Smart and Functional Polymers

Smart and Functional Polymers

Special Issue Editors

Jianxun Ding
Yang Li
Mingqiang Li

MDPI • Basel • Beijing • Wuhan • Barcelona • Belgrade

MDPI

Special Issue Editors

Jianxun Ding
Changchun Institute of Applied
Chemistry, Chinese Academy of Sciences
China

Yang Li
Boston Children's Hospital
USA

Mingqiang Li
Sun Yat-sen University
China

Editorial Office
MDPI
St. Alban-Anlage 66
4052 Basel, Switzerland

This is a reprint of articles from the Special Issue published online in the open access journal *Molecules* (ISSN 1420-3049) from 2018 to 2019 (available at: https://www.mdpi.com/journal/molecules/special_issues/SFP)

For citation purposes, cite each article independently as indicated on the article page online and as indicated below:

LastName, A.A.; LastName, B.B.; LastName, C.C. Article Title. *Journal Name* **Year**, *Article Number*, Page Range.

ISBN 978-3-03921-590-4 (Pbk)
ISBN 978-3-03921-591-1 (PDF)

Contents

About the Special Issue Editors

Jianxun Ding received his B.S. degree in Polymer Chemistry from University of Science and Technology of China in 2007 and obtained his Ph.D. degree at Changchun Institute of Applied Chemistry (CIAC), Chinese Academy of Sciences (CAS), in 2013 under the supervision of Dr. Xuesi Chen. During 2017–2019, he worked with Dr. Omid C. Farokhzad and Dr. Jinjun Shi from Brigham and Women's Hospital, Harvard Medical School, as a Postdoctoral Research Fellow. During his study, he was awarded over 10 awards, such as the Tang Aoqing Chemistry Scholarship in 2011, and the Graduate National Scholarship and the President Excellence Award of Chinese Academy of Sciences in 2012. He works now as an Associate Professor at CIAC, CAS. His research focuses on the synthesis of functional biodegradable polymers, the development of smart polymer platforms for controlled drug delivery, the exploitation of polymer-based adjuvants for immunotherapy, and the preparation of polymer scaffolds for regenerative medicine. He was invited to serve as the Editorial Board Members of *Molecules, Current Pharmaceutical Design, Medicine, Current Molecular Pharmacology, Sci,* and *Heliyon*, and as a Young Editorial Board Member of *Journal of Functional Polymers*. In addition, he served as a Guest Editor for several journals, including *Frontiers in Pharmacology, Current Pharmaceutical Design, Current Pharmaceutical Biotechnology, Polymers,* and *Journal of Nanomaterials.* Heretofore, he has published more than 160 academic articles in mainstream journals, including *Advanced Materials, Nano Letters, ACS Nano, Nature Communications, Advanced Functional Materials, Advanced Sciences, Biomaterials, Journal of Controlled Release,* and so forth, with over 6000 citations. Moreover, he has applied for over 70 patents in China, of which over 60 have been authorized. Among them, two patents were awarded the Invention Contest of Jilin Province and the Excellence Award of Chinese Patent Award in 2013. Meanwhile, owing to his significant accomplishments, he has won more than 10 awards, such as the Natural Science Award of Jilin Province, the Science and Technology Progress Award of Jilin Province, and the Natural Science Academic Achievement Award of Jilin Province. He was selected for the Fifth Batch Outstanding Innovative Talents in Jilin Province in 2015. He was selected in the Young Talents Promotion Project of Jilin Province and joined the CIAC Yinghua Young Innovation Promotion Association, CAS, in 2018, and joined the CAS Young Innovation Promotion Association in 2019. Based on previous studies and accomplishments, he has undertaken more than 10 grants for groups, such as the National Natural Science Foundation of China and the Science and Technology Development Program of Jilin Province.

Yang Li obtained his B.S. degree (2011) in Polymer Chemistry from University of Science and Technology of China. He earned his Ph.D. degree (2017) in Biomedical Science from University of Texas Southwestern Medical Center under the supervision of Dr. Jinming Gao. Currently, he works as a postdoctoral research fellow in the research group of Dr. Daniel Kohane at Boston Children's Hospital, Harvard Medical School. His research interest focuses on developing stimuli-responsive nanomaterials for biological sensing and drug delivery. He is also interested in the mechanistic investigation of supramolecular self-assembly.

Mingqiang Li received his B.S. degree in Polymer Chemistry from University of Science and Technology of China in 2009. He carried out his graduate studies under Dr. Xuesi Chen at Changchun Institute of Applied Chemistry, Chinese Academy of Sciences, and postdoctoral training with Dr. Kam W. Leong at Columbia University. His current research is mainly focused on biomaterials, nanomedicines, and microfluidics.

Preface to "Smart and Functional Polymers"

Smart and functional polymers have attracted increasing interest in recent years thanks to their full range of applications in various fields, such as industry, bioengineering, and medicine. Polymers with different physicochemical properties can be synthesized through the polymerization of functional monomers or the modification of different active groups after polymerization. Compared with small molecules, these polymers have relatively large surface-to-volume ratio, adjustable size and surface, enhanced sensitivity and specificity, excellent targeting binding affinity, and stimuli-responsive ability. With the development of synthesis and characterization techniques, a series of smart and functional polymer materials, including shape memory polymers, drug release polymer nanosystems, polymer gels, polymer films, polymer fibers, and so forth, have been successfully designed and achieved rapid progress. However, the synthesis technology, structure and performance, stability, and biosafety of smart and functional polymers need to be further improved to meet the increasing needs of different fields.

This book is based on the Special Issue of the journal Molecules on "Smart and Functional Polymers". The collected research and review articles focus on the synthesis and characterizations of advanced functional polymers, polymers with specific structures and performances, current improvements in advanced polymer-based materials for various applications, and the opportunities and challenges in the future. The topics cover the emerging synthesis and characterization technologies of smart polymers, core–shell structure polymers, stimuli-responsive polymers, anhydrous electrorheological materials fabricated from conducting polymers, reversible polymerization systems, and biomedical polymers for drug delivery and disease theranostics.

In summary, this book provides a comprehensive overview of the latest synthesis approaches, representative structures and performances, and various applications of smart and functional polymers. It will serve as a useful reference for all researchers and readers interested in polymer sciences and technologies.

Jianxun Ding, Yang Li, and Mingqiang Li
Special Issue Editors

$$\text{DLC} = \frac{\text{Weight of drug in PUM/PTX}}{\text{Weight of drug} - \text{loaded micelle}} \times 100\% \tag{1}$$

$$\text{DLE} = \frac{\text{Weight of drug in PUM/PTX}}{\text{Weight of feeding drug}} \times 100\% \tag{2}$$

PTX was successfully loaded into mPEG−poly(**1/3**) micelle with DLC of 8.7% and DLE of 87.5%. The morphological characteristic of PUM/PTX was observed by TEM (Figure 2c), and the average nanoparticle size was around 43 nm. The R_h was 44.7 ± 11.6 nm by the DLS test, showing a similar result with TEM. The results suggested that PUM/PTX was a monodisperse micelle and had a uniform particle size distribution.

2.2. PTX Release, In Vitro Cell Uptake, and Cell Proliferation Inhibition

The release characteristics of PTX from PUM/PTX were detected in PBS. As depicted in Figure 2d, the release profile showed no apparent burst release of PTX from PUM/PTX in PBS at pH 7.4 within 24 h. The amount of PTX released from PUM/PTX was lower than 45% during the first 12 h. At 48 h, over 65% of PTX was released. Since the tumoral microenvironment is more acidic, the PTX release behavior was also tested in PBS at pH 6.8 (Figure 2e). The amount of PTX released from PUM/PTX was 46% in the first 12 h. At 60 h, about 69% of PTX was released. These two release rates indicated that the hydrolytic rate of PUM was faster in the acidic condition. This controlled release pattern indicated that polyurea could be used as a suitable drug delivery system.

Drug release was a complicated process [13]. To just explain the nature of drug release behaviors, a classic empirical equation was established by Peppas et al. [17]. The equations were written as:

$$\frac{M_t}{M_\infty} = k\,t^n \tag{3}$$

$$\lg\left(\frac{M_t}{M_\infty}\right) = \lg k + n \lg t \tag{4}$$

In Equations (3) and (4), M_t and M_∞ were the cumulative drug release at time t and infinite time, respectively; k was the proportionality constant, and n was the release exponent that was related to the release mechanism of payloads. In the study of drug release, an increase in n indicates that the release was more influenced in a swelling-controlled way. n was calculated using the Equations (3) and (4), and the values for a pH of 6.8 and 7.4 were 0.32 and 0.21, respectively. The value of n was larger at pH 6.8 than pH 7.4, which was attributed to the faster hydrolysis of polyurea in an acidic environment.

Coumarin-6 (C_6) as a hydrophobic model fluorescence molecule was used for cell uptake study [18]. C_6 was loaded into mPEG−poly(**1/3**) micelle to form PUM/C_6. The internalization of PUM/C_6 by 4T1 cells was monitored through flow cytometry (FCM) and confocal laser scanning microscopy (CLSM). As shown in Figure 3a, after 1 h incubation, the control group had no fluorescence signal of C_6, while in the PUM/C_6 group the fluorescence signals were significantly increased, and fluorescence intensity was further increased at 6 h. Similarly, CLSM images in Figure 3b,c showed that the highest fluorescence intensity was detected when PUM/C_6 was incubated with 4T1 cells for 6 h. The results demonstrated that the PUM micelle could efficiently deliver PTX into 4T1 cells.

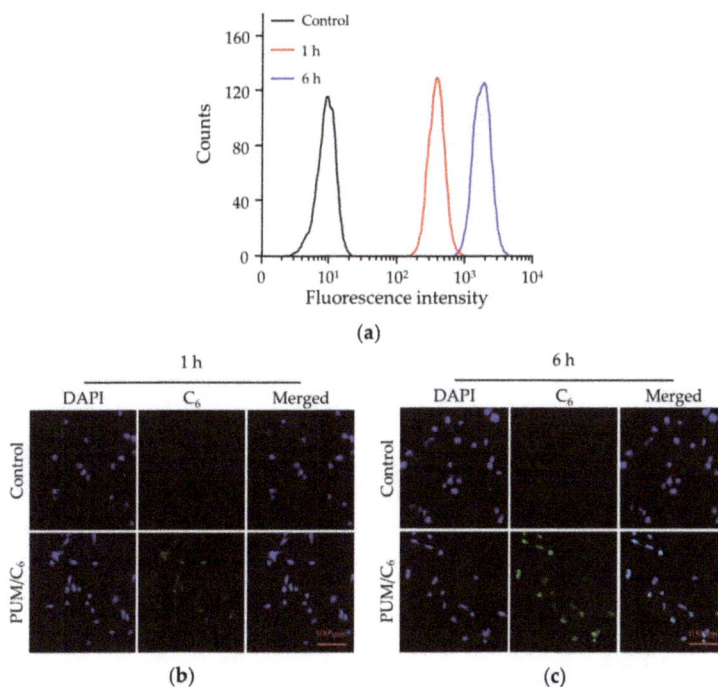

Figure 3. Cell uptake of PUM/C_6 after incubation with 4T1 cells detected by (**a**) flow cytometry (FCM) and (**b,c**) confocal laser scanning microscopy (CLSM) analyses in 1 h (**b**) and 6 h (**c**).

PUM with HUBs was synthesized by the reversible reaction between cyclohexyl diisocyanate and *tert*-butyl diamine. With the hydrolysis of PUM, the degradation products were cyclohexane-1,3-diyldimethanamine and *tert*-butyl diamine. The cytotoxicity of the hydrolytic products of HUBs to 4T1 and L929 cells was tested by methyl thiazolyl tetrazolium (MTT) assays (Figure 4a, b). After incubation with PUM at the concentration of 100.0 µg mL^{-1} for 72 h, the viability of L929 cells was kept around 93%, indicating negligible toxicity of PUM to normal cells. After incubation with PUM for 72 h at a concentration of 50.0 µg mL^{-1}, the cell viability of 4T1 cells was 86.97%, indicating that the polymer had little effect on the growth of tumor cells. MTT assays compared the toxicity of free PTX and PUM/PTX toward 4T1 cells. Both free PTX and PUM/PTX inhibited the growth of 4T1 cells. The cell viability was reduced to 51.9% after incubation with free PTX for 48 h at a concentration of 10.0 µg mL^{-1}, while the cells treated with an equivalent dose of PUM/PTX showed viability of 46.6% (Figure 4c). As shown in Figure 4d, the cell proliferation was further suppressed at 72 h with cell viability reduced to around 45.9%. The data above confirmed that PUM/PTX could be efficiently endocytosed by 4T1 cells and release PTX to perform the antitumor effect. In vitro experiments proved that the polymer could make a proper drug delivery vehicle.

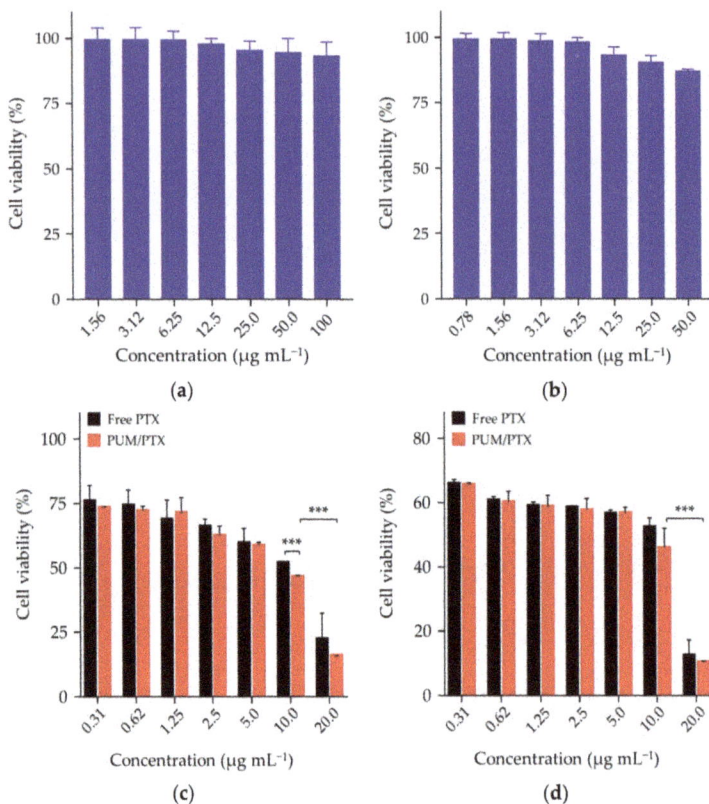

Figure 4. In vitro cytotoxicity of (**a**) L929 and (**b**) 4T1 cells in 72 h; (**c**) PUM/PTX in 48 h and (**d**) PUM/PTX in 72 h after incubation with 4T1 cells. The error bars represent the standard deviation ($n = 3$; * $p < 0.05$, ** $p < 0.01$, *** $p < 0.001$).

2.3. In Vivo Antitumor Efficacy

The antitumor activity was investigated in a BALB/c model of mice bearing allograft orthotopic murine 4T1 breast tumors. The antitumor efficacy of PUM/PTX was further detected in vivo. Free PTX was dissolved in the mixture of castor oil and ethanol [19], and diluted with PBS. The mice were treated with free PTX or PUM/PTX at a dosage of 5.0 mg kg^{-1} PTX, and the mice treated with PBS were set as a control group. The treatment started at the time when the tumor volumes reached approximately 200 mm^3. As shown in Figure 5a, free PTX showed modest antitumor efficacy compared with the control group within 18 days, whereas PUM/PTX showed the best inhibition rate of 32.7%. This result suggests that PUM/PTX could well inhibit tumor growth. The photograph of the tumors had a similar result (Figure 5b).

3.13. *Histopathological and Immunofluorescence Analyses*

The mice were sacrificed two days after the last injection. According to the protocol reported in previous studies, the lung is the most common organ for breast cancer metastasis, while the spleen is the primary immune organ that should be observed at the end of the experiments [25]. The tumors and major organs (i.e., the lungs and spleens) were collected, fixed in 4% (*W/V*) PBS-buffered paraformaldehyde overnight, and then embedded in paraffin. The paraffin-embedded tissues were cut into ~5 μm slices for H&E staining and ~3 μm sheets for immunofluorescence analyses (i.e., Ki-67 and caspase-3). The histological and immunofluorescence alterations were detected by a microscope (Nikon Eclipse Ti, Optical Apparatus Co., Ardmore, PA, USA).

4. Conclusions

In this study, we synthesized four kinds of hydrolyzable polyureas with different hydrolytic rates by changing the chemical groups on the polyureas. Among them, poly(**1/3**) from cyclohexyl diisocyanate and *tert*-butyl diamine showed the fastest hydrolytic rate. After modification by hydrophilic mPEG, the amphiphilic mPEG–poly(**1/3**) was synthesized for delivery of PTX. The PTX was successfully encapsulated by mPEG–poly(**1/3**) micelle with a DLC and DLE of 8.75% and 87.5%, respectively. PUM/PTX could be efficiently internalized by murine breast cancer 4T1 cells and released PTX along with the hydrolysis of polyurea. The results showed that PUM/PTX drastically suppressed the proliferation of tumor cells in vitro and significantly inhibited tumor growth in an orthotopic 4T1 breast tumor model in vivo. Therefore, the hydrolyzable PEGylated polyureas with adjustable degradation might become a promising platform for controlled drug delivery.

Author Contributions: J.D., W.X. and Y.Y. conceived and designed the experiments; M.C. performed the experiments; M.C., X.F. and Z.J. analyzed the data; M.C. wrote the draft manuscript; X.F., Z.J., Y.W. and J.D. revised the manuscript. All the authors confirmed the content of the manuscript and approved the submission.

Funding: This research was financially supported by the National Natural Science Foundation of China (Grant Nos. 51873207, 51803006, 51833010, 51673190, 51603204, 51673187, and 51520105004), the Science and Technology Development Program of Jilin Province (Grant No. 20190201068JC), and the National Key Research and Development Program of China (Grant No. 2016YFC1100701).

Conflicts of Interest: The authors declare no conflict of interest.

References

1. Cai, K.M.; Ying, H.Z.; Cheng, J.J. Dynamic ureas with fast and pH-independent hydrolytic kinetics. *Chem. Eur. J.* **2018**, *24*, 7345–7348. [CrossRef] [PubMed]
2. Ying, H.Z.; Zhang, Y.F.; Cheng, J.J. Dynamic urea bond for the design of reversible and self-healing polymers. *Nat. Commun.* **2014**, *5*, 3218–3226. [CrossRef] [PubMed]
3. Hutchby, M.; Houlden, C.E.; Ford, J.G.; Tyler, S.N.G.; Gagne, M.R.; Lloyd-Jones, G.C.; Booker-Milburn, K.I. Hindered ureas as masked isocyanates: Facile carbamoylation of nucleophiles under neutral conditions. *Angew. Chem. Int. Ed.* **2009**, *48*, 8721–8724. [CrossRef]
4. Rocas, P.; Cusco, C.; Rocas, J.; Albericio, F. On the importance of polyurethane and polyurea nanosystems for future drug delivery. *Curr. Drug Delivery* **2018**, *15*, 37–43. [CrossRef]
5. Ying, H.Z.; Cheng, J.J. Hydrolyzable polyureas bearing hindered urea bonds. *J. Am. Chem. Soc.* **2014**, *136*, 16974–16977. [CrossRef]
6. Florez-Grau, G.; Rocas, P.; Cabezon, R.; Espana, C.; Panes, J.; Rocas, J.; Albericio, F.; Benitez-Ribas, D. Nanoencapsulated budesonide in self-stratified polyurethane-polyurea nanoparticles is highly effective in inducing human tolerogenic dendritic cells. *Int. J. Pharm.* **2016**, *511*, 785–793. [CrossRef] [PubMed]
7. John, J.V.; Seo, E.J.; Augustine, R.; Jang, I.H.; Kim, D.K.; Kwon, Y.W.; Kim, J.H.; Kim, I. Phospholipid end-capped bioreducible polyurea micelles as a potential platform for intracellular drug delivery of doxorubicin in tumor cells. *ACS Biomater. Sci. Eng.* **2016**, *2*, 1883–1893. [CrossRef]

8.   Valerio, A.; Feuser, P.E.; Bubniak, L.D.; Dos Santos-Silva, M.C.; de Araujo, P.H.H.; Sayer, C. In vitro biocompatibility and macrophage uptake assays of poly(urea-urethane) nanoparticles obtained by miniemulsion polymerization. *J. Nanosci. Nanotechnol.* **2017**, *17*, 4955–4960. [CrossRef]

9.   Shoaib, M.; Bahadur, A.; Rahman, M.S.U.; Iqbal, S.; Arshad, M.I.; Tahir, M.A.; Mahmood, T. Sustained drug delivery of doxorubicin as a function of pH, releasing media, and NCO contents in polyurethane urea elastomers. *J. Drug Deliv. Sci. Tec.* **2017**, *39*, 277–282. [CrossRef]

10. Morral-Ruiz, G.; Melgar-Lesmes, P.; Lopez-Vicente, A.; Solans, C.; Garcia-Celma, M.J. Biotinylated polyurethane-urea nanoparticles for targeted theranostics in human hepatocellular carcinoma. *Nano Res.* **2015**, *8*, 1729–1745. [CrossRef]

11. Zhang, Y.F.; Ying, H.Z.; Hart, K.R.; Wu, Y.X.; Hsu, A.J.; Coppola, A.M.; Kim, T.A.; Yang, K.; Sottos, N.R.; White, S.R.; et al. Malleable and recyclable poly(urea-urethane) thermosets bearing hindered urea bonds. *Adv. Mater.* **2016**, *28*, 7646–7651. [CrossRef] [PubMed]

12. Jiang, Z.Y.; Chen, J.J.; Cui, L.G.; Zhuang, X.L.; Ding, J.X.; Chen, X.S. Advances in stimuli-responsive polypeptide nanogels. *Small Methods* **2018**, *2*, 307–321. [CrossRef]

13. Gao, S.; Tang, G.; Hua, D.; Xiong, R.; Han, J.; Jiang, S.; Zhang, Q.; Huang, C. Stimuli-responsive bio-based polymeric systems and their applications. *J. Mater. Chem. B* **2019**, *7*, 709–729. [CrossRef]

14. Jiang, Z.; Liu, Y.; Feng, X.; Ding, J. Functional polypeptide nanogels. *J. Funct. Polym.* **2019**, *32*, 13–27.

15. Tang, B.Q.; Zaro, J.L.; Shen, Y.; Chen, Q.; Yu, Y.L.; Sun, P.P.; Wang, Y.Q.; Shen, W.C.; Tu, J.S.; Sun, C.M. Acid-sensitive hybrid polymeric micelles containing a reversibly activatable cell-penetrating peptide for tumor-specific cytoplasm targeting. *J. Controlled Release* **2018**, *279*, 147–156. [CrossRef] [PubMed]

16. Wang, J.; Xu, W.; Li, S.; Qiu, H.; Li, Z.; Wang, C.; Wang, X.; Ding, J. Polylactide-cholesterol stereocomplex micelle encapsulating chemotherapeutic agent for improved antitumor efficacy and safety. *J. Biomed. Nanotechnol.* **2018**, *14*, 2102–2113. [CrossRef] [PubMed]

17. Shi, F.H.; Ding, J.X.; Xiao, C.S.; Zhuang, X.L.; He, C.L.; Chen, L.; Chen, X.S. Intracellular microenvironment responsive PEGylated polypeptide nanogels with ionizable cores for efficient doxorubicin loading and triggered release. *J. Mater. Chem.* **2012**, *22*, 14168–14179. [CrossRef]

18. Chu, B.Y.; Qu, Y.; Huang, Y.X.; Zhang, L.; Chen, X.X.; Long, C.F.; He, Y.Q.; Ou, C.W.; Qian, Z.Y. PEG-derivatized octacosanol as micellar carrier for paclitaxel delivery. *Int. J. Pharm.* **2016**, *500*, 345–359. [CrossRef]

19. Cheng, Y.L.; He, C.L.; Ding, J.X.; Xiao, C.S.; Zhuang, X.L.; Chen, X.S. Thermosensitive hydrogels based on polypeptides for localized and sustained delivery of anticancer drugs. *Biomaterials* **2013**, *34*, 10338–10347. [CrossRef]

20. Zhang, Y.; Cai, L.L.; Li, D.; Lao, Y.H.; Liu, D.Z.; Li, M.Q.; Ding, J.X.; Chen, X.S. Tumor microenvironment-responsive hyaluronate-calcium carbonate hybrid nanoparticle enables effective chemotherapy for primary and advanced osteosarcomas. *Nano Res.* **2018**, *11*, 4806–4822. [CrossRef]

21. Wang, Y.P.; Gao, W.Q.; Shi, X.Y.; Ding, J.J.; Liu, W.; He, H.B.; Wang, K.; Shao, F. Chemotherapy drugs induce pyroptosis through caspase-3 cleavage of a gasdermin. *Nature* **2017**, *547*, 99–103. [CrossRef]

22. Chen, J.J.; Ding, J.X.; Wang, Y.C.; Cheng, J.J.; Ji, S.X.; Zhuang, X.L.; Chen, X.S. Sequentially responsive shell-stacked nanoparticles for deep penetration into solid tumors. *Adv. Mater.* **2017**, *29*, 170–177. [CrossRef]

23. Wang, Y.; Jiang, Z.; Xu, W.; Yang, Y.; Zhuang, X.; Ding, J.; Chen, X. Chiral Polypeptide Thermogels Induce Controlled Inflammatory Response as Potential Immunoadjuvants. *ACS Appl. Mater. Interfaces* **2019**, *11*, 8725–8730. [CrossRef] [PubMed]

24. Xu, W.G.; Ding, J.X.; Xiao, C.S.; Li, L.Y.; Zhuang, X.L.; Chen, X.S. Versatile preparation of intracellular-acidity-sensitive oxime-linked polysaccharide-doxorubicin conjugate for malignancy therapeutic. *Biomaterials* **2015**, *54*, 72–86. [CrossRef] [PubMed]

25. Lv, Y.; Xu, C.; Zhao, X.; Lin, C.; Yang, X.; Xin, X.; Zhang, L.; Qn, C.; Han, X.; Yang, L.; et al. Nanoplatform assembled from a CD44-targeted prodrug and smart liposomes for dual targeting of tumor microenvironment and cancer cells. *ACS Nano* **2018**, *12*, 1519–1536. [CrossRef] [PubMed]

Sample Availability: Samples of the compounds are available from the authors.

molecules

MDPI

Article

Purification and Glutaraldehyde Activation Study on HCl-Doped PVA–PANI Copolymers with Different Aniline Concentrations

Jorge M. Guerrero [1], Amanda Carrillo [2,*], María L. Mota [2,3], Roberto C. Ambrosio and Francisco S. Aguirre [1,*]

[1]   Centro de Investigación en Materiales Avanzados, S.C., Alianza Norte 202, Parque de Investigación e Innovación Tecnológica, Apodaca, NL C.P. 66600, Mexico; jorge.guerrero@cimav.edu.mx
[2]   Instituto de Ingeniería y Tecnología, Universidad Autónoma de Ciudad Juárez, Av. Del Charro 610, Ciudad Juárez, CHIH C.P. 32310, Mexico; mdllmotago@conacyt.mx
[3]   CONACYT, Universidad Autónoma de Ciudad Juárez, Ciudad Juárez, CHIH C.P. 32310, Mexico
[4]   Facultad de Electrónica, Benemérita Universidad Autónoma de Puebla, Puebla C.P 72000, Mexico; roberto.ambrosio@correo.buap.mx
*   Correspondence: amanda.carrillo@uacj.mx (A.C.); servando.aguirre@cimav.edu.mx (F.S.A.); Tel.: +52-1-656-688-4800 (A.C.); +52-1-811-721-6676 (F.S.A.)

Academic Editors: Jianxun Ding, Yang Li and Mingqiang Li
Received: 30 November 2018; Accepted: 21 December 2018; Published: 25 December 2018

Abstract: In this work, we report the synthesis and purification of polyvinyl alcohol-polyaniline (PVA–PANI) copolymers at different aniline concentrations, and their molecular (^{1}H-NMR and FTIR), thermal (TGA/DTG/DSC), optical (UV–Vis-NIR), and microstructural (XRD and SEM) properties before and after activation with glutaraldehyde (GA) in order to obtain an active membrane. The PVA–PANI copolymers were synthesized by chemical oxidation of aniline using ammonium persulfate (APS) in an acidified (HCl) polyvinyl alcohol matrix. The obtained copolymers were purified by dialysis and the precipitation–redispersion method in order to eliminate undesired products and compare changes due to purification. PVA–PANI products were analyzed as gels, colloidal dispersions, and thin films. ^{1}H-NMR confirmed the molecular structure of PVA–PANI as the proposed skeletal formula, and FTIR of the obtained purified gels showed the characteristic functional groups of PVA gels with PANI nanoparticles. After exposing the material to a GA solution, the presence of the FTIR absorption bands at 1595 cm^{-1}, 1650 cm^{-1}, and 1717 cm^{-1} confirmed the activation of the material. FTIR and UV–Vis-NIR characterization showed an increase of the benzenoid section of PANI with GA exposure, which can be interpreted as a reduction of the polymer with the time of activation and concentration of the solution.

Keywords: polymerization dispersion method; polyaniline; polyvinyl alcohol; glutaraldehyde; chemical activation

---

1. Introduction

Conductive polymers have been studied since their discovery in 1977 [1]. Among all known conductive polymers, polyaniline (PANI) has been one of the most studied due its high environmental stability, straightforward control of its chemical and physical properties through doping, and relatively low cost of development in comparison to other conductive polymers. PANI has been the subject in numerous studies for applications such as membranes [2], anticorrosive coatings [3], biosensors [4–7], and electronic devices [8,9]. However, the implementation of PANI has been limited because of its poor mechanical properties and its poor solubility in most organic solvents. This insolubility results in heterogeneous solutions, where the presence of microparticles hinders the formation of homogeneous PANI thin films at low cost by physical methods such as spin or dip coating. To overcome these drawbacks, several methods have been studied, such as the possibility of processing PANI in the form of mixtures with electrically insulating polymers, improving the presented deficiencies, and opening a range of potential applications. In recent years, several publications report on the use of polystyrene (PS) [10], polyvinyl chloride (PVC) [11], and polyvinyl alcohol (PVA) [12], to stabilize PANI nanoparticles and improve the processing and formation of thin films.

The preparation of PVA–PANI copolymers by polymerization dispersion has already been reported showing a stable colloidal dispersion without sedimentation [13]. Also, with the increase of PVA content the stability of the colloidal dispersion increases, improving its mechanical properties [14]. Nevertheless, PVA addition affects the conductivity of the material. Work has been done to correlate the mechanical and electrical properties of PVA–PANI as function of aniline concentration [15].

Both PANI and PVA–PANI have been activated with glutaraldehyde (GA) to obtain PANI-G and PVA–PANI-G materials that can be used as biological platforms for the detection of different analytes such as enzymes and proteins [16,17]. The biocompatibility of PVA is already known, but PVA–PANI obtained by polymerization dispersion requires further purification in order to be applied in the biomedical field because of the presence of undesired byproducts produced by the chemical polymerization of aniline [18], which can also affect the activation of the material.

The present work is focused on the study of purified PVA–PANI copolymers at different concentrations of aniline. The molecular, thermal, optical, and microstructural properties before and after PVA–PANI copolymer activation with GA have been studied. The aim of this work is to provide information about the effects of purified PANI in a PVA matrix with GA activation.

2. Results and Discussion

2.1. Molecular Structure Variation and Modification of PVA–PANI Copolymers

2.1.1. Proton Nuclear Magnetic Resonance Spectroscopy (^{1}H-NMR)

To obtain the molecular structure of PVA–PANI copolymers, ^{1}H-NMR analysis (500 MHz, TMS) was carried out using deuterium oxide (D_2O) as solvent. The obtained spectrum for PVA–PANI is shown in Figure 1 where at $\delta = 7.3$ ppm and $\delta = 7.5$ ppm there is evidence of a doublet of doublets (d, $J = 6.5$ Hz, C_6H_4) corresponding to the benzenoid and quinoid sections of PANI in emeraldine phase, then $\delta = 4.7$ ppm (s, D_2O). The doublet located at $\delta = 3.9$ ppm (d, $J = 28.5$ Hz) corresponds to the characteristic signal of α proton of oxygen binding PVA–PANI, while the multiplet located at $\delta = 1.5$ ppm (m, $J = 18.7$ Hz) suggests the integration of methylene protons from the PVA backbone [19,20].

The smooth baseline of the spectrum suggests absence of impurities in the obtained material (Figure 2). After water suppression the actual PVA:PANI ratio can be calculated by integrating the signal corresponding to PVA and PANI, obtaining a 2:1 ratio (PVA:PANI). Moreover, two sharp singlets can be clearly seen at $\delta = 2$ ppm and $\delta = 3.25$ ppm corresponding to absorbed solvent in the copolymer. The $\delta = 4.7$ ppm (d, $J = 19$ Hz), can correspond to the proton of PVA hydroxyl (–OH), pendant groups or PANI secondary amine (–NH) groups within the polymer chain. The doublet located at $\delta = 3.9$ ppm (d, $J = 28.5$ Hz) corresponds to the characteristic signal of α proton of oxygen binding PVA–PANI, this completes the molecular structure of the PVA–PANI copolymers as proposed from the skeletal formula in Figure 2. Is important to mention that the signals could be clearly seen only after the water suppression and is considered to be masked by the solvent peak in Figure 1.

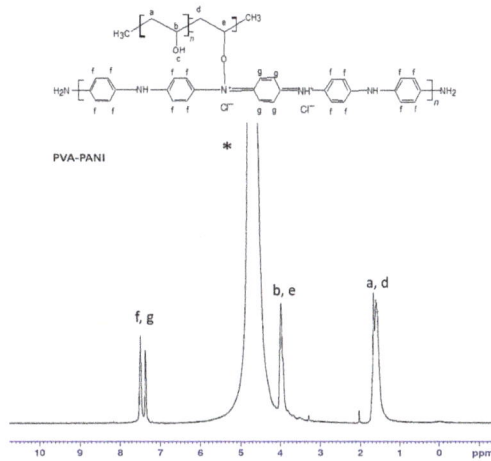

Figure 1. ^{1}H-NMR spectra for PVA–PANI in D_2O (*).

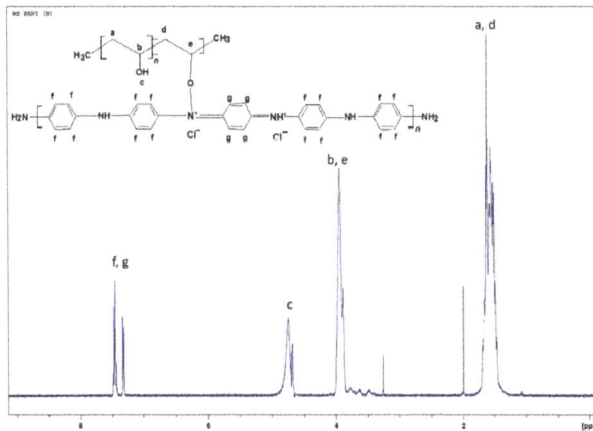

Figure 2. ^{1}H-NMR spectra for PVA–PANI with water suppression.

2.1.2. Fourier Transform Infrared Spectroscopy (FTIR)

To confirm the structure of PVA–PANI and evaluate chemical changes upon activation with GA, an FTIR analysis was carried out. First, a comparison of both purification methods was made through a study of the functional groups present (Figure 3). It was observed that PVA–PANI purified by the precipitation–redispersion method presents the characteristic bands as follows, O–H between 3600 and 3000 cm^{-1}, aliphatic groups for the PVA backbone between 2900 and 2800 cm^{-1}, and stretching of C=O and C–O acetate groups from hydrolyzed PVA in the fingerprint region [21]. The presence of the bands at 1179, 1031, 739, and 689 cm^{-1} corresponding to C–N and C–H vibrations from the benzenoid and quinoid section of PANI confirm the presence of the polymer in the PVA matrix. On the other hand, the spectrum for the dialysis purified material shows the characteristic bands for PVA–PANI at a medium–high concentration of aniline in a PVA matrix, according to previous studies [15]. The N-H, C=C, C–C, and C–N vibrations centered at 3383, 1600, 1423, and 1297 cm^{-1}, respectively, can be seen along with the C–H in plane and out of plane vibrations from the quinoid and benzenoid rings [22]. Also, from Figure 3, it is seen that both purification methods do not show any uncoupled or overlapped bands by the content of unreacted material, especially in the fingerprint region of the spectrum.

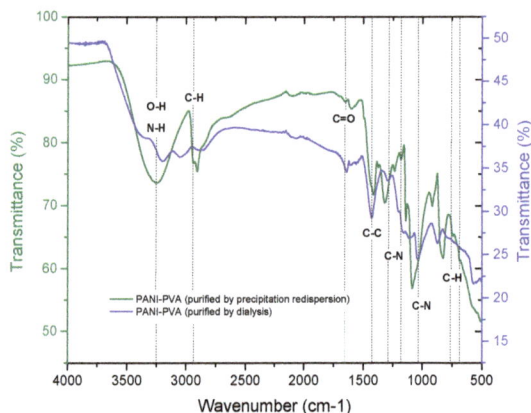

Figure 3. FTIR spectra of PVA–PANI purified by dialysis (-) and by precipitation–redispersion (-).

The implementation of dialysis has been used for the purification of PANI, since it provides a better dispersion of particles and minimizes product loss. However, filtration of the product obtained from precipitation–redispersion is highly recommended, since it takes a short time and allows the obtaining of only the desired copolymer in gel form through precipitation. Figure 4 shows the corresponding bands of PVA gels modified with PANI (ES) nanoparticles obtained by the precipitation–redispersion method with different concentrations of aniline (Table 1). An increase in the absorption bands with the increase of monomer in the PVA matrix was observed.

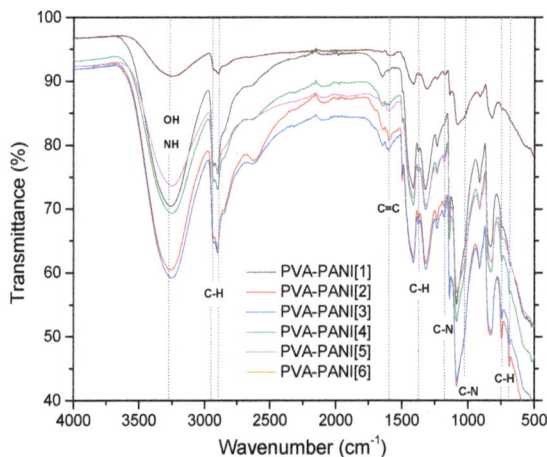

Figure 4. FTIR spectra of PVA–PANI gels at different PANI concentrations.

The absorption bands located between 3600 and 3000 cm^{-1}, attributed to O–H and N–H groups of PVA and PANI, respectively, show an intensity increase which is due to the functional groups in this range and widening due to minimum wavenumber displacements. Subsequently, in the range of 2350 to 1750 cm^{-1}, small absorption bands can be observed corresponding to the presence of overtones from PVA–PANI. These bands show an intensity decrease with increasing aniline in the polymer matrix. The latter can be attributed to chain length variation of the obtained PANI for each experiment. An increase in the absorption band located at 1586 cm^{-1} indicates an increase in C=C bonds associated to quinoid rings which can also be correlated to significant production of PANI in the

PVA matrix. The absorption bands found at 1416, 1369, and 1320 cm^{-1} correspond to the absorption of C–C stretching of the quinoid ring in PANI, the flexion of the C–H group, and the stretching vibration of the C–N bond of the amine in the aromatic ring, respectively. Absorption bands due to the vibration of the C–N bond and the stretching of the C–H group from the aromatic ring within the plane are shown at 1230 and 1130 cm^{-1}, respectively. The band located at 1090 cm^{-1} corresponds to the stretching vibration of the C–O bond which tends to be much higher in PVA–PANI composites, showing a more pronounced shoulder in 1031 cm^{-1} with the increase of monomer, demonstrating a cross-linking between the materials [23]. The FTIR spectra obtained for the activated material with 1% GA solution for an activation time of 5, 15, and 30 min can be observed in Figure 5a–c, respectively.

Table 1. Experiment of aniline concentration variation against PVA 5 wt %.

Concentration	Sample	PVA 5 wt %	Aniline	PVA–PANI (*wt/wt*)
Low	PVA–PANI-1	0.0202 mmol	3.5203 mmol	12.5 wt %
	PVA–PANI-2	0.0202 mmol	7.0640 mmol	25 wt %
Medium	PVA–PANI-3	0.0202 mmol	14.1281 mmol	37.5 wt %
	PVA–PANI-4	0.0202 mmol	17.6601 mmol	50 wt %
	PVA–PANI-5	0.0202 mmol	21.1921 mmol	62.5 wt %
High	PVA–PANI-6	0.0202 mmol	24.7241 mmol	75 wt %

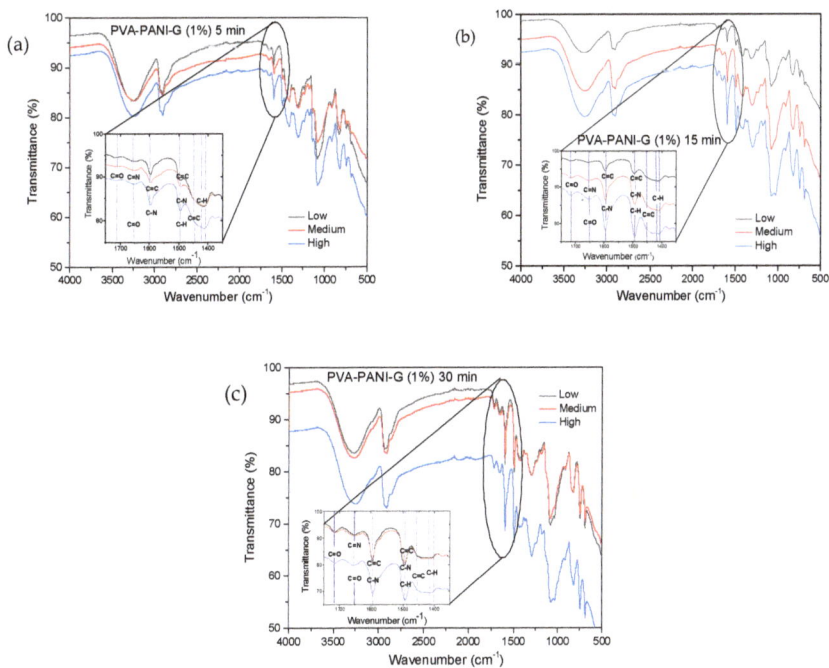

Figure 5. FTIR spectra of PVA–PANI gels at low, medium and high concentration of PANI activated with GA at 1% for: (**a**) 5 min, (**b**) 15 min, and (**c**) 30 min.

In Figure 5, a reduction in the intensity of the absorption bands corresponding to the O–H and N–H groups of PVA–PANI is shown, indicating an interaction between the activating solution and the polymer [24]. The inset shows the range from 1550 to 1750 cm^{-1}, where the presence of three absorption bands can be observed at 1595, 1650, and 1717 cm^{-1}, corresponding to the presence of C=N (Schiff base) bonds formed by the linkage of primary amines with C=O groups, and the presence of free C=O groups [24,25]. As the activation time so does the absorption band at 1595 cm^{-1}, correlated to the vibration of the C=N (imine) bond, overlapping the vibration of the C=C bond of the quinoid section of PANI. Moreover, an increase in the absorption band located at 1490 cm^{-1} can be observed, which is associated to the benzenoid ring stretching [22], along with a reduction of the intensity of the C–C stretching from the quinoid section, which can indicate a reduction of the purified material with the increase of activation time [26]. Further activation of PVA–PANI with a more concentrated solution (2.5% GA) confirmed the reduction of the material as well as a branching between both materials (Figure S1).

2.1.3. UV–Visible-Near-Infrared Spectroscopy (UV–Vis-NIR)

UV–Vis-NIR absorption spectra and its normalized form for PVA–PANI films at low, medium, and high concentrations of PANI can be observed in Figure 6. Absorption bands corresponding to π–π* transitions from the excitation of benzenoid segments (~330 nm), polaron-π* transitions associated with benzenoid and quinoid ring (~430 nm), and π-polaron transitions from the excitation of quinoid rings from doped PANI (~800 nm) can be seen in each experiment [3,27]. The absorption bands corresponding to π–π* transitions show a bathochromic effect as a function of the increase in aniline concentration in the PVA matrix, attributed to an increase in the molecular weight of the produced PANI [28]. Furthermore, a clear hyperchromic effect is observed, associated to the increase in concentration of the material, as established by Beer–Lambert's law.

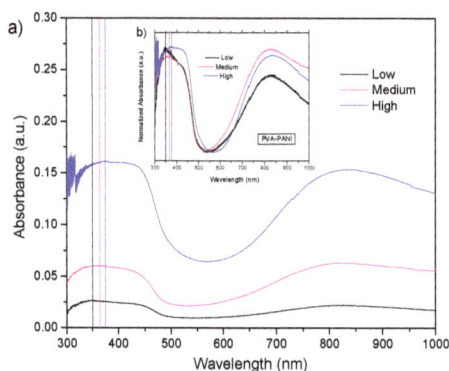

Figure 6. (a) UV–Vis-NIR absorbance and (b) normalized UV–Vis-NIR absorbance spectra of PVA–PANI thin films at low, medium, and high concentrations of PANI.

According to the results of activated PVA–PANI obtained by FTIR, an immersion time of 30 min was used for the PVA–PANI-G films. The UV–Vis-NIR absorption spectra and its normalized form for PVA–PANI-G films at low, medium, and high aniline concentrations are shown in Figure 7.

It can be seen that the absorption bands of the obtained PVA–PANI-G films present an hypsochromic effect for low and medium aniline concentrations, and a bathochromic effect for medium and high aniline concentrations, which can be attributed to a change in polarity of the molecule when interacting with the aldehyde groups of GA. The characteristic absorption bands for the electronic transitions of PANI, as well as the characteristic shift due to the increase in aniline concentration, according to the PVA–PANI experiment shown in Table 1, are also seen in Figure 7. A hyperchromic effect can be seen in the absorption band located at 380 nm, which can be related to an increase of the benzenoid section of PANI due to the activation time and concentration of GA. Comparison of bandgap variations due to concentration and activation of the material are shown in Figure S2.

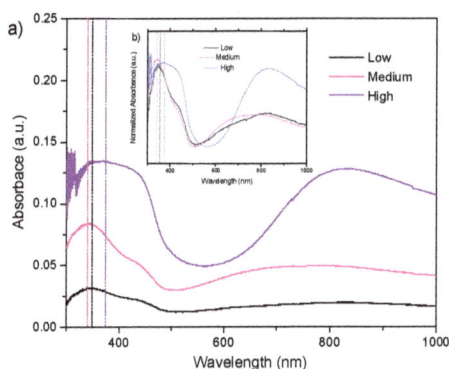

Figure 7. (a) UV–Vis-NIR absorbance spectra and (b) normalized UV–Vis-NIR absorbance spectra of PVA–PANI thin films activated with 1% GA for 30 min.

2.2. Purity, Thermal Stability and Cross-Linking of PVA–PANI Copolymers

Thermogravimetric Analysis, Derivative Thermogravimetric Analysis, and Differential Scanning Calorimetry (TGA/DTG/DSC)

In order to evaluate the thermal stability, the purity of the obtained materials by both purification methods, and to correlate an increase in hydroxyl and amine functional groups, TGA-DTG analysis were carried out. DSC study was performed to observe the cross-linking effects on PVA–PANI and PVA–PANI-G in order to evaluate the sensitivity of the material towards GA. First, to confirm purification of the material, an analysis of the transition temperatures and molecular weight for the obtained byproducts and sideproducts of the synthesis route for PVA–PANI (Figure 8) was attained as shown in Figure 9.

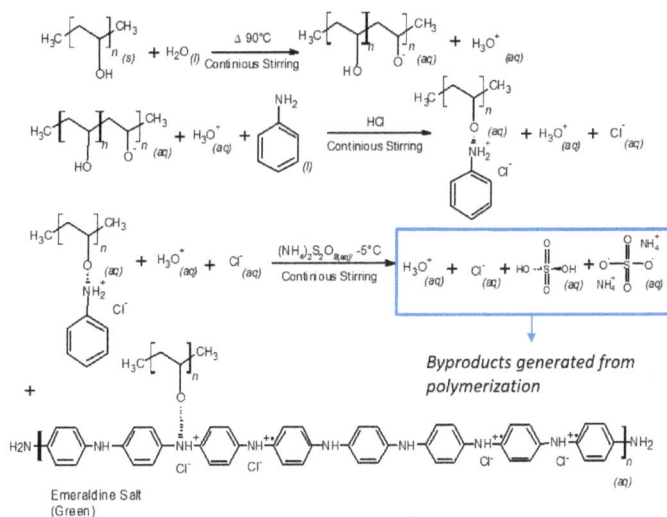

Figure 8. Synthesis route of PVA–PANI and generated byproducts.

H₃C... CH₃ ... 130,000 g/mol (200°C) ÓH

NH₂ ... 93.13 g/mol (184.1°C)

HCl 35.45 g/mol (48°C)

H₂O 19.02 g/mol (100°C)

O NH₄⁺ 132.14 g/mol (257°C) NH₄⁺ O

H₂N— —NH₂ 184.24 g/mol (400°C)

NH₂ 184.24 g/mol (354°C)

O 98.08 g/mol HO—S—OH (337°C) O

NH₄⁺ O—S—O O—S—O NH₄⁺ 228.18 g/mol (120°C)

HN ... 184.24 g/mol (354°C)

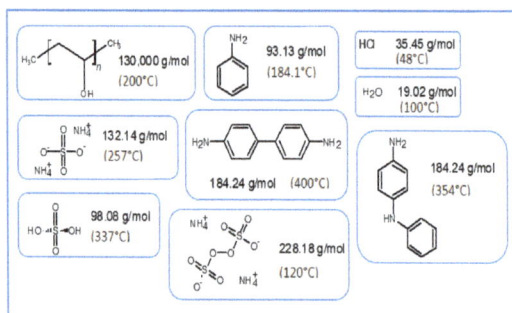

Figure 9. Molecular weight and transition temperatures of byproducts, sideproducts, and reagents present during synthesis.

We expected to observe a weight loss for each undesired product along with an exothermic peak in the temperature range of the thermograms presented in Figure 10. TGA and DTG of the material purified by precipitation–redispersion method (Figure 10a) show a loss of 8.41 wt % between 25 to 172 °C, which can be attributed to solvent and moisture loss at the surface of the material. The absence of an exothermic peak in this range and a gradual weight loss can be related to the absence of undesired products that can present a transition in this range of temperature, such as APS and HCl. Subsequently, the presence of two exothermic peaks located at 173.9 and 193.3 °C is associated to the removal of bound water, moisture, and dopant molecules inside the polymer matrix, as well as the loss of free hydroxyl functional groups due to the decomposition of PVA [29] (mass loss of 21.97 wt %). At ~250 °C an exothermic peak is shown as a shoulder corresponding to the breaking of free amine functional groups associated to the decomposition of the PANI polymeric chain, which continues to gradually decrease along with the dehydroxylation of the copolymer up to 395 °C (total mass loss of 44 wt %). In the range of 400 to 500 °C several peaks associated with the degradation of similar materials can be observed, which can be related to a variation in distribution of different molecular weights of PANI. On the other hand, PVA–PANI purified by dialysis (Figure 10b) shows a weight loss of 12.37%, which is attributed to moisture and the removal of HCl from the material [12], showing an exothermic peak at 52.94 °C. In the same way as the material purified by precipitation–redispersion, the decomposition of PVA and PANI is shown at 199.55 °C and 250.74 °C, respectively, to which the gradual loss of hydroxyl and amine functional groups can be related, having a total weight loss of 20.6% of the material. The presence of a small exothermic band at 355.25 °C can be attributed to a cross-linking reaction in the polymer matrix and not to the presence of oligomers in the material, which can be confirmed with the presence of a single exothermic peak at 489.15 °C associated with the degradation of the copolymer with a loss of 16.95 wt %. Also, from Figure 10, it can be seen that PVA–PANI purified by dialysis shows improved thermal stability and molecular integration which is evidenced by the presence of peaks correlated to decomposition and degradation of PVA and PANI only; this can be attributed to the processing of the material during the purification process.

After analyzing the thermal stability and purity of the material obtained by both purification methods, the effect of aniline concentration was studied. Figure 11 shows three weight losses. The first weight loss in the temperature range from 25 to 160 °C can be attributed to the loss of moisture and free acid, corresponding to a weight loss of 9.76%. There is another weight loss between 160 and 400 °C which is attributed to degradation of short polymer chains along with the removal of hydrogen bonds and Coulomb interactions from O–H and N–H functional groups, corresponding to a weight loss of 40.24% [30,31]. With regard to the increase of aniline, a shift to higher temperatures of 21.14 °C can be observed starting from a low concentration level to a high concentration of PANI from room temperature to 400 °C, which corresponds to a weight loss of 27%, 29%, and 30% for low, medium, and high concentrations, respectively, indicating that an increase in molecular weight of

PANI contributes to a higher thermal stability, but also a mayor weight loss because of the presence of more hydroxyl and amine functional groups in the polymer chain. The decomposition of the purified PVA–PANI gels tends to be gradual until 400 °C where there is another pronounced weight loss of 36.3% due to the degradation of the material, leaving a residue of 14.5 wt % with a variation of 1.21% between each concentration.

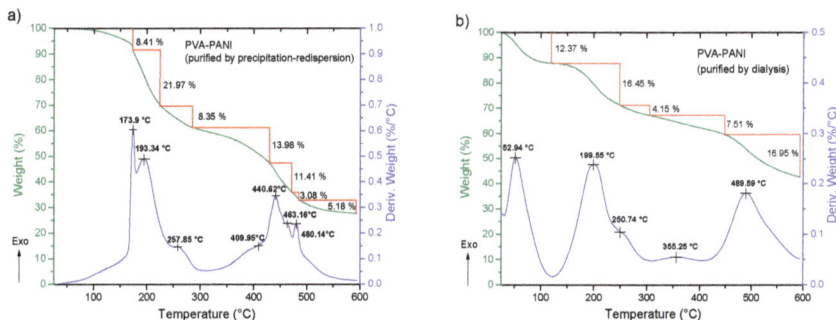

Figure 10. TGA-DTG of PVA–PANI purified by (**a**) precipitation–redispersion and (**b**) dialysis.

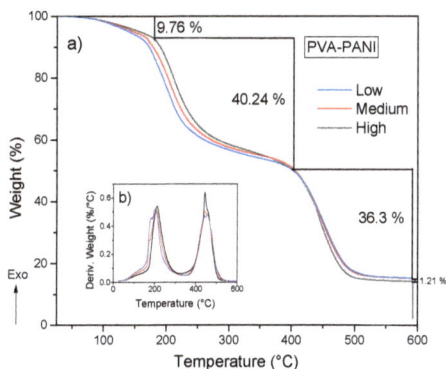

Figure 11. (**a**) TGA and (**b**) DTG of purified PVA–PANI gels at low, medium, and high concentrations of PANI.

Correlation of previously obtained thermal stability and cross-linking was studied by DSC, in order to observe energy variations associated to interactions of the polymer matrix with the PANI nanoparticles. According to the obtained DSC results (Figure 12), the presence of an endothermic peak at ~200 °C can be attributed to a cross-linking reaction between PVA and PANI at that temperature, where enthalpy changes are inversely proportional to the concentration of chemically linked PANI [32]; indicating that PVA–PANI at low concentration of monomer shows a better cross-linking between the materials. Also, it can be seen that no other prominent transition peaks are shown in the thermogram, indicating that the purified material is amorphous.

The obtained TGA-DTG results for PVA–PANI-G gels are shown in Figure 13. First, a weight loss of 6 to 9% can be observed, which corresponds to the loss of absorbed solvent and moisture. Subsequently there is a weight loss of 33 to 41% corresponding to the loss of pendant functional groups and hydroxyl and/or amino terminal groups. Starting at 400 °C, the degradation of the copolymer is seen, which no longer presents a gradual and constant decomposition as shown in Figure 11. At this range of temperature, the material shows two stages of weight loss, one with an almost linear loss with respect to the increase in temperature and another with gradual degradation, which can be attributed to a modification in the main chain of the polymer caused by the interaction with GA. PVA–PANI at high concentration activated for 30 min showed the best thermal properties.

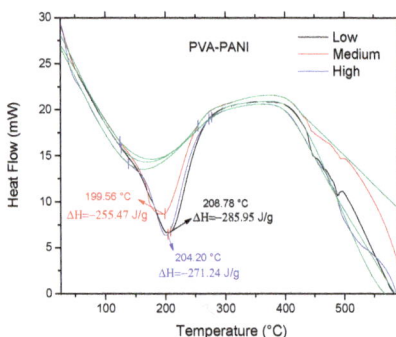

Figure 12. DSC of purified PVA–PANI gels at low, medium, and high concentrations of PANI.

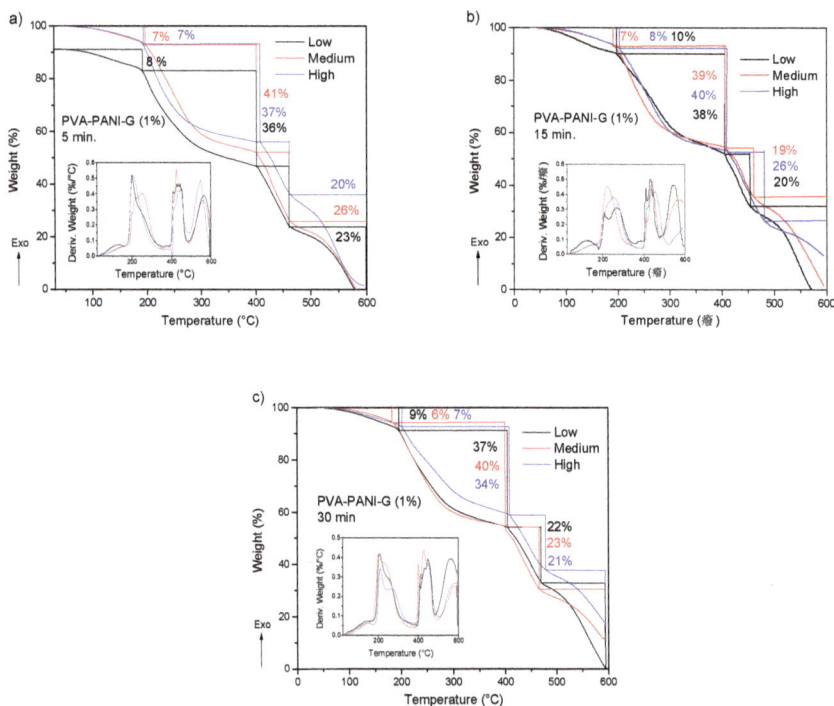

Figure 13. TGA and DTG thermograms of purified PVA–PANI-G gels, activated with 1% GA for (**a**) 5 min, (**b**) 15 min, and (**c**) 30 min.

The DSC analysis for the previous gels is shown in Figure 14. Comparing the thermogram from Figure 12 to the obtained results for the activated copolymers, it is seen that a decrease in transition temperatures is obtained for low aniline concentration along with a decrease in enthalpy from -285.95 J/g to a minimum of -320.86 J/g at 15 min of activation, which indicates that cross-linking is favored with an increase of GA for this particular concentration. However, with an increase in activation time, the material shows an increase of enthalpy up to a maximum of -56.69 J/g for a high concentration of PANI at 30 min of activation, meaning that cross-linking between PVA–PANI and GA is disfavored, which can be related to a molecular disarrangement produced by branching of PVA–PANI, making it difficult to cross-link. Results for a higher concentration of GA are provided in Figure S3.

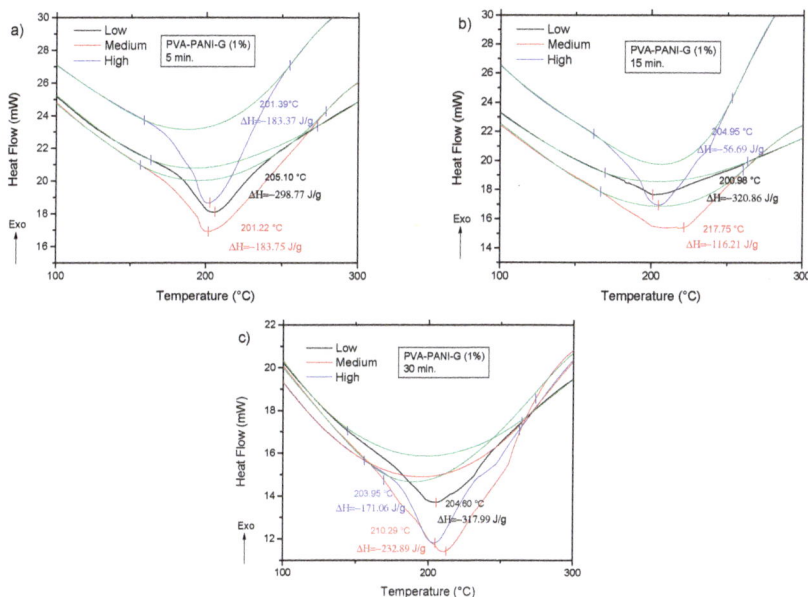

Figure 14. DSC thermograms of purified PVA–PANI-G gels, activated with 1% GA for (**a**) 5 min, (**b**) 15 min, and (**c**) 30 min.

2.3. Microstructural Variation of PVA–PANI Copolymers

2.3.1. X-ray Diffraction (XRD)

XRD analysis was carried out to observe if the material crystallinity is affected by the purification process or the concentration of PANI nanoparticles in the PVA matrix. In Figure 15, it is observed that both purification processes seem to reduce the crystallinity of the thin films showing a broad peak between 16 and 40° (2θ), where PVA–PANI thin films purified by dialysis tend to show a better molecular order than PVA–PANI purified by the precipitation–redispersion method, presenting a diffraction peak located at $2\theta = 19.77°$ which corresponds to the (101) plane of semicrystalline PVA with $d_{101} = 4.485$ Å [29,33]. These results were compared to the XRD results obtained from unpurified PVA–PANI thin films with high PVA concentration, showing the discussed diffraction peak without any trace of PANI peaks at $2\theta \approx 25°$, which indicates that the obtained PANI is amorphous.

From the data shown in Figure 16a, it can be observed that the purified PVA–PANI doped with HCl 1 M (pH = 1) has a low atomic order regardless of aniline concentration, showing a broad peak with a maximum intensity found at $2\theta = 23.29°$, which is in agreement with the reported for PANI doped with HCl [34,35]. It is expected to have a lower crystallinity of the copolymer with GA activation due to the disarrangement produced by cross-linking, increasing molecular dispersion, and broadening of the XRD peaks [36]. This behavior can be observed for PVA–PANI-G thin films at low, medium, and high concentrations of aniline (Figure 16b).

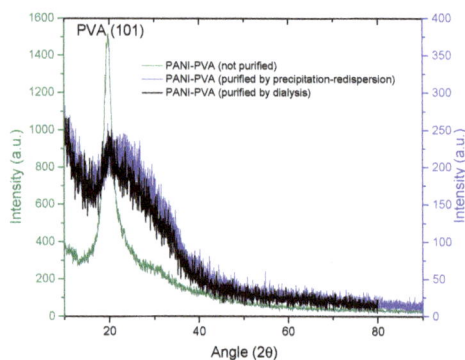

Figure 15. XRD diffractogram comparison for unpurified and purified PVA–PANI by precipitation –redispersion and dialysis.

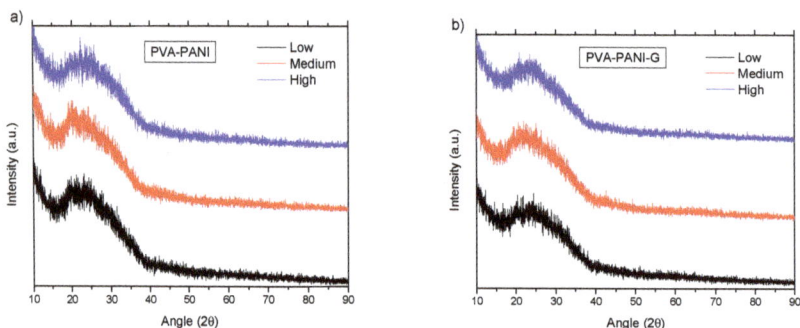

Figure 16. XRD diffractogram of (**a**) PVA–PANI thin films at different concentrations of aniline and (**b**) its comparison with PVA–PANI-G thin films.

2.3.2. Scanning Electron Microscopy (SEM)

Figure 17 shows the SEM micrographs obtained for PVA–PANI films purified by the precipitation–redispersion method for low, medium, and high aniline concentrations with respect to the PVA polymer matrix. Similar results were obtained for PVA–PANI-G thin films (Figure S4).

The morphology of the particles presents as needles which changed with increasing aniline concentration, generating particles in granular form which tend to agglomerate in different nucleation points, according to the reports of Gangopadhyay et al. [12]. On the other hand, the morphology observed for both PVA–PANI and PVA–PANI-G purified by dialysis maintained a spherical shape with an average size of 156 nm (Figure 18), which has been reported for various PANI synthesis processes [37]. This indicates that the precipitation–redispersion process (Figure 17) influences morphological changes in the material, along with pH, temperature, and solvents used.

Even though a significant difference in morphology is shown between both purification methods, we found no significant difference in the crystallinity of the material according to Figure 15, where PVA–PANI purified by dialysis tends to show a better molecular order due to the stabilizing effect of the PVA, promoting better control in the growth of nucleation sites for the obtained films.

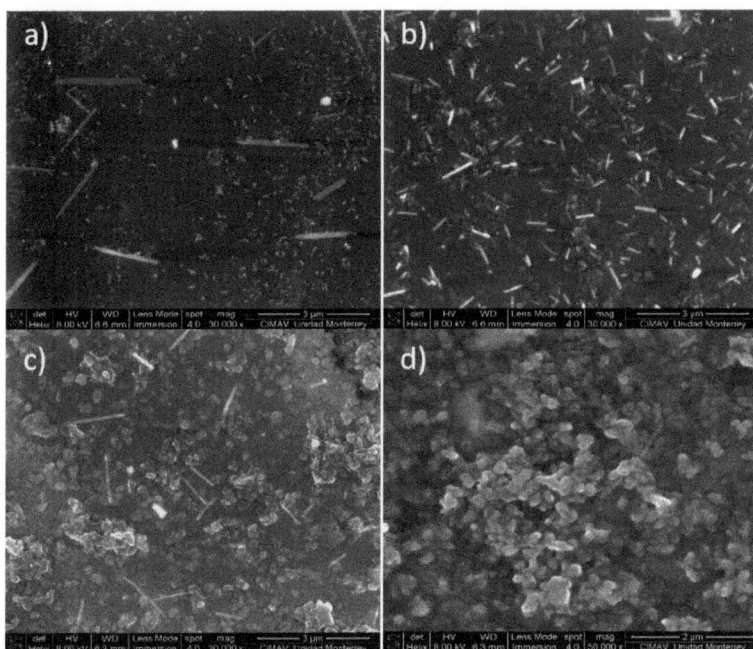

Figure 17. SEM micrographs of purified PVA–PANI thin films and its morphologic variations at (**a**) low concentration, (**b**) medium concentration, and (**c,d**) high concentration.

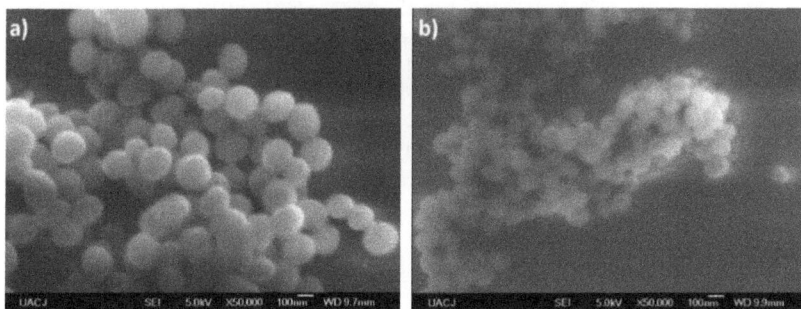

Figure 18. SEM micrographs of (**a**) PVA–PANI and (**b**) PVA–PANI-G purified by dialysis.

3. Materials and Methods

3.1. Materials

The reagents used for the synthesis of PVA–PANI copolymers are as follows. Polyvinyl alcohol (PVA) (130,000 MW, 99% hydrolyzed), high purity aniline monomer ($\geq$99.5%), and ammonium persulfate (APS) (ACS $\geq$ 98%) (Sigma Aldrich Co., Toluca, Edo. Mex., Mexico). Hydrochloric acid (36–38%) (J.T. Baker, Phillipsburg, NJ, USA) and glutaraldehyde solution (50 wt % in H_2O) (Sigma Aldrich Co.) for the activation of the obtained PVA–PANI blends. Aniline was stored in a dark environment and under refrigeration, all reagents were used as acquired.

3.2. Synthesis of PVA–PANI

First a solution of PVA 5 wt % was made by dissolving 0.0202 mmol (2.6315 g) of PVA (MW = 130,000 g/mol) in 50 mL of deionized water (Milli-Q) under constant magnetic stirring at 80 °C until a clear solution was obtained. Then, the PVA 5 wt % was used as a polymer matrix for the polymerization of aniline at different ratios with respect to the PVA as shown in Table 1.

Subsequently, the pH of the matrix was adjusted to a value of ≤2, and it was taken immediately to an ice bath (T ≤ 5 °C) where ammonium persulfate was added drop wise at a ratio of 1:1 M to aniline according to Table 2. After a few minutes, the color change of the colloidal dispersion was observed, obtaining a dark green dispersion without sedimentation as reported in the literature [13].

Table 2. Monomer and intermediary molar ratios.

PVA (mmol)	Aniline (mmol)	Aniline (M)	HCl (mmol)	APS (mmol)	APS (M)
0.0202	3.532	0.070	100	3.532	0.070
0.0202	7.064	0.141	100	7.064	0.141
0.0202	10.596	0.212	100	10.596	0.212
0.0202	14.128	0.282	100	14.128	0.282
0.0202	17.660	0.353	100	17.660	0.353
0.0202	21.192	0.423	100	21.192	0.423

3.3. Purification Process

When carrying out the PVA–PANI synthesis by the polymerization dispersion method, a number of byproducts related to doping, oxidant material (APS) and oligomers are generated (Figure 8). Therefore, these byproducts must be eliminated by means of a purification process, in order to obtain the PVA–PANI material without any other product that influences the effects produced by the material in future characterizations.

The purification of PVA–PANI copolymers was carried out mainly by two methods: precipitation–redispersion based on a solvent system $CH_3OH:H_2O$ (5:1) in which the copolymer precipitates, then this precipitate is filtered, dried in vacuum, and redispersed in deionized water (Milli-Q) under temperature and continuous agitation. As well as a dialysis process, which is based on the implementation of membranes with a cut out molecular weight (MWCO) of 12,000 Da (Obtained from Sigma Aldrich Co.), where a certain amount of PVA–PANI is placed against deionized water (Milli-Q) for 48 h so that most byproducts permeate through the membrane. Prior to its chemical and thermal characterization, the obtained PVA–PANI was vacuum-filtered and dried.

3.4. Thin Film Development

PVA–PANI thin films were obtained by dip coating (for precipitation–redispersion purification) and spin coating (for dialysis purification) physical methods using Corning glass substrates previously cleaned with acetone, isopropanol, and deionized water (Milli-Q). Once the thin films were obtained they were washed, dried with N_2 gas, and kept in a desiccator before being analyzed.

3.5. Activation of PVA–PANI Thin Films

Activation of PVA–PANI films with GA was carried out by immersing them in a 1% dilution of GA for a time of 5, 15, and 30 min at room temperature. Followed by a wash with deionized water (Milli-Q) to remove any agent that did not reacted. Later they were placed in a desiccator for storage.

3.6. Characterization Methods

Purity and molecular structure of the obtained PVA–PANI blend was confirmed by ^{1}H-NMR using a Brucker NMR equipment (Brucker, Billerica, MA., USA) at a frequency of 500 MHz, using tetramethyl silane (TMS) as standard at room temperature. To complement molecular structure and observe the present functional groups in the material an FTIR analysis was carried out using a Thermo Scientific Nicolet iS10 FT-IR spectrometer (ATR) in the region from 400 to 4000 cm^{-1} in air atmosphere and at room temperature. Optical properties and transitions of the developed PVA–PANI thin films were analyzed using a Jenway 6850 UV–Vis-NIR spectrophotometer (Bibby Scientific, Staffs., UK) in the region from 1000 to 300 nm in absorption mode. The thermal stability, composition, purity, and cross-linking effect of the PVA–PANI gels were analyzed using a TGA/DSC SDT-Q600 implementing a temperature program from 25 °C to 600 °C with a heating speed of 10 °C/min and controlled N$_2$ atmosphere. Microstructural characteristics for the thin films were obtained by XRD using a Panalytical diffractometer with a Cu kα radiation with a 1.54 wavelength used at a glancing angle. And the morphological variations of the material were observed with a Nova NanoSEM 200 equipment (FEI, Tokyo, Japan) using secondary electrons.

4. Conclusions

The purification of HCl-doped PVA–PANI copolymers at different aniline concentrations and the effect of their activation with GA was studied. ^{1}H-NMR and FTIR analysis was used to confirm the molecular structure of the material as proposed from the skeletal formula before carrying out further characterizations. Regarding the activation of the material, it was observed that GA has a reduction effect on the PVA–PANI copolymer (increasing the number of benzene units of PANI), which increases proportionally with immersion time and concentration as confirmed by FTIR and UV–Vis-NIR. The activation of the material was confirmed by FTIR showing the characteristic absorption bands in the 1550 to 1750 cm^{-1} range, where the C=N (Schiff base) bond corresponding to the formation of imines can be found. Further analysis of cross-linking between PVA–PANI and GA was carried out by DSC, showing that the interaction of GA towards the material is favored at low aniline concentration, where minor variations where observed due to doping level. The purity of the material was confirmed by TGA-DTG, which showed bands corresponding to weight losses and exothermic peaks of PVA–PANI only. Moreover, it was seen that the precipitation–redispersion method seems to affect the structural and morphological properties of the PVA–PANI at different aniline concentrations, even after GA activation, as shown from the XRD and SEM results. Therefore, the dialysis purification method is recommended for upcoming experiments. We propose, as future work, to carry out experiments immobilizing an antibody on the activated material to test it as a platform for biological detection, in order to apply it as an active membrane in future in vitro diagnostic devices.

Supplementary Materials: Supplementary materials are available online.

Author Contributions: Conceptualization, J.M.G., A.C., and F.S.A.; Methodology, J.M.G., A.C., and M.L.M.; Formal Analysis, J.M.G., A.C., M.L.M., and F.S.A.; Investigation, J.M.G., A.C., M.L.M., R.C.A., and F.S.A; Resources, A.C. and F.S.A.; Data Curation, J.M.G.; Writing—Original Draft Preparation, J.M.G; Writing—Review and Editing, J.M.G., R.C.A., A.C., M.L.M., and F.S.A.; Visualization, J.M.G; Supervision, A.C., M.L.M., R.C.A., and F.S.A.; Validation, Project Administration, and Funding Acquisition, F.S.A., A.C., and CONACYT.

Funding: This research was funded by CONACYT, México, Grant Number 294690.

Acknowledgments: The authors would like to acknowledge to José Guadalupe Téllez Padilla (^{1}H-NMR), Alberto Toxqui Terán (TGA/DTG/DSC), Nayely Pineda Aguilar (SEM), and Francisco Enrique Longoria Rodríguez (XRD) for assisting in the presented characterization, and personnel from UACJ (Chihuahua, México), Biotechnology Incubator (Nuevo León, México), and CIMAV (Nuevo León, México) for technical support and for providing the infrastructure to develop this work.

Conflicts of Interest: The authors declare no conflicts of interest.

References

1. Shirakawa, H.; Louis, E.J.; Chiang, C.K.; Heeger, A.J. Synthesis of Electrically Conducting Organic Polymers: Halogene Derivatives of Polyacetylene, (CH)x. *J. C. S. Chem. Comm.* **1977**, *16*, 578–580. [CrossRef]
2. Blinova, N.V.; Stejskal, J.; Trchová, M.; Ćirić-Marjanović, G.; Sapurina, I. Polymerization of aniline on polyaniline membranes. *J. Phys. Chem. B* **2007**, *111*, 2440–2448. [CrossRef] [PubMed]

3. Gao, X.Z.; Liu, H.J.; Cheng, F.; Chen, Y. Thermoresponsive polyaniline nanoparticles: Preparation, characterization, and their potential application in waterborne anticorrosion coatings. *Chem. Eng. J.* **2016**, *283*, 682–691. [CrossRef]

4. Wang, P.; Liu, M.; Kan, J. Amperometric phenol biosensor based on polyaniline. *Sensors Actuators B Chem.* **2009**, *140*, 577–584. [CrossRef]

5. Parente, A.H.; Marques, E.T.A.; Azevedo, W.M.; Filho, J.L.L. Glucose Biosensor Using Glucose Oxidase Immobilized in Polyaniline. *Appl. Biochem. Biotechnol.* **1992**, *37*, 267–273. [CrossRef]

6. Tiwari, A.; Shukla, S.K. Chitosan-g-polyaniline: A creatine amidinohydrolase immobilization matrix for creatine biosensor. *Express Polym. Lett.* **2009**, *3*, 553–559. [CrossRef]

7. Chowdhury, A.D.; De, A.; Chaudhuri, C.R.; Bandyopadhyay, K.; Sen, P. Label free polyaniline based impedimetric biosensor for detection of *E. coli* O157:H7 Bacteria. *Sensors Actuators B Chem.* **2012**, *171–172*, 916–923. [CrossRef]

8. Sawangphruk, M.; Suksomboon, M.; Kongsupornsak, K.; Khuntilo, J.; Srimuk, P.; Sanguansak, Y.; Klunbud, P.; Suktha, P.; Chiochan, P. High-performance supercapacitors based on silver nanoparticle-polyaniline- graphene nanocomposites coated on flexible carbon fiber paper. *J. Mater. Chem. A* **2013**, *1*, 9630–9636. [CrossRef]

9. Bhadra, J.; Sarkar, D. Field effect transistor fabricated from polyaniline-polyvinyl alcohol nanocomposite. *Indian J. Phys.* **2010**, *84*, 693–697. [CrossRef]

10. Qin, Y. Alginate fibers: An overwiew of the production processes and applications in wound management. *Polym. Int.* **2008**, *57*, 171–180. [CrossRef]

11. Ameen, S.; Ali, V.; Zulfequar, M.; Mazharul, M.; Husain, M. Synthesis and characterization of polyaniline-polyvinyl chloride blends doped with sulfamic acid in aqueous tetrahydrofuran. *Cent. Eur. J. Chem.* **2006**, *4*, 565–577. [CrossRef]

12. Gangopadhyay, R.; De, A.; Ghosh, G. Polyaniline-poly(vinyl alcohol) conducting composite: Material with easy processability and novel application potential. *Synth. Met.* **2001**, *123*, 21–31. [CrossRef]

13. Bhadra, J.; Sarkar, D. Self-assembled polyaniline nanorods synthesized by facile route of dispersion polymerization. *Mater. Lett.* **2009**, *63*, 69–71. [CrossRef]

14. Mirmohseni, A.; Wallace, G.G. Preparation and characterization of processable electroactive polyaniline-polyvinyl alcohol composite. *Polymer* **2003**, *44*, 3523–3528. [CrossRef]

15. Bhadra, J.; Al-Thani, N.J.; Madi, N.K.; Al-Maadeed, M.A. Effects of aniline concentrations on the electrical and mechanical properties of polyaniline polyvinyl alcohol blends. *Arab. J. Chem.* **2015**, *10*, 664–672. [CrossRef]

16. Purcena, L.L.A.; Caramori, S.S.; Mitidieri, S.; Fernandes, K.F. The immobilization of trypsin onto polyaniline for protein digestion. *Mater. Sci. Eng. C* **2009**, *29*, 1077–1081. [CrossRef]

17. Caramori, S.S.; De Faria, F.N.; Viana, M.P.; Fernandes, K.F.; Carvalho, L.B. Trypsin immobilization on discs of polyvinyl alcohol glutaraldehyde/ polyaniline composite. *Mater. Sci. Eng. C* **2011**, *31*, 252–257. [CrossRef]

18. Stejskal, J.; Hajná, M.; Kašpárková, V.; Humpolíček, P.; Zhigunov, A.; Trchová, M. Purification of a conducting polymer, polyaniline, for biomedical applications. *Synth. Met.* **2014**, *195*, 286–293. [CrossRef]

19. Alves, M.; Young, C.; Bozzetto, K.; Poole-Warren, L.A.; Martens, P.J. Degradable, poly(vinyl alcohol) hydrogels: Characterization of degradation and cellular compatibility. *Biomed. Mater.* **2012**, *7*, 024106. [CrossRef]

20. Nagy, M.; Szollosi, L.; Keki, S.; Faust, R.; Zsuga, M. Poly(vinyl alcohol)-based amphiphilic copolymer aggregates as drug carrying nanoparticles. *J. Macromol. Sci. Part A Pure Appl. Chem.* **2009**, *46*, 331–338. [CrossRef]

21. Andrade, G.; Barbosa-Stancioli, E.F.; Piscitelli Mansur, A.A.; Vasconcelos, W.L.; Mansur, H.S. Design of novel hybrid organic-inorganic nanostructured biomaterials for immunoassay applications. *Biomed. Mater.* **2006**, *1*, 221–234. [CrossRef] [PubMed]

22. Trchová, M.; Stejskal, J. Polyaniline: The infrared spectroscopy of conducting polymer nanotubes (IUPAC Technical Report). *Pure Appl. Chem.* **2011**, *83*, 1803–1817. [CrossRef]

23. Bhadra, J.; Sarkar, D. Size variation of polyaniline nanoparticles dispersed in polyvinyl alcohol matrix. *Bull. Mater. Sci.* **2010**, *33*, 519–523. [CrossRef]

24. Mansur, H.S.; Sadahira, C.M.; Souza, A.N.; Mansur, A.A.P. FTIR spectroscopy characterization of poly (vinyl alcohol) hydrogel with different hydrolysis degree and chemically crosslinked with glutaraldehyde. *Mater. Sci. Eng. C* **2008**, *28*, 539–548. [CrossRef]

25. Lin-Vien, D.; Colthup, N.B.; Fateley, W.G.; Grasselli, J.G. *The Handbook of Infrared and Raman Characteristic Frequencies of Organic Molecules*, 1st ed.; Academic Press: San Diego, CA, USA, 1991; p. 503. ISBN 9780080571164.

26. De Melo, J.V.; Bello, M.E.; de Azevêdo, W.M.; de Souza, J.M.; Diniz, F.B. The effect of glutaraldehyde on the electrochemical behavior of polyaniline. *Electrochim. Acta* **1999**, *44*, 2405–2412. [CrossRef]

27. Gomes, E.C.; Oliveira, M.A.S. Chemical Polymerization of Aniline in Hydrochloric Acid (HCl) and Formic Acid (HCOOH) Media. Differences Between the Two Synthesized Polyanilines. *Am. J. Polym. Sci.* **2012**, *2*, 5–13. [CrossRef]

28. Yang, D.; Lu, W.; Goering, R.; Mattes, B.R. Investigation of polyaniline processibility using GPC/UV-vis analysis. *Synth. Met.* **2009**, *159*, 666–674. [CrossRef]

29. Honmute, S.; Ganachari, S.V. Studies on Polyaniline-Polyvinyl Alcohol (PANI-PVA) Interpenetrating Polymer Network (IPN) Thin Films. *Int. J. Sci. Res.* **2012**, *1*, 102–106.

30. Bhadra, J.; Madi, N.K.; Al-Thani, N.J.; Al-Maadeed, M.A. Polyaniline/polyvinyl alcohol blends: Effect of sulfonic acid dopants on microstructural, optical, thermal and electrical properties. *Synth. Met.* **2014**, *191*, 126–134. [CrossRef]

31. Mahato, M.; Adhikari, B. Vapor phase sensing response of doped polyaniline-poly (vinyl alcohol) composite membrane to different aliphatic alcohols. *Synth. Met.* **2016**, *220*, 410–420. [CrossRef]

32. Vargas, L.R.; Poli, A.K.; Lazzarini Dutra, R.; Brito de Souza, C.; Ribeiro Baldan, M.; Sarmento Goncalves, E. Formation of Composite Polyaniline and Graphene Oxide by Physical Mixture Method. *J. Aerosp. Technol. Manag.* **2017**, *9*, 29–38. [CrossRef]

33. Meftah, A.M.; Gharibshahi, E.; Soltani, N.; Mat Yunus, W.M.; Saion, E. Structural, optical and electrical properties of PVA/PANI/Nickel nanocomposites synthesized by gamma radiolytic method. *Polymers* **2014**, *6*, 2435–2450. [CrossRef]

34. Mansour, F.; Elfalaky, A.; Maged, F.A. Synthesis, Characterization and Optical properties of PANI/ PVA Blends. *IOSR J. Appl. Phys.* **2015**, *7*, 37–45.

35. Geethalakshmi, D.; Muthukumarasamy, N.; Balasundaraprabhu, R. Effect of dopant concentration on the properties of HCl-doped PANI thin films prepared at different temperatures. *Optik* **2014**, *125*, 1307–1310. [CrossRef]

36. Hu, H.; Xin, J.H.; Hu, H.; Chan, A.; He, L. Glutaraldehyde-chitosan and poly (vinyl alcohol) blends, and fluorescence of their nano-silica composite films. *Carbohyd. Polym.* **2013**, *91*, 305–313. [CrossRef] [PubMed]

37. Zhang, L.; Ma, H.; Cong, C.; Su, Z. Nonaqueos synthesis of uniform polyaniline nanospheres via cellulose acetate template. *J. Polym. Sci. Part A Polym. Chem.* **2012**, *50*, 912–917. [CrossRef]

Sample Availability: Samples of the compounds are not available from the authors.

molecules

MDPI

Article

Synthesis of Carbohydrate-Grafted Glycopolymers Using a Catalyst-Free, Perfluoroarylazide-Mediated Fast Staudinger Reaction

William Ndugire, Bin Wu and Mingdi Yan *

Department of Chemistry, University of Massachusetts Lowell, 1 University Ave., Lowell, MA 01854, USA;
William_Ndugire@student.uml.edu (W.N.); Bin_Wu1@student.uml.edu (B.W.)
* Correspondence: mingdi_yan@uml.edu; Tel.: +978-934-3647

Received: 5 December 2018; Accepted: 30 December 2018; Published: 3 January 2019

Abstract: Glycopolymers have gained increasing importance in investigating glycan-lectin interactions, as drug delivery vehicles and in modulating interactions with proteins. The synthesis of these glycopolymers is still a challenging and rigorous exercise. In this regard, the highly efficient click reaction, copper (I)-catalyzed alkyne-azide cycloaddition, has been widely applied not only for its efficiency but also for its tolerance of the appended carbohydrate groups. However, a significant drawback of this method is the use of the heavy metal catalyst which is difficult to remove completely, and ultimately toxic to biological systems. In this work, we present the synthesis of carbohydrate-grafted glycopolymers utilizing a mild and catalyst-free perfluorophenyl azide (PFPA)-mediated Staudinger reaction. Using this strategy, mannose (Man) and maltoheptaose (MH) were grafted onto the biodegradable poly(lactic acid) (PLA) by stirring a PFAA-functionalized PLA with a phosphine-derivatized Man or MH in DMSO at room temperature within an hour. The glycopolymers were characterized by ^{1}H-NMR, ^{19}F-NMR, ^{31}P-NMR and FTIR.

Keywords: Glycopolymer; post-polymerization functionalization; perfluoroaryl azides; Staudinger reaction

---

1. Introduction

Carbohydrates are not only the core of metabolism in many biological systems, but are also integral in many cells as structural [1], communication [2], and recognition [3] elements. Synthetic carbohydrate-functionalized polymers, i.e., glycopolymers, has become an important tool in fundamental glycobiology research, and in biomedical applications such as sensing [4,5], drug delivery [6], and cryopreservation [7,8]. Synthetic glycopolymers can be categorized broadly into two types based on the mode of synthesis: (1) polymerization of carbohydrate-derivatized monomers (such as allyl [9], acrylamide [10] and acrylate [11]), and (2) grafting of carbohydrates onto a polymer backbone [12–17]. The latter technique, typically referred to as post-polymerization modification, allows facile synthesis of the polymer backbone and control over carbohydrate grafting density. To successfully graft carbohydrates onto the polymer, the conjugation reaction must be highly efficient, of high yield, and tolerant of the functional groups on the polymer. The popular 'click' type of reactions, particularly the copper(I)-catalyzed alkyne-azide cycloaddition (CuAAC), meet these requirements and have been applied in the synthesis of glycopolymers by post-polymerization and in glycosylation of surfaces [18–20]. This reaction tolerates a wide variety of carbohydrate side groups. Analytically, it offers the advantage in the form of the distinct chemical shift of the triazole proton in ^{1}H-NMR that facilitates straightforward characterization [21,22]. However, a drawback of this reaction is the necessity of Cu (I) catalyst that has proved difficult to be removed completely from the products. While at least 2–5 mol% of Cu (I) is required to achieve a respectable reaction rate, even small amounts of residual metal in the final product has been shown to denature proteins and cause oxidative damage to cells [21,23,24]. Other click-type reactions that do not require metal catalysts include SPAAC (strain-promoted azide-alkyne cycloaddition), azide-aryne based "benzyne click" reaction, and iEDDA (inverse electron demand Diels–Alder) reaction [25–27], but these have not been applied directly in the synthesis of glycopolymers.

Another click-type reaction is the phosphine-azide Staudinger reaction [28]. The recent improvement by Bertozzi via the introduction of an ester trap [29] rendered this bioorthogonal reaction suitable for in vivo conjugations [30,31] and in oligosaccharide metabolic engineering for cell-surface labelling [29,32]. Despite its versatility, the Staudinger ligation suffers from a slow reaction rate. To overcome this issue, we have recently shown that by using an electron-deficient perfluoroaryl azide (PFPA), the rate of reaction can be increased up to four orders of magnitude to give an iminophosphorane product [33]. The rate enhancement can be attributed to the highly electronegative F atoms lowering the LUMO of PFPA thus accelerating the reactions with dienophiles and nucleophiles [34–37]. In addition, the PFPA-iminophosphorane formed in this reaction was stable and not readily hydrolyzed in vivo, as demonstrated in the metabolic labeling of A549 cells [33].

The fast reaction rate and high yield make this PFPA-Staudinger reaction an excellent candidate for polymerization. In an earlier work, we demonstrated that the reaction of bis-PFPA and a bisphosphene occurred at room temperature in 30 min to yield poly(iminophosphorane) having molecular weight of over 59,000 and a narrow dispersity of 1.1–1.2 [38]. In this work, we applied this reaction in the post-polymerization synthesis of carbohydrate–grafted glycopolymers. By taking advantage of the fast reaction rate, mild reaction conditions, and high chemoselectivity of the PFPA-Staudinger reaction, we conjugated maltoheptaose and mannose onto poly(lactic acid) (PLA), a biocompatible and biodegradable glycopolymer.

2. Results and Discussion

While the availability of pendant functional groups like the hydroxyl on the polymer would be suited for condensation with a carboxyl-modified sugar, this strategy is less desirable as it requires the use of coupling agents that can be difficult to remove like DCC/DMAP [39] or strong acid [40] that would hydrolyze the polymer.

Our design for grafting carbohydrates to PLA was to functionalize PLA with PFPA followed by reaction with a phosphine-derivatized carbohydrate. A PLA-*co*-PLA-PFPA copolymer is synthesized so that the density of PFPA and the carbohydrate can be varied and controlled. A hydroxy-functionalized polylactide copolymer **4** (Scheme 1A) was synthesized to conjugate PFPA and subsequently graft the carbohydrate. The benzyl-derivatized lactide monomer **1** was synthesized according to Scheme 1B. Ring-opening copolymerization of lactide **1** and lactide **2** in toluene using stannous octoate as the catalyst gave polylactide copolymer **4** in 74% yield [5]. The ratio of the two monomers can be varied so as to control the grafting density of carbohydrate on the PLA polymer. In this work, the polymer obtained had a monomer ratio (m:n) of 1:22 for **1** and **2** after copolymerization. Deprotection of the benzyl group on **3** gave the hydroxy-functionalized polylactide copolymer **4** in 63% yield [5]. The m:n ratio was calculated from the ^{1}H-NMR spectrum of **4** (Figure 1A) by taking the ratio of the methine protons (H-a and H-c) that overlap at 5.2 ppm. The integral value of the H-a is equivalent to methylene protons H-b/2. Subtraction of this value from the overlapped (H-a + H-c) gives the integration of H-c and therefore H-a/H-c can be calculated.

(A)

(B)

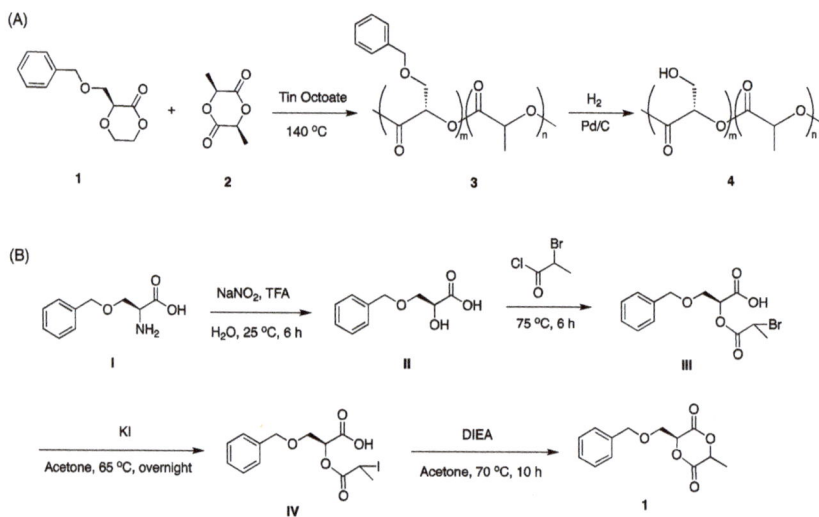

Scheme 1. Synthesis of (**A**) hydroxy-functionalized polylactide copolymer **4**, and (**B**) lactide monomer **1**.

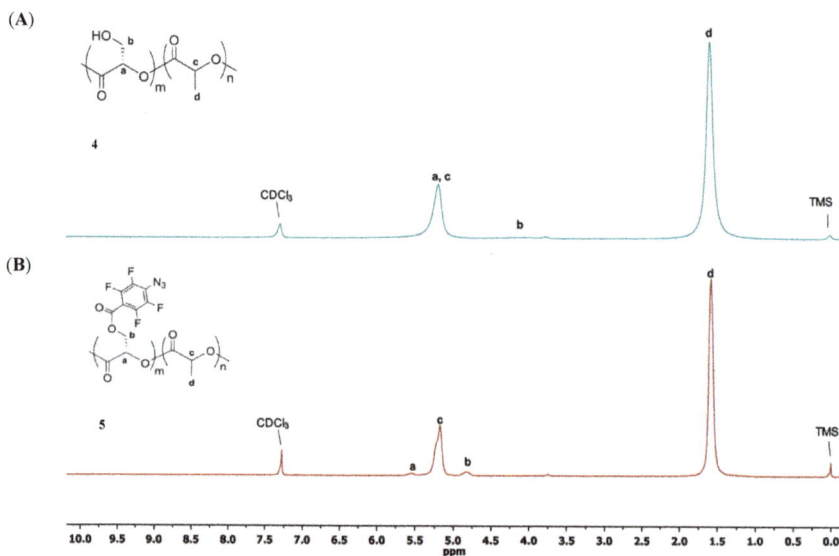

(A)

(B)

Figure 1. ^{1}H-NMR spectra of copolymers **4** (**A**) and **5** (**B**) in CDCl$_3$.

PFPA-functionalized PLA copolymer **5** was prepared by esterification of **4** with carboxy-derivatized PFPA using DCC and a catalytic amount of DMAP giving **5** in 86% yield (Scheme 2).

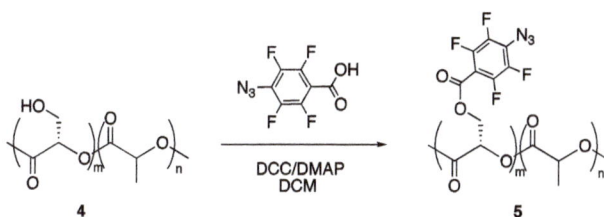

Scheme 2. Synthesis of PFPA-PLA copolymer **5**.

The ^{1}H-NMR spectrum of **5** showed a near complete reaction as demonstrated by a marked shift of the methylene proton H-b from 4.05 ppm to 4.83 ppm (Figure 1). No degradation of the copolymers to the free lactic acid monomer was observed during the reaction, as evidenced by the absence of peaks at ~1.2 ppm which belongs to the free lactic acid. In addition, the characteristic azide peak at 2133 cm^{-1} appeared in the Fourier transform-infrared spectroscopy (FTIR) spectrum of copolymer 5 (Figure 2).

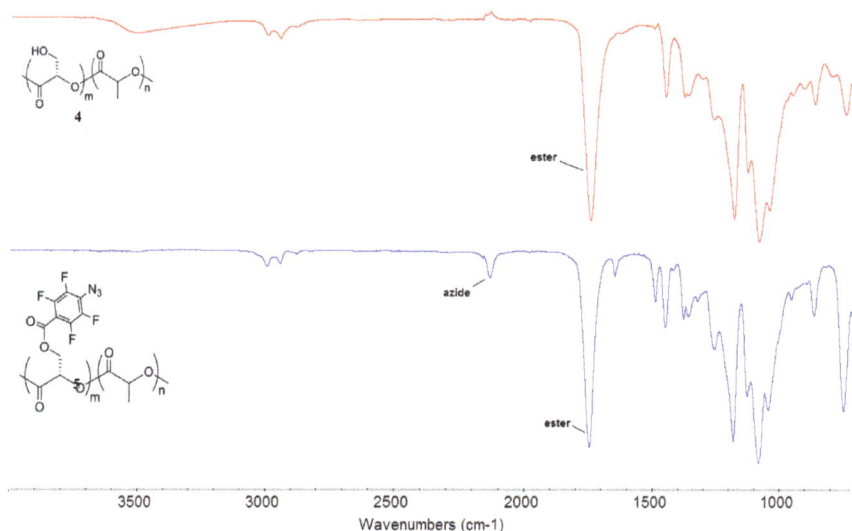

Figure 2. FTIR spectra of PLA copolymers **4** and **5**.

The phosphine-derivatized carbohydrates were prepared from an amine-derivatized carbohydrate and the NHS-functionalized phosphine (Scheme 3). A monosaccharide, D-mannose (Man), and an oligosaccharide, D-maltoheptaose (MH), were used as model carbohydrates in this study. The amine-Man **6** was synthesized according to previously reported procedure (see Scheme S2 and detailed procedures in SI) [41]. The amine-MH **7** was synthesized following the procedure in Scheme S1 (see detailed procedures in SI). Reaction of NHS-functionalized triphenylphosphine with excess of **6** or **7** in DMSO at room temperature gave the triphenylphosphine-derivatized Man (**8**) or MH (**9**).

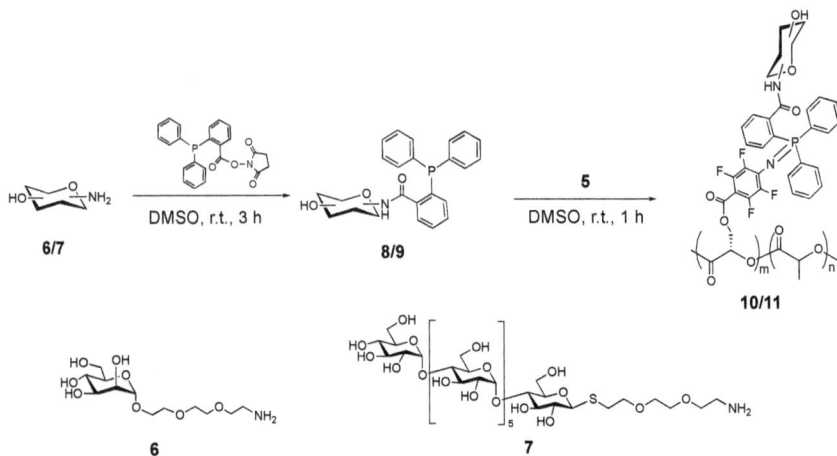

Scheme 3. Synthesis of phosphine-derivatized mannose **8** and maltoheptaose **9**, and subsequent grafting to PFPA-PLA copolymer **5** to yield mannose-polymer **10** and maltoheptaose-polymer **11**, respectively.

To prepare Man- or MH-grafted PLA **10** and **11**, copolymer **5** was added directly to the reaction mixture of **8** or **9** in DMSO and stirred at room temperature for 1 h. The products were purified by dialysis for 48 h and dried by lyophilization. After the carbohydrate was grafted, the aromatic and the carbohydrate peaks appeared in the ^{1}H-NMR spectra of **10** and **11** at 7.5–7.7 ppm and 3.0–6.0 ppm, respectively (Figure 3). In the ^{31}P-NMR spectra of the products (Figure 4), a new peak was observed at 13 ppm after conjugation of the phosphine onto the copolymer. The absence of any peaks higher than 13 ppm confirmed the absence of byproducts resulting from the oxidation of phosphine.

The yield of conjugation was obtained from the ^{1}H-NMR spectrum by taking the ratio of peak integration of the phenyl protons (Ph) at 7.5–7.7 ppm and the methyl protons H-b at 1.48 ppm, together with the previously calculated monomer ratio of m:n = 1:22 to give the formula:

$$(\%)coupling\ yield = \left(\frac{\frac{Ph}{14} \times 22}{\frac{H-b}{3}} \right) \times 100 \tag{1}$$

Using this equation, the yields were calculated to be 25% and 35% for the Man-grafted PLA copolymer **10** and MH-grafted PLA copolymer **11**, respectively (see Figures S5 and S6 for peak integrations).

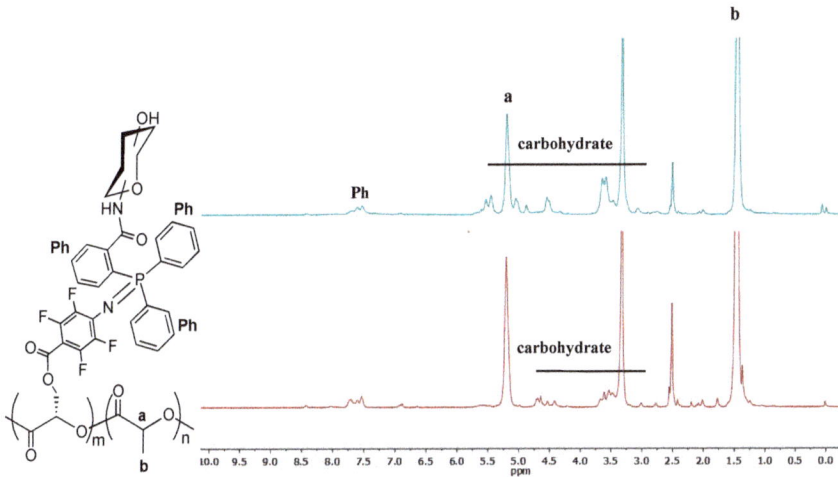

Figure 3. ^{1}H-NMR spectra of Man- and MH-grafted copolymers **10** (**top**) and **11** (**bottom**) in DMSO-d$_6$.

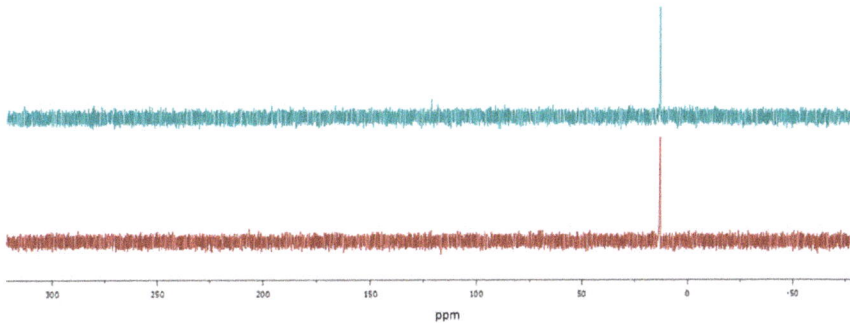

Figure 4. ^{31}P-NMR spectra of copolymer **10** (**top**) and **11** (**bottom**) in DMSO-d$_6$.

3. Conclusions

In this work, we utilized an electron-deficient perfluorophenyl azide-mediated fast Staudinger reaction to efficiently synthesize carbohydrate grafted-polylactide glycopolymers by post-polymerization modification. A polylactide copolymer was synthesized by stannous octoate-catalyzed cationic ring-opening copolymerization, which was subsequently modified with PFPA. Conjugation with a phosphine-derivatized carbohydrate yield mannose- and maltoheptaose-grafted polylactide glycopolymers in 35% and 25% yields. These yields are comparable to those obtained by other groups in post-polymerization synthesis of glycopolymers using other techniques [42,43]. The main limiting factor of efficient grafting being steric hindrance between ligands and the polymer backbone restricting access to the reactive sites. In our work, this occurs between carbohydrate-phosphines reaction with the pendant PFPA groups, which is further complicated by the difference in polarity between the PLA and D-mannose/maltoheptaose. However, the grafting reaction is fast, carried out under mild conditions without the use of any catalyst. This metal catalyst-free approach to glycopolymer synthesis is significant as it eliminates the concerns over the potential toxicity of heavy metals, making these glycopolymers attractive for biomedical applications.

4. Experimental Procedures

4.1. Materials and Instruments

All chemicals and solvents were purchased from Sigma-Aldrich (St. Louis, MO, USA), TCI America (Portland, Oregon, USA) or Fisher Scientific (Hampton, NH, USA), and were used without further purification unless otherwise noted. Dichloromethane (DCM), dimethylformamide (DMF), ethyl acetate (EtOAc) and dimethyl sulfoxide (DMSO) were purified by distillation over CaH_2. Deuterated solvents were acquired from Cambridge Isotope Lab., Inc. (Tewksbury, MA, USA). Amberlite IRC-120H$^+$ resin was activated by washing with NaOH and HCl, followed by water, ethanol and toluene.

Nuclear magnetic resonance spectroscopy data were collected on either a Bruker 500 MHz spectrometer (^{1}H-NMR) or a Bruker 200 MHz spectrometer (^{19}F- and ^{31}P-NMR) (Bruker Corporation, Billerica, MA, USA). FT-IR spectra were recorded on a Nicolet 6700 FT-IR spectrometer (Thermo Fisher Scientific, Waltham, MA, USA).

4-Azido-2,3,5,6-tetrafluorobenzoic acid (PFPA-COOH) was synthesized following our previously developed protocol [44,45]. The detailed synthesis of 1-(2-(2-(2-aminoethoxy)ethoxy)ethoxy-α-D-mannopyranoside (**6**, Scheme S2) [41,46] and 1-(2-(2-(2-aminoethoxy)ethoxy)ethoxy-maltoheptaoside (**7**, Scheme S1) can be found in the Supporting Information.

4.2. Synthesis of 3-benzyloxy-2-hydroxypropionic acid (II)

Synthesized according to literature procedure [47]. To a 200 mL of 0.7 M aqueous solution of trifluoroacetic acid, H-Ser(benzyl)-OH (**I**, 10.0 g, 52 mmol) was added, and the mixture was stirred at room temperature until all solids were dissolved. Then, 50 mL of aqueous $NaNO_2$ (5.3 g, 77 mmol) was added dropwise with a syringe pump under Ar protection and the reaction was stirred for another 3 h. After confirming of the consumption of starting material by TLC, NaCl (10 g) was added and the mixture was extracted with ethyl acetate three times followed by washing with brine and dried over $MgSO_4$. After passing through a flash column using CH_2Cl_2/MeOH/AcOH ($v/v/v$ 100:8:1), compound **II** was obtained (7.2 g, 72%).^{3} ^{1}H-NMR (500 MHz, DMSO-d$_6$) δ 7.35–7.26 (m, 5H, Ar-H), 4.60 (s, 2H, PhCH$_2$O-), 4.38 (s, 1H, -OCH$_2$CH(COOH)O-) 3.83 (d, J = 15 Hz, 2H, -OCH$_2$CH-). IR: 3324, 3070, 3032, 2925, 2871, 2645, 2537, 1693, 1495, 1455, 1412, 1379, 1274, 1225, 1202, 1161, 1019, 1001, 973, 926, 827, 791, 732, 610, 529 cm^{-1}.

4.3. Synthesis of 3-(benzyloxy)-2-(2-bromopropanoyloxy)propanoic acid (III)

Synthesized per literature [47]. Compound **II** (10.0 g, 51 mmol) and 2-bromopropionyl chloride (6.3 mL, 61.2 mmol) were mixed in a 50-mL round-bottom flask backfilled with Ar. The reaction mixture was heated to 70 °C and stirred for 6 h. Upon the completion of the reaction, the crude product was heated at 60 °C under reduced pressure to remove unreacted 2-bromopropionyl chloride and 2-bromopropionyl acid. After cooling to room temperature, the residue was washed with water and extracted with ethyl acetate 3 times. The combined organic phase was washed with brine and dried over $MgSO_4$. Compound **III** was obtained as a brown oil after further purification by flash column CH_2Cl_2/MeOH/AcOH ($v/v/v$ 100:2:0.5) (12.6 g, 75%). ^{1}H-NMR (500 MHz, CDCl$_3$): 7.35–7.25 (m, 5H, Ar-H), 5.34–5.32 (m, 1H, -OCH$_2$CH(COOH)O-), 4.62–4.38 (m, 3H, PhCH$_2$O- and -OCCH(Br)CH$_3$), 3.97–3.85 (m, 2H, -OCH$_2$CH-), 1.88–1.83 (m, 3H, -CH(Br)CH$_3$). IR: 3442, 3031, 2928, 2871, 1732, 1452, 1362, 1211, 1155, 1097, 1070, 985, 910, 738, 697, 610 cm^{-1}.

4.4. Synthesis of 3-(benzyloxy)-2-(2-iodopropanoyloxy)propanoic acid (IV)

Prepare following literature procedures [47]. Compound **III** (8.0 g, 24 mmol) and potassium iodide (40 g, 0.24 mol) were mixed with 100 mL of anhydrous acetone. The mixture was heated at 60 °C overnight under Ar. The solid salt was removed by passing through a layer of Celite$^®$ and the filtrate was concentrated under vacuum. Ethyl acetate was added to the oily residue and the solution was filtered again to remove trace potassium iodide/potassium bromide. The organic phase was washed with 2 M aq. $Na_2S_2O_3$ for 3 times and dried over $MgSO_4$. After removing the solvent from the filtrate, the product **IV** was obtained and used directly in the next step without purification (7.5 g, 82%). ^{1}H-NMR (500 MHz, CDCl$_3$): δ 10.26 (s, 1H, -COO*H*), 7.50–7.16 (m, 5H, Ar-H), 5.39 (m, 1H, -OC*H*$_2$C*H*(COOH)O-), 4.84–4.41 (m, 3H, PhC*H*$_2$O- and –OCC*H*(Br)CH$_3$), 3.94 (dd, J = 48.3, 10.4 Hz, 2H, -OC*H*$_2$CH-), 2.29–1.87 (m, 3H, -CH(I)C*H*$_3$). IR: 2923, 1728, 1496, 1452, 1362, 1197, 1124, 1094, 1043, 1026, 977, 909, 737, 697, 633, 603, 582, 570, 563, 554, 548, 538, 529, 526 cm^{-1}.

4.5. Synthesis of 3-(benzyloxymethyl)-6-methyl-1,4-dioxane-2,5-dione (Monomer 1)

Synthesized as described in literature [7]. A solution of compound **IV** (10.0 g, 26.8 mmol) in dry CH$_2$Cl$_2$ (100 mL) was added dropwise to refluxing dry acetone (1 L) containing DIEA (8.8 mL, 53.6 mmol) under Ar. It took 10 h to finish the addition and the reaction was refluxed for another hour. The solvents were removed under reduced pressure and ether was added to dissolve the crude product. Insoluble ammonium iodide was filtered and the filtrate was concentrated. After purification by flash column chromatography using hexanes/ethyl acetate (*v*/*v* 4:1), the title compound **1** was obtained as a yellow oil (2.1 g, 31%). The diastereomers were used directly without separation. ^{1}H-NMR (500 MHz, CDCl$_3$): (SS) δ 7.34–7.26 (m, 5H, Ar H), 5.24–5.04 (2H; -OCC*H*(CH$_2$O-)O- and - OCC*H*(CH$_3$)O-), 4.61–4.56 (2H; ArC*H*$_2$O-), 3.97 (2H; -OC*H*$_2$CH-), 1.63 (3H; -CHC*H*$_3$). IR: 2993 (w, ν$_s$(aromatic C-H)), 2942 (w, ν$_{as}$(-CH$_2$- and –CH$_3$)), 2872 (w, ν$_s$(-CH$_2$- and aliphatic -CH-)), 1748 (vs, ν$_s$(ester C=O)), 1453 (m, δ$_s$(-CH$_2$-)), 1365 (w), 1268 (w), 1182 (s), 1084 (vs), 1046 (m), 865 (w), 739 (m, ω(aromatic C-H)), 698 (m, τ(aromatic ring)) cm^{-1}.

4.6. Synthesis of PLA copolymer 3

Monomer **1** (1.0 g, 4.0 mmol), recrystallized L-lactide (**2**, 1.0 g, 6.9 mmol) and Sn(Oct)$_2$ (10 mg in 1 mL anhydrous toluene) were added into a 5-mL round-bottom flask. The mixture was heated to 70 °C under vacuum for 1 h. Ar was filled in the flask and the temperature was increased to 140 °C. The mixture was stirred until the stir bar stopped moving. After cooling to room temperature, the solid was dissolved in CH$_2$Cl$_2$ and hexanes was added. The precipitate was re-dissolved in CH$_2$Cl$_2$, and this dissolution/precipitation was repeated three times, and the precipitate was finally dried under vacuum to give copolymer **1c** as a dark brown solid (1.48 g, 74%). ^{1}H-NMR (500 MHz, CDCl$_3$) δ 5.20 (-OC*H*$_2$CH(COOH)O- and -OC*H*(CH$_3$)CO-), 4.59 (PhC*H*$_2$O-), 3.90 (-OC*H*$_2$CH-), 1.56 (-CHC*H*$_3$). IR (ATR) 2942, 1747, 1497, 1452, 1365, 1192, 1084, 1046, 865, 739, 698 cm^{-1}.

4.7. Synthesis of PLA copolymer 4

Following literature synthesis [47]. Copolymer **3** (1.0 g) was dissolved in 50 mL ethyl acetate/methanol (3:1), and catalytic amount of Pd/C was added. The mixture was purged with Ar for 20 min and filled with H$_2$ under vigorous stirring. After 12 h, the solution was passed through a pile of Celite to remove Pd/C and the filtrate was dried under vacuum to give copolymer **4** as a light brown solid (630 mg, 63%). ^{1}H-NMR (500 MHz, CDCl$_3$): δ 5.0–5.5 (-OC*H*(CH$_2$OH)CO- and -OC*H*(CH$_3$)CO-), 3.78 (HOC*H*$_2$CH-), 1.49 (-CHC*H*$_3$). IR (ATR): ~3500 (br, w), 2995 (m), 2945 (m), 1744 (s), 1452 (m), 1380 (w), 1182 (s), 1129 (s), 1084 (s), 864 (m), 743 (m) cm^{-1}.

4.8. Synthesis of PFPA-grafted PLA copolymer 5

Copolymer **4** (100 mg) and PFPA-COOH (20 mg, 0.09 mmol) were added together with DCC (41 mg, 0.2 mmol) and DMAP (2.4 mg, 0.02 mmol) to 10 mL anhydrous dichloromethane under Ar, and the mixture was stirred at room temperature overnight. The solution was then concentrated to 3 mL and was poured into 50 mL of hexane/methanol (*v*/*v* 9:1) to precipitate the crude product. The yellow precipitate was dissolved in dichloromethane and was precipitated in hexane/methanol. This dissolution/ precipitation was repeated for a total of 3 times. Finally, the precipitate was dried under vacuum to give PFPA-grafted PLA copolymer **5** as a bright yellow solid (104 mg, 86%). ^{1}H-NMR (500 MHz, CDCl$_3$): δ 5.56 (-O*CH*(-CH$_2$OCOPFPA)CO-), 5.19 (-O*CH*(CH$_3$)CO), 4.83 (-O*CH$_2$*(OCOPFPA)CH-), 1.59 (-CH*CH$_3$*). ^{19}F-NMR (188 MHz, CDCl$_3$): δ −137.5 (doublet), −149.62 (singlet). FTIR (ATR): 2995 (m), 2945 (m), 2133 (m), 1747 (s), 1648 (w), 1563 (m), 1490 (s), 1381 (m), 1363 (m), 1258 (s), 1182 (s), 1128 (s), 1082 (s), 1044 (s), 957 (w), 865 (m), 751 (s), 667 (m) cm^{-1}.

4.9. Synthesis of Man- or MH-grafted PLA glycopolymers **10** *or* **11**

General Procedures

The mole ratio of *N*-succinimidyl 2-(diphenylphosphanyl)benzoate:amine-carbohydrate: PFPA in polymer **10** or **11** was set as 4:6:1. *N*-Succinimidyl 2-(diphenylphosphanyl) benzoate and amine-Man **6** or amine-MH **7** were added to 5 mL anhydrous DMSO, and the solution was stirred for 3 h. Then, the mixture was added to a DMSO solution containing copolymer **5**. Afterwards, the mixture was stirred for another hour. The mixture was transferred into a dialysis tube (molecule cutoff: 3500) and dialyzed in water for 2 days. Finally, the product was obtained after lyophilization.

Polymer **10**: yellow powder (yield: 92%). ^{1}H-NMR (500 MHz, DMSO-d$_6$): δ 7.70–7.20 (Ar-H), 5.21 (-O*CH*(CH$_3$)CO), 5.80–3.00 (carbohydrate), 1.48 (-CH*CH$_3$*); IR (ATR): 3362, 2945, 1747, 1651, 1489, 1451, 1381, 1515, 1184, 1082, 1043, 865, 749, 695 cm^{-1}.

Polymer **11**: a white powder (yield: 94%). ^{1}H-NMR (500 MHz, DMSO-d$_6$): 7.75–7.40 (Ar-H), 5.21 (-O*CH*(CH$_3$)CO-), 4.90–3.00 (carbohydrate), 1.48 (-CH*CH$_3$*); IR (ATR): 3371, 2945, 1748, 1651, 1503, 1452, 1381, 1133, 1084, 865, 752, 695 cm^{-1}.

Supplementary Materials: Supporting information including detailed synthetic protocols, ^{1}H, ^{19}F and ^{31}P NMR and IR spectra are available online.

Author Contributions: The authors contributed as follows M.Y., B.W. and W.N. in conceptualization, visualization and methodology and synthesis, W.N. and M.Y. in writing—original draft preparation, M.Y., B.W. and W.N. in writing—review and editing; supervision, project administration, and funding acquisition, M.Y.

Funding: This study was supported in part by the National Institute of Health (R15GM128164) and National Science Foundation (CHE-1808671).

Acknowledgments: The authors thank Xuan Chen for his help in the synthesis.

Conflicts of Interest: The authors declare no conflict of interest.

References

1. Caroff, M.; Karibian, D. Structure of bacterial lipopolysaccharides. *Carbohydr. Res.* **2003**, *338*, 2431–2447. [CrossRef] [PubMed]
2. Ramon, M.; Rolland, F.; Sheen, J. Sugar sensing and signaling. *Arabidopsis Book* **2008**, *6*, e0117. [CrossRef] [PubMed]
3. Brandley Brian, K.; Schnaar Ronald, L. Cell—Surface Carbohydrates in Cell Recognition and Response. *J. Leukocyte Biol.* **1986**, *40*, 97–111. [CrossRef]
4. Jin, Y.; Wong, K.H.; Granville, A.M. Enhancement of Localized Surface Plasmon Resonance polymer based biosensor chips using well-defined glycopolymers for lectin detection. *J. Colloid Interface Sci.* **2016**, *462*, 19–28. [CrossRef] [PubMed]
5. Chen, X.; Wu, B.; Jayawardana, K.W.; Hao, N.; Jayawardena, H.S.N.; Langer, R.; Jaklenec, A.; Yan, M. Magnetic Multivalent Trehalose Glycopolymer Nanoparticles for the Detection of Mycobacteria. *Adv. Healthc. Mater* **2016**, *5*, 2007–2012. [CrossRef] [PubMed]
6. Zhou, W.-J.; Kurth, M.J.; Hsieh, Y.-L.; Krochta, J.M. Synthesis and thermal properties of a novel lactose-containing poly(N-isopropylacrylamide-co-acrylamidolactamine) hydrogel. *J. Polym. Sci. A Polym. Chem.* **2000**, *37*, 1393–1402. [CrossRef]
7. Newman, Y.M.; Ring, S.G.; Colaco, C. The Role of Trehalose and Other Carbohydrates in Biopreservation. *Biotechnol. Genet. Eng. Rev* **1993**, *11*, 263–294. [CrossRef] [PubMed]

8. Beattie, G.M.; Crowe, J.H.; Lopez, A.D.; Cirulli, V.; Ricordi, C.; Hayek, A. Trehalose: A Cryoprotectant That Enhances Recovery and Preserves Function of Human Pancreatic Islets After Long-Term Storage. *Diabetes* **1997**, *46*, 519. [CrossRef]

9. Roy, R.; Tropper, F. Syntheses of copolymer antigens containing 2-acetamido-2-deoxy-α-orβ-d-glucopyranosides. *Glycoconjugate J.* **1988**, *5*, 203–206. [CrossRef]

10. Cao, S.; Roy, R. Synthesis of glycopolymers containing GM3-saccharide. *Tetrahedron Lett.* **1996**, *37*, 3421–3424. [CrossRef]

11. Desport, J.; Moreno, M.; Barandiaran, M. Fructose-Based Acrylic Copolymers by Emulsion Polymerization. *Polymers* **2018**, *10*, 488. [CrossRef]

12. Spain, S.G.; Gibson, M.I.; Cameron, N.R. Recent advances in the synthesis of well-defined glycopolymers. *J. Polym. Sci. A Polym. Chem.* **2007**, *45*, 2059–2072. [CrossRef]

13. Varma, A.J.; Kennedy, J.F.; Galgali, P. Synthetic polymers functionalized by carbohydrates: A review. *Carbohydr. Polym.* **2004**, *56*, 429–445. [CrossRef]

14. Wang, Q.; Dordick, J.S.; Linhardt, R.J. Synthesis and Application of Carbohydrate-Containing Polymers. *Chem. Mater.* **2002**, *14*, 3232–3244. [CrossRef]

15. Yang, L.; Sun, H.; Liu, Y.; Hou, W.; Yang, Y.; Cai, R.; Cui, C.; Zhang, P.; Pan, X.; Li, X.; et al. Self-Assembled Aptamer-Grafted Hyperbranched Polymer Nanocarrier for Targeted and Photoresponsive Drug Delivery. *Angew. Chem. Int. Ed.* **2018**, *57*, 17048–17052. [CrossRef] [PubMed]

16. Tang, H.; Tsarevsky, N.V. Preparation and functionalization of linear and reductively degradable highly branched cyanoacrylate-based polymers. *J. Polym. Sci. A Polym. Chem.* **2016**, *54*, 3683–3693. [CrossRef]

17. Ding, J.; Xiao, C.; Li, Y.; Cheng, Y.; Wang, N.; He, C.; Zhuang, X.; Zhu, X.; Chen, X. Efficacious hepatoma-targeted nanomedicine self-assembled from galactopeptide and doxorubicin driven by two-stage physical interactions. *J. Control. Release* **2013**, *169*, 193–203. [CrossRef] [PubMed]

18. Dondoni, A. Triazole: The Keystone in Glycosylated Molecular Architectures Constructed by a Click Reaction. *A. Chem. Asian J.* **2007**, *2*, 700–708. [CrossRef]

19. Pérez-Balderas, F.; Ortega-Muñoz, M.; Morales-Sanfrutos, J.; Hernández-Mateo, F.; Calvo-Flores, F.G.; Calvo-Asín, J.A.; Isac-García, J.; Santoyo-González, F. Multivalent Neoglycoconjugates by Regiospecific Cycloaddition of Alkynes and Azides Using Organic-Soluble Copper Catalysts. *Org. Lett.* **2003**, *5*, 1951–1954. [CrossRef] [PubMed]

20. Fazio, F.; Bryan, M.C.; Blixt, O.; Paulson, J.C.; Wong, C.-H. Synthesis of Sugar Arrays in Microtiter Plate. *J. Am. Chem. Soc.* **2002**, *124*, 14397–14402. [CrossRef]

21. Aragão-Leoneti, V.; Campo, V.L.; Gomes, A.S.; Field, R.A.; Carvalho, I. Application of copper(I)-catalysed azide/alkyne cycloaddition (CuAAC) 'click chemistry' in carbohydrate drug and neoglycopolymer synthesis. *Tetrahedron* **2010**, *66*, 9475–9492. [CrossRef]

22. Slavin, S.; Burns, J.; Haddleton, D.M.; Becer, C.R. Synthesis of glycopolymers via click reactions. *Eur. Polym. J.* **2011**, *47*, 435–446. [CrossRef]

23. Baskin, J.M.; Prescher, J.A.; Laughlin, S.T.; Agard, N.J.; Chang, P.V.; Miller, I.A.; Lo, A.; Codelli, J.A.; Bertozzi, C.R. Copper-free click chemistry for dynamic in vivo imaging. *Proc. Natl. Acad. Sci. USA* **2007**, *104*, 16793–16797. [CrossRef] [PubMed]

24. Wu, P.; Feldman, A.K.; Nugent, A.K.; Hawker, C.J.; Scheel, A.; Voit, B.; Pyun, J.; Fréchet, J.M.J.; Sharpless, K.B.; Fokin, V.V. Efficiency and Fidelity in a Click-Chemistry Route to Triazole Dendrimers by the Copper(I)-Catalyzed Ligation of Azides and Alkynes. *Angew. Chem. Int. Ed.* **2004**, *43*, 3928–3932. [CrossRef]

25. Becer, C.R.; Hoogenboom, R.; Schubert, U.S. Click Chemistry beyond Metal-Catalyzed Cycloaddition. *Angew. Chem. Int. Ed.* **2009**, *48*, 4900–4908. [CrossRef]

26. Chang, P.V.; Prescher, J.A.; Sletten, E.M.; Baskin, J.M.; Miller, I.A.; Agard, N.J.; Lo, A.; Bertozzi, C.R. Copper-free click chemistry in living animals. *Proc. Natl. Acad. Sci. USA* **2010**, *107*, 1821–1826. [CrossRef]

27. Campbell-Verduyn, L.; Elsinga, P.H.; Mirfeizi, L.; Dierckx, R.A.; Feringa, B.L. Copper-free 'click': 1,3-dipolar cycloaddition of azides and arynes. *Org. Biomol. Chem.* **2008**, *6*, 3461–3463. [CrossRef]

28. Staudinger, H.; Meyer, J. Über neue organische Phosphorverbindungen III. Phosphinmethylenderivate und Phosphinimine. *Helv. Chim. Acta* **1919**, *2*, 635–646. [CrossRef]

29. Saxon, E.; Bertozzi, C.R. Cell Surface Engineering by a Modified Staudinger Reaction. *Science* **2000**, *287*, 2007. [CrossRef]

30. Köhn, M.; Breinbauer, R. The Staudinger Ligation—A Gift to Chemical Biology. *Angew. Chem. Int. Ed.* **2004**, *43*, 3106–3116. [CrossRef]
31. Lin, F.L.; Hoyt, H.M.; van Halbeek, H.; Bergman, R.G.; Bertozzi, C.R. Mechanistic Investigation of the Staudinger Ligation. *J. Am. Chem. Soc.* **2005**, *127*, 2686–2695. [CrossRef] [PubMed]
32. Prescher, J.A.; Dube, D.H.; Bertozzi, C.R. Chemical remodelling of cell surfaces in living animals. *Nature* **2004**, *430*, 873–877. [CrossRef] [PubMed]
33. Sundhoro, M.; Jeon, S.; Park, J.; Ramström, O.; Yan, M. Perfluoroaryl Azide Staudinger Reaction: A Fast and Bioorthogonal Reaction. *Angew. Chem.* **2017**, *129*, 12285–12289. [CrossRef]
34. Xie, S.; Ramström, O.; Yan, M. N,N-Diethylurea-Catalyzed Amidation between Electron-Deficient Aryl Azides and Phenylacetaldehydes. *Org. Lett.* **2015**, *17*, 636–639. [CrossRef] [PubMed]
35. Xie, S.; Zhang, Y.; Ramström, O.; Yan, M. Base-catalyzed synthesis of aryl amides from aryl azides and aldehydes. *Chem. Sci.* **2016**, *7*, 713–718. [CrossRef] [PubMed]
36. Xie, S.; Lopez, S.A.; Ramström, O.; Yan, M.; Houk, K.N. 1,3-Dipolar Cycloaddition Reactivities of Perfluorinated Aryl Azides with Enamines and Strained Dipolarophiles. *J. Am. Chem. Soc.* **2015**, *137*, 2958–2966. [CrossRef] [PubMed]
37. Xie, S.; Fukumoto, R.; Ramström, O.; Yan, M. Anilide Formation from Thioacids and Perfluoroaryl Azides. *J. Org. Chem.* **2015**, *80*, 4392–4397. [CrossRef]
38. Sundhoro, M.; Park, J.; Wu, B.; Yan, M. Synthesis of Polyphosphazenes by a Fast Perfluoroaryl Azide-Mediated Staudinger Reaction. *Macromolecules* **2018**, *51*, 4532–4540. [CrossRef]
39. Duchateau, J.; Lutsen, L.; Guedens, W.; Cleij, T.J.; Vanderzande, D. Versatile post-polymerization functionalization of poly(p-phenylene vinylene) copolymers containing carboxylic acid substituents: Development of a universal method towards functional conjugated copolymers. *Polym. Chem.* **2010**, *1*, 1313–1322. [CrossRef]
40. Ting, S.R.S.; Chen, G.; Stenzel, M.H. Synthesis of glycopolymers and their multivalent recognitions with lectins. *Polym. Chem.* **2010**, *1*, 1392–1412. [CrossRef]
41. Kong, N.; Shimpi, M.R.; Park, J.H.; Ramström, O.; Yan, M. Carbohydrate conjugation through microwave-assisted functionalization of single-walled carbon nanotubes using perfluorophenyl azides. *Carbohydr. Res.* **2015**, *405*, 33–38. [CrossRef] [PubMed]
42. Peng, K.-Y.; Hua, M.-Y.; Lee, R.-S. Amphiphilic polyesters bearing pendant sugar moieties: Synthesis, characterization, and cellular uptake. *Carbohydr. Polym.* **2014**, *99*, 710–719. [CrossRef] [PubMed]
43. Disney, M.D.; Zheng, J.; Swager, T.M.; Seeberger, P.H. Detection of Bacteria with Carbohydrate-Functionalized Fluorescent Polymers. *J. Am. Chem. Soc.* **2004**, *126*, 13343–13346. [CrossRef]
44. Liu, L.; Yan, M. A General Approach to the Covalent Immobilization of Single Polymers. *Angew. Chem.* **2006**, *118*, 6353–6356. [CrossRef]
45. Wang, X.; Ramström, O.; Yan, M. Dynamic light scattering as an efficient tool to study glyconanoparticle–lectin interactions. *Analyst* **2011**, *136*, 4174–4178. [CrossRef] [PubMed]
46. Deng, L.; Norberg, O.; Uppalapati, S.; Yan, M.; Ramström, O. Stereoselective synthesis of light-activatable perfluorophenylazide-conjugated carbohydrates for glycoarray fabrication and evaluation of structural effects on protein binding by SPR imaging. *Org. Biomol. Chem.* **2011**, *9*, 3188–3198. [CrossRef] [PubMed]
47. Gerhardt, W.W.; Noga, D.E.; Hardcastle, K.I.; García, A.J.; Collard, D.M.; Weck, M. Functional Lactide Monomers: Methodology and Polymerization. *Biomacromolecules* **2006**, *7*, 1735–1742. [CrossRef]

Sample Availability: Samples of the compounds are not currently available from the authors.

molecules

MDPI

Article

Candidate Polyurethanes Based on Castor Oil (*Ricinus communis*), with Polycaprolactone Diol and Chitosan Additions, for Use in Biomedical Applications

Yomaira L. Uscátegui [1,2], Luis E. Díaz [3], José A. Gómez-Tejedor [4,5], Ana Vallés-Lluch [4], Guillermo Vilariño-Feltrer [4], María A. Serrano [4] and Manuel F. Valero [2,*]

1 Doctoral Program of Biosciences, Universidad de La Sabana, Chía 140013, Colombia; yomairausma@unisabana.edu.co
2 Energy, Materials and Environment Group, Faculty of Engineering, Universidad de La Sabana, Chía 140013, Colombia
3 Bioprospecting Research Group, Faculty of Engineering, Universidad de La Sabana, Chía 140013, Colombia; luis.diaz1@unisabana.edu.co
4 Centre for Biomaterials and Tissue Engineering, Universitat Politècnica de València, Camino de Vera, s/n, 46022 Valencia, Spain; jogomez@fis.upv.es (J.A.G.-T.); avalles@ter.upv.es (A.V.-L.); guivifel@upv.es (G.V.-F.); mserranj@fis.upv.es (M.A.S.)
5 Biomedical Research Networking Center in Bioengineering, Biomaterials, and Nanomedicine (CIBER-BBN), 46022 Valencia, Spain
* Correspondence: manuelvv@unisabana.edu.co; Tel.: +57-1-8615555 (ext. 25224)

Received: 14 December 2018; Accepted: 4 January 2019; Published: 10 January 2019

Abstract: Polyurethanes are widely used in the development of medical devices due to their biocompatibility, degradability, non-toxicity and chemical versatility. Polyurethanes were obtained from polyols derived from castor oil, and isophorone diisocyanate, with the incorporation of polycaprolactone-diol (15% w/w) and chitosan (3% w/w). The objective of this research was to evaluate the effect of the type of polyol and the incorporation of polycaprolactone-diol and chitosan on the mechanical and biological properties of the polyurethanes to identify the optimal ones for applications such as wound dressings or tissue engineering. Polyurethanes were characterized by stress-strain, contact angle by sessile drop method, thermogravimetric analysis, differential scanning calorimetry, water uptake and in vitro degradation by enzymatic processes. In vitro biological properties were evaluated by a 24 h cytotoxicity test using the colorimetric assay MTT and the LIVE/DEAD kit with cell line L-929 (mouse embryonic fibroblasts). In vitro evaluation of the possible inflammatory effect of polyurethane-based materials was evaluated by means of the expression of anti-inflammatory and proinflammatory cytokines expressed in a cellular model such as THP-1 cells by means of the MILLIPLEX® MAP kit. The modification of polyols derived from castor oil increases the mechanical properties of interest for a wide range of applications. The polyurethanes evaluated did not generate a cytotoxic effect on the evaluated cell line. The assessed polyurethanes are suggested as possible candidate biomaterials for wound dressings due to their improved mechanical properties and biocompatibility.

Keywords: castor oil; biomedical devices; polyurethanes; polycaprolactone-diol; chitosan

---

1. Introduction

Polyurethanes (PUs) are widely used in the preparation of medical devices due to their biocompatibility, degradability, and non-toxicity when compared to polymers such as polylactic acid (PLA), polycarbonate, polycaprolactone, among others [1–4]. Examples of PU applications in the biomedical field are implants, artificial heart valves, sutures, catheters, artificial heart, vascular prostheses, wound coatings, blood compatible coatings, drug delivery systems, porous supports for tissue regeneration, among others [5–9].

Since the mechanical, thermal, chemical and biological properties of PUs can be varied during the synthesis process [10–14], the addition of polymers, such as polycaprolactone diol (PCL) or chitosan (Ch) can modify the properties of PUs such as biocompatibility [15] and the antimicrobial activity. PCL is an attractive polymer for the development of biomaterials due to its properties such as biocompatibility, biodegradability, ease in the processing of biomaterials, among others [16]. Ch is a polysaccharide that is obtained from renewable sources because it is part of the structure of some crustaceans. Ch is mainly characterized by being biocompatible, biodegradable, bioadhesive, non-the toxic, and has antimicrobial properties, among others [15,17,18]. These properties have allowed PCL and Ch to be used in some applications such as wound dressings, surgical sutures, scaffolds in tissue engineering, among others [19]. The use of Ch and PCL is expected to increase the biocompatibility of the PUs synthesized with polyols derived from castor oil. Additionally, the filler effect is expected to increase the mechanical properties such as tensile strength. And when using the mixture of chitosan with polycaprolactone, it is sought to evaluate if there is a possible synergistic effect or not to obtain biocompatible materials with antimicrobial properties, or if both mechanical and biological properties are affected.

Biocompatibility is interpreted as a series of interactions that occur at the tissue/material interface, allowing the identification of those materials with surface characteristics and/or more biocompatible polymer chemistry; these interactions are influenced by the intrinsic characteristics of the material. Some biocompatibility tests involve analytical tests or observations of physiological phenomena, reactions or surface properties attributable to a specific application [12].

Cell cultures are ideal systems for the study and observation of a specific cell type under specific conditions since these systems do not have the complexity that an in vivo system entails, due to a large number of variables that interact. In vitro tests assess the morphology, cytotoxicity and secretory functions of different cell types. The tests can be by direct contact of the cells and the material or indirect, adding an extract of the material to the cell culture [20,21].

Monocytes and macrophages are part of the innate immune system because they are cells that are involved in inflammatory processes with the ability to synthesize and secrete pro and anti-inflammatory cytokines [22]. Cytokines correspond to a diverse group of extracellular, water-soluble proteins, which influence the production and activity of other cytokines by increasing (proinflammatory) or decreasing (anti-inflammatory) the inflammatory response [23].

The human monocyte cell line THP-1 is widely used in research thanks to the ability of monocytes to differentiate into macrophages [24]. THP-1 monocytes have a round shape in suspension; when they differentiate upon stimulation by phorbol 12-myristate-13-acetate (PMA), the cells adhere to the culture plates, gaining phenotypic and functional characteristics similar to primary human macrophages [25,26]. The immune response is assessed by measuring cytokines in the cell culture medium [22].

The inflammatory response of macrophages is activated by invading pathogens, particles, lipopolysaccharides (LPSs), and other stimuli [27]. LPSs are part of the outer membrane of Gram-negative bacteria and can cause tissue damage and the release of multiple pro-inflammatory cytokines [27]. Therefore, when an inflammatory response is induced, pro-inflammatory cytokines, such as interleukin-1 beta (IL-1β), tumor necrosis factor alpha (TNF-α), and interleukin-6 (IL-6), can be released. Likewise, anti-inflammatory cytokines, such as interleukin-10 (IL-10) [28], can be released. Biomaterials, such as high molecular weight polyethylene, can activate macrophages to secrete pro-inflammatory cytokines, including TNF-α and IL-1β, among others, as a response to material implantation [29].

PU applications in biomedicine are diverse due to Pus' various properties. Using aliphatic chains derived from vegetable oils creates flexible PUs, and cyclic diisocyanates provide greater mechanical strength [30]. Therefore, it is necessary to specifically characterize each synthesized material to determine its functionality and suitability for biomedical devices. The aim of this research was to determine the physicochemical, mechanical, morphological, biodegradability, and in vitro biocompatibility characteristics of the PUs, in addition to the possible inflammatory effects of the materials synthesized with castor oil polyols, depending on segment structure and PU cross-linking density. In this study, different PUs were synthesized with castor oil (chemically modified or not) and isophorone diisocyanate (IPDI) by adding polycaprolactone diol (PCL) (15% *w*/*w*) and chitosan (Ch) (3% *w*/*w*). In vitro degradation was determined in acidic and basic media, and enzymatic degradation was carried out with pig liver esterase. The in vitro cell viability was determined using L929 mouse fibroblasts (ATCC® CCL-1), human fibroblasts (MRC-5) (ATCC® CCL-171™), and adult human dermal fibroblasts (HDFa) (ATCC® PCS-201-012™) with the PUs. The viability was also determined by a live/dead kit for the L929 mouse fibroblasts. Pro- and anti-inflammatory responses were evaluated by cytokine expression (IFN-γ, IL-1β, IL-2, IL-4, IL-5, IL-6, IL-8, IL-10, and TNF-α) of the THP-1 cells with and without stimulation by LPS. The present paper serves as a screening of the immunomodulatory effects of PU materials synthesized with castor oil.

2. Results and Discussion

2.1. Obtaining Polyols

The reaction to obtaining polyols derived from the castor oil is presented in Scheme 1. The hydroxyl number of the polyols transesterified with pentaerythritol was determined. The values of the hydroxyl index for each polyol (P.1, P.2 and P.3) were 160, 191 and 236 mg KOH per g of castor oil sample, respectively. According to the results of the hydroxyl index, it is noted that the chemical modification of the castor oil increases as the pentaerythritol content increases. The increment of the hydroxyl index can be related to the increase of the crosslinking reactions and to the gain of the bulk density of the polymeric materials. An increase in the crosslink density would generate an improvement in the mechanical properties of the polymer matrices.

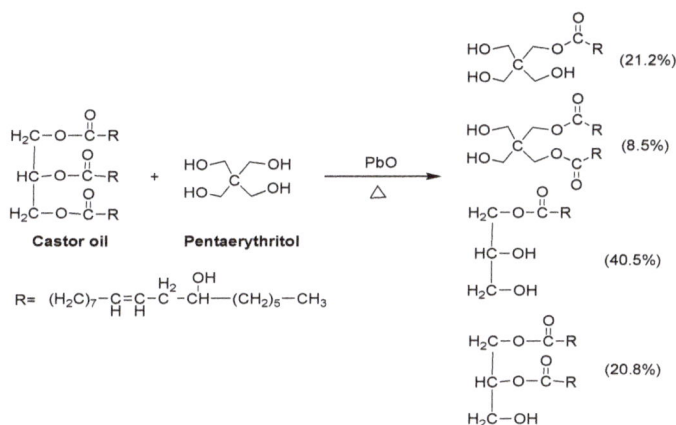

Scheme 1. Reaction scheme for obtaining polyols.

2.2. Mechanical Properties of PUs

Aromatic diisocyanates are the most commonly used in the synthesis of PUs due to their mechanical properties, but they can produce carcinogenic and mutagenic diamines upon degradation [31], therefore, an aliphatic diisocyanate was used in this research to avoid the side effects of the raw materials on living tissues.

Mechanical testing is essential to establish the use of a biomaterial because it allows obtaining the load parameters required for a tissue of interest [32]. The mechanical properties of 12 PU matrices were evaluated by determining the stress-strain curves from which the tensile strength and the elongation at the break were obtained. Figure 1 shows the results of the mechanical properties depending on the polyol used in the synthesis.

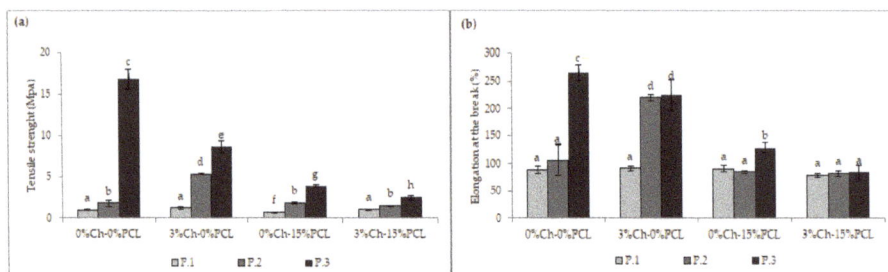

Figure 1. Mechanical properties of the synthesized PUs. (**a**) Maximum stress of PUs; (**b**) Percent elongation of PUs. The data are expressed as the mean $\pm$ SD (n = 3). Bars with different letters (a–h) indicate significant differences ($p < 0.05$) between the polyols.

Physical and mechanical properties depend on the atomic and molecular structure of the materials used in the synthesis. The nature of the bonds and subunits of the structure affects the mechanical properties and therefore the stress-strain properties, which are of interest for the evaluation of biomaterials [32]. The highest tensile strength (16.86 MPa) was obtained with polyol P.3 (P.3-0%Ch-0%PCL). Figure 1a shows an increase in the tensile strength of all the materials synthesized with polyol P.3, which has the highest cross-linking. This is how the chemical modification of the polyols increases the maximum stress—via the increase of physical cross-linking of the polymer matrix [33].

The mechanical properties of PUs are attributed to presence of hard and soft segment domains [34]. The hard segment generally refers to the combination of the chain extender and the diisocyanate components, while the soft segments refer to polymeric diols. Depending on the structure of the hard and soft segments, crystalline and amorphous domains can be formed, which determine the stiffness and stability of the material [35]. Hydrogen bond cause strong interactions, so the polar nature of the hard segment causes a strong attraction, forming the domains [36]. Therefore, when using polyol P.3 in the synthesis, the values of the mechanical properties increased because the soft segments had a higher number of hydroxyl groups, increasing the cross-linking density, and the hard segments formed a ring, providing greater resistance.

Regarding the percent elongation at break (Figure 1b), the analysis showed significant differences ($p < 0.05$) of polyol P.1 compared with the other polyols (P.2 and P.3), with the percentage increasing as the polyol was modified. The mechanical properties of PUs depend on many factors, including molecular weight, chemical bonds, cross-linking, crystallinity of the polymer, and the size, shape, and interactions of the hard segment present in the structure [37]. Thus, PUs with a higher degree of cross-linking have higher values of tensile strength and percent elongation at break. Increased cross-linking produces a more compact structure [37]. An increase in strength is attributed to the content of intermolecular hydrogen bonds and cross-linking density [38].

When analyzing the influence of the additives used in the synthesis on the mechanical properties, significant differences are found when 3% Ch was added to polyol P.2 (P.2-3%Ch-0%PCL), obtaining a higher value than the other materials synthesized with P.1 and P.3. For polyol P.3, the additives decreased the maximum stress compared with the material without additives.

Chen et al. (2018) synthesized PUs with PCL as a polyol, IPDI, and polylactic acid (PLA), obtaining tensile strength values between 41 and 60 MPa when using PLA- and PCL-based PU ratios in the range of 80/20 to 95/5. The authors attributed the decrease in tensile strength as the ratio of PCL increased to the plasticizing effect of certain non-cross-linked PCL polyols and to a possible decrease in compatibility as the PCL content increased [39].

A similar effect was observed for percent elongation at break because it decreased when PCL and Ch were added to polyol P.3 (P.3-3%Ch-15%PCL). The other materials did not differ with the additives used from the material without additives. The flexibility of PU may be due to the long oil hydrocarbon chain present in the polymer chain [37]. This agrees with the results reported by Park et al. (2013), who synthesized PUs with polycaprolactone, hexamethylene diisocyanate, and isosorbide, with silk added, and determined that a higher silk content increased the stiffness and decreased the maximum stress. The authors stated that the design of flexible and soft polymers allows for the production of a wide range of biomaterials to regenerate soft tissues such as muscles and ligaments [6]. Additionally, Vannozzi et al. reported that in general, soft and deformable substrates are key features for skeletal muscle tissue engineering [33].

2.3. Fourier-Transform Infrared Spectroscopy (FTIR)

FTIR was used to determine the efficiency of the synthesis process by the identification of characteristic functional groups of PU and the absence of characteristic peaks of the monomers used in the synthesis process. Figure 2 shows the infrared spectra of the synthesized PUs.

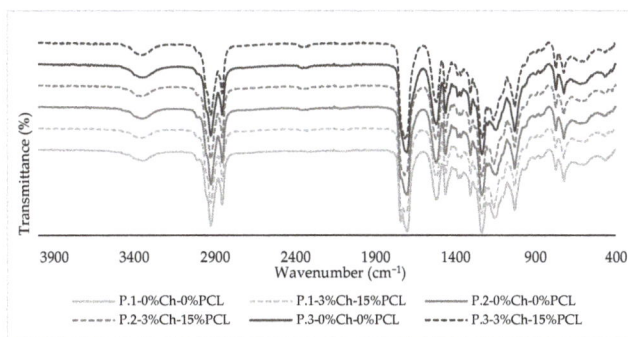

Figure 2. FTIR spectra of the synthesized PUs.

Figure 2 shows that all of the FTIR spectra had similar peaks, independent of the polyol or additive used in the synthesis, and the peaks observed corresponded to the expected PU matrices. The absence of the stretching peak of the $-N=C=O$ bond of the diisocyanates at 2250 cm^{-1} [31], indicates there were no unreacted free isocyanate groups in the synthesized PU matrices, showing that the reaction was complete.

Spectral peaks characteristic of PUs can be seen in the spectra. Thus, around 3330 cm^{-1}, the characteristic bands for the stretching vibrations of the $-N-H$ bonds are observed [40] from which it can be inferred that they correspond to the urethane bonds present in the matrices. Near 2923 cm^{-1}, the stretching peak of the methyl group can be observed, and at around 2855 cm^{-1}, the symmetric stretching of the C–H bond is present. At around 1700 cm^{-1}, an intense band is present due to the C=O bond stretching [40] thus indicating the formation of the urethane group. Around 1250 cm^{-1}, the C–N bond stretching is observed; at about 1140 cm^{-1}, the stretching vibrations of the C–O bond appear [37].

2.4. Thermal Analysis

2.4.1. Thermogravimetric Analysis

Thermogravimetric analysis was performed by producing thermograms of the PUs. Figure 3 shows the results for the synthesized matrices.

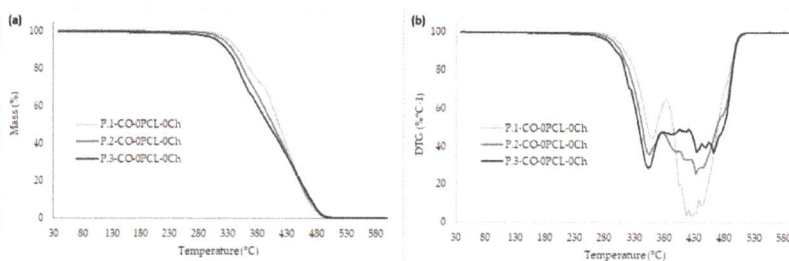

Figure 3. Thermograms of the synthesized PUs. (**a**) Thermogravimetric (TG) curve of PUs; (**b**) Derivative of the thermogravimetric (DTG) curve of PUs.

Figure 3 shows a trend regarding thermal behavior of the synthesized PUs, showing no displacements of the degradation temperatures with the use of modified polyols. When obtaining the curves derived from thermogravimetry, several peaks were observed, which agrees with other authors stating that the mechanism of degradation of PUs is complex due to the formation of various compounds in the process [41].

For each polymer matrix a thermogram was obtained from which the stability of the synthesized PUs in this study were determined. It was observed that the polyol type and additives did not affect the thermal stability of the materials when compared with the synthesized material without additives. The thermograms showed that all PUs were stable at temperatures below 300 °C and showed complete degradation at temperatures near 600 °C, which coincides with the research conducted by Jutrzenka et al. (2018), who synthesized PUs based on a glycerin derivative, polyethylene-butylene, and diphenylmethane diisocyanate. They determined that the materials were stable up to 300 °C [36]. In a study on PUs synthesized with castor oil and isophorone diisocyanate that were proposed as surgical adhesives, the PUs had the same degradation values, and the authors stated that the values related to the degradation temperatures do not affect the biomedical application since physiological temperature is lower ($\approx$37 °C) [42].

Three degradation regions were detected for the polymer matrices. This is consistent with the study performed with polymeric matrices synthesized with castor oil and isophorone diisocyanate by adding different concentrations of PCL and Ch [43]. The first stage of degradation was observed in the range of 250–370 °C and corresponds to the thermal degradation of the urethane bonds formed in the hard segments, characterized by being technically unstable [2]. The second stage was between 375–430 °C and corresponds to the degradation of the soft segments [36]. The last stage was in the range of 425–500 °C and corresponds to the thermal degradation of the double bonds of remaining fatty acids from castor oil [44,45].

2.4.2. Differential Scanning Calorimetry (DSC)

Table 1 shows the glass transition temperature (T_g) values for the synthesized materials, as determined by the DSC curves of the materials. Table 1 shows that the polyol type and the additives significantly influence the T_g value. This agrees with what is understood from the mechanical properties of the molecular structure produced by modifying the polyol derived from castor oil. The values of T_g for P.1, P.2, and P.3 without additives were −14.8 °C, −12.8 °C and 14 °C, respectively. P.1 had the lowest values of T_g (between −14 °C and −25 °C, approximately).

Table 1. DSC thermograms of the synthesized PUs depending on the polyol used.

Polymeric Material	T_g (°C)		
	P.1	P.2	P.3
0%Ch-0%PCL	−14.8	−12.8	14.8
3%Ch-0%PCL	−13.3	−1.0	13.7
0%Ch-15%PCL	−25.3	−10.2	4.7
3%Ch-15%PCL	−25.4	−14.8	2.8

The observed trend is a decrease in T_g as the hydroxyl index decreases, which agrees with the cross-linking of the synthesized PUs, indicating that greater energy is needed for reordering the structure by a change in intermolecular forces. This may be due to secondary interactions resulting from the hyperbranched structure [46]. The thermal properties of PUs depend on the number of urethane bonds present in the structure because they can tolerate a considerable amount of heat [47]. Saénz-Pérez et al. (2016) synthesized PUs with polytetramethylene glycol and diphenylmethane or toluene diisocyanates, and butanediol. In the determination of T_g, they found that the values increased as the amount of chain extender increased. Therefore, the authors state that the increase in T_g was caused by the reduction in the mobility of chain segments due to the increase of hard segments [48].

The ricinoleic acid triglyceride from castor oil used as a polyol contains an ordered structure in which hydroxyl groups are uniformly distributed within the chain, helping to obtain a PU with a uniform cross-linked structure, achieving high mechanical properties and thermal stability [30]. In general, all of the synthesized matrices had a single value of T_g, indicating that all the materials showed homogeneous segment dispersion. The T_g values were similar to those reported for PUs based on polyethylene glycol, poly (ε-caprolactone-co-D,L-lactide), and diurethane diisocyanate (with hexamethylene diisocyanate and butanediol), where the authors found values near −33 °C. Likewise, the authors did not find exothermic peaks because the materials were amorphous [49]. With the above information, it can be generalized that Pus are thermally stable and that they can be used in various biomedical applications, for example, as materials for non-absorbable sutures.

2.5. Hydrophilic Character

2.5.1. Contact Angle

To evaluate the hydrophilic nature of PUs, the water contact angle on their surfaces was determined. Figure 4 shows the results of the synthesized PUs.

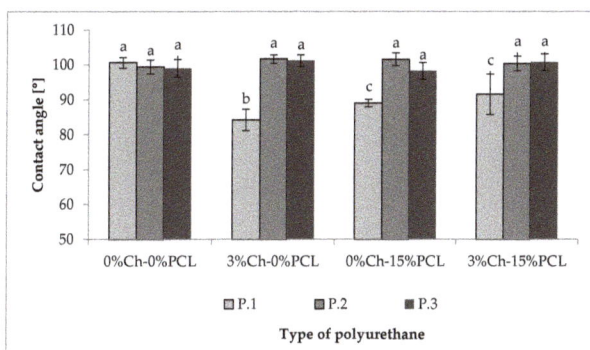

Figure 4. Contact angle of the synthesized PUs. The data are expressed as the mean ± SD (n = 10). Bars with different letters (a–c) indicate significant differences ($p < 0.05$).

According to the statistical analysis in Figure 4, it is observed that for polyols P.2 and P.3, there are no significant differences in contact angle values. For P.1, significant differences are observed by using additives because there is a reduction of angle values, indicating a decrease in the hydrophilic character. Values near 100 degrees were found, so it can be deduced that the materials tend to be hydrophobic. Mi et al. determined the contact angles of PCL-based thermoplastic PUs and different chain extenders; the values were near 90 degrees. They attributed the results of the contact angle to the hydrophilic functional groups present in the chain extenders that were used. The authors state that although PUs maintain the same chemical structure, monomer variation in the synthesis can affect the wettability and therefore vary the degradation behavior [50].

Likewise, Gossart et al. evaluated the contact angle of PUs synthesized with L-lysine diisocyanate, hydroxyethyl methacrylate, and poly (hexamethylene-carbamate) and found values above 80 degrees. The authors state that the common values reported for PU matrices were in the range of 80 to 90 degrees depending on the structure of the PU and interactions with the surfaces [51]. The results of the contact angle shown in Figure 4 are greater than those reported; this possibly results from the cross-linked network generated by the monomers used in the synthesis, as the materials tended to be hydrophobic.

2.5.2. Water Absorption Rate

Figure 5 shows the rates of water absorption for the synthesized PUs during 72 h of testing.

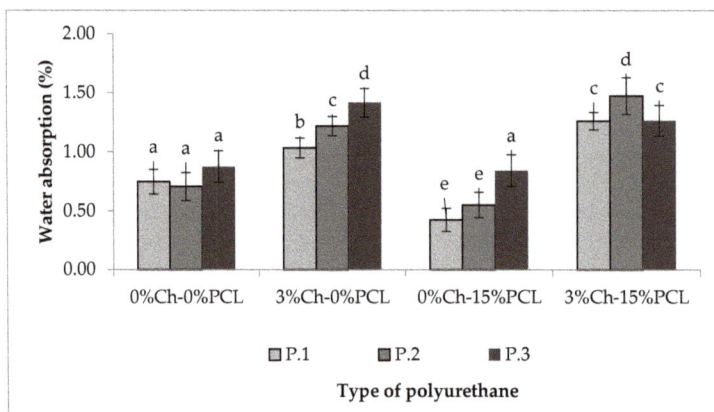

Figure 5. Water absorption rate over 72 h. Absorption results are expressed as the mean $\pm$ SD (n = 3). Bars with different letters (a–d) indicate significant differences ($p < 0.05$).

The weights of the materials were monitored until constant weight over 24, 48, 72, and 144 h, but the data are not shown because no significant differences were found after 48 h. The results show that the water absorption rates range between 0.5 and 1.5%. As seen in Figure 5, adding PCL and Ch increases the rate of absorption compared to the material without additives, although the difference in rates was not greater than one. It is likely that by increasing the amount of additive more functional groups became available to interact with the medium, which is also polar. However, by increasing the functionality of the polyol, the effect generated is inverse, showing a reduction in the rate of absorption compared to PUs without the additive. Internal interactions (hydrogen bonds) increase the barrier effect by preventing fluid diffusion. In addition, the additive function causes the chains to reorganize, showing a reduction of volumetric defects or voids in which the water can be deposited [32].

Marques et al. evaluated a bioadhesive synthesized from lactic acid in which the water absorption was 10%. The authors noted that moderate rates of water absorption improved the hemostatic character of the materials [52].

The contact angle and the rate of absorption provide information about hydrophobicity/hydrophilicity and could be an indirect indicator of surface molecular mobility.

Surface wettability can affect protein adsorption on the surface and biocompatibility [53]. This is one reason why it is indispensable to perform a study of material biocompatibility to determine its possible acceptance by the human body. Therefore, the PUs synthesized in this study can be considered suitable for biomedical applications, such as materials for non-absorbable sutures, considering the water absorption rates under the test conditions.

2.6. Density Determination

Figure 6 shows the density of the materials synthesized with IPDI and the additives (PCL and Ch).

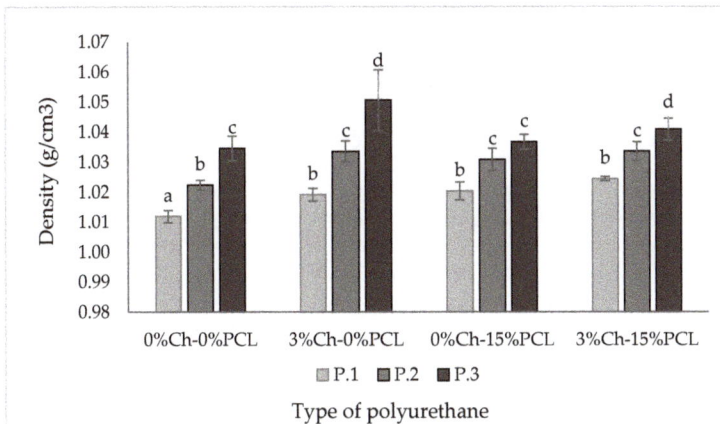

Figure 6. Density of the synthesized PUs. The data are expressed as the mean $\pm$ SD (n = 3). Bars with different letters (a–d) indicate significant differences ($p < 0.05$).

As seen in Figure 6, the density of PUs depends on the type of polyol, which agrees with the results shown for the mechanical and thermal properties. As the cross-linking density increased, the density of the resulting polymeric material significantly increased, although the difference between the highest and lowest values was less than 5%. This agrees with the results of Conejero-García et al. who synthesized polyglycerol sebacate, with a different degree of cross-linking, as a material for various applications in tissue engineering. The density values reported by the authors ranged between 1.13 and 1.14 g mL^{-1} [54].

The high densities of polymeric materials can be related to a higher hydroxyl (OH) content due to increased cross-linking reactions [55]. In a study conducted by Carriço et al. they found that increasing the castor oil content in the formulation of foam increased the apparent density, suggesting that the polymer chains were more packed, with less free volume and smaller cells, increasing the stiffness of these materials [55].

2.7. Dynamo-Mechanical Thermal Analysis (DMTA)

Figure 7 shows the dynamic behavior, DMTA, in a tension mode of PUs corresponding to the evolution of the modulus and the loss factor versus temperature. With the variation of the storage modulus and the loss factor, it was possible to observe the displacements suffered by T_g for the evaluated materials. It can be seen that T_g increases when polyols P.2 and P.3 are used; this is probably due to the stiffness of the structure and greater generation of hydrogen bonds because of the high hydroxyl index present in the polyol [38,39]. According to the above results, it can be inferred that material compatibility decreases when the polyol without modifications, and therefore with less cross-linking, is used.

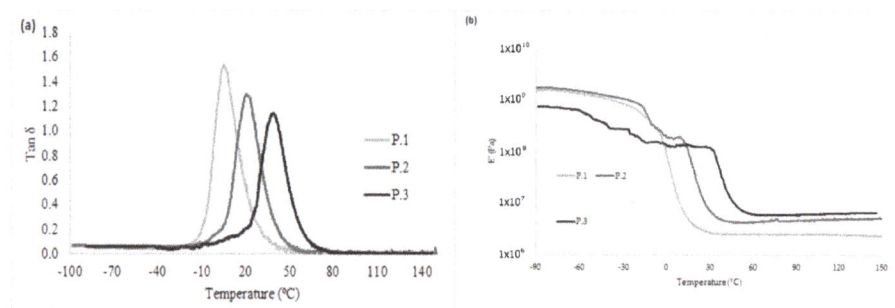

Figure 7. DMTA thermograms of the synthesized PUs. (**a**) Loss factor; (**b**) Storage modulus.

Results with a similar trend were found by Chen et al. on PU matrices synthesized with PCL as a polyol, IPDI, and PLA. The authors reported that when the mobility of the chain decreased, the value of T_g increased. Therefore, they observed that phase compatibility decreased when the PCL content increased due to a possible plasticizing effect of non-cross-linked material [39].

A similar behavior occurred with the modulus results. The lowest values corresponded to the materials synthesized with P.1 (polyol without modification) compared with polyols P.2 and P.3, which have a higher content of hydroxyl groups. An increase in hydroxyl groups results in an increase in cross-linking, hindering polymer chain mobility and thereby increasing the storage modulus. A decrease in polymer chain mobility can limit energy transfer and diffusion, which could decrease the absorption capacity of impact resistance and deformation [39].

The relationship between hard and soft segments is important as they simultaneously act as a physical cross-linking agent and as a high-modulus filler. When there is an organization of the hard and soft segments in the respective domains, pre-polymers tend to have two T_g values. One temperature will be negative corresponding to the soft segments, while the other will be positive, corresponding to the hard segments [38]. When a single event of T_g occurs, it could be inferred that there is a homogeneous phase distribution [38]. According to the results obtained by DSC, it can be observed that the trend of T_g is similar, that is, a single value of T_g is present, increasing as the polyol is modified. The presence of a single transition can be related to the existence of a dominant phase, so it can be inferred that there is a uniform distribution of the components [56]. The differences in T_g between the polyols used may be due to the cross-linking density because this would cause a greater compatibility between hard and soft segments [38].

2.8. Field-Emission Scanning Electron Microscopy (FESEM)

Figure 8 shows the morphology of PUs synthesized with IPDI as a function of the polyol. The FESEM micrographs showed a uniform distribution of PUs, but it was not possible to differentiate the hard segments from the soft segments. Similarly, no differences were observed related to the type of polyol used for the polymeric matrix. These results can be correlated with the calorimetric results because if there is only a single T_g, it is probable that there is a homogeneous phase distribution. This agrees with the results reported by Thakur et al. on PUs synthesized with castor oil and toluene diisocyanate for the coating of materials [47].

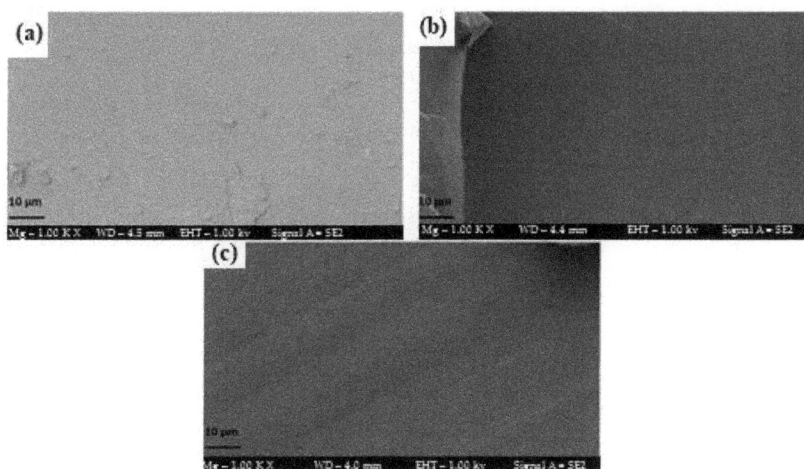

Figure 8. FESEM micrographs of PUs synthesized at 100×. (**a**) P.1; (**b**) P.2; (**c**) P.3.

2.9. In Vitro Biodegradability Assays

The biodegradability of PUs was evaluated in the presence of different media (HCl, NaOH, and enzymatic) over a specific period of time, the results are shown in Figure 9.

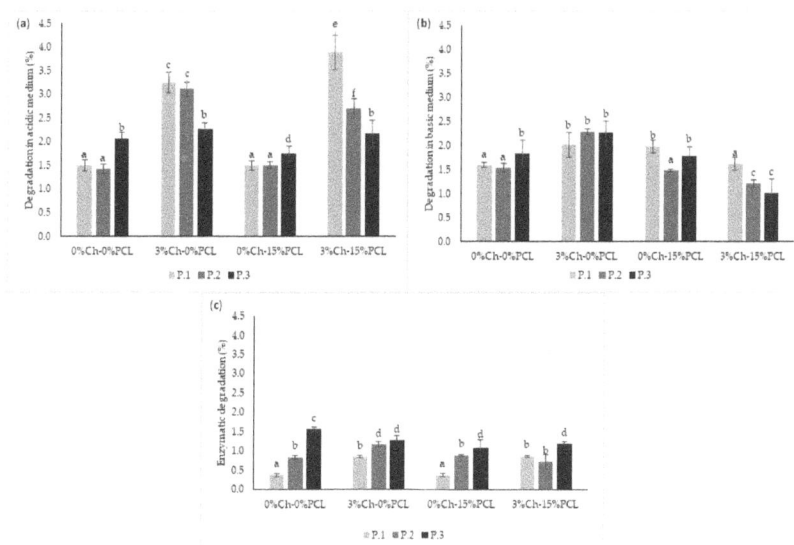

Figure 9. Degradation rate of PUs in different media. (**a**) Acidic medium (0.1 M HCl) for 90 days; (**b**) Basic medium (0.1 M NaOH) for 90 days; (**c**) Enzyme medium (esterase) for 21 days. The data are expressed as the mean $\pm$ SD (n = 3). Bars with different letters (a–f) indicate significant differences ($p <$ 0.05) between the polyols.

PCL was used as a positive control in enzymatic degradation, showing a degradation of 9.08 $\pm$ 0.30% after 21 days. It can be inferred that the linear structure of PCL facilitates the diffusion of the cleaved chains throughout the polymer and their release into the medium, producing a higher degradation rate. For the synthesized PUs, characterized by being cross-linked materials, the chain mobility is lower, thus hindering the diffusion of the cleaved chains.

The highest value of the degradation rate after 90 days of testing (acidic and basic media) was obtained with the acidic medium under the test conditions for the material synthesized with polyol P.1 by adding the additives (3%Ch-15%PCL). FTIR of the degraded materials (Figure 10) helped determine whether the degradation corresponded to one of the functional groups of the material.

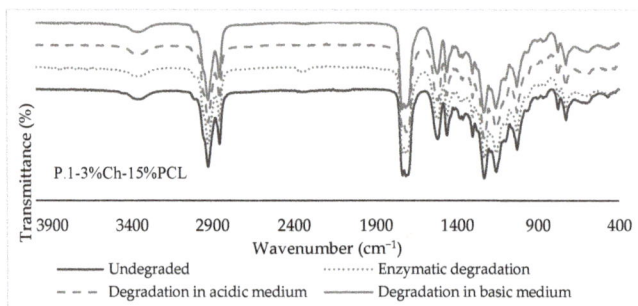

Figure 10. FTIR spectra of PUs degraded in acidic, basic, and enzymatic media.

Figure 10 shows that the functional groups characteristic of PUs are conserved compared with the undegraded material, therefore, it can be inferred that degradation occurs at the surface level. Surface images of the degraded materials were also taken using FESEM, observing a possible surface degradation of the materials (Figure 11). When observing that the degradation rate is less than 4% in the evaluated media, it can be said this is due to the hydrophobic character of PUs, which agrees with the results found for the rate of water absorption and contact angle.

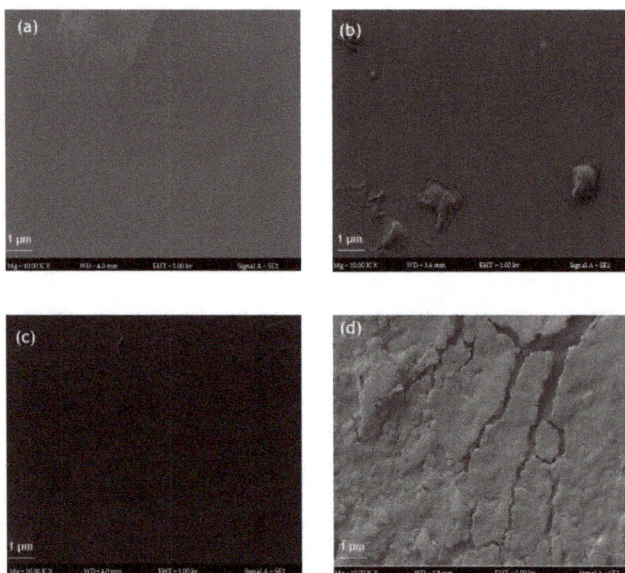

Figure 11. FESEM micrographs of the PU synthesized with P.2-3%Ch-15%PCL after the degradation process in different media (10,000×). (**a**) Undegraded material; (**b**) Acidic medium (0.1 M HCl) after 90 days; (**c**) Basic medium (0.1 M NaOH) after 90 days; (**d**) Enzyme medium (esterase) after 21 days.

The results obtained in this research show a similar trend with those reported by Thakur et al., who evaluated the chemical resistance of PUs in acidic and basic media, finding that in general, PUs show chemical resistance, and PUs based on castor oil showed higher resistance to the basic medium because the oil contains hydrolyzable functional groups [37]. This may be the reason why the biodegradation of the PUs in basic medium was lower than in acidic medium.

The degradation of poly (ester-urethane) occurs mainly by a hydrolytic attack of the ester and urethane bond [31]. Das et al. attributed the degradation in acidic and basic media to different types of strong interactions in the structures of PUs [57]. The authors stated that low resistance to alkalis is due to the hydrolyzable ester bonds in the monoglyceride residues and PCL of polymers [57]. After 90 days of testing, it was observed that the highest values of the degradation rate under the test conditions were relatively low (3.9% in acidic medium and 2.3% in basic medium). These results indicate that the polymeric materials are resistant to degradation due to the structure of PUs, that is, to the cross-linked matrix with high mechanical properties and a hydrophobic character.

Regarding enzymatic degradation (Figure 12c), the highest rate observed was 1.6% after 21 days. This degradation process was greater than those obtained with the other treatments evaluated. This behavior can be attributed to an increase of hydroxyl groups in the polyols, producing an increase in the physical cross-linking of the polymer, and therefore, an increase in urethane groups. Urethane bonds are similar to amides and may be hydrolyzed by enzymes such as the esterase used in this research [58,59]. This agrees with what was stated by Gogoi et al. who reported that amide and urea bonds present in the branched polymer structure facilitate degradation [60]. According to Cherng et al., the degradation of PUs is due to cleavage of the hydrolytically weak bonds that are characteristic of the soft segments; therefore, they concluded that the in vitro degradation rate depends mainly on the type of polyol used in the synthesis due to the ester bonds in the structure [58].

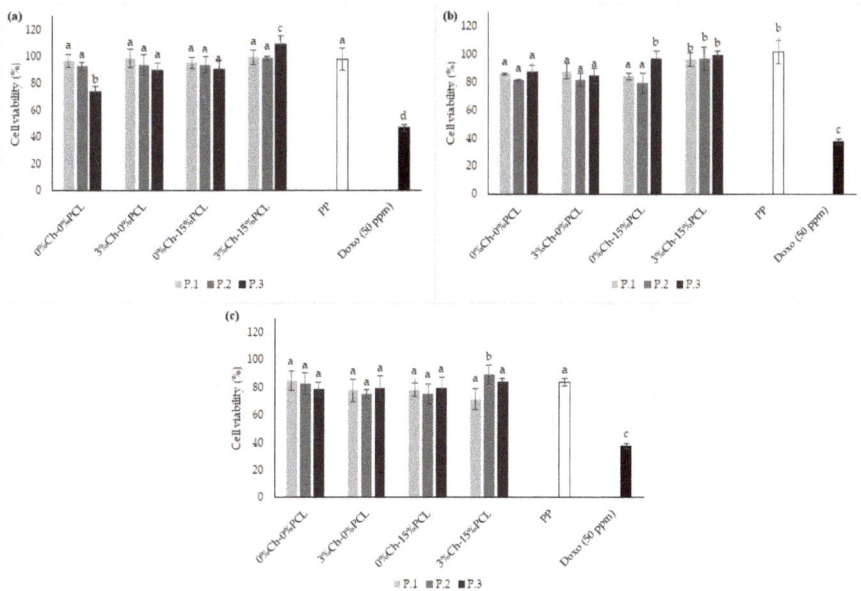

Figure 12. Percentage of cell viability at 24 h. (**a**) L929 Mouse fibroblasts; (**b**) Human fibroblasts (MRC-5); (**c**) Human dermal fibroblasts (HDFa). The data are expressed as the mean ± SD (n = 3). Bars with different letters (a–d) indicate significant differences (*p* < 0.05) between the polyols.

2.10. In Vitro Cell Viability Assay by the MTT Method

As part of the in vitro biological evaluation of PUs, the percentage of cell viability was determined on three fibroblast cell lines. The results are shown in Figure 12.

Figure 12 shows that all polymers had a cell viability of greater than 70% for the three cell types evaluated. According to the ISO/CD 10993-5 standard, values greater than 70% can be considered suitable for biomaterials since they would be non-cytotoxic. As a control material, polypropylene (PP) was used because it is a biocompatible and non-absorbable biomaterial. As a negative control, doxorubicin (Doxo) was used. Figure 12a shows the cell viability results of the PUs on L929 fibroblasts. The percent viability of all matrices, except for the PU synthesized with P.3 without additive, did not show significant differences ($p < 0.05$) compared to PP. For the PU synthesized with P.3, significant differences were observed with the polymer synthesized with P.1 and with the control, but the cell viabilities are suitable to propose the synthesized materials as possible biomaterials. For the L929 fibroblasts, as the cross-linking of PU increased, the percentage of cell viability decreased. It is likely that the high degree of cross-linking produces a low availability of functional groups, such as −OH groups, in the cross-linked polymer. These groups would decrease the cell adhesion to the surface [33] mediated by proteins from the medium.

Bakhshi et al. evaluated PUs synthesized by adding quaternary ammonium salts in the epoxidation of soybean oil and found that PUs show a cell viability between 78–108% with L929 mouse fibroblasts, indicating that there was no toxicity of the synthesized polymers, coinciding with the results of this study [2].

Similar results were found by Calvo-Correas et al. in a preliminary study of in vitro cytotoxicity with murine L929 cells. These authors determined that PUs synthesized from castor oil and lysine diisocyanate had a cell viability greater than 100% in the first 24 h of the assay when following the ISO 10993-12 standard, indicating that the synthesized PUs were non-toxic and could have a potential use in biomedical applications [61]. Likewise, the results from the current study agree with those reported by Reddy et al., who synthesized a PU from lysine diisocyanate, PCL, and 1,4-butane-diamide and found that the polymers did not show toxicity when in contact with NIH/3T3 mouse fibroblasts [62].

Figure 12b shows the cell viability results of the polymers on human fibroblasts (MRC-5). For this cell line, no trend was observed related to the polyol type used in the synthesis of the PUs. The viability was decreased compared to that of the control material (PP); nevertheless, the materials are suitable for use in the design of biomaterials because they show a cell viability of greater than 80%.

Figure 12c shows the cell viability results of PUs on human fibroblasts (HDFa), and the materials show greater than 80% cell viability. Here, there were no statistically significant differences ($p < 0.05$) between the polyols used in the synthesis of PUs. Similarly, there were no significant differences between the synthesized materials and the PP control material; therefore, it can be inferred that the synthesized PUs can be considered suitable for the design of biomaterials for non-absorbable sutures.

The above results are in accordance with those reported by Coakley et al., who evaluated HDFa cells in vitro from a construct derived from porcine urinary bladder as a potential scaffold in tissue engineering. The study demonstrated the viability of the human dermal fibroblasts. The results showed cell viability values of greater than 90% [63].

Cytotoxicity may be related to the relative cell viability of controls, where values lower than 30% indicate severe toxicity of the materials, values between 30 and 60% indicate moderate toxicity, values between 60 and 90% indicate slight toxicity, and values above 90% indicate that the materials are non-toxic [64]. Therefore, the fibroblast cell viability results shown in Figure 12 demonstrate that the PUs are non-toxic for all of the lines evaluated.

2.11. Immunocytochemical Techniques

2.11.1. In Vitro Cell Viability Assay by a Live/Dead Kit

L929 cells that were cultured directly on the material and evaluated by the MTT method showed no toxic effects. When evaluating the cells by a live/dead viability kit, it was possible to observe that the cells adhered to the material, showed viability, and proliferated during the exposure time with the material. This kit allows determination of the number of live and dead cells during the test. In this experiment, viable cells were determined. Figure 13 shows the fluorescence units of the viability results as evaluated by the live/dead kit.

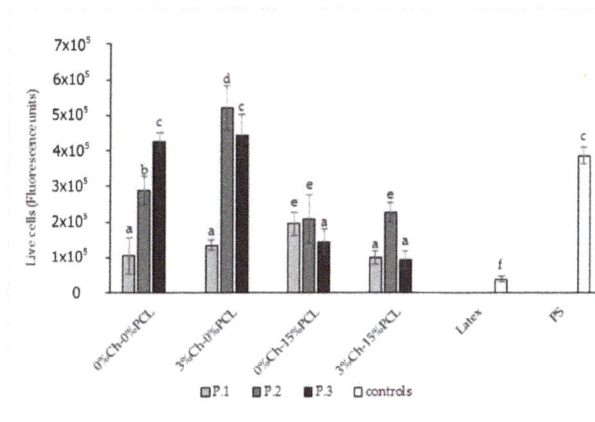

Figure 13. Fluorescence unit results of the in vitro cell viability test after 48 h with the live/dead kit. Polystyrene (PS) positive control; Latex negative control. The data are expressed as the mean $\pm$ SD ($n = 3$). Bars with different letters (a–f) indicate significant differences ($p < 0.05$) between polyols.

The results shown in Figure 13 indicate that PUs synthesized with modified polyols by transesterification with and without the addition of Ch show values of living cells similar to and higher than the material used as a positive control, PS. The materials synthesized with modified polyols by adding Ch showed statistically significant differences compared to the other PUs after 48 h of testing. Likewise, all the materials showed higher values than the negative control used in the test.

All the materials showed statistically significant differences compared to the negative control (latex). When comparing to the results of the positive control (PS), no significant differences were observed between the materials synthesized with P.3 without additive and by adding 3% Ch. This indicated that these materials had a similar cell viability behavior to that obtained with the PS reference material, which is widely used to grow and proliferate cells in vitro. Similarly, it was observed that one material evaluated showed significant differences compared to PS, with the viability values being higher than the positive control. This PU corresponds to the one synthesized with polyol P.2 with 3% Ch. These results indicate that Ch has a positive effect because it improves the biocompatibility of the PU.

Chitosan has a primary amino group and two free hydroxyl groups for each glucose unit, which is beneficial for biomedical applications [15]. The presence of free amino groups causes an increase in the positive charge of the polymer; therefore, there is a greater interaction between Ch and the cells [65].

These results agree with those reported by Laube et al., who performed a live/dead staining assay of NIH/3T3 fibroblasts grown on PU foams that were synthesized with lysine diisocyanate. They found that the cells were viable and proliferated, showing an increased cell density over time. Therefore, the authors proposed these materials as possible substitutes for soft tissue fillers [31].

2.11.2. Fixation and Morphological Analysis with Phalloidin and DAPI

In this assay, the adhesion and proliferation of L929 cells were observed by morphological analysis and were compared to the latex negative control. PS was used as a positive control, which is a common biomaterial used for cell culturing. Figure 14 shows the images obtained from the cells fixed and stained after 24 h of contact with the PUs by polyol type.

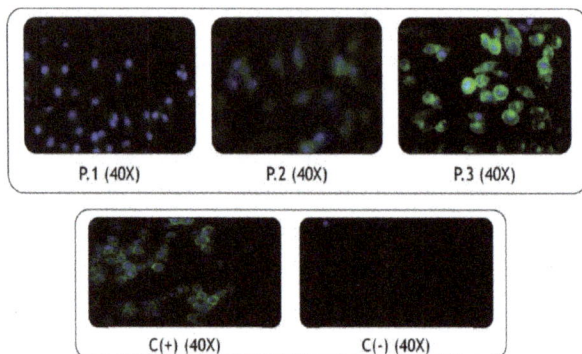

Figure 14. Optical microscopy images of phalloidin (green) and DAPI (blue) staining for representative PUs evaluated on L929 fibroblasts for 24 h. C(+) polystyrene positive control, C(−) latex negative control. Images were analyzed by ImageJ software.

This technique stains the F-actin fibers (a main component of the cytoskeleton) with phalloidin, producing a green color [66], and DAPI staining produces a blue color for the fibroblast nuclei. As observed in the images, the structure of the L929 fibroblasts was conserved when compared with the positive control. Likewise, varying type of polyol or additive did not produce conformational changes in cell structure. As seen in the images, fibroblasts could adhere to the polymeric material, as evaluated after 24 h of contact. Cell morphology can indicate substrate-cell interactions. Flat and extended cell forms indicate strong adhesion from cell to substrate, while a rounded morphology generally means that the cells have difficulty in adhering to the substrate [67]. With the above, it can be inferred that the PUs evaluated in direct contact with cells allow cell proliferation and do not affect the cell morphology after 24 h of testing. Therefore, the synthesized PUs in this research may be suitable as materials for the design of non-absorbable sutures.

2.12. Evaluation of Inflammatory Processes

Twelve materials synthesized with castor oil polyols and IPDI with 3% Ch and 15% PCL were evaluated to determine the in vitro inflammation processes produced after contact with the materials. Inflammation was evaluated by the release of different cytokines to the medium of THP-1 cells that were differentiated into macrophages, with and without LPS stimulation, and in contact with the PUs. Nine inflammatory markers were analyzed, including pro- and anti-inflammatory cytokines (IFN-γ, IL-1β, IL-2, IL-4, IL-5, IL-6, IL-8, IL-10, and TNF-α). The first test group (group A) was used to determine the possible effect of PUs on the inflammatory process of macrophages.

Figure 15 shows the concentrations of anti-inflammatory cytokines expressed by the macrophages after 24 h of exposure to the PUs. Figure 16 shows the concentration of pro-inflammatory cytokines expressed in the culture medium. The reference material was a polypropylene (PP) biomaterial used in non-absorbable biomedical sutures. The control was the culture medium without polymer. Figures 15 and 16 show cytokine expression in the monocytes (the cells not treated with PMA). The second test group (group B) was formed to determine if the synthesized materials had anti-inflammatory activity against cells in which inflammation had been stimulated by LPS before direct contact with the PUs.

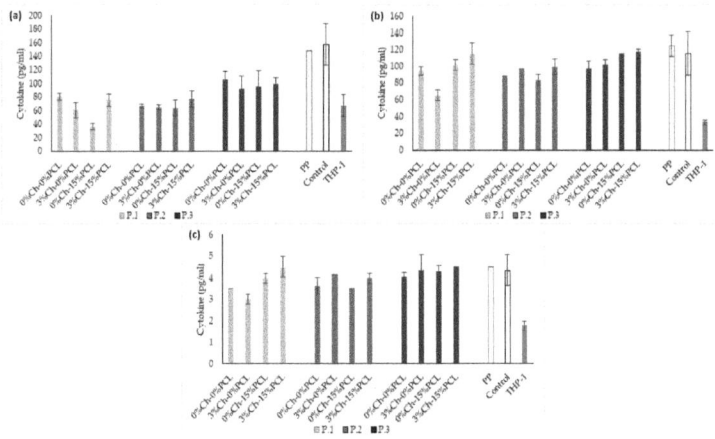

Figure 15. Anti-inflammatory cytokines without LPS stimulation. (**a**) IL-10; (**b**) IL-5; (**c**) IL-4. PP: polypropylene as a reference material; Control: culture medium without PUs; THP-1: monocyte cell line. The concentration of cytokines is expressed as the mean ± SD (n = 3).

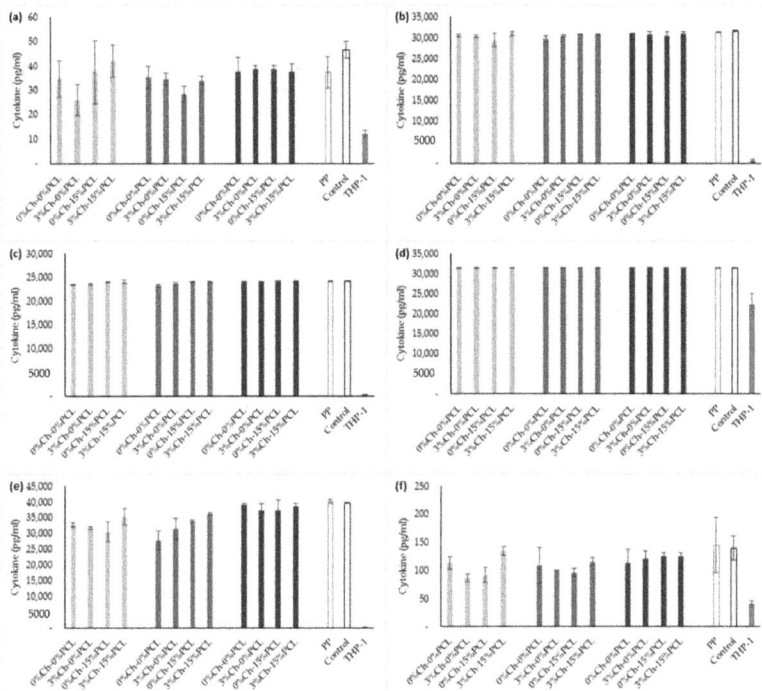

Figure 16. Pro-inflammatory cytokines without LPS stimulation. (**a**) INF-γ; (**b**) TNF-α; (**c**) IL-1β; (**d**) IL-8; (**e**) IL-6; (**f**) IL-2. PP: polypropylene as a reference material; Control: culture medium without PUs; THP-1: monocyte cell line. The concentration of cytokines is expressed as the mean ± SD (n = 3).

The results of the concentration of pro- and anti-inflammatory cytokines from test B are shown in Figures 17 and 18, respectively. PP was tested as a reference material, and culture medium without polymer was used as a control.

As observed in the results of groups A and B, there were no significant differences between the concentrations of the cytokines expressed by the cells that were stimulated with LPS and those that were not stimulated. This may be related to the protocol that was used. A standardized protocol to differentiate THP-1 monocytes into macrophages with PMA is not yet available. Therefore, a wide variation of PMA concentrations is found—between 6 and 600 nM. The time of stimulation with PMA can vary from 3 to 72 h, and the recovery periods vary between 0 and 10 days [25,68]. It is probable that high concentrations of PMA (higher than 100 ng mL^{-1}) increase the levels of expression of genes associated with inflammation and cause an increase in the secretion of pro-inflammatory cytokines, such as TNF or IL-8 [25].

As stated by Park et al. (2007), high concentrations of PMA can induce the expression of some genes during the differentiation process, which can mask the effect of post-differentiation stimuli [69]. The above results indicate that the concentrations of PMA used during the differentiation process of THP-1 monocytes into macrophages triggered the release of pro- and anti-inflammatory cytokines, and this process was prior to exposing the cells to LPS stimulation.

The results obtained for group A (Figures 15 and 16) show that the PUs do not produce an inflammatory process in THP-1 macrophages. This is based on the control results (medium without PU) and the PP reference material, which showed higher values (although not always significant) in the concentrations of the evaluated pro- and anti-inflammatory cytokines.

An acute inflammation in an organism occurs due to a protective response against tissue damage caused by infectious or foreign agents. The first cytokines formed in response to bacterial lipopolysaccharides, tissue injury or infection are TNF-α and IL-1β, which act directly on specific receptors to trigger a cascade of other effectors, such as cytokines and chemokines, among others [23,25,70]. TNF-α and IL-1β have a synergic effect on inflammation, also promoted by IFN-γ through the increase of TNF-α [70]. Tumor necrosis factor (TNF-α) is a pro-inflammatory cytokine that is stimulated early in the inflammatory response [23].

One cytokine evaluated was interleukin-4 (IL-4), characterized by having a potent anti-inflammatory activity and the capacity to inhibit the synthesis of pro-inflammatory cytokines [71]. It acts on activated macrophages to reduce the effects of cytokines IL-1, TNF-α, IL-6, and IL-8 [23]. Therefore, by expressing a pro- or anti-inflammatory cytokine, the expression of its opposite is inhibited. Figure 17 shows IL-4 values near 5 pg mL^{-1}, and its opposites such as TNF-α, IL-6, and IL-8 show values over 20,000 pg mL^{-1}, as observed in Figure 15. Interleukin-10 (IL-10), which exerts anti-inflammatory actions on monocytes or macrophages, was also evaluated [71]. It inhibits the pro-inflammatory cytokines IL-1, TNF-α, and IL-6, stimulating the endogenous production of anti-inflammatory cytokines [23].

TNF-α plays an important role in the inflammatory response of an organism, as previously mentioned. Upon LPS stimulation, a systemic inflammatory state occurs that is characterized by increased levels of the pro-inflammatory cytokines TNF-α and IL-1β, and decreased levels of anti-inflammatory cytokines, such as IL-10 [26]. This behavior was observed in the values obtained from group B because in Figure 18a, the values obtained for TNF-α showed concentrations near 30,000 pg mL^{-1} and IL-1β had values of 25,000 pg mL^{-1}, but for IL-10, mean values of 80 pg mL^{-1} were found.

Other evaluated cytokines were interleukins 2 and 6. Interleukin-2 (IL-2) is a pro-inflammatory cytokine characterized by the generation and propagation of the antigen-specific immune response [23]. Interleukin-6 (IL-6) is a cytokine with pro-inflammatory properties, and high levels are associated with severity in septic processes. IL-6 is sometimes considered an anti-inflammatory cytokine because it can induce beneficial proteins in septic shock [71].

For the two groups (A and B), the cytokines of the THP-1 monocytes that were not transformed to macrophages were evaluated to observe the basal expression levels of each cytokine. According to Chanput et al., the basal levels of cytokines of THP-1 monocytes and macrophages have values near 20 and 30 pg mL^{-1} [22]. For the study performed with PUs, cytokines IFN GAMA, IL-2, IL-5, and IL-6 had values that were within the range mentioned by the study of Chanput et al. for THP-1 monocytes, as observed in Figures 15–18. After the differentiation of monocytes into macrophages, it was observed (Figures 17 and 18) that the expression levels of all cytokines increased compared to the basal levels of the monocytes. Anti-inflammatory cytokines IL-4, IL-5, and IL-10 had increases of 55, 65, and 35%, respectively. Pro-inflammatory cytokines IL-2, IL-6, IL-8, IL-1β, IFN-γ, and TNF-α had increases of 65, 122, 29, 100, 70, and 97%, respectively. Regarding the results of the macrophages stimulated with LPS, the percent increases in the concentrations of the cytokines were similar to these.

According to the results from the reference biomaterial (PP) and the medium without PU, the behavior does not show statistically significant differences of the PUs evaluated in group B. Therefore, it can be inferred that the PUs do not have anti-inflammatory activity.

The results show that, for the differentiation of THP-1 monocytes into macrophages, in vitro standardization and optimization is necessary to achieve adequate cytokine expression results. THP-1 monocytes and macrophages could be an adequate and reliable model to evaluate the inflammatory response before conducting a more detailed study with human-derived cells [22].

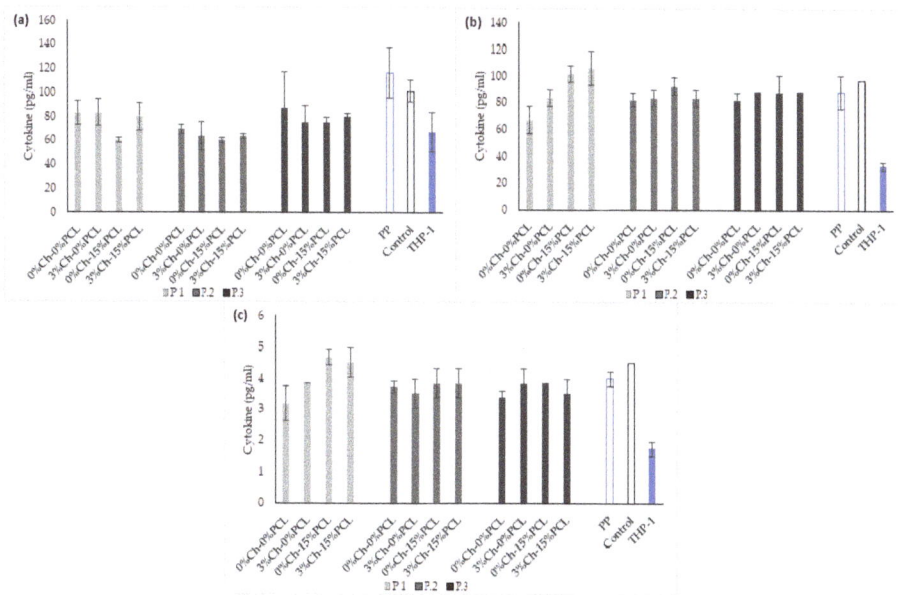

Figure 17. Anti-inflammatory cytokines with LPS stimulation. (a) IL-10; (b) IL-5; (c) IL-4. PP: polypropylene as a reference material; Control: culture medium without PUs; THP-1: monocyte cell line. The concentration of cytokines is expressed as the mean ± SD (n = 3).

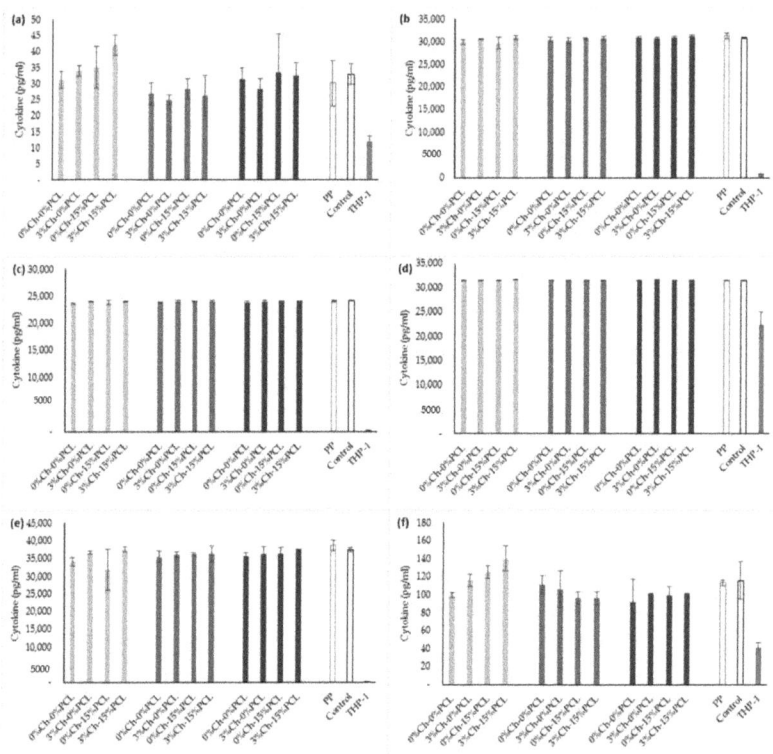

Figure 18. Pro-inflammatory cytokines with LPS stimulation. (**a**) INF-γ; (**b**) TNF-α; (**c**) IL-1β; (**d**) IL-8; (**e**) IL-6; (**f**) IL-2. PP: polypropylene as a reference material; Control: culture medium without PUs; THP-1: monocyte cell line. The concentration of cytokines is expressed as the mean ± SD (n = 3).

3. Materials and Methods

3.1. Reagents

Castor (*Ricinus communis*) oil was purchased from Químicos Campota and Co. Ltd. (Bogotá, Colombia). Isophorone diisocyanate (IPDI), polycaprolactone diol (PCL) (average molecular weight of 2000 g mol^{-1}), low molecular weight chitosan (Ch) (with a percentage of deacetylation between 75–85%), *n*-octane, 0.1 M hydrochloric acid (HCl), 0.1 M NaOH, porcine liver esterase (18 units mg^{-1}), 4′,6-diamidino-2-phenylindole dihydrochloride (DAPI), and phorbol 12-myristate-13-acetate (PMA) were acquired from Sigma-Aldrich Chemical Co. (St. Louis, MO, USA). Pentaerythritol was from Merck KGaA (Darmstadt, Germany). Phosphate-buffered saline (PBS: Dulbecco's phosphate-buffered saline), 3-(4,5-dimethyl-2-thiazolyl)-2,5-diphenyl-2H-tetrazolium bromide (MTT), 2.5% trypsin (10×), penicillin-streptomycin (10,000 units of penicillin and 10,000 μg of streptomycin per milliliter), Dulbecco's modified Eagle medium (DMEM, 1×), Roswell Park Memorial Institute (RPMI) 1640, Live/dead® Viability/Cytotoxicity Kit, Alexa Flour™ 488 phalloidin, and lipopolysaccharide (LPS, *Escherichia coli* 026:B6) were obtained from Gibco/Invitrogen (Paisley, UK). Fetal bovine serum (FBS) was from Eurobio (Les Ulis, France).

3.2. Biological Material

Human acute monocytic leukemia cell line, THP-1 (ATCC® TIB-202™). Mouse subcutaneous connective tissue fibroblasts L929 (ATCC® CCL-1). Human lung fibroblasts MRC-5 (ATCC® CCL-171™). Adult human dermal fibroblast HDFa (ATCC® PCS-201-012™). All biological materials were acquired from the strain library of the Universidad de La Sabana (Chía, Colombia).

3.3. Obtaining Polyols

The polyols were derived from castor oil (P.1, P.2, and P.3). P.1 corresponded to commercial unmodified castor oil. Polyols P.2 and P.3 were obtained by transesterification with pentaerythritol [72]. For the reaction, the castor oil temperature was raised to 120 °C for 10 min. The temperature was then further increased to 210 °C, adding pentaerythritol (1.32% and 2.64% mol pentaerythritol/mol castor oil for P.2 and P.3, respectively) and 0.05% lead oxide as a catalyst. The mixture was maintained at 210 °C for 2 h. After the reaction time, the catalyst was removed by decantation for 24 h and filtration. The hydroxyl number of the polyols were determined according to ASTM D1957-86 [73].

3.4. Synthesis of Polyurethanes

The PUs were synthesized by the pre-polymer method. The polyol was brought to 60 °C, and the diisocyanate was added the reactor at a constant NCO:OH ratio (1:1) [74], and maintained at 300 rpm for 5 min. Them the PCL (15% *w/w* of oil polyol weight) and Ch (3% *w/w* of oil polyol weight) were then added and maintained at 300 rpm for 5 min additional. The formation of the PU sheets was carried out by pouring the prepolymer into a steel mold. Twelve PU matrices were synthesized in sheets (15 cm × 9 cm × 0.3 cm (length × width × height)). Curing was performed at 110 °C for 12 h [72]. The synthesized PUs are identified with the following nomenclature: Pn-xCh-yPCL, where n represents the polyol used (1 for unmodified castor oil, 2 for castor oil with 1.32% pentaerythritol/mol, and 3 for the castor oil with 2.64% pentaerythritol/mol), and x and y represent the percentage of Ch and PCL, respectively.

3.5. Mechanical Tests

A universal testing machine EZ-LX (Shimadzu, Kyoto, Japan) was used to determine maximum stress, percent elongation, and Young's modulus of the polyurethanes (following the ASTM D638-10 standard). A load cell of 5 kN with a crosshead speed of 25 mm min^{-1} was used [75,76]. Three samples of 40 mm × 6 mm × 3 mm (length × width × thickness) were tested.

3.6. Fourier-Transform Infrared Spectroscopy (FTIR)

Chemical structures were evaluated by an ATR-FTIR spectrometer (Bruker Alpha, Billerica, MA, USA), in the range from 400 to 4000 cm^{-1}. The spectrum corresponds to the average of 24 scans at a spectral resolution of 4 cm^{-1} [40,77].

3.7. Thermal Analysis

3.7.1. Thermogravimetric Analysis

Thermal behavior was evaluated in a TGA/DCS1 thermogravimetric analyzer coupled to DSC (Mettler-Toledo Inc., Schwerzenbach, Switzerland). According to ASTM D6370, the conditions used were as follows: heating speed of 25 °C min^{-1}, a temperature range of 25–600 °C, nitrogen atmosphere, and samples of 15 ± 2 mg [48].

3.7.2. Differential Scanning Calorimetry

Glass transition temperatures were determined by a DSC 3+ analyzer (Mettler-Toledo, Columbus, OH, USA). The conditions were as follows: temperature range from −70 °C to 150 °C, nitrogen atmosphere with 20 mL min^{-1} flow, and sample weights of 10 ± 2 mg [78].

3.8. Hydrophilic Character

3.8.1. Contact Angle

A "Drop Shape Analysis—DSA" device (GH11, Krüss, Hamburg, Germany) was used to measure the contact angle. According to ASTM-D7334-08 (2013), it was measured with the sessile drop method using 10 μL of distilled water at 20 °C [79]. Ten measurements of each material were performed.

3.8.2. Water Absorption Rate

To remove uncross-linked chains and monomer residues after synthesis, the PU materials were washed with ethanol for two days, renewing the ethanol every day. The solvent was then replaced by deionized water for two more days. The materials were dried in a vacuum chamber (0.01 mm Hg) at 37 °C for 24 h [54]. The rate of water absorption was determined by immersing the sample in distilled water until constant weight. The residual water was then removed from the samples with dry filter paper, and the samples were weighed [80]. All tests were performed in triplicate. The rate of water absorption at equilibrium was calculated by comparing the mass of the sample (m_t) after obtaining constant weight with the initial mass (m_i) of the sample using Equation (1):

$$\% \text{ Water absortion of PUs} = (m_i - m_t)/m_i \tag{1}$$

3.9. Determination of Density

The material density was determined using the Archimedes immersion technique using a Mettler AX 205 balance (Mettler-Toledo Inc.) with a sensitivity of 0.01 mg and a Mettler ME-33360 density determination kit. Dry samples of 5 mm × 5 mm × 3 mm were used to calculate the density in triplicate. The PUs were weighed in the air (m_{air}) and then immersed in n-octane with a known density ($\rho_{n\text{-}octane}$), and the immersion weight ($m_{n\text{-}octane}$) was obtained at 20 °C. The density was calculated by the following Equation (2) [54]:

$$\text{Density of PUs} = (m_{aire} \times \rho_{n-octano})/(m_{aire} - m_{n-octano}) \tag{2}$$

3.10. Dynamic Mechanical Thermal Analysis (DMTA)

The effect of temperature on the mechanical properties was evaluated by a DMTA test. A DMA 8000 thermomechanical analyzer (Perkin-Elmer, Waltham, MA, USA) was used at a frequency of 1 Hz, a deformation of 0.1%, and a temperature program between −90 °C and 150 °C, with a heating rate of 5 °C min^{-1}. The storage modulus and the loss factor, tanδ [38], were determined.

3.11. Field Emission Scanning Electron Microscopy (FESEM)

Morphological characterization of the PUs was performed with a field-emission scanning electron microscope (FESEM, ZEISS ULTRA 55 from Oxford Instruments (Abingdon, UK). The PUs were washed with ethanol for two days, renewing the ethanol every day. The solvent was then replaced by deionized water for two more days. The materials were dried in a vacuum chamber (0.01 mm Hg) at 37 °C for 24 h [54]. The samples were coated with platinum to observe the morphology of the materials with an accelerating voltage of 5 kV [81].

3.12. In Vitro Biodegradability Tests

Biodegradability tests were performed in independent tests in triplicate following ASTM F1635-11. To remove uncross-linked chains and monomer residues, the PUs were washed with ethanol for two days, renewing the ethanol each day. The solvent was then replaced by deionized water for two more days. The materials were dried in a vacuum chamber at 37 °C for 24 h [54]. Samples of 5 mm × 5 mm × 3 mm were placed in the biodegradation media and incubated at 37 °C for three months, except for the enzyme medium (10 units mg^{-1} solid), which was for 21 days. After this time, the samples were washed with distilled water, dried in a vacuum chamber, and weighed. The reagents used were: 0.1 M HCl medium, 0.1 M NaOH medium, and porcine liver esterase enzymatic medium. The biodegradation rate was calculated by comparing the dry weight (w_f) of the sample after degradation during the predetermined time with the initial dry weight (w_i) of the sample using Equation (3) [81,82]:

$$\% \text{ Degradation of PUs} = (w_i - w_t)/w_i \qquad (3)$$

3.13. In Vitro Cell Viability Assay by the MTT Method

L929 mouse fibroblasts and MRC-5 and HDFa human fibroblasts were cultured in DMEM supplemented with 10% FBS and 1% penicillin-streptomycin in T-75 cell culture flasks. They were grown at 37 °C and 5% CO_2. The cell culture medium was changed every 48 h [83]. At 100% confluence, the cells were trypsinized (trypsin-EDTA) for viability analysis.

The effect of polyurethanes on cell viability was evaluated by the MTT method defined by ISO/CD 10993-5. Cells were seeded in 96-well plates at a concentration of 4.0×10^4 cells per well in supplemented medium and cultured at 37 °C and 5% CO_2 for 24 h. Subsequently, PU cylinders of 3 mm × 2 mm (diameter × height) (previously sterilized under ultraviolet UV light (260 nm) for 30 min on each side [33]) were placed in 100 µL of supplemented medium. The materials were left in contact with the polymers for 24 h at 37 °C and 5% CO_2. The supernatant and polymers were then removed, and the MTT solution (12 mM in PBS) was added to a total volume of 100 µL and incubated for 4 h at 37 °C. The supernatant was removed, and 100 µL of dimethyl sulfoxide was added and incubated for 15 min at 37 °C. The optical density was then determined in a plate reader (Bio-Tek ELx800 Microplate Reader, Highland Park, Winooski, VT, USA) at 570 nm. A polypropylene biomaterial (PP) was used as a positive control for cell viability, and doxorubicin (DOXO) was used as a negative control. All tests were performed in triplicate. Cell viability was determined according to Equation (4):

$$\% \text{ Cell viability of PUs} = \text{Abs}_{sample}/\text{Abs}_{control} \qquad (4)$$

where Abs_{sample} corresponds to the absorbance value of the cells after treatment with the PU, and $\text{Abs}_{control}$ corresponds to the cells without treatment.

3.14. Immunocytochemical Techniques

In vitro cell viability assay by the live/dead kit. Cell viability was also evaluated by the live/dead kit based on plasmatic membrane integrity and intracellular esterase activity [84]. L929 mouse fibroblasts were cultured in RPMI supplemented with 10% FBS and 1% penicillin-streptomycin in T-75 cell culture flasks at 37 °C and 5% CO_2. The cell culture medium was changed every 48 h [83]. At 100% confluence, the cells were trypsinized (trypsin-EDTA) for viability analysis.

Prior to the assay, the PUs were washed with ethanol and water for 3 days and dried in a vacuum chamber at 37 °C for 24 h [54]. PU sheets with a diameter of 4 mm (previously sterilized under UV light (260 nm) for 30 min on each side) were placed in 96-well plates [33]. Five microliters of cells, at a concentration of 1.0×10^4 cells mL^{-1}, were then seeded onto each material using supplemented medium and incubated at 37 °C and 5% CO_2 for 30 min. Next, 95 µL of supplemented medium was added and incubated for 24 h at 37 °C and 5% CO_2 [54]. The supernatant was removed, and 2 µL of calcein AM (live cell marker) staining solution and 4 µL of ethidium homodimer (dead cell marker) were added and incubated for 20 min in the dark with stirring [84]. Fluorescence measurement was performed for calcein in a Victor 1420 Multilabel Counter spectrophotometer (Perkin-Elmer, Waltham, MA, USA) at 485 nm to determine the concentration of living cells. PS was used as a positive control and latex as a negative control. All tests were performed in triplicate.

Fixation and morphological analysis with phalloidin and DAPI. L929 fibroblasts in contact with the material after 24 h of incubation were washed with PBS and fixed for 20 min with paraformaldehyde 4% *w/v* at 20 °C. They were then washed twice with PBS at 4 °C. Phalloidin (Alexa Flour™ 488 phalloidin) was added and incubated for 20 min to reveal F-actin. The samples were then washed twice with PBS. The cell nuclei were stained for 5 min with DAPI at a 1/5000 dilution in PBS, followed by washing twice with PBS. Next, 0.05% sodium azide was added. The cell structure was observed using an Eclipse 80i optical microscope (Nikon, Tokyo, Japan) equipped with a Nikon Intensilight Illuminator [54]. The images were analyzed by ImageJ software version 1.45k (Bethesda, MD, USA).

3.15. Differentiation into Macrophages and Inflammation Stimulation

A human acute monocytic leukemia cell line, THP-1, was used. The cells were cultured at 37 °C and 5% CO_2 in RPMI medium supplemented with 10% FBS and 1% penicillin-streptomycin in T-75 cell culture flasks. For cell differentiation into macrophages, the cells were centrifuged at 2000 rpm for 10 min, resuspended in 3 mL of RPMI culture medium and seeded into 24-well plates [69] by adding 200 mM of PMA and incubating for 48 h [27]. The differentiated cells (macrophages) were washed twice with PBS to remove undifferentiated cells [26]. The macrophages that had been differentiated with PMA were stimulated with 700 ng mL^{-1} of LPS for 3 h [22].

3.16. Evaluation of Inflammatory Processes

With the THP-1 cell line and PUs, two independent tests were performed to evaluate the inflammatory processes of the materials in contact with the cells. The first test (group A) consisted in the direct contact of PUs on the cells differentiated into macrophages with PMA, for 24 h, to determine whether the PUs caused inflammation after contact with the cell line. The next test group (group B) consisted of PUs in contact for 24 h with macrophages differentiated with PMA and stimulated with LPS to determine whether the materials had anti-inflammatory activity.

3.17. Immunoassay

The MILLIPLEX® MAP kit for flow cytometry was used to evaluate cytokine production (IFN-γ, IL-1β, IL-2, IL-4, IL-5, IL-6, IL-8, IL-10, and TNF-α) and release to the supernatant of the macrophages cultured in contact with PUs for 24 h, following the instructions provided by the manufacturers. A mixture of beads with different fluorescence intensities, coated with antibodies for the mentioned cytokines, was pre-incubated in the dark for one hour on an orbital shaker. Concentration standards of 3.2, 16, 80, 400, 2000 and 10,000 pg mL^{-1} were used to determine the concentration curve of each analyte. In a 96-well plate, 200 μL of wash buffer was added, the plate was shaken for 10 min, and the supernatant was discarded. Twenty-five microliters of the standards, 25 μL of running buffer, 25 μL of the stock solution, 25 μL of supernatant from cells in the appropriate wells, and 25 μL of the bead mixture were added. The plate was incubated overnight with shaking at 4 °C. The supernatant was then removed, and the samples were washed twice with 200 μL of wash buffer. Next, 25 μL of antibody was added for detection, and the samples were incubated for 1 h in the dark and with shaking. Twenty-five microliters of the streptavidin phycoerythrin binding protein was then added and incubated for 30 min with shaking in the dark. The supernatant was discarded, and the wells were washed twice with 200 μL of wash buffer. One hundred and fifty microliters of the coating liquid were then added, and the plate was read on a MAGPIX flow cytometer from MILLIPLEX MAP (Darmstadt, Germany). The results of the samples were analyzed using the xPONENT MAGPIX software (Madison, WI, USA).

3.18. Statistical Analysis

The results were expressed as mean values ± standard deviation (SD). The data were analyzed by means of an analysis of variance (ANOVA) and the significant differences were determined for $p < 0.05$. For the comparison between samples the *t*-Student test was used with the SPSS Statistics Software v.23 (IBM, Armonk, NY, USA).

4. Conclusions

PUs synthesized with castor oil polyols, isophorone diisocyanate, 15% *w/w* polycaprolactone and 3% *w/w* chitosan were used to evaluate their mechanical, physicochemical, morphological, biodegradability, and biocompatibility characteristics, along with their possible inflammatory effects. The type of polyol and additive showed a significant impact on the maximum stress, percent elongation, and contact angle of the material. Chemical modification of the polyols improves the evaluated properties due to the increased cross-linking of the resulting materials. The resulting mechanical properties of the PUs verified their dependence on the presence of hydrogen bonds, the number of hydroxyl groups in the polyols, and the interactions between the hard and soft segments of the matrix. The polyol with the highest hydroxyl index showed the highest values for the mechanical and thermal properties. The percent elongation, with a maximum of 265%, can be used to obtain resistant materials with flexibility, which allows designing biomaterials that do not cause injuries to soft tissues. The degradation rates under the study conditions showed values of less than 4%, so it can be inferred that cross-linking of the synthesized PUs hinders the degradation process. In vitro cell viability was determined using L929 mouse fibroblasts, human fibroblasts (MRC-5) and adult human dermal fibroblast (HDFa). The cell viability results of the PUs in contact with the three cell lines demonstrated that there was no effect on cell viability; therefore, it is likely that these materials could be used for direct contact with the skin without causing damage to surrounding cells. Cell viability was also determined by the live/dead kit for the L929 mouse fibroblasts. Due to the biocompatible properties of the PUs evaluated in this study, it can be inferred that they are suitable for use in biomedical applications as materials for non-absorbable biomedical sutures.

Author Contributions: Conceptualization, M.F.V., Y.L.U., L.E.D., J.A.G.-T., and A.V.-L.; Methodology, Y.L.U.; Formal analysis, Y.L.U.; Investigation, Y.L.U., G.V.-F. and M.A.S.; Resources, M.F.V.; Writing—original draft preparation, Y.L.U.; Visualization, Y.L.U.; Supervision, M.F.V. and L.E.D.; Project administration, M.F.V.; Funding acquisition, M.F.V. and L.E.D.

Funding: This research was funded by the UNIVERSIDAD DE LA SABANA, grant number ING-202-2018 and by COLCIENCIAS under scholarship grant 617-2-2014. CIBER-BBN is an initiative funded by the VI National R&D&I Plan 2008–2011, Iniciativa Ingenio 2010, Consolider Program. CIBER Actions are financed by the Instituto de Salud Carlos III with assistance from the European Regional Development Fund. J.A.G.-T. and A.V.-LL. acknowledge the support of the Spanish Ministry of Economy and Competitiveness (MINECO) through project DPI2015-65401-C3-2-R (including FEDER financial support).

Acknowledgments: The authors thank the Universidad de La Sabana for financing research project ING-202-2018 which is part of this research, Colciencias for the doctoral scholarship under grant 617-2 of 2014, and the Universitat Politècnica de València for assistance and advice with the equipment.

Conflicts of Interest: The authors declare no conflict of interest.

References

1. Alishiri, M.; Shojaei, A.; Abdekhodaie, M.J.; Yeganeh, H. Synthesis and characterization of biodegradable acrylated polyurethane based on poly(ε-caprolactone) and 1,6-hexamethylene diisocyanate. *Mater. Sci. Eng. C* **2014**, *42*, 763–773. [CrossRef] [PubMed]
2. Bakhshi, H.; Yeganeh, H.; Yari, A.; Nezhad, S.K. Castor oil-based polyurethane coatings containing benzyl triethanol ammonium chloride: Synthesis, characterization, and biological properties. *J. Mater. Sci.* **2014**, *49*, 5365–5377. [CrossRef]

3. Kucinska-Lipka, J.; Gubanska, I.; Janik, H.; Sienkiewicz, M. Fabrication of polyurethane and polyurethane based composite fibres by the electrospinning technique for soft tissue engineering of cardiovascular system. *Mater. Sci. Eng. C Mater. Biol. Appl.* **2015**, *46*, 166–176. [CrossRef] [PubMed]
4. Tsai, M.-C.; Hung, K.-C.; Hung, S.-C.; Hsu, S. Evaluation of biodegradable elastic scaffolds made of anionic polyurethane for cartilage tissue engineering. *Colloids Surf. B Biointerfaces* **2015**, *125*, 34–44. [CrossRef] [PubMed]
5. Rocco, K.A.; Maxfield, M.W.; Best, C.A.; Dean, E.W.; Breuer, C.K. In vivo applications of electrospun tissue-engineered vascular grafts: A review. *Tissue Eng. Part B* **2014**, *20*, 628–640. [CrossRef]
6. Park, H.; Gong, M.-S.; Park, J.-H.; Moon, S.-I.; Wall, I.B.; Kim, H.-W.; Lee, J.H.; Knowles, J.C. Silk fibroin-polyurethane blends: Physical properties and effect of silk fibroin content on viscoelasticity, biocompatibility and myoblast differentiation. *Acta Biomater.* **2013**, *9*, 8962–8971. [CrossRef]
7. Rajan, K.P.; Al-ghamdi, A.; Parameswar, R.; Nando, G.B. Blends of thermoplastic polyurethane and polydimethylsiloxane rubber: Assessment of biocompatibility and suture holding strength of membranes. *Int. J. Biomater.* **2013**, *2013*. [CrossRef]
8. Rodríguez-Galán, A.; Franco, L.; Puiggal, J. Biodegradable Polyurethanes and Poly(ester amide)s. In *Handbook of Biodegradable Polymers: Synthesis, Characterization and Applications*; Lendlein, A., Sisson, A., Eds.; John Wiley & Sons: Hoboken, NJ, USA, 2011; pp. 133–154.
9. Adolph, E.J.; Pollins, A.C.; Cardwell, N.L.; Davidson, J.M.; Guelcher, S.A.; Nanney, L.B. Biodegradable lysine-derived polyurethane scaffolds promote healing in a porcine full-thickness excisional wound model. *J. Biomater. Sci. Polym. Ed.* **2014**, *25*, 1973–1985. [CrossRef]
10. Shourgashti, Z.; Khorasani, M.T.; Khosroshahi, S.M.E. Plasma-induced grafting of polydimethylsiloxane onto polyurethane surface: Characterization and in vitro assay. *Radiat. Phys. Chem.* **2010**, *79*, 947–952. [CrossRef]
11. Qiu, H.; Li, D.; Chen, X.; Fan, K.; Ou, W.; Chen, K.C.; Xu, K. Synthesis, characterizations, and biocompatibility of block poly(ester-urethane)s based on biodegradable poly(3-hydroxybutyrate-co-4-hydroxybutyrate) (P3/4HB) and poly(ε-caprolactone). *J. Biomed. Mater. Res. A* **2013**, *101*, 75–86. [CrossRef]
12. Morral-Ruíz, G.; Melgar-Lesmes, P.; García, M.L.; Solans, C.; García-Celma, M.J. Polyurethane and polyurea nanoparticles based on polyoxyethylene castor oil derivative surfactant suitable for endovascular applications. *Int. J. Pharm.* **2014**, *461*, 1–13. [CrossRef] [PubMed]
13. Dulińska-Molak, I.; Lekka, M.; Kurzydłowski, K.J. Surface properties of polyurethane composites for biomedical applications. *Appl. Surf. Sci.* **2013**, *270*, 553–560. [CrossRef]
14. Chan-Chan, L.H.; Solis-Correa, R.; Vargas-Coronado, R.F.; Cervantes-Uc, J.M.; Cauich-Rodríguez, J.V.; Quintana, P.; Bartolo-Pérez, P. Degradation studies on segmented polyurethanes prepared with HMDI, PCL and different chain extenders. *Acta Biomater.* **2010**, *6*, 2035–2044. [CrossRef] [PubMed]
15. Usman, A.; Zia, K.M.; Zuber, M.; Tabasum, S.; Rehman, S.; Zia, F. Chitin and chitosan based polyurethanes: A review of recent advances and prospective biomedical applications. *Int. J. Biol. Macromol.* **2016**, *86*, 630–645. [CrossRef] [PubMed]
16. Wu, C.-S. Enhanced antibacterial activity, antioxidant and in vitro biocompatibility of modified polycaprolactone-based membranes. *Int. J. Polym. Mater. Polym. Biomater.* **2016**, *65*, 872–880. [CrossRef]
17. Anirudhan, T.S.; Nair, S.S.; Nair, A.S. Fabrication of a bioadhesive transdermal device from chitosan and hyaluronic acid for the controlled release of lidocaine. *Carbohydr. Polym.* **2016**, *152*, 687–698. [CrossRef]
18. Kaur, G.; Mahajan, M.; Bassi, P. Derivatized Polysaccharides: Preparation, characterization, and application as bioadhesive polymer for drug delivery. *Int. J. Polym. Mater.* **2013**, *62*, 475–481. [CrossRef]
19. Wu, H.; Williams, G.R.; Wu, J.; Wu, J.; Niu, S.; Li, H.; Wang, H.; Zhu, L. Regenerated chitin fibers reinforced with bacterial cellulose nanocrystals as suture biomaterials. *Carbohydr. Polym.* **2018**, *180*, 304–313. [CrossRef]
20. Aranguren, M.I.; González, J.F.; Mosiewicki, M.A. Biodegradation of a vegetable oil based polyurethane and wood flour composites. *Polym. Test.* **2012**, *31*, 7–15. [CrossRef]
21. Guelcher, S.; Srinivasan, A.; Dumas, J. Synthesis, mechanical properties, biocompatibility, and biodegradation of polyurethane networks from lysine polyisocyanates. *Biomaterials* **2008**, *29*, 1762–1775. [CrossRef]
22. Chanput, W.; Mes, J.; Vreeburg, R.A.M.; Savelkoul, H.F.J.; Wichers, H.J. Transcription profiles of LPS-stimulated THP-1 monocytes and macrophages: A tool to study inflammation modulating effects of food-derived compounds. *Food Funct.* **2010**, *1*, 254–261. [CrossRef] [PubMed]
23. Oliveira, C.M.; Sakata, R.K.; Issy, A.M.; Gerola, L.R. Citocinas y dolor. *Rev. Bras. Anestesiol.* **2011**, *61*, 137–142. [CrossRef]

24. Small, A.; Lansdown, N.; Al-Baghdadi, M.; Quach, A.; Ferrante, A. Facilitating THP-1 macrophage studies by differentiating and investigating cell functions in polystyrene test tubes. *J. Immunol. Methods* **2018**, *461*, 73–77. [CrossRef] [PubMed]
25. Lund, M.E.; To, J.; O'Brien, B.A.; Donnelly, S. The choice of phorbol 12-myristate 13-acetate differentiation protocol influences the response of THP-1 macrophages to a pro-inflammatory stimulus. *J. Immunol. Methods* **2016**, *430*, 64–70. [CrossRef] [PubMed]
26. Ballerini, P.; Diomede, F.; Petragnani, N.; Cicchitti, S.; Merciaro, I.; Cavalcanti, M.F.X.B.; Trubiani, O. Conditioned medium from relapsing-remitting multiple sclerosis patients reduces the expression and release of inflammatory cytokines induced by LPS-gingivalis in THP-1 and MO3.13 cell lines. *Cytokine* **2017**, *96*, 261–272. [CrossRef] [PubMed]
27. Dreskin, S.C.; Thomas, G.W.; Dale, S.N.; Heasley, L.E. Isoforms of Jun kinase are differentially expressed and activated in human monocyte/macrophage (THP-1) cells. *J. Immunol.* **2001**, *166*, 5646–5653. [CrossRef] [PubMed]
28. Dash, B.C.; Thomas, D.; Monaghan, M.; Carroll, O.; Chen, X.; Woodhouse, K.; Brien, T.O.; Pandit, A. An injectable elastin-based gene delivery platform for dose- dependent modulation of angiogenesis and inflammation for critical limb ischemia. *Biomaterials* **2015**, *65*, 126–139. [CrossRef] [PubMed]
29. Lin, T.H.; Yao, Z.; Sato, T.; Keeney, M.; Li, C.; Pajarinen, J.; Yang, F.; Egashira, K.; Goodman, S.B. Suppression of wear-particle-induced pro-inflammatory cytokine and chemokine production in macrophages via NF-κB decoy oligodeoxynucleotide: A preliminary report. *Acta Biomater.* **2014**, *10*, 3747–3755. [CrossRef] [PubMed]
30. Zhang, C.; Garrison, T.F.; Madbouly, S.A.; Kessler, M.R. Recent advances in vegetable oil-based polymers and their composites. *Prog. Polym. Sci.* **2017**, *71*, 91–143. [CrossRef]
31. Laube, T.; Weisser, J.; Berger, S.; Börner, S.; Bischoff, S.; Schubert, H.; Gajda, M.; Bräuer, R.; Schnabelrauch, M. In situ foamable, degradable polyurethane as biomaterial for soft tissue repair. *Mater. Sci. Eng. C* **2017**, *78*, 163–174. [CrossRef]
32. Temenoff, J.S.; Mikos, A.G. *Biomaterials: The Intersection of Biology and Materials Science*, 8th ed.; Prentice Hall, Inc.: Upper Saddle River, NJ, USA, 2008.
33. Vannozzi, L.; Ricotti, L.; Santaniello, T.; Terencio, T.; Oropesa-Nunez, R.; Canale, C.; Borghi, F.; Menciassi, A.; Lenardi, C.; Gerges, I. 3D porous polyurethanes featured by different mechanical properties: Characterization and interaction with skeletal muscle cells. *J. Mech. Behav. Biomed. Mater.* **2017**, *75*, 147–159. [CrossRef] [PubMed]
34. Chashmejahanbin, M.R.; Daemi, H.; Barikani, M.; Salimi, A. Noteworthy impacts of polyurethane-urea ionomers as the efficient polar coatings on adhesion strength of plasma treated polypropylene. *Appl. Surf. Sci.* **2014**, *317*, 688–695. [CrossRef]
35. Braun, U.; Lorenz, E.; Weimann, C.; Sturm, H.; Karimov, I.; Ettl, J.; Meier, R.; Wohlgemuth, W.A.; Berger, H.; Wildgruber, M. Mechanic and surface properties of central-venous port catheters after removal: A comparison of polyurethane and silicon rubber materials. *J. Mech. Behav. Biomed. Mater.* **2016**, *64*, 281–291. [CrossRef] [PubMed]
36. Jutrzenka Trzebiatowska, P.; Santamaria Echart, A.; Calvo Correas, T.; Eceiza, A.; Datta, J. The changes of crosslink density of polyurethanes synthesised with using recycled component. Chemical structure and mechanical properties investigations. *Prog. Org. Coat.* **2018**, *115*, 41–48. [CrossRef]
37. Thakur, S.; Karak, N. Castor oil-based hyperbranched polyurethanes as advanced surface coating materials. *Prog. Org. Coat.* **2013**, *76*, 157–164. [CrossRef]
38. Gurunathan, T.; Mohanty, S.; Nayak, S.K. Isocyanate terminated castor oil-based polyurethane prepolymer: Synthesis and characterization. *Prog. Org. Coat.* **2015**, *80*, 39–48. [CrossRef]
39. Chen, H.; Yu, X.; Zhou, W.; Peng, S.; Zhao, X. Highly toughened polylactide (PLA) by reactive blending with novel polycaprolactone-based polyurethane (PCLU) blends. *Polym. Test.* **2018**, *70*, 275–280. [CrossRef]
40. Shah, S.A.A.; Imran, M.; Lian, Q.; Shehzad, F.K.; Athir, N.; Zhang, J.; Cheng, J. Curcumin incorporated polyurethane urea elastomers with tunable thermo-mechanical properties. *React. Funct. Polym.* **2018**, *128*, 97–103. [CrossRef]
41. Cakić, S.M.; Ristić, I.S.; Cincović, M.M.; Nikolić, N.C.; Nikolić, L.; Cvetinov, M.J. Synthesis and properties biobased waterborne polyurethanes from glycolysis product of PET waste and poly (caprolactone) diol. *Prog. Org. Coat.* **2017**, *105*, 111–122. [CrossRef]

42. Ferreira, P.; Pereira, R.; Coelho, J.F.J.; Silva, A.F.M.; Gil, M.H. Modification of the biopolymer castor oil with free isocyanate groups to be applied as bioadhesive. *Int. J. Biol. Macromol.* **2007**, *40*, 144–152. [CrossRef] [PubMed]
43. Arévalo, F.; Uscategui, Y.L.; Diaz, L.; Cobo, M.; Valero, M.F. Effect of the incorporation of chitosan on the physico-chemical, mechanical properties and biological activity on a mixture of polycaprolactone and polyurethanes obtained from castor oil. *J. Biomater. Appl.* **2016**, *31*, 708–720. [CrossRef] [PubMed]
44. Corcuera, M.A.; Rueda, L.; Fernandez d'Arlas, B.; Arbelaiz, A.; Marieta, C.; Mondragon, I.; Eceiza, A. Microstructure and properties of polyurethanes derived from castor oil. *Polym. Degrad. Stab.* **2010**, *95*, 2175–2184. [CrossRef]
45. Uscátegui, Y.L.; Arévalo-Alquichire, S.J.; Gómez-Tejedor, J.A.; Vallés-Lluch, A.; Díaz, L.E.; Valero, M.F. Polyurethane-based bioadhesive synthesized from polyols derived from castor oil (*Ricinus communis*) and low concentration of chitosan. *J. Mater. Res.* **2017**, *32*, 3699–3711. [CrossRef]
46. Saikia, A.; Karak, N. Renewable resource based thermostable tough hyperbranched epoxy thermosets as sustainable materials. *Polym. Degrad. Stab.* **2017**, *135*, 8–17. [CrossRef]
47. Thakur, S.; Hu, J. Polyurethane: A Shape Memory Polymer (SMP). In *Aspects of Polyurethanes*; Yilmaz, F., Ed.; InTechOpen: London, UK, 2017; pp. 53–71.
48. Sáenz-Pérez, M.; Lizundia, E.; Laza, J.M.; García-Barrasa, J.; Vilas, J.L.; León, L.M. Methylene diphenyl diisocyanate (MDI) and toluene diisocyanate (TDI) based polyurethanes: Thermal, shape-memory and mechanical behavior. *RSC Adv.* **2016**, *6*, 69094–69102. [CrossRef]
49. Hou, Z.; Zhang, H.; Qu, W.; Xu, Z.; Han, Z. Biomedical segmented polyurethanes based on polyethylene glycol, poly(ε-caprolactone-co-D,L-lactide), and diurethane diisocyanates with uniform hard segment: Synthesis and properties. *Int. J. Polym. Mater. Polym. Biomater.* **2016**, *65*, 947–956. [CrossRef]
50. Mi, H.Y.; Jing, X.; Hagerty, B.S.; Chen, G.; Huang, A.; Turng, L.S. Post-crosslinkable biodegradable thermoplastic polyurethanes: Synthesis, and thermal, mechanical, and degradation properties. *Mater. Des.* **2017**, *127*, 106–114. [CrossRef]
51. Gossart, A.; Battiston, K.G.; Gand, A.; Pauthe, E.; Santerre, J.P. Mono vs multilayer fibronectin coatings on polar/hydrophobic/ionic polyurethanes: Altering surface interactions with human monocytes. *Acta Biomater.* **2018**, *66*, 129–140. [CrossRef]
52. Marques, D.S.; Santos, J.M.C.; Ferreira, P.; Correia, T.R.; Correia, I.J.; Gil, M.H.; Baptista, C.M.S.G. Photocurable bioadhesive based on lactic acid. *Mater. Sci. Eng. C* **2016**, *58*, 601–609. [CrossRef]
53. Sheikh, Z.; Khan, A.S.; Roohpour, N.; Glogauer, M.; Rehman, I.U. Protein adsorption capability on polyurethane and modified-polyurethane membrane for periodontal guided tissue regeneration applications. *Mater. Sci. Eng. C* **2016**, *68*, 267–275. [CrossRef]
54. Conejero-García, Á.; Gimeno, H.R.; Sáez, Y.M.; Vilariño-Feltrer, G.; Ortuño-Lizarán, I.; Vallés-Lluch, A. Correlating synthesis parameters with physicochemical properties of poly(glycerol sebacate). *Eur. Polym. J.* **2017**, *87*, 406–419. [CrossRef]
55. Carriço, C.S.; Fraga, T.; Pasa, V.M.D. Production and characterization of polyurethane foams from a simple mixture of castor oil, crude glycerol and untreated lignin as bio-based polyols. *Eur. Polym. J.* **2016**, *85*, 53–61. [CrossRef]
56. Fuentes, L.E.; Pérez, S.; Martínez, S.I.; García, Á.R. Redes poliméricas interpenetradas de poliuretano a partir de aceite de ricino modificado y poliestireno: Miscibilidad y propiedades mecánicas en función de la composición. *Revisata Ion.* **2011**, *24*, 45–50.
57. Das, B.; Konwar, U.; Mandal, M.; Karak, N. Sunflower oil based biodegradable hyperbranched polyurethane as a thin film material. *Ind. Crops Prod.* **2013**, *44*, 396–404. [CrossRef]
58. Cherng, J.Y.; Hou, T.Y.; Shih, M.F.; Talsma, H.; Hennink, W.E. Polyurethane-based drug delivery systems. *Int. J. Pharm.* **2013**, *450*, 145–162. [CrossRef] [PubMed]
59. Spontón, M.; Casis, N.; Mazo, P.; Raud, B.; Simonetta, A.; Ríos, L.; Estenoz, D. Biodegradation study by Pseudomonas sp. of flexible polyurethane foams derived from castor oil. *Int. Biodeterior. Biodegradation* **2013**, *85*, 85–94. [CrossRef]
60. Gogoi, S.; Barua, S.; Karak, N. Biodegradable and thermostable synthetic hyperbranched poly(urethane-urea)s as advanced surface coating materials. *Prog. Org. Coat.* **2014**, *77*, 1418–1427. [CrossRef]

61. Calvo-Correas, T.; Santamaria-Echart, A.; Saralegi, A.; Martin, L.; Valea, Á.; Corcuera, M.A.; Eceiza, A. Thermally-responsive biopolyurethanes from a biobased diisocyanate. *Eur. Polym. J.* **2015**, *70*, 173–185. [CrossRef]
62. Reddy, T.T.; Kano, A.; Maruyama, A.; Takahara, A. Synthesis, characterization and drug release of biocompatible/biodegradable non-toxic poly(urethane urea)s based on poly(epsilon-caprolactone)s and lysine-based diisocyanate. *J. Biomater. Sci. Polym. Ed.* **2010**, *21*, 1483–1502. [CrossRef] [PubMed]
63. Coakley, D.N.; Shaikh, F.M.; O'Sullivan, K.; Kavanagh, E.G.; Grace, P.A.; McGloughlin, T.M. In vitro evaluation of acellular porcine urinary bladder extracellular matrix—A potential scaffold in tissue engineered skin. *Wound Med.* **2015**, *10–11*, 9–16. [CrossRef]
64. Shahrousvand, M.; Sadeghi, G.M.M.; Shahrousvand, E.; Ghollasi, M.; Salimi, A. Superficial physicochemical properties of polyurethane biomaterials as osteogenic regulators in human mesenchymal stem cells fates. *Colloids Surf. B Biointerfaces* **2017**, *156*, 292–304. [CrossRef] [PubMed]
65. Aranaz, I.; Mengíbar, M.; Harris, R.; Paños, I.; Miralles, B.; Acosta, N.; Galed, G.; Heras, Á. Functional characterization of chitin and chitosan. *Curr. Chem. Biol.* **2009**, *3*, 203–230.
66. Ortuno-Lizarán, I.; Vilarino-Feltrer, G.; Martinez-Ramos, C.; Pradas, M.M.; Vallés-Lluch, A. Influence of synthesis parameters on hyaluronic acid hydrogels intended as nerve conduits. *Biofabrication* **2016**, *8*, 1–12. [CrossRef] [PubMed]
67. Jing, X.; Mi, H.Y.; Huang, H.X.; Turng, L.S. Shape memory thermoplastic polyurethane (TPU)/poly(ε-caprolactone) (PCL) blends as self-knotting sutures. *J. Mech. Behav. Biomed. Mater.* **2016**, *64*, 94–103. [CrossRef] [PubMed]
68. Starr, T.; Bauler, T.J.; Malik-Kale, P.; Steele-Mortimer, O. The phorbol 12-myristate-13-acetate differentiation protocol is critical to the interaction of THP-1 macrophages with Salmonella Typhimurium. *PLoS ONE* **2018**, *13*, 1–13. [CrossRef] [PubMed]
69. Park, E.K.; Jung, H.S.; Yang, H.I.; Yoo, M.C.; Kim, C.; Kim, K.S. Optimized THP-1 differentiation is required for the detection of responses to weak stimuli. *Inflamm. Res.* **2007**, *56*, 45–50. [CrossRef] [PubMed]
70. Gómez Estrada, H.A.; González Ruiz, K.N.; Medina, J.D. Actividad antiinflamatoria de productos naturales. *Bol. Latinoam. Caribe Plantas Med. Aromat.* **2011**, *10*, 182–217.
71. González, R.; Zamora, Z.; Alonso, Y. Citocinas anti-inflamatorias y sus acciones y efectos en la sepsis y el choque séptico. *REDVET Rev. Electrónica Vet.* **2009**, *10*, 1–11.
72. Valero, M.F.; Ortegón, Y. Polyurethane elastomers-based modified castor oil and poly(e-caprolactone) for surface-coating applications: Synthesis, characterization, and in vitro degradation. *J. Elastomers Plast.* **2015**, *47*, 360–369. [CrossRef]
73. Valero, M.F.; Pulido, J.E.; Ramírez, Á.; Cheng, Z. Determinación de la densidad de entrecruzamiento de poliuretanos obtenidos a partir de aceite de ricino modificado por transesterificación. *Polímeros* **2009**, *19*, 14–21. [CrossRef]
74. Valero, M.F. Poliuretanos elastoméricos obtenidos a partir de aceite de ricino y almidón de yuca original y modificado con anhídrido propiónico: Síntesis, propiedades fisicoquímicas y fisicomecánicas. *Quim. Nov.* **2010**, *33*, 850–854. [CrossRef]
75. Simón-Allué, R.; Pérez-López, P.; Sotomayor, S.; Peña, E.; Pascual, G.; Bellón, J.M.; Calvo, B. Short- and long-term biomechanical and morphological study of new suture types in abdominal wall closure. *J. Mech. Behav. Biomed. Mater.* **2014**, *37*, 1–11. [CrossRef] [PubMed]
76. Yoshida, K.; Jiang, H.; Kim, M.J.; Vink, J.; Cremers, S.; Paik, D.; Wapner, R.; Mahendroo, M.; Myers, K. Quantitative evaluation of collagen crosslinks and corresponding tensile mechanical properties in mouse cervical tissue during normal pregnancy. *PLoS ONE* **2014**, *9*. [CrossRef] [PubMed]
77. Mekewi, M.A.; Ramadan, A.M.; ElDarse, F.M.; Abdel Rehim, M.H.; Mosa, N.A.; Ibrahim, M.A. Preparation and characterization of polyurethane plasticizer for flexible packaging applications: Natural oils affirmed access. *Egypt. J. Pet.* **2017**, *26*, 9–15. [CrossRef]
78. Hormaiztegui, M.E.V.; Aranguren, M.I.; Mucci, V.L. Synthesis and characterization of a waterborne polyurethane made from castor oil and tartaric acid. *Eur. Polym. J.* **2018**, *102*, 151–160. [CrossRef]
79. Kanmani, P.; Rhim, J.-W. Physical, mechanical and antimicrobial properties of gelatin based active nanocomposite films containing AgNPs and nanoclay. *Food Hydrocoll.* **2014**, *35*, 644–652. [CrossRef]
80. Członka, S.; Bertino, M.F.; Strzelec, K. Rigid polyurethane foams reinforced with industrial potato protein. *Polym. Test.* **2018**, *68*, 135–145. [CrossRef]

81. Basak, P.; Adhikari, B. Effect of the solubility of antibiotics on their release from degradable polyurethane. *Mater. Sci. Eng. C* **2012**, *32*, 2316–2322. [CrossRef]

82. Wang, Y.; Yu, Y.; Zhang, L.; Qin, P.; Wang, P. One-step surface modification of polyurethane using affinity binding peptides for enhanced fouling resistance. *J. Biomater. Sci. Polym. Ed.* **2015**, *26*, 459–467. [CrossRef]

83. Rezvanain, M.; Ahmad, N.; Mohd Amin, M.C.I.; Ng, S.F. Optimization, characterization, and in vitro assessment of alginate-pectin ionic cross-linked hydrogel film for wound dressing applications. *Int. J. Biol. Macromol.* **2017**, *97*, 131–140. [CrossRef]

84. Vilariño Feltrer, G.; Martínez Ramos, C.; Monleon De La Fuente, A.; Vallés Lluch, A.; Moratal Pérez, D.; Barcia Albacar, J.; Monleón Pradas, M. Schwann-cell cylinders grown inside hyaluronic-acid tubular scaffolds with gradient porosity. *Acta Biomater.* **2016**, *30*, 199–211. [CrossRef] [PubMed]

Sample Availability: Samples of the compounds are not available from the authors.

molecules

MDPI

Article

Immobilizing Polyether Imidazole Ionic Liquids on ZSM-5 Zeolite for the Catalytic Synthesis of Propylene Carbonate from Carbon Dioxide

Liying Guo [1,*], Xianchao Jin [1], Xin Wang [2], Longzhu Yin [1,2], Yirong Wang [1] and Ying-Wei Yang [2,*]

[1] School of Petrochemical Engineering, Shenyang University of Technology, Liaoyang 111003, China; xcjin1993@163.com (X.J.); yinlz18@mails.jlu.edu.cn (L.Y.); yrwang2000@163.com (Y.W.)

[2] State Key Laboratory of Inorganic Synthesis and Preparative Chemistry, International Joint Research Laboratory of Nano-Micro Architecture Chemistry (NMAC), College of Chemistry, Jilin University, 2699 Qianjin Street, Changchun 130012, China; xinwangjlu@163.com

* Correspondence: lyguo1981@163.com (L.G.); ywyang@jlu.edu.cn (Y.-W.Y.); Tel.: +86-151-4099-8399 (L.G.); +86-431-8516-8468 (Y.-W.Y.)

Received: 21 September 2018; Accepted: 20 October 2018; Published: 21 October 2018

Abstract: Traditional ionic liquids (ILs) catalysts suffer from the difficulty of product purification and can only be used in homogeneous catalytic systems. In this work, by reacting ILs with co-catalyst ($ZnBr_2$), we successfully converted three polyether imidazole ionic liquids (PIILs), i.e., HO-[Poly-epichlorohydrin-methimidazole]Cl (HO-[PECH-MIM]Cl), HOOC-[Poly-epichlorohydrin-methimidazole]Cl (HOOC-[PECH-MIM]Cl), and H_2N-[Poly-epichlorohydrin-methimidazole]Cl (H_2N-[PECH-MIM]Cl), to three composite PIIL materials, which were further immobilized on ZSM-5 zeolite by chemical bonding to result in three immobilized catalysts, namely ZSM-5-HO-[PECH-MIM]Cl/[$ZnBr_2$], ZSM-5-HOOC-[PECH-MIM]Cl/[$ZnBr_2$], and ZSM-5-H_2N-[PECH-MIM]Cl/[$ZnBr_2$]. Their structures, thermal stabilities, and morphologies were fully characterized by Fourier-transform infrared spectroscopy (FT-IR), X-ray diffractometry (XRD), thermogravimetric analysis (TGA), and scanning electron microscopy (SEM). The amount of composite PIIL immobilized on ZSM-5 was determined by elemental analysis. Catalytic performance of the immobilized catalysts was evaluated through the catalytic synthesis of propylene carbonate (PC) from CO_2 and propylene oxide (PO). Influences of reaction temperature, time, and pressure on catalytic performance were investigated through the orthogonal test, and the effect of catalyst circulation was also studied. Under an optimal reaction condition (130 °C, 2.5 MPa, 0.75 h), the composite catalyst, ZSM-5-HOOC- [PECH-MIM]Cl/[$ZnBr_2$], exhibited the best catalytic activity with a conversion rate of 98.3% and selectivity of 97.4%. Significantly, the immobilized catalyst could still maintain high heterogeneous catalytic activity even after being reused for eight cycles.

Keywords: catalyst; CO_2; heterogeneous catalysis; molecular sieve; polyether imidazole ionic liquid

---

1. Introduction

Carbon dioxide (CO_2) is a rich carbon resource in nature. Over the years, experts and scholars around the world have been working on the catalytic conversion and utilization of CO_2 [1–5]. Although research results have shown that the chemically speaking CO_2 is extremely inactive, careful selection of proper catalysts could make CO_2 become a low-cost and widely used resource. Therefore, the development of efficient catalysts is the key to achieve chemical fixation and conversion of CO_2 under mild conditions.

Ionic liquids (ILs), as environmentally friendly chemical materials, have been widely used in various fields including organic synthesis, biomass dissolution, catalysis, and composite materials preparation due to their unique properties [6–10]. In recent years, studies on the conversion of CO_2 to cyclic carbonate using ILs as catalysts have been widely reported [11–13], demonstrating that the catalytic performance of ILs as a single component in catalysts is unfavorable, and the addition of Lewis acids or other co-catalysts in the catalysts is necessary to achieve better catalytic activity. However, these studies are still associated with several disadvantages including limitation of the intermittent tank reaction system, catalyst separation, and product purification [14]. These defects make ILs difficult to be applied in industrial processes. To solve these problems, a structural design concept based on ILs-immobilized catalysts has gained considerable attention. Yin and coworkers [15] immobilized 3-(2-hydroxyethyl)-1-propylimidazolium bromide ILs on SBA-15, Al-SBA-15, and SiO_2, but the yield declined obviously after three cycles of catalysis. Zhang's group [16] and Xiong's group [17] immobilized ILs on chitosan and coconut shell activated carbon supporter, respectively. However, gradual loss of active components in the immobilized catalysts also occurred in the catalytic process. Therefore the utilization of chemical bonding between ILs and supporters is key to realize the resource transformation of CO_2. Compared with other catalytic systems, such as metal [18], metal complexes [19], and imidazolium salt [20], ILs with the advantages of low energy consumption, high recycling efficiency, and less corrosion to equipment have become a hotspot [16]. Polyether polymer catalysts contain macromolecular chains that can form a synergistic and base barrier effect to enhance catalytic activity [21].

In this work, three polyether imidazole ionic liquids (PIILs) were selected to react with co-catalyst ($ZnBr_2$) to obtain three composite PIILs, followed by further immobilization on ZSM-5 zeolite to result in three immobilized PIILs catalysts (Figure 1 and Supplementary Materials). The chemical bonding between the composite PIILs and ZSM-5 was determined and the phase transitions of PIILs were also achieved. The immobilization is stable, so that the prepared catalyst can still maintain high catalytic activity after eight cycles.

Molecules **2018**, 23, 2710

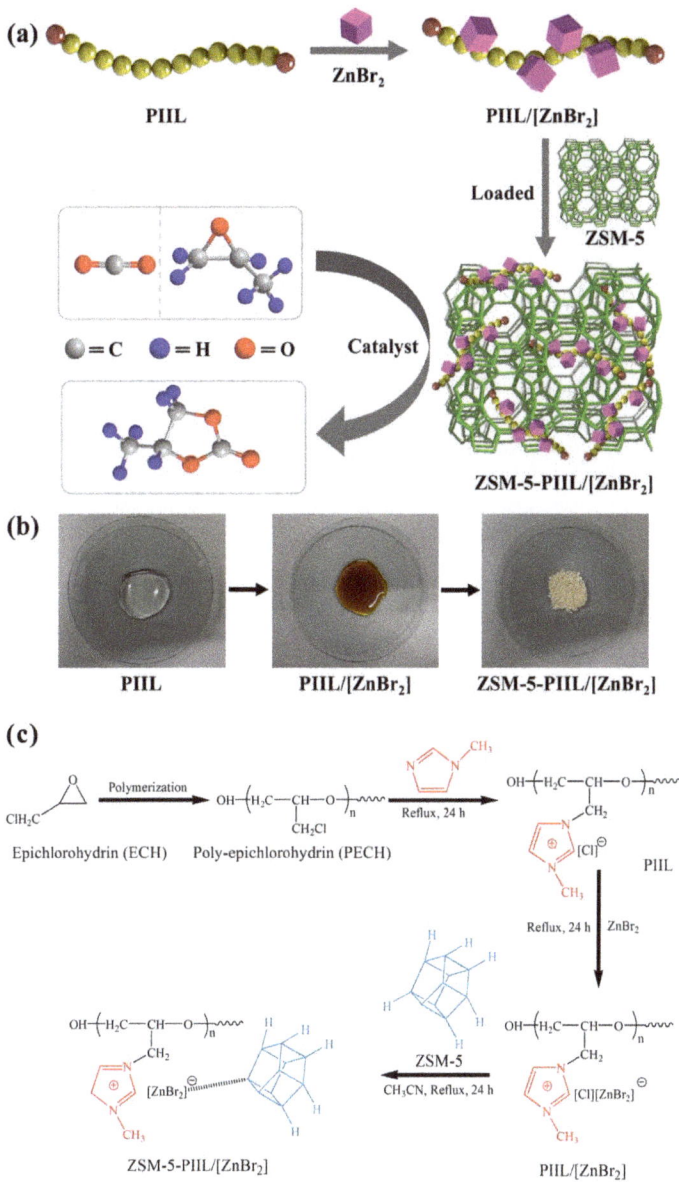

Figure 1. (**a**) The diagram for the preparation of immobilized catalysts for the synthesis of propylene carbonates from carbon dioxide; (**b**) the photos of the synthesized materials showing the color change and phase transition; and (**c**) the chemical structure of the catalyst and its synthesis process, taking ZSM-5-HO-[PECH-MIM]Cl/[ZnBr2] as an example.

2. Results and Discussion

2.1. Fourier-Transform Infrared Spectroscopy (FT-IR)

FT-IR studies on zeolite ZSM-5 and its immobilized catalysts were carried out (Figure 2). As in pattern (a), the peak at 1066 cm^{-1} corresponds to the Si–O characteristic peak of ZSM-5 [22]. Patterns (b, c, d) contain all the characteristic absorption peaks of pattern (a), and meanwhile, the characteristic peaks at ca. 1655 cm^{-1} and 626 cm^{-1}, which are associated with the stretching frequency of the imidazole ring, were also found. The peaks at ca. 1275 cm^{-1} in patterns (b, c, d) could be assigned to the enhanced stretching vibration of C–O–C. The strong absorption peak at 3386 cm^{-1} of pattern (b) corresponds to the stretching vibration of the hydroxyl group. When polyether imidazole molecular chains are linked to ZSM-5, hydrogen chloride is released [15,23,24]. However, a small portion of HOOC-[PECH-MIM]Cl/[ZnBr$_2$] still exists as deduced from the peak at 1720 cm^{-1} with low intensity in pattern (c), which is due to the physical adsorption on the surface of ZSM-5. In the immobilization process of H$_2$N-[PECH-MIM]Cl/[ZnBr$_2$], there are hydrogen atoms on the surface and inside of the molecular sieve [25]. A part of hydrogen on ZSM-5 reacted with the chloride ions of H$_2$N-[PECH-MIM]Cl/[ZnBr$_2$] thus releasing hydrogen chloride, and the other part of the hydrogen easily forms a hydrogen bond with oxygen in the polyether chain [15,23]. The hydrogen bonds caused the appearance of a wide and strong hydroxyl peak at 3396 cm^{-1} in pattern (d), thus the low amino peak was covered. The above results indicated the successful anchorage of composite PIILs on ZSM-5.

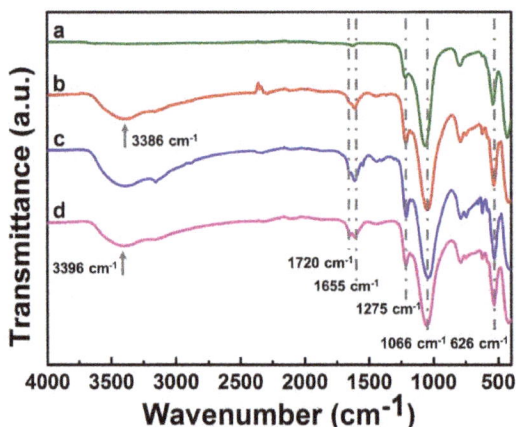

Figure 2. FT-IR spectra of zeolite ZSM-5 and its immobilized catalysts: (**a**) ZSM-5; (**b**) ZSM-5-HO-[PECH-MIM]Cl/[ZnBr$_2$]; (**c**) ZSM-5-HOOC-[PECH-MIM]Cl/[ZnBr$_2$]; and (**d**) ZSM-5-H$_2$N-[PECH-MIM]Cl/[ZnBr$_2$].

2.2. Scanning Electron Microscopy (SEM)

SEM images of ZSM-5 and its immobilized catalysts are provided in Figure 3. After the immobilization, the surfaces of the catalysts (b, c, d) were covered with a white translucent substance, and the surfaces became smoother. The particles became larger in size and their profiles became clearer, indicating that a part of the PIILs was immobilized on the surface of ZSM-5. SEM results further suggest the successful immobilization of PIILs catalysts on ZSM-5 and the solid state of the obtained immobilized catalysts. See Figure 1b for ordinary photos of the phase transition.

Figure 3. SEM images of zeolite ZSM-5 and its immobilized catalysts: (**a**) ZSM-5; (**b**) ZSM-5-HO-[PECH-MIM]Cl/[ZnBr$_2$]; (**c**) ZSM-5-HOOC-[PECH-MIM]Cl/[ZnBr$_2$]; and (**d**) SM-5-H$_2$N-[PECH-MIM]Cl/[ZnBr$_2$].

2.3. X-Ray Diffractometry (XRD)

The XRD patterns of ZSM-5 and its immobilized catalysts are shown in Figure 4. In pattern (a), the peaks of high intensity at 23.1°, 23.9°, and 24.4° are the characteristic diffraction peaks of ZSM-5, indicating good crystallinity of our synthesized ZSM-5. Compared with pattern (a), patterns (b, c, d) exhibit all the diffraction peaks of ZSM-5, and the shape and intensity of the diffraction peaks have negligible changes, indicating that the prepared catalysts maintained the good crystallinity of ZSM-5 after immobilization of the three composite PIILs onto ZSM-5.

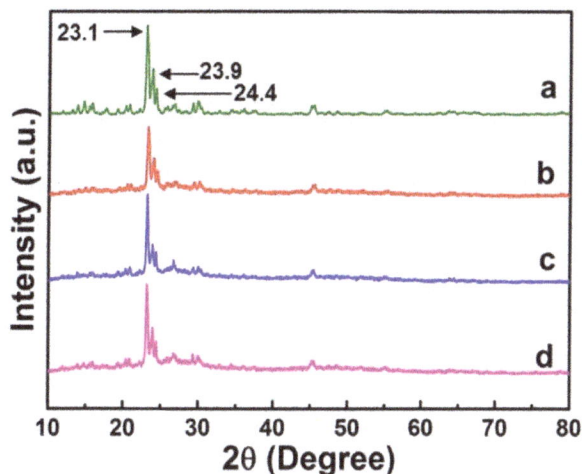

Figure 4. XRD patterns of zeolite ZSM-5 and its immobilized catalysts: (**a**) ZSM-5; (**b**) ZSM-5-HO-[PECH-MIM]Cl/[ZnBr$_2$]; (**c**) ZSM-5-HOOC-[PECH-MIM]Cl/[ZnBr$_2$]; and (**d**) ZSM-5-H$_2$N-[PECH-MIM]Cl/[ZnBr$_2$].

2.4. Thermogravimetry (TG) and Derivative Thermogravimetry (DTG)

Molecular sieve ZSM-5 and its immobilized catalysts were further characterized by thermogravimetric analyzer (TGA) to explore their thermal stability with a heating rate of 10 °C/min. TG and DTG results are shown in Figure 5. As seen in Figure 5a, ZSM-5 and the three catalysts have slight weight loss before 150 °C, which is caused by the evaporation of water. As the temperature rises, the curve of ZSM-5 in Figure 5b approximates to a straight line, where the weight loss rate does not change and ZSM-5 does not break down. From Figure 5a, the three immobilized catalysts were broken down in two steps. The first loss (230–550 °C) is attributed to the loss of grafted composite PIILs [26–28], and the second slight loss (550–700 °C) is due to the decomposition of some molecular sieve [29–31]. The three catalysts with functional groups of –OH, –COOH, and –NH$_2$ have maximum decomposition rate at 490 °C, 460 °C, and 475 °C, respectively, and their residual rates are maintained above 50%. The decomposition speed and thermal stability of the three immobilized catalysts with different functional groups are different, and among them, ZSM-5-HOOC-[PECH-MIM]Cl/[ZnBr$_2$] possesses the fastest decomposition and the highest residual rate. The thermal stability of the other two immobilized catalysts shows little difference, and the weight loss processes are similar. TG and DTG results suggest that the immobilized catalysts have good thermal stability, and that the immobilized catalysts would not decompose under 230 °C.

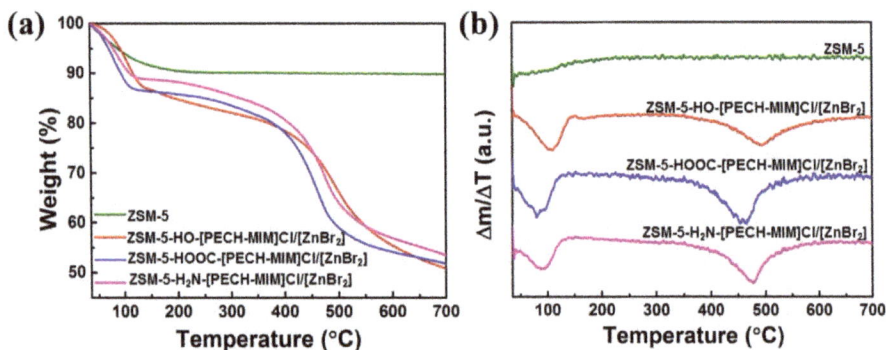

Figure 5. (a) TG and (b) DTG of zeolite ZSM-5 and its immobilized catalysts: ZSM-5-HO-[PECH-MIM]Cl/[ZnBr$_2$]; ZSM-5-HOOC-[PECH-MIM]Cl/[ZnBr$_2$]; and ZSM-5-H$_2$N-[PECH-MIM]Cl/[ZnBr$_2$].

2.5. Effects of Different Catalysts on the Catalytic Performance

The catalytic performance of the above mentioned three PIILs, three composite PIILs, ZSM-5, and three immobilized catalysts for the synthesis of propylene carbonate (PC) from CO$_2$ and propylene oxide (PO) was investigated (Table 1). Three PIILs showed catalytic activity (entries 1–3), among them, HOOC-[PECH-MIM]Cl has the best catalytic performance because the terminal carboxyl group has acidity and can also act as donor of a hydrogen bond, which could add a synergistic effect with the anion. The synergistic effect contributes to the activation and ring opening of PO and improves the catalytic activity [15,24]. Zinc bromide has a catalytic effect when used alone as a co-catalyst (entry 4). The catalytic performance of PIILs were further improved after addition of zinc bromide (entries 5–7) [26,27], where the conversion rate reached 100%, and the selectivity and yield both reached above 97%. The improvement of the catalytic performance can be attributed to the addition of zinc bromide that increased the amount of bromide ions. The Br$^-$ attacked the β-carbon atom of PO, and therefore increased the conversion rate. The polyether macromolecular chain has a basic isolation effect and an infinite dilution effect, which could not only inhibit the formation of by-products but also improve the selectivity [32,33]. The catalytic activity of ZSM-5 is poor (entry 8). The excellent catalytic performance of the three immobilized catalysts is very promising, although there is no big difference between them (entries 9–11). This is because the structures of the immobilized catalysts were similar after three composite PIILs were immobilized on the molecular sieve by chemical bonding. The chemical bonding occurs through the covalent bond formation by condensation between the chlorine of PIILs and the hydrogen on ZSM-5 [25]. All of the three immobilized catalysts have polyether macromolecular chains as the main chains and imidazole rings as the branched chains. ZSM-5-HOOC-[PECH-MIM]Cl/[ZnBr$_2$] has better catalytic activity than the other two immobilized catalysts, because there is still a small amount of HOOC-[PECH-MIM]Cl/[ZnBr$_2$] physically adsorbed on the surface of ZSM-5. It still has acidity, which is good for ring opening [15,24], consistent with the FT-IR study. The catalytic activity of the immobilized catalysts is slightly lower than that of the composite PIILs since the amount of composite PIILs is reduced after immobilization. However, the conversion rate and selectivity both reached 95%. Experimental results showed an excellent catalytic performance of the three immobilized catalysts, among which ZSM-5-HOOC-[PECH-MIM]Cl/[ZnBr$_2$] represents the best.

Table 1. Effects of different catalysts on catalytic performance [1].

	Catalyst	Conversion Rate (%)	Selectivity (%)	Yield (%)	Purity (%)
1	HO-[PECH-MIM]Cl	90.4	96.4	87.1	99.1
2	HOOC-[PECH-MIM]Cl	95.1	96.8	92.1	99.4
3	H$_2$N-[PECH-MIM]Cl	93.3	95.3	88.9	98.9
4	ZnBr$_2$	40.8	68.2	27.8	95.2
5	HO-[PECH-MIM]Cl/[ZnBr$_2$]	100	97.1	97.1	99.5
6	HOOC-[PECH-MIM]Cl/[ZnBr$_2$]	100	98.9	98.9	99.7
7	H$_2$N-[PECH-MIM]Cl/[ZnBr$_2$]	100	97.6	97.6	99.4
8	ZSM-5	9.4	41.2	3.87	96.4
9	ZSM-5-HO-[PECH-MIM]Cl/[ZnBr$_2$]	95.8	96.5	92.4	98.3
10	ZSM-5-HOOC-[PECH-MIM]Cl/[ZnBr$_2$]	96.3	96.8	93.2	99.2
11	ZSM-5-H$_2$N-[PECH-MIM]Cl/[ZnBr$_2$]	96.1	96.4	92.6	99.0

[1] Reaction condition: catalyst, 2.5 wt%, temperature, 120 °C; pressure, 2.5 MPa; time, 1 h.

2.6. Effects of Reaction Conditions on the Catalytic Performance

The effect of reaction conditions on the catalytic performance of ZSM-5-HOOC-[PECH-MIM]Cl/[ZnBr$_2$] was studied. To define the optimal reaction conditions, the orthogonal test of L9 (3^4) was used to investigate the effects of reaction pressure, temperature, and time on the conversion rate and selectivity (Table 2). K_{nj} (n = 1,2,3) indicated the sums of three conversion rates or selectivity when the values of the corresponding factors were the same for each time. R was the range. The results show that reaction pressure has the greatest effect on conversion rate and selectivity, followed by reaction temperature and time. When the reaction pressure is 2.5 MPa, the temperature is 130 °C, and the time is 0.75 h, the immobilized catalyst exhibited the best catalytic performance with a conversion rate of 98.3% and a selectivity of 97.4%. When the reaction pressure is increased, more CO$_2$ can be dissolved in PO, therefore, the conversion rate and selectivity can be increased. If the pressure was too high or the reaction time prolonged, the selectivity declined, which is consistent with the previous literature reports [16,34–36]. Excessive temperature could cause a decrease in selectivity due to the exothermic nature of the cycloaddition reaction that could lead to the production of some byproduct such as from the polymerization of PC [15,37].

Table 2. Orthogonal test-effects of reaction conditions on catalytic performance.

	Factor			Conversion Rate (%)	Selectivity (%)
	Pressure (MPa)	Temperature (°C)	Time (h)		
1	2.0	110	0.75	82.3	92.4
2	2.0	120	1.0	90.6	90.1
3	2.0	130	1.25	95.4	85.2
4	2.5	110	1.0	95.9	98.2
5	2.5	120	1.25	97.6	97.5
6	2.5	130	0.75	98.3	97.4
7	3.0	110	1.25	92.1	94.6
8	3.0	120	0.75	97.9	90.3
9	3.0	130	1.0	99.5	84.2
K_{1j}	268.3/267.7	270.3/285.2	278.5/280.1	—	—
K_{2j}	291.8/293.1	286.1/277.9	286.0/272.5	—	—
K_{3j}	289.5/269.1	293.2/266.8	285.1/277.3	—	—
R	23.5/25.4	22.9/18.4	7.5/7.6	—	—

2.7. Effect of the Catalyst Circulation on the Catalytic Performance

The effect of catalyst circulation on catalytic performance was investigated under an optimized reaction condition (2.5 MPa, 130 °C, 0.75 h), with ZSM-5-HOOC-[PECH-MIM]Cl/[ZnBr$_2$] as the catalyst. As in Figure 6, the experimental results suggest an excellent reusability of the immobilized catalyst. The conversion rate and yield are 98.3% and 97.4%, respectively, in the first use. In the next two cycles, the catalytic performance decreased. Elemental analysis was used to measure the fresh catalyst, three and eight cycle catalysts, respectively (Table 3). The results showed that the grafted rate of composite PIILs catalyst decreased over 1 to 3 cycles, and there was negligible change after three cycles. The reason is that a small amount of physically adsorbed PIILs is easily falls off from ZSM-5 zeolite during the 1st to 3rd cycles. From the 4th cycle, the grafted rate of PIILs, conversion rate and yield showed small changes, which indicate that the chemical bonding between the composite PIILs and the zeolite is steady, and therefore the catalytic performance of the immobilized catalyst remains stable. The conversion rate is 86.7% and the yield is 77.9% after the 8th cycle. The results in Figure 6 are consistent with those in Table 3. The amount of the grafted PIIL is 1.121 mmol/g, the chemical graft is about 0.803 mmol/g, and physical graft is about 0.318 mmol/g.

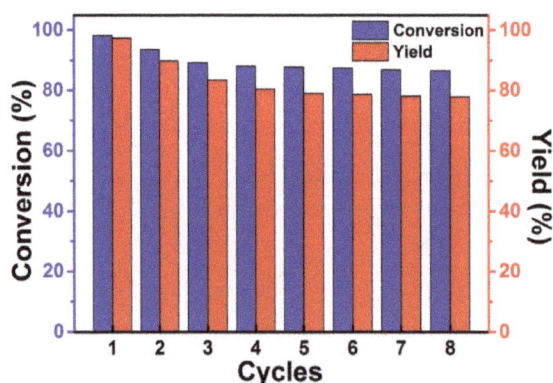

Figure 6. Effect of the catalyst circulation on catalytic performance.

Table 3. Elemental analysis results of the catalysts.

Catalyst	N (wt%)	C (wt%)	H (wt%)	PIILgrafted (mmol/g)	Standard Error (mmol/g)
ZSM-5-HOOC-[PECH-MIM]Cl/[ZnBr$_2$] [1]	3.149	7.744	2.000	1.121	1.780×10^{-3}
ZSM-5-HOOC-[PECH-MIM]Cl/[ZnBr$_2$] [2]	2.275	7.131	1.694	0.810	1.472×10^{-3}
ZSM-5-HOOC-[PECH-MIM]Cl/[ZnBr$_2$] [3]	2.256	7.059	1.681	0.803	1.633×10^{-3}

[1] Fresh catalyst; [2] Catalyst reused 2 times; [3] Catalyst reused 8 times.

3. Materials and Methods

3.1. Reagents and Instruments

Three PIILs, i.e., HO-[PECH-MIM]Cl, HOOC-[PECH-MIM]Cl, and H$_2$N-[PECH-MIM]Cl, were prepared according to our published procedures [26–28]. Zinc bromide was purchased from Tianjin Guangfu chemical reagents factory (Tianjin, China). Acetonitrile was received from Tianjin Guangfu Fine Chemical Research Institute and used without further purification. PO was obtained from Jiangsu Yonghua fine chemicals Research Company (Suzhou, China). ZSM-5 zeolite, CO$_2$, and nitrogen were supplied by Liaoyang Petrochemical Industries Company (Liaoning, China).

Instruments used in this study included a MAGNA-IR750 Fourier Transform Infrared Spectrometer (FT-IR, Thermo Nicolet Corporation, Markham, ON, Canada); TM3000 Scanning Electron Microscope (SEM, Keyence, Osaka, Japan); D/max-2400 Automatic X-ray Diffractometer (XRD, RKC Instrument Inc., Tokyo, Japan); TGA4000 Thermogravimetric Analyzer (TGA, PerkinElmer, Waltham, MA, USA); PerkinElmer-2400 Element Analyzer (EA, PerkinElmer, Waltham, MA, USA); PARR4523 Catalytic Device (PARR, Moline, IL, USA); 1790F Gas Chromatograph (GC) (Agilent Technologies, Inc., Santa Clara, CA, USA); D08-8C Carbon Dioxide Flowmeter (Beijing Sevenstar Electronics Co., Ltd., Beijing, China); DF-101S Magnetic Stirrer (Gongyi Yuhua Instrument Co., Ltd., Gongyi, China); SFX-2L Rotary Evaporator (Taiwan Xinyue Instrument and Meter Co., Ltd., Xiamen, China); DZF-6050 Vacuum Drying box (Gongyi Yuhua Instrument Co., Ltd., Gongyi, China); 2-XZ-4 rotary vane vacuum pump (Wenling City Yangan Electromechanical Co., Ltd., Wenling, China).

3.2. Preparation of Immobilized Catalysts

Three PIILs, i.e., HO-[PECH-MIM]Cl, HOOC-[PECH-MIM]Cl, and H$_2$N-[PECH-MIM]Cl, were prepared according to our published procedures [26–28]. Briefly, epichlorohydrin (ECH) was polymerized to form poly-epichlorohydrin (PECH) possessing –OH groups, then chloroacetic acid and ammonia were used to react with PECH to obtain products with –COOH and –NH$_2$ groups, respectively. Then three types of PECH samples with –OH, –COOH, and –NH$_2$ were reacted with imidazole to obtain three PIILs. Next, a given amount of ZnBr$_2$ was added into the three PIILs, respectively, and the resulting mixtures were heated at reflux under stirring for 24 h. As the reaction proceeded, the mixtures became more viscous with a dark color change. When the reactions were complete, three reddish brown liquids were obtained, namely three composite PIILs, i.e., HO-[PECH-MIM]Cl/[ZnBr$_2$], HOOC-[PECH-MIM]Cl/[ZnBr$_2$], and H$_2$N-[PECH-MIM]Cl/[ZnBr$_2$]. Then, the composite PIILs were added, respectively, to the suspension of ZSM-5 in acetonitrile, and the resulting mixtures were stirred at 70 °C for 24–36 h to obtain milk-white liquids. Acetonitrile was removed from the obtained liquids by rotary evaporation at 65 °C. The afforded three samples of white powder were then dried in vacuum at 65 °C for 24 h under 0.08 MPa to remove residual acetonitrile, leading to the final products of immobilized PIILs catalysts, i.e., ZSM-5-HO-[PECH-MIM]Cl/[ZnBr$_2$], ZSM-5-HOOC-[PECH-MIM]Cl/[ZnBr$_2$] and, ZSM-5-H$_2$N-[PECH-MIM]Cl/[ZnBr$_2$] (Figure 1c).

3.3. Typical Procedure for the Synthesis of PC from PO and CO$_2$

The immobilized catalyst (2.5% mass fraction of PO) was added in the stainless-steel autoclave. After the atmosphere was replaced by nitrogen, PO and CO$_2$ were filled into the autoclave. When the

Molecules **2018**, *23*, 2710

flow of CO_2 was 0, the reaction was complete. The crude product was purified to obtain the refined PC, and its purity was determined by gas chromatography. Specific steps refer to our reported papers [26,27].

4. Conclusions

Three immobilized PIILs catalysts were newly developed, and their catalytic performance studied. The results showed that the immobilization of composite PIILs on ZSM-5 was mainly based on chemical bonding, while a small amount of physical adsorption also existed. ZSM-5-HOOC-[PECH-MIM]Cl/[ZnBr$_2$] exhibited the highest catalytic activity in the synthesis of PC under the optimized condition: reaction temperature of 130 °C, pressure of 2.5 MPa, and time of 0.75 h, where the conversion rate reached 98.3% and the selectivity 97.4%. Significantly, the catalyst still maintained a good catalytic activity after eight cycles. We envisage that the newly prepared catalysts can solve the problems of traditional ILs catalysts including the difficulty in production purification and short service life in the catalytic process. We believe that our research can provide industrial continuous conversion of CO_2 in a packed bed reactor with acceptable performance.

Supplementary Materials: The supplementary materials are available online.

Author Contributions: Conceptualization, L.G.; Investigation, L.G., X.J., X.W., L.Y., and Y.W.; Supervision, L.G. and Y.-W.Y.; Writing—original draft, L.G. and X.J.; Writing—review and editing, L.G. and Y.-W.Y.

Funding: This work was supported by the National Natural Science Foundation of China (21706163), Department of Education of Liaoning Province (LQGD2017020), and Jilin Province-University Cooperative Construction Project—Special Funds for New Materials (SXGJSF2017-3).

Conflicts of Interest: The authors declare no conflict of interest. The funding sponsors had no role in the design of the study; in the collection, analyses, or interpretation of data; in the writing of the manuscript, and in the decision to publish the results.

References

1. Rulev, Y.A.; Gugkaeva, Z.; Maleev, V.I.; North, M.; Belokon, Y.N. Robust bifunctional aluminium-salen catalysts for the preparation of cyclic carbonates from carbon dioxide and epoxides. *Beilstein J. Org. Chem.* **2015**, *11*, 1614–1623. [CrossRef] [PubMed]
2. Kember, M.R.; Buchard, A.; Williams, C.K. Catalysts for CO_2/epoxide copolymerisation. *Chem. Commun.* **2011**, *47*, 141–163. [CrossRef] [PubMed]
3. Yang, L.H.; Wang, H.M. Recent advances in carbon dioxide capture, fixation, and activation by using N-heterocyclic carbenes. *ChemSusChem* **2014**, *7*, 962–998. [CrossRef] [PubMed]
4. Chen, S.M.; Liu, Y.; Guo, J.P.; Li, P.Z.; Huo, Z.Y.; Ma, P.T.; Niu, J.Y.; Wang, J.P. A multi-component polyoxometalate and its catalytic performance for CO_2 cycloaddition reactions. *Dalton Trans.* **2015**, *44*, 10152–10155. [CrossRef] [PubMed]
5. Li, Z.X.; Na, W.; Wang, H.; Gao, W.G. Direct syntheses of Cu-Zn-Zr/SBA-15 mesoporous catalysts for CO_2 hydrogenation to methanol. *Chem. J. Chin. Univ.* **2014**, *35*, 2616–2623.
6. Agatemor, C.; Ibsen, K.N.; Tanner, E.E.L.; Mitragotri, S. Ionic liquids for addressing unmet needs in healthcare. *Bioeng. Transl. Med.* **2018**, *3*, 7–25. [PubMed]
7. Earle, M.J.; Seddon, K.R. Ionic liquids. Green solvents for the future. *Pure Appl. Chem.* **2009**, *72*, 1391–1398. [CrossRef]
8. Sheldon, R. Catalytic reactions in ionic liquids. *Chem. Commun.* **2001**, *23*, 2399–2407. [CrossRef]
9. Seddon, K.R. Ionic liquids for clean technology. *J. Chem. Technol. Biotechnol.* **1997**, *68*, 351–356. [CrossRef]
10. Shi, F.; Gu, Y.L.; Zhang, Q.H.; Deng, Y.Q. Development of ionic liquids as green reaction media and catalysts. *Catal. Surv. Asia* **2004**, *8*, 179–186. [CrossRef]
11. Peng, J.J.; Deng, Y.Q. Formation of propylene carbonate catalyzed by room temperature ionic liquids. *Chin. J. Catal.* **2001**, *22*, 598–600.
12. Li, F.W.; Xiao, L.F.; Xia, C.G. Synthesis of cyclic carbonates catalyzed by ionic liquid mediated ZnBr$_2$ catalytic system. *Chem. J. Chin. Univ.* **2005**, *26*, 343–345.

13. Guo, L.Y.; Deng, L.L.; Jin, X.C.; Wu, H.; Yin, L.Z. Composite ionic liquids immobilized on MCM-22 as efficient catalysts for the cycloaddition reaction with CO_2 and propylene oxide. *Catal. Lett.* **2017**, *147*, 2290–2297. [CrossRef]

14. Wang, Y.H.; Li, W.; Luo, S.; Liu, S.X.; Ma, C.H.; Li, J. Research advances on the applications of immobilized ionic liquids functional materials. *Acta Chim. Sin.* **2018**, *76*, 85–94. [CrossRef]

15. Dai, W.L.; Chen, L.; Yin, S.F.; Luo, S.L.; Au, C.T. 3-(2-Hydroxyl-Ethyl)-1-propylimidazolium bromide immobilized on SBA-15 as efficient catalyst for the synthesis of cyclic carbonates via the coupling of carbon dioxide with epoxides. *Catal. Lett.* **2010**, *135*, 295–304. [CrossRef]

16. Sun, J.; Wang, J.Q.; Cheng, W.G.; Zhang, J.X.; Li, X.H.; Zhang, S.J.; She, Y.B. Chitosan functionalized ionic liquid as a recyclable biopolymer-supported catalyst for cycloaddition of CO_2. *Green Chem.* **2012**, *14*, 654–660. [CrossRef]

17. Zhang, Y.L.; Tan, Z.T.; Liu, B.L.; Mao, D.S.; Xiong, C.R. Coconut shell activated carbon tethered ionic liquids for continuous cycloaddition of CO_2 to epichlorohydrin in packed bed reactor. *Catal. Commun.* **2015**, *68*, 73–76. [CrossRef]

18. Zhuo, C.W.; Qin, Y.S.; Wang, X.H.; Wang, F.S. Temperature-responsive catalyst for the coupling reaction of carbon dioxide and propylene oxide. *Chin. J. Chem.* **2018**, *36*, 299–305. [CrossRef]

19. Della Monica, F.; Maity, B.; Pehl, T.; Buonerba, A.; De Nis, A.; Monari, M.; Grassi, A.; Rieger, B.; Cavallo, L.; Capacchione, C. [OSSO]-type iron(III) complexes for the low-pressure reaction of carbon dioxide with epoxides: Catalytic activity, reaction kinetics, and computational study. *ACS Catal.* **2018**, *8*, 6882–6893. [CrossRef]

20. Bobbink, F.D.; Vasilyev, D.; Hulla, M.; Chamam, S.; Menoud, F.; Laurenczy, G.; Katsyuba, S.; Dyson, P.J. Intricacies of cation–anion combinations in imidazolium salt-catalyzed cycloaddition of CO_2 into epoxides. *ACS Catal.* **2018**, *8*, 2589–2594. [CrossRef]

21. Wang, G.J.; Wang, D.H.; Qiu, J.; Zhao, L.Q. *Functional Polymer Materials*, 1st ed.; East China University of Science and Technology Press: Shanghai, China, 2006; pp. 283–285.

22. Sang, Y.; Liu, H.X.; He, S.C.; Li, H.S.; Jiao, Q.Z.; Wu, Q.; Sun, K.N. Catalytic performance of hierarchical H-ZSM-5/MCM-41 for methanol dehydration to dimethyl ether. *J. Energy Chem.* **2013**, *22*, 769–777. [CrossRef]

23. Valkenberg, M.H.; deCastro, C.; Hölderich, W.F. Immobilisation of ionic liquids on solid supports. *Green Chem.* **2002**, *4*, 88–93. [CrossRef]

24. Cheng, W.G.; Chen, X.; Sun, J.; Wang, J.Q.; Zhang, S.J. SBA-15 supported triazolium-based ionic liquids as highly efficient and recyclable catalysts for fixation of CO_2 with epoxides. *Catal. Today* **2013**, *200*, 117–124. [CrossRef]

25. Mao, D.S.; Guo, Q.S.; Meng, T. Effect of magnesium oxide modification on the catalytic performance of nanoscale HZSM-5 zeolite for the conversion of methanol to propylene. *Acta Phys-Chim. Sin.* **2010**, *26*, 2242–2248.

26. Guo, L.Y.; Ma, X.Y.; Li, C.Y.; Deng, L.L.; Bai, S.Y. Preparation and catalytic properties of chloride 1-carboxyl polyether-3-mtthyl imidazole ionic liquid. *Acta Pet. Sin. (Pet. Process Sect.)* **2017**, *33*, 342–348.

27. Guo, L.Y.; Ma, X.Y.; Wang, L.Y.; Deng, L.L.; Jin, X.C. Preparation and catalytic properties of chloride 1-amino polyether-3-methyl imidazole ionic liquid. *Chem. Ind. Eng. Prog.* **2017**, *36*, 581–587.

28. Guo, L.Y.; Zhang, B.; Li, C.Y.; Ma, X.Y.; Chang, X.T.; Gao, X. Synthesis of polyether ionic liquids and its influence on properties of phenolic resin. *Plastics.* **2015**, *44*, 57–59.

29. Yuan, S.B.; She, L.Q.; Liu, X.Y.; Li, X.W.; Pang, L.; Huang, H.Z.; Zhou, Y. Characterization and aromazization activity of gallium modified HZSM-5 catalysts. *Chin. J. Catal.* **1988**, *1*, 25–31.

30. Leng, Y.; Lu, D.; Jiang, P.P.; Zhang, C.J.; Zhao, J.W.; Zhang, W.J. Highly cross-linked cationic polymer microspheres as an efficient catalyst for facile CO_2 fixation. *Catal. Commun.* **2016**, *74*, 99–103. [CrossRef]

31. Saptal, V.B.; Bhanage, B.M. Bifunctional ionic liquids for the multitask fixation of carbon dioxide into valuable chemicals. *ChemCatChem* **2016**, *8*, 244–250. [CrossRef]

32. Han, L.; Li, H.Q.; Choi, S.J.; Park, M.S.; Lee, S.M.; Kim, Y.J.; Park, D.W. Ionic liquids grafted on carbon nanotubes as highly efficient heterogeneous catalysts for the synthesis of cyclic carbonates. *Appl. Catal. A* **2012**, *429*, 67–72. [CrossRef]

33. Fujita, S.; Nishiura, M.; Arai, M. Synthesis of styrene carbonate from carbon dioxide and styrene oxide with various zinc halide-based ionic liquids. *Catal. Lett.* **2010**, *135*, 263–268. [CrossRef]

34. Kim, M.I.; Choi, S.J.; Kim, D.W.; Park, D.W. Catalytic performance of zinc containing ionic liquids immobilized on silica for the synthesis of cyclic carbonates. *J. Ind. Eng. Chem.* **2014**, *20*, 3102–3107. [CrossRef]

35. Takahashi, T.; Watahiki, T.; Kitazume, S.; Yasuda, H.; Sakakura, T. Synergistic hybrid catalyst for cyclic carbonate synthesis: Remarkable acceleration caused by immobilization of homogeneous catalyst on silica. *Chem. Commun.* **2006**, *37*, 1664–1666. [CrossRef] [PubMed]

36. Sun, J.; Cheng, W.G.; Fan, W.; Wang, Y.H.; Meng, Z.Y.; Zhang, S.J. Reusable and efficient polymer-supported task-specific ionic liquid catalyst for cycloaddition of epoxide with CO_2. *Catal. Today* **2009**, *148*, 361–367. [CrossRef]

37. Appaturi, J.N.; Adam, F. A facile and efficient synthesis of styrene carbonate via cycloaddition of CO_2 to styrene oxide over ordered mesoporous MCM-41-Imi/Br catalyst. *Appl. Catal. B* **2013**, *136*, 150–159. [CrossRef]

Sample Availability: Samples of the PIILs and the immobilized catalysts are available from the authors.

![molecules logo]

Article

Structure Characterization and Otoprotective Effects of a New Endophytic Exopolysaccharide from Saffron

Juan Li [1], Guimei Wu [1], Cuiying Qin [1], Wuhai Chen [1], Gang Chen [1,2,3,*] and Lu Wen [1,*]

[1] School of Pharmacy, Guangdong Pharmaceutical University, Guangzhou 510006, China;
smilewinnie@126.com (J.L.); gdpuwgm@126.com (G.W.); gdpuwen@126.com (C.Q.);
gdpucwh@126.com (W.C.)

[2] Guangdong Provincial Key Laboratory of Advanced Drug Delivery, Guangdong Pharmaceutical University,
Guangzhou 510006, China

[3] Guangdong Provincial Engineering Center of Topical Precise Drug Delivery System,
Guangdong Pharmaceutical University, Guangzhou 510006, China

* Correspondence: cg753@126.com (G.C.); gdpuwen@gdpu.edu.cn (L.W.)

Received: 28 January 2019; Accepted: 18 February 2019; Published: 19 February 2019

Abstract: Saffron, a kind of rare medicinal herb with antioxidant, antitumor, and anti-inflammatory activities, is the dry stigma of *Crocus sativus* L. A new water-soluble endophytic exopolysaccharide (EPS-2) was isolated from saffron by anion exchange chromatography and gel filtration. The chemical structure was characterized by FT-IR, GC-MS, and 1D and 2D-NMR spectra, indicating that EPS-2 has a main backbone of (1→2)-linked α-D-Manp, (1→2, 4)-linked α-D-Manp, (1→4)-linked α-D-Xylp, (1→2, 3, 5)-linked β-D-Araf, (1→6)- linked α-D-Glcp with α-D-Glcp-(1→ and α-D-Galp-(1→ as sidegroups. Furthermore, EPS-2 significantly attenuated gentamicin-induced cell damage in cultured HEI-OC1 cells and increased cell survival in zebrafish model. The results suggested that EPS-2 could protect cochlear hair cells from ototoxicity exposure. This study could provide new insights for studies on the pharmacological mechanisms of endophytic exopolysaccharides from saffron as otoprotective agents

Keywords: hearing loss; saffron; endophytic exopolysaccharide; gentamicin; cochlear hair cell

1. Introduction

Hearing loss is a global problem. To date, more than 466 million people have moderate to severe or greater hearing loss, and one-third of them are over 65 years old. As the world's population ages, it is estimated that approximately 900 million people (or one in every 10 people) will suffer hearing loss by 2050. World Health Organization (WHO) estimates that an annual global cost of hearing loss will be US $750 billion, including health sector costs (excluding hearing equipment), education support costs, productivity losses and social costs [1]. Unfortunately, hearing loss has not received sufficient attention by the pharmaceutical industry, and until now there have been few Food and Drug Administration (FDA)-approved drugs to treat or prevent different types of hearing loss [2]. Most hearing loss is caused by permanent loss of hair cells in the inner ear. One of the most likely causes of hair cell death is exposure to ototoxic agents, including aminoglycoside antibiotics such as gentamicin (GM) and neomycin, and cisplatin anticancer agents [3,4]. Additionally, it is estimated that aminoglycoside antibiotics generate hearing thresholds of almost 50% [5]. However, these aminoglycoside drugs also continue to be used in view of their cost and effectiveness, and their ototoxicity usually limits the dose range of drugs [4,6]. GM is a cationic aminoglycoside that enters cells via endocytosis and forms a complex with iron, which drives the formation of free radicals and directly promotes the formation of ROS [7]. Recent reports show that antioxidant drugs can benefit patients with hearing loss, because hearing loss is produced by excessive generation of reactive oxygen species (ROS) in cells of the cochlea [8,9]. Therefore, it is highly possible to find potential drugs in antioxidants to prevent hair cell loss due to ototoxicity exposure and attenuate hearing loss.

With today's interest in new renewable sources of polymers, polysaccharides are macromolecular carbohydrates that play a vital role in the growth and development of living organisms, and serve as important biological response modifiers. Furthermore, it is well known that the polysaccharides can be used in various pharmaceutical formulations due to their sustainability, biodegradability, and biosafety [10–12]. Polysaccharides are widely distributed in plants, animals, and microorganisms. Endophytic fungi symbiotically live in plant tissues and all or part of their entire life cycles is spent in and/or between plant cells, often without causing apparent symptoms of diseases. Moreover, these microorganisms play important roles as components of plant ecosystems [13]. Endophytic fungi benefit their hosts by enhancing resistance to disease, abiotic stress and plant growth, and they have been widely recognized as a rich, potential and novel source of natural bioactive substances in agricultural, pharmaceutical and food industries [14]. Nowadays, endophytic fungi, as a renewable resource, are of growing interest. They often produce exopolysaccharides with unique structures and diverse biological activities, which have become the most promising group of antioxidants [15,16].

Saffron (*Crocus sativus* L.) is a perennial stemless herb of the Iridaceae family grown in Turkey, Iran, and China, and it has a variety of biological activities, such as antitumor, antioxidant, and anti-inflammatory effects [17,18]. As the main medicinal part of saffron, the stigma extract has antioxidant properties, which can scavenge oxygen radicals and effectively inhibit the activity of oxygen radicals [19]. Our previous studies have shown that crude exopolysaccharide extracted from fermentation mycelia of saffron exhibited excellent scavenging activities against 1, 1-diphenyl-2-picrylhydrazyl, hydroxyl and superoxide anion radicals [20]. However, the purification of exopolysaccharide and its further activity have not been researched. The present research describes the purification and characterization of a water-soluble exopolysaccharide (EPS-2) from the fermentation culture of endophytic fungus *Penicillium citreonigrum* CSL-27 of saffron and investigates its protective role in the damaged cochlear hair cells.

2. Results and Discussion

2.1. Isolation and Purification of Exopolysaccharide

The main exopolysaccharide fraction EPS-2 was collected according to the detection curve of phenol-sulphuric acid colorimetry. After purification with a DEAE-52 cellulose column (Figure 1a), one major fractional peak was obtained. According to Figure 1b, the fraction appeared as a single and symmetrical peak after being further purified by Sephadex G-75 column chromatography. EPS-2 appeared as a single and symmetrical peak in the high performance gel permeation chromatography (HPGPC) (Figure 1c), indicating homogeneity. By comparison of the retention times of EPS-2 with the molecular standards, the molecular weight of EPS-2 was estimated to be 40.4 kDa. The colorimetric analysis has shown that EPS-2 contains 88.9% total carbohydrate, and no sulfate ester, protein, or uronic acid is detected. Monosaccharide composition analysis indicated that EPS-2 was mainly composed of mannose, glucose, galactose, xylose, and arabinose with a molar ratio of 51.77:36.76:5.76:3.16:6.94. The results of SEM have shown that EPS-2 has a frizzy shape, and the surface has a scaly texture. The irregular aggregation has determined that EPS-2 is an amorphous solid. The molecular aggregation may be attributed to a repulsive force between the polysaccharides and the side chains.

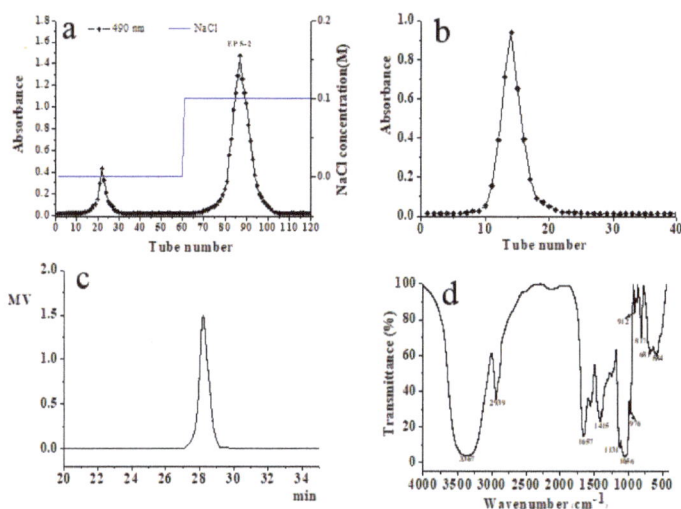

Figure 1. Chromatograms of EPS-2 from *P. citreonigrum* CSL-27 from (**a**) DEAE-52 cellulose column chromatography, (**b**) Sephadex G-75 chromatography, (**c**) HPGPC, and (**d**) FT-IR spectrum.

2.2. Fourier Transformed Infrared (FT-IR) Spectroscopy Analysis

The IR spectrum of EPS-2 displayed the characteristic peaks of polysaccharide (Figure 1d). The strong and broad absorption peak at 3367 cm^{-1} was related to the stretch vibration of O–H (hydroxyl group) bond existing in all polymers. The strong peak at 2939 cm^{-1} was due to the C–H stretching vibration in the sugar ring and the strong absorption peak at 1657 cm^{-1} represented the stretching vibration of C=O and carboxyl group. Another peak at 1418 cm^{-1} could be attributed to the symmetric stretching of the COO-group [21]. The presence of strong absorbance in the region 1200–950 cm^{-1} indicated the polysaccharide nature of EPS-2. The strong absorption at 1131 and 1056 cm^{-1} in the range of 1200–1000 cm^{-1}, which is anomeric region, was attributed to C–O–C and C–O groups in the polysaccharide, suggesting that the monosaccharide in the EPS-2 has a pyranose ring [22]. Moreover, the band at 912 cm^{-1} indicated the pyranose form of the glucosyl residue and absorption peak at 817 cm^{-1} as well as the existence of glycosidic linkages of the EPS-2. Moreover, the weak absorption at 912 and 817 cm^{-1} was assigned to the coexistence of α and β glycosidic bonds [23].

2.3. Methylation Analysis

Methylation is an essential method to analyze linkages that form methoxyl groups on the free hydroxyl groups of polysaccharides [24]. After methylation of EPS-2, a series of methylated derivatives were identified based on a gas chromatography coupled with mass spectrometry (GC-MS) analysis. According to the retention time and the ion fragment characteristics in the GC-MS spectra, 1,2,5-tri-O-acetyl-3,4,6-tri-O-methyl-mannitol (residue A), 1,2,4,5-tri-O-acetyl-3,6-tri-O-methyl-mannitol (residue B), 1,5-di-O-acetyl-2,3,4,6-tetra-O-methyl-D-glucitol (residue C), 1,5,6-tri-O-acetyl-2,3, 4-tri-O-methyl-D-glucitol (residue D), 1, 5-di-O-acetyl-2, 3, 4, 6-tetra-O-methyl-galactitol (residue E), 1,4,5-tri-O-acetyl-2, 3-di-O-methyl-D-xylitol (residue F), and 2,3,4,5-tetra-O-acetyl-D-arabinitol (residue G) were detected (summarized in Table 1), indicating the presence of →2)–Manp-(1→, →2, 4)–Manp-(1→, Glc-(1→, →6)-Glcp-(1→, Gal-(1→, →4)-α-D-Xyl-(1→ and →2, 3, 5)–Ara-(1→, respectively. The molar ratio was 7.2:2.3:3.3:1.2:1.0:1.1:1.1, which was in agreement with the results of the GC analysis of monosaccharide components.

Table 1. Glycosidic linkage composition of methylated EPS-2.

PMAA	Mass Fragments (*m/z*)	Molar Ratio	Linkage Type
3,4,6-Me$_3$-Manp [a]	87, 101, 117, 129, 161, 189, 233	7.2	→2)-D-Manp-(1→
3,6-Me$_2$-Manp	43, 71, 87, 129, 173, 189, 233	2.3	→2, 4)-D-Manp-(1→
2,3,4,6-Me$_4$-Glcp	45, 71, 87, 101, 117, 129, 145, 161, 205	3.3	D-Glcp-(1→
2,3,4-Me$_3$-Glcp	58, 71, 87, 99, 101, 117, 129, 161, 189, 233	1.2	→6)-D-Glcp-(1→
2,3,4,6-Me$_4$-Galp	87, 101, 117, 129, 189, 233	1.0	D-Galp-(1→
2,3-Me$_2$-Xylp	43, 87, 101, 117, 129, 189	1.1	→4)-D-Xylp-(1→
Araf	73, 85, 115, 145, 158, 187, 217	1.1	→2, 3, 5)-D-Araf-(1→

[a] 3, 4, 6-Me$_3$-Manp = 1, 2, 5-Tri-*O*-acetyl-3,4,6-tri-*O*-methyl-mannitol.

2.4. Nuclear Magnetic Resonance (NMR) Spectroscopy Analysis

The structure of EPS-2 was further analyzed by NMR spectroscopy. The ^{1}H, ^{13}C, HSQC, and HMBC spectra of EPS-2 are shown in Figure 2. Several anomeric proton signals (5.31–5.03 ppm) appeared in the ^{1}H-NMR spectrum. Other proton signals were located in the region of about 4.25–3.36 ppm, which were attributed to the protons of the C-2–C-6 of hexosyl glycosidic ring. The corresponding anomeric carbon signals (102.4–98.2 ppm) were identified in the ^{13}C-NMR and HSQC spectra of EPS-2. These signals corresponded to seven types of residues (residue A–G, respectively), and this result was consistent with the GC-MS result.

The carbon and proton signals of residues A–G were grouped by comprehensive analysis, comparison of the NMR spectra, GC-MS data of EPS-2 and published literature [25–28]. The α/β configurations of residues were judged by the chemical shift and coupling constant of the anomeric proton [29,30]. In the HSQC spectrum, the anomeric signals at 100.5/5.30, 98.2/5.16, 98.3/5.07, 102.4/5.10, 102.1/5.06, 102.0/5.03, and 102.3/5.07 ppm corresponded to residues A–G, respectively, which belonged to →2)-α-D-Manp-(1→, →2, 4)-α-D-Manp-(1→, α-D-Glcp-(1→, →6)-α-D-Glcp-(1→, α-Galp-(1→, →4)-α-D-Xylp-(1→, and →2, 3, 5)-β-D-Araf-(1→, respectively. Additionally, the adjacent carbon and hydrogen signals of each residue were assigned according to the HMBC spectrum. The data on carbon and hydrogen for EPS-2 are shown in Table 2.

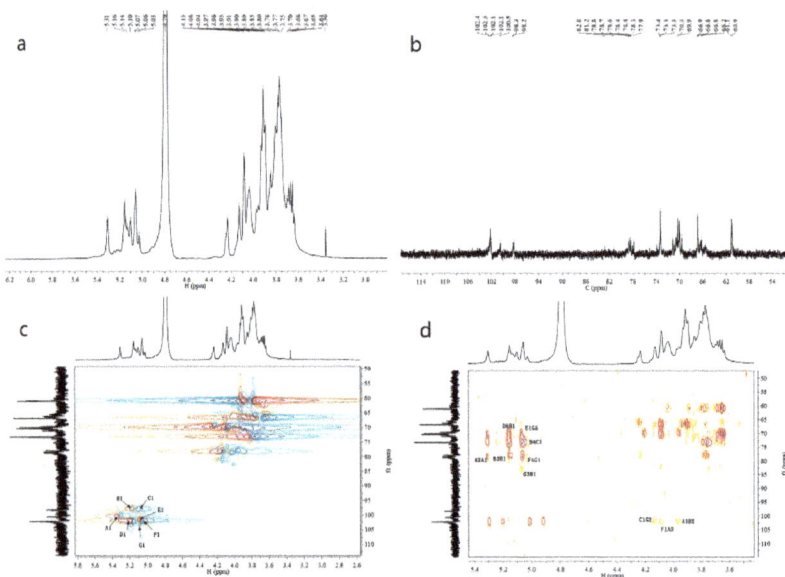

Figure 2. *Cont.*

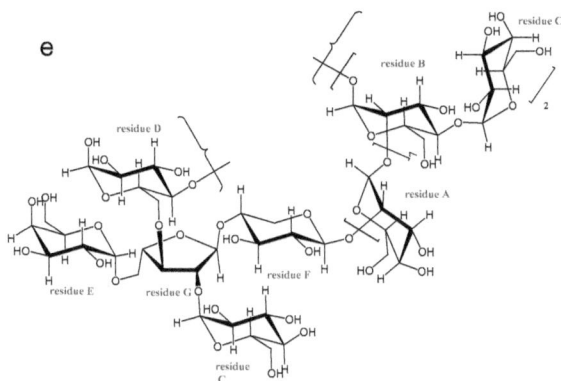

Figure 2. ^{1}H-NMR (**a**), ^{13}C-NMR (**b**), HSQC (**c**), HMBC spectrum (**d**), and predicted repetitive structural unit (**e**) of EPS-2.

Table 2. Assignment of ^{13}C-NMR and ^{1}H-NMR chemical shifts of EPS-2.

No.	Glycosyl Residues	Chemical Shifts, δ (ppm)					
		C1/H1	C2/H2	C3/H3	C4/H4	C5/H5	C6/H6
A	→2)–α-D-Manp-(1→	100.5	78.5	70.6	66.8	73.9	63.4
		5.30	4.11	3.97	3.65	3.80	3.91
B	→2, 4)–α-D-Manp-(1→	98.2	78.3	70.1	70.3	73.6	66.9
		5.16	3.96	4.07	3.89	3.76	3.85, 3.77
C	α-D-Glcp-(1→	98.3	70.4	73.3	76.7	70.1	61.0
		5.07	3.80	3.85	4.08	3.90	3.81, 3.75
D	→6)–α-D-Glcp-(1→	102.4	73.3	73.5	70.4	76.8	66.2
		5.10	3.36	3.60	3.62	4.08	3.75
E	α-D-Galp-(1→	102.1	70.2	73.2	71.2	73.3	61.1
		5.06	3.82	4.13	4.02	4.04	3.72
F	→4)–α-D-Xylp-(1→	102.0	75.3	76.3	78.4	65.5	
		5.03	3.82	3.95	3.73	3.65	
G	→2, 3, 5)–β-D-Araf-(1→	102.3	81.1	82.8	81.1	66.6	
		5.07	4.25	4.02	4.23	3.96, 3.88	

Some points existed in HSQC and also showed in HMBC were removed, and the remained linkage sites between the residues were determined by analyzing the HMBC spectrum of EPS-2. In the HMBC spectrum, the peak at δ 78.5/5.30 ppm (AC2/AH1) suggested that there is a recurring C-2 linked O-1 of residue A structure in EPS-2. The signal at δ 78.3/5.16 ppm (B C2/B H1) has shown that C-2 is linked to O-1 of residue B, and the signal at δ 100.5/3.96 ppm (A C1/B H2) has indicated that C-2 of residue B is linked to O-1 of recurring residue A. Likewise, the linkages of residue F O-1 with residue A C-2, residue C O-1 with residue B C-4, and residue B O-1 with residue D C-6 were deduced by the signals at δ 102.0/4.11 ppm (FC1/AH2), 70.3/5.07 ppm (BC4/CH1), 66.2/5.16 ppm (DC6/BH1), respectively. In addition, signals at δ δ98.3/4.25 ppm (CC1/GH2), δ 82.8/5.10 ppm (GC3/DH1), δ 66.6/5.06 ppm (GC5/EH1), and δ 78.4/5.07 ppm (FC4/GH1) illustrate that O-1, C-2, C-3 and C-5 of residue G were linked to the C-4 of residue F, O-1 of residue C, O-1 of residue D, O-1 of residue E, respectively. The possible repetitive structure unit of EPS-2 was inferred and is shown in Figure 2e.

2.5. Effects of EPS-2 on the Viability of House Ear Institute-Organ of Corti 1 (HEI-OC1) Cells Treated with GM

As evidenced by the MTT assay, the exposure to GM for 24 h decreased the cell viability in a dose-dependent manner. Cell viability was reduced by ca. 50% by 10 mM GM (Figure 3a).

Thus, 10 mM GM was used subsequently. Cells were pretreated with 50, 100, 200, 400, and 800 µg/mL EPS-2 for 1 h before adding 10 mM GM. Control cells were treated with a vehicle (0.1% DMSO). Cell survival was determined after 24 h. In the control group, no cytotoxicity was observed at 0.1% DMSO. The exposure of HEI-OC1 cells to GM resulted in a significant reduction of cell viability, but cell pretreatment with EPS-2 significantly inhibited GM-mediated cytotoxicity in a dose-dependent manner, as shown in Figure 3b. In cells treated with GM only, EPS-2 at 50 µg/mL increased cell viability by 50% EDA (40 µM), a positive control also significantly increased cell survival.

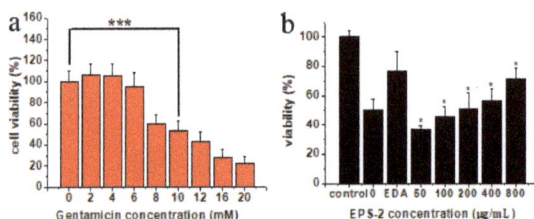

Figure 3. Effect of EPS-2 on GM-mediated decrease in viability of HEI-OC1 cells. (**a**) GM decreased cell viability dose dependently. GM (10 mM) treatment for 24 h decreased the cell viability by ca. 50%. (**b**) EPS-2 protected HEI-OC1 cells following GM (10 mM) treatment. * $p < 0.05$, *** $p < 0.001$ vs. control group. Results are shown as the mean $\pm$ SD ($n = 6$).

2.6. Protective Effect of EPS-2 on Hair Cells in Neuromasts

Using the zebrafish lateral line as a model of hair cell death, we tested EPS-2 prevented GM-induced hair cell death. DASPEI was performed to stain mitochondria, and the mean number of hair cells in the four neuromasts (SO1, SO2, O1, and OC1) of the zebrafish larvae was calculated to quantitatively assess the changes. Aminoglycoside treatment caused nuclear fragmentation and reduced neuromast fluorescence, while protective compound could prevent fragmentation and preserve labeling intensity.

From two independent screens, we found that EPS-2 protected hair cells from GM (Figure 4). We first determined whether 1 h of exposure to EPS-2 alone caused hair cell death, and the 400 µg/mL concentration of EPS-2 was revealed to have toxicity for hair cells compared to negative control, whereas 200 µg/mL EPS-2 had no toxicity for hair cells compared to the negative control group. Therefore, EPS-2 at a concentration of 200 µg/mL was expected to be applied as a maximal concentration in the following study. Pretreatment with 25, 50, 100, and 200 µg/mL EPS-2 resulted in significant protection of hair cells exposed to 100 µM GM compared to GM alone. The mean ($\pm$ SD) number of hair cells in the neuromasts in the negative control group was 36.62 $\pm$ 1.85, and the viability of hair cells was set to 100%. In the model group (100 µM GM), the viability of hair cells was 50%; in the positive control group (0.5 µM EDA), the viability of hair cells was 82%; the protective effect of pre-treated EPS-2 increased in a dose-dependent manner until 200 µg/mL (the viability of hair cells was up to 72%) of concentration with a significantly large number of viable hair cells than that in the model group.

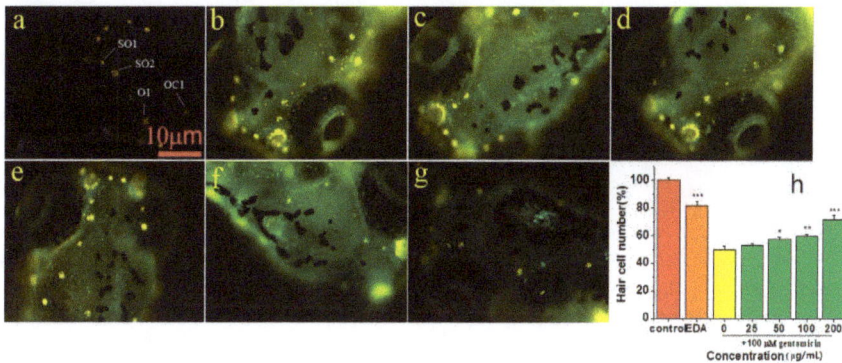

Figure 4. Analysis of hair cell damage by DASPEI assay (×10). (**a**) negative control: 36.62 ± 1.85 cells; (**b**) GM 100 μM treatment: 18.37 ± 2.26 cells; (**c**) EPS-2 25 μg/mL + GM 100 μM treatment: 19.37 ± 1.41 cells; (**d**) EPS-2 50 μg/mL + GM 100 μM treatment: 21.00 ± 1.85cells; (**e**) EPS-2 100 μg/mL + GM 100 μM treatment: 21.87 ± 1.55 cells; (**f**) EPS-2 200 μg/mL + GM 100 μM treatment: 26.25 ± 3.01 cells; (**g**) positive control: 30.00 ± 3.16 cells; and (**h**) EPS-2 protected hair cells following GM (100 μM) treatment. * $p < 0.05$, ** $p < 0.01$, *** $p < 0.001$ vs. control group). Results are shown as the mean ± SD ($n = 8$ fish per treatment). Scale bar = 10 μm.

3. Experimental

3.1. Chemicals and Reagents

3-Aminobenzoic acid ethyl ester methanesulfonate (MS-222, CAS no. 886-86-2), 3-methy-1-pheny-2-pyrazoline-5-one (EDA, CAS no. 89-25-8), 2-(4-(dimethylamino)styryl)-*N*- ethylpyridinium iodide (DASPEI, CAS no. 3785-01-1) and 3-(4, 5-dimethylthiazol-2-yl)-2, 5-diphenyltetrazolium bromide (MTT, CAS no. 298-93-1) were obtained from Sigma-Aldrich (St Louis, MO, USA). Gentamycin (GM, CAS no.1405-41-0) was purchased from Dalilan Meilun Biotechnology Co., Ltd. (Dalian, China). The Hepes-buffered DMEM solution (pH 7.4, Gibco, 1×, sterile, CAS no. 21063-029) used in this study contained neither phenol red nor sodium pyruvate was purchased from Life Technologies Co. (Grand Island, NY, USA). High-glucose Dulbecco's modified Eagle medium (DMEM) and Fetal bovine serum (FBS) were purchased from Thermo Fisher Scientific (Waltham, MA, USA). DEAE-cellulose 52 was purchased from Beijing Dingguo Changsheng Biotechnology Co., Ltd. (Beijing, China). Sephadex G-75 gel filtration medium was purchased from Shanghai Lanji Biotechnology Co., Ltd. (Shanghai, China). Trifluoroacetic acid (TFA), 1-phenyl-3-methyl-5-pyrazolone (PMP), standard monosaccharides (D-mannose, L-rhamnose, D-glucuronic acid, D-galacturonic acid, D-glucose, D-galactose, D-xylose, L-arabinose, and D-fucose), T-series dextrans of different molecular weights (T-5.2, T-11.6, T-23.8, T-48.6, T-148, T-273, T-410, T-668, and T-1400) and dialysis tubing (molecular weight cut off, 8000–14,000 Da) were obtained from Sigma-Aldrich (St. Louis, MO, USA). The reagents used in high-performance liquid chromatography (HPLC) and GC-MS were of chromatograph grade, and all other chemicals and reagents used were of analytical grade (AR).

3.2. Fungal Material Microbial Strain and Culture Conditions

The strain CSL-27 was isolated from the corm of saffron and identified as *Penicillium citreonigrum* by Beijing Dingguo Changsheng Biotechnology Co. Ltd. The strain was stored at China Center for Type Culture Collection (CCTCC) (Wuhan, China). The strain was activated on potato dextrose agar (PDA) slants, and then cultivated on a rotary shaker (TCYQ, Taicang Laboratory Equipment Factory, Jiangsu Province, China) constantly at 120 rpm and 28 °C for 14 days. The liquid culture medium contained 10 g/L glucose, 2 g/L peptone, 1 g/L yeast extract and 1 g/L NaCl with a pH of 6.5.

3.3. Cell Culture

HEI-OC1 cell line, derived from the organ of Corti was obtained from the House Ear Institute (Los Angeles, CA, USA). Cells were cultured in high-glucose DMEM, supplemented with 10% FBS at 33 °C and 10% CO_2 in a humidified atmosphere without antibiotics. The cell incubator (HERAcell 150i) was derived from Thermo Fisher Scientific (Waltham, MA, USA).

3.4. Zebrafish Husbandry

Zebrafish (*Danio rerio*) embryos were produced by paired matings of AB wild-type adult fish from School of Pharmaceutical Sciences, Sun Yat-sen University and maintained in zebrafish facilities at School of Pharmacy, Guangdong Pharmaceutical University. Experiments were performed on 5–6 day old larval zebrafish maintained at 28 °C in a defined embryo medium (EM) containing 1 mM $MgSO_4$, 120 μM KH_2PO_4, 74 μM Na_2HPO_4, 1 mM $CaCl_2$, 500 μM KCl, 15 mM NaCl, and 500 μM $NaHCO_3$ in distilled water at pH 7.2. This age range was selected due to the fact that hair cells in 5 day-old fish show mature responses to ototoxic insult, and the small fish size allows for high throughput screening of compounds in small volumes [31]. All procedures were approved by the appropriate Institutional Animal Care and Use Committee at Guangdong Pharmaceutical University.

3.5. Isolation and Purification of EPS-2

The supernatant from the culture of strain CSL-27 was collected and concentrated by a vacuum rotary evaporator (EYELA, Japan). The concentrated solution was treated with four volumes of cold 95% EtOH and kept overnight at 4 °C. The precipitate was separated and collected by centrifugation at 3000 rpm for 20 min and dissolved in distilled water and deproteinated by the Sevag method [32]. Finally, the precipitate was dialyzed in distilled water for 48 h at 4 °C and then freeze-dried to obtain a crude polysaccharide. The sugar content in the EPS was analyzed using phenol-sulfuric acid method with glucose as the standard [33].

The crude exopolysaccharide was purified using a column (2.6 × 95 cm) packed with Macroporous resin AB-8. Distilled water was employed as the mobile phase. The flow rate was 2 mL/min. Each fraction (10 mL) was collected and analyzed with the phenol-sulfuric acid reagent at 490 nm using a spectrophotometer [33]. The fractions, which coincided with the major peak, were collected together, concentrated at 60 °C with a rotary evaporator under vacuum, dialyzed (Mw cut off: 8000 Da) and lyophilized. The exopolysaccharide samples obtained by lyophilizing were dissolved in distilled water and fractionated on a pre-equilibrated DEAE-52 cellulose column (2.6 × 60 cm) equilibrated with distilled water and then eluted with aqueous NaCl solution (0.1 mol/L) at a flow rate of 1 mL/min. All the fractions were assayed for carbohydrate content by the phenol-sulfuric acid method and the fraction representing only one sharp peak was collected, dialyzed, concentrated and further purified using a Sephadex G-75 gel-filtration column (1.6 × 70 cm) by eluting with distilled water at a flow rate of 0.2 mL/min. Consequently, a fine exopolysaccharide, named EPS-2, was obtained. After freeze-drying, EPS-2 was available for use in the subsequent experiments.

3.6. Analysis of Physicochemical Characteristics

The molecular weight of EPS-2 was assessed by HPGPC on TSK-5000PWXL and TSK G-3000 PWXL gel columns (1.8 × 300 mm) in series (Tosoh Biosep, Tokyo, Japan). The columns were calibrated with dextran standards and a refractive index detector (Waters 2414, Milford, MA, USA), and eluted with 0.02 M KH_2PO_4 solution at a flow rate of 0.6 mL/min and column temperature of 35 °C. The molecular weight was estimated by reference to a calibration curve made by a set of standards dextran (Mw: 1400, 668, 410, 273, 148, 48.6, 23.8, and 5.2 kDa) [34]. The sample was dissolved in 1 mL of distilled water and mixed with an equal volume of 4.0 M TFA. The sample was allowed to stand still for 4 h at 100 °C and the acid-hydrolyzed sample was filtered through a 0.45 μm syringe filter and the residual acid was removed through decompression and distillation with methanol for thrice [35]. The resulting monosaccharide compositions were determined by HPLC after precolumn derivatization with PMP using a Shimadzu HPLC system fitted with Phenomenex GEMINI-NX C_{18} HPLC column (4.6 nm × 250 mm) and Shimadzu prominence diode array detector. The sugar was identified by comparison with reference sugars (L-rhamnose, L-arabinose, D-fucose, D-xylose, D-mannose, D-galactose, D-glucose, D-glucuronic acid, and D-galacturonic acid). Calculation of the molar ratio of the monosaccharide was carried out on the basis of the peak area of the monosaccharide [24].

The morphology of EPS-2 was observed under a low vacuum scanning electron microscope (SEM, Philips Quanta-400, Netherlands). The dried exopolysaccharide powder was placed on a specimen holder with the help of double-sided adhesive tapes and then sputtered with the gold powder using a sputter coater. The sample was observed at magnifications of 800× and 1600× at an accelerating potential of 20 kV under low vacuum conditions.

3.7. FT-IR Analysis

FT-IR spectroscopy was used to determine the functional groups of the purified EPS. Infrared spectra of the purified EPS fraction were recorded in the 4000–400 cm^{-1} region using a FT-IR system (Perkin Elmer Spectrometer 100, Wellesley, MA, USA). The sample (10 mg) was mixed with 100 mg of dried potassium bromide (KBr) and compressed to prepare as a salt disc (10 mm diameter) for reading the spectrum further. The determinations were performed in two independent replicates and are reported as the mean with standard deviations.

3.8. Methylation Analysis

A methylation analysis was performed by the method of Hakomori with some modifications [15]. In brief, polysaccharide in dimethyl sulfoxide (DMSO) was methylated using NaH and iodomethane. After 6 h total hydrolysis with 2 M TFA at 105 °C, the methylated sugar residues were converted to partially methylated alditol acetates by reduction with $NaBH_4$, followed by acetylation with acetic anhydride. The derived sugar residues were dissolved in 100 µL chloroform. Subsequently, partial methylated alditol acetates (PMAAs) were analyzed by GC-MS on the Shimadzu NTST system equipped with a TG WAXMS capillary column (30.0 m × 0.25 mm × 0.25 µm) (ThermoFinnigan, Silicon valley, CA, USA). The temperature was set to 50 °C, maintained for 3 min, then increased to 240 °C at a rate of 15 °C/min, and maintained at 240 °C for 20 min. Helium acted as the carrier gas, with the flow rate maintained at 1.0 mL/min. PMAAs were identified by the retention times and fragmentation patterns.

3.9. NMR Spectroscopy Analysis

^{1}H-NMR and ^{13}C-NMR spectra were recorded using a Bruker AVANCE IIIT600 NMR spectrometer at 25 °C. The sample (35 mg) was deuterium-exchanged by lyophilization two times with 99.97% D_2O, and then was dissolved in 1.0 mL of 99.97% D_2O. Acetone was taken as the internal standard (2.225 ppm for ^{1}H and 31.07 ppm for ^{13}C). ^{1}H-NMR, ^{13}C-NMR, ^{1}H-^{1}H COSY, HSQC, and HMBC were performed using the standard Agilent software.

3.10. Protection Assay

Hair cells act as sensory receptors for the auditory and vestibular systems in all vertebrates. Conventional vertebrate experimental animal models are well suited to detect and solve problems related to hair cell death and survival, but they are not suitable for drug screening. This is mainly because the inner ear is inaccessible and the alive time of inner ear cochlear tissue is very short in vitro [36]. As present, HEI-OC1 cell line is a mature and immortalized cell lines derived from the cochlea and vestibular tissue that has been shown to be sensitive to some known ototoxins, such as GM and cisplatin [37]. In addition, Hair cells in the inner ear of mammals are similarity with the pathway activated by ototoxicity exposure in zebrafish lateral line, but hair cells in zebrafish regenerate following ototoxicity exposure unlike mammals. Therefore, the zebrafish lateral line is an excellent model for drug screening that modulates hair cell survival, an intractable approach in mammalian systems [3]. In this experiment, HEI-OC1 cell line and zebrafish model were effectively combined for the activity assay of EPS-2.

3.10.1. Cell Viability Assay

Cell viability was measured using the MTT assay as described previously [38]. HEI-OC1 cells were seeded at a density of 1×10^4 cells/well in a 96-well plate and cultured overnight. To investigate the effect of EPS-2 on cell viability, HEI-OC1 cells were treated with 50, 100, 200, 400, and 800 μg/mL EPS-2 for 1 h, before being exposed to GM. When the cells were confluent, the culture medium was replaced with medium containing GM, which was the calculated half-maximal inhibitory concentration (IC_{50}). After one day of incubation, 100 μL MTT was added to each culture well and the 96-well plate was incubated at 33 °C in an atmosphere of 10% CO_2 for 4 h. EDA (40 μM) was used as a positive control. The effect of EPS-2 on viability at each concentration was calculated as a percentage of the control activity from the absorbance values. Absorbance at 490 nm was measured using a Microplate spectrophotometer (RT-2100C, Shenzhen Rayto Life Science Co., Ltd., China) for cell viability and the average OD in control cells was taken as 100% of viability. A final concentration of 10 mM GM was selected to damage the HEI-OC1 cells in the following experiments.

$$\text{Cell relative viability (\%)} = OD_{experiment}/OD_{control} \times 100\% \ (OD_{blank} \text{ was used to zero})$$

Six wells were used for each EPS-2 concentration and three independent experiments were performed.

3.10.2. Assay of Zebrafish Neuromast Hair Cell Protection

At 5–6 days post-fertilization (dpf) AB zebrafish larvae were raised at 28.5 °C in Petri dishes and transferred to cell culture baskets placed in 96-well culture plates in groups of 3–4 fish per basket. The larvae were exposed to EPS-2 at the following concentrations: 25, 50, 100, 200, and 400 μg/mL for 1 h for the experimental group. A negative control group with no additional sample was also established. The larvae were then washed with the EM three times and anesthetized using 40 μg/mL MS-222 for 5 min as described in previous publications [39]. The mean count of hair cells was calculated within four neuromasts (SO1, SO2, O1, and OC1) on one side of each fish at a $10\times$ magnification using a Zeiss inverted fluorescence microscope (Carl ZEISS AG, Germany) for each group ($n = 8$). All zebrafish were alive and no abnormal developments were observed.

Then the 5–6 dpf zebrafish larvae were pretreated with EPS-2 for 1 h at concentrations of 25, 50, 100, and 200 μg/mL followed by treatment with 100 μM GM, respectively. Following 1 h GM exposure, the larvae were rinsed briefly with EM, and incubated in a staining agent (0.005% DASPEI) for 15min, rinsed three times with fresh EM and anesthetized in 40 μg/mL MS-222 for 5 min. The hair cells within the above-mentioned four neuromasts were examined. Each neuromast was scored for presence of a normal compliment of hair cells, with reduced or absent DASPEI staining indicating a reduction in the number of hair cells. Composite scores were calculated for the larvae in each treatment group, normalized to the control group and expressed as % hair cell survival. Negative controls were treated with GM while positive controls were treated with 0.5 μM EDA.

3.10.3. Statistical Analysis

All data were presented as the mean $\pm$ standard deviation. One-way analysis of variance (ANOVA) was used for multiple comparisons; $p < 0.05$ was considered statistically significant. Statistical analysis was performed with IBM SPSS 21.0 for Windows (IBM, Armonk, NY, USA).

4. Conclusions

Hearing loss is the one of the most common sensory disorders in humans, and a large number of cases are due to hair cell damage caused by ototoxicity drugs such as GM. Therefore, it is great significance to identify agents and their mechanisms that protect hair cells from ototoxicity damage [40]. Saffron has strong biological activities, and the active part is concentrated in the stigma, and the amount is too small to be detrimental to further research. Thus, endophytic fungus is a good substitute for studying saffron. Moreover, many literatures have reported that endophytic exopolysaccharides have unique charms and effective activities. We previously reported the antioxidant activity of crude exopolysaccharide extracted from fermentation mycelia of saffron. The characterization of polysaccharides of the endophytic fungus had great significance for the further structure-function relationship study, and the development and application of the endophytic polysaccharide. Therefore, we purified the polysaccharide with DEAE-52 cellulose and Sephadex G-75 columns, and a new water-soluble endophytic polysaccharide EPS-2 with a molecular weight of 40.4 kDa was obtained. The results of monosaccharide composition, FT-IR spectroscopy, GC-MS and NMR analyses suggested that EPS-2 is composed of →2)–Manp-(1→, →2, 4)–Manp-(1→, Glc-(1→, →6)- Glcp-(1→, Gal-(1→, →4)-α-D-Xyl-(1→, and →2, 3, 5)–Ara-(1→. The possible repetitive structural unit of EPS-2 was inferred. The most effective concentration of EPS-2 for attenuating GM-induced HEI-OC1 cell damage was 200 µg/mL (50% cell viability), and EPS-2 protected hair cells from a concentration of 25 µg/mL (50% hair cell number) in a zebrafish model. In conclusion, the study reports the systematic purification, structural identification, and the in-vitro testing of its protective effects on hair cells against GM toxicity of EPS-2 and notes its potential as a natural candidate lead for new drugs to combat hearing loss that can be utilized in the pharmaceutical and healthcare industries.

Author Contributions: L.W. and G.C. conceived, designed the experiment and elaborated the manuscript; J.L. performed activity assays and analyzed the data; G.W. performed structure elucidation and provided 1D- and 2D-NMR spectra; C.Q. cultured the fungus, prepared a crude fungal extract and assisted in structure elucidation; W.C. isolated and purified the exopolysaccharide; and all authors approved the final paper.

Funding: This research was financially supported by the National Natural Science Foundation of China (Nos. 81502944 and 81573618).

Conflicts of Interest: The authors declare no conflicts of interest.

References

1. WHO Media centre. Deafness and hearing loss. World Health Organization. Available online: http://www.who.int/mediacentre/factsheets/fs300/en/ (accessed on 15 March 2018).
2. Thomas, A.J.; Wu, P.; Raible, D.W.; Rubel, E.W.; Simon, J.A.; Ou, H.C. Identification of small molecule inhibitors of cisplatin-induced hair cell death: Results of a 10,000 compound screen in the zebrafish lateral line. *Otol. Neurotol.* **2015**, *36*, 519–525. [CrossRef] [PubMed]
3. Kruger, M.; Boney, R.; Ordoobadi, A.J.; Sommers, T.F.; Trapani, J.G.; Coffin, A.B. Natural bizbenzoquinoline derivatives protect zebrafish lateral line sensory hair cells from aminoglycoside toxicity. *Front Cell Neurosci.* **2016**, *10*, 83–98. [CrossRef]
4. Esterberg, R.; Coffin, A.B.; Ou, H.; Simon, J.A.; Raible, D.W.; Rubel, E.W. Fish in a dish: drug discovery for hearing habilitation. *Drug Discov. Today* **2013**, *10*, 23–29. [CrossRef] [PubMed]
5. Fausti, S.A.; Henry, J.A.; Schaffer, H.I.; Olson, D.J.; Frey, R.H.; McDonald, W.J. High-frequency audiometric monitoring for early detection of aminoglycoside ototoxicity. *J. Infect. Dis.* **1992**, *165*, 1026–1032. [CrossRef] [PubMed]
6. Noack, V.; Pak, K.; Jalota, R.; Kurabi, A.; Ryan, A.F. An antioxidant screen identifies candidates for protection of cochlear hair cells from gentamicin toxicity. *Front Cell Neurosci.* **2017**, *11*, 242–252. [CrossRef]
7. Draz, E.I.; Abdin, A.A.; Sarhan, N.I.; Gabr, T.A. Neurotrophic and antioxidant effects of silymarin comparable to 4-methylcatechol in protection against gentamicin-induced ototoxicity in guinea pigs. *Pharmacol. Rep.* **2015**, *67*, 317–325. [CrossRef] [PubMed]

8. Oishi, N.; Kendall, A.; Schacht, J. Metformin protects against gentamicin-induced hair cell death in vitro but not ototoxicity in vivo. *Neurosci. Lett.* **2014**, *583*, 65–69. [CrossRef] [PubMed]

9. Le Prell, C.G.; Ojano-Dirain, C.; Rudnick, E.W.; Nelson, M.A.; DeRemer, S.J.; Prieskorn, D.M.; Miller, J.M. Assessment of nutrient supplement to reduce gentamicin-induced ototoxicity. *J. Assoc. Res. Oto.* **2014**, *15*, 375–393. [CrossRef] [PubMed]

10. Zhang, Y.; Wang, F.; Li, M.; Yu, Z.; Qi, R.; Ding, J.; Zhang, Z.; Chen, X. Self-stabilized hyaluronate nanogel for intracellular codelivery of doxorubicin and cisplatin to osteosarcoma. *Adv. Sci.* **2018**, *5*, 1700821. [CrossRef]

11. Chen, J.; Ding, J.; Xu, W.; Sun, T.; Xiao, H.; Zhuang, X.; Chen, X. Receptor and microenvironment dual-recognizable nanogel for targeted chemotherapy of highly metastatic malignancy. *Nano Lett.* **2017**, *17*, 4526–4533. [CrossRef]

12. Li, D.; Han, J.; Ding, J.; Chen, L.; Chen, X. Acid-sensitive dextran prodrug: A higher molecular weight makes a better efficacy. *Carbohydr. Polym.* **2017**, *161*, 33–41. [CrossRef] [PubMed]

13. Rodriguez, R.J.; White, J.F., Jr.; Arnold, A.E.; Redman, R.S. Fungal endophytes: diversity and functional roles. *New Phytol.* **2009**, *182*, 314–330. [CrossRef] [PubMed]

14. Park, Y.H.; Chung, J.Y.; Ahn, D.J.; Kwon, T.R.; Lee, S.K.; Bae, I.; Yun, H.K.; Bae, H. Screening and characterization of endophytic fungi of *Panax ginseng* Meyer for biocontrol activity against *Ginseng pathogens*. *Biol. Control* **2015**, *91*, 71–81. [CrossRef]

15. Yan, M.X.; Mao, W.J.; Liu, X.; Wang, S.Y.; Xia, Z.; Cao, S.J.; Li, J.; Qin, L.; Xian, H.L. Extracellular polysaccharide with novel structure and antioxidant property produced by the deep-sea fungus *Aspergillus versicolor* N2bc. *Carbohydr. Polym.* **2016**, *147*, 272–281. [CrossRef]

16. Kavitake, D.; Devi, P.B.; Singh, S.P.; Shetty, P.H. Characterization of a novel galactan produced by *Weissella confusa* KR780676 from an acidic fermented food. *Int. J. Biol. Macromol.* **2016**, *86*, 681–689. [CrossRef] [PubMed]

17. Hatziagapiou, K.; Kakouri, E.; Lambrou, G.I.; Bethanis, K.; Tarantilis, P.A. Antioxidant properties of *Crocus sativus* L. and its constituents and relevance to neurodegenerative diseases; focus on Alzheimer's and Parkinson's disease. *Curr. Neuropharmacol.* **2018**, *16*, 1–26. [CrossRef] [PubMed]

18. Lopresti, A.L.; Drummond, P.D.; Inarejos-García, A.M.; Prodanov, M. Affron®, a standardised extract from saffron (*Crocus sativus* L.) for the treatment of youth anxiety and depressive symptoms: a randomised, double-blind, placebo-controlled study. *J. Affect Disorders* **2018**, *232*, 349–357. [CrossRef] [PubMed]

19. Christodoulou, E.; Kadoglou, N.P.E.; Stasinopoulou, M.; Konstandi, O.A.; Kenoutis, C.; Kakazanis, Z.I.; Rizakou, A.; Kostomitsopoulos, N.; Valsami, G. *Crocus sativus* L. aqueous extract reduces atherogenesis, increases atherosclerotic plaque stability and improves glucose control in diabetic atherosclerotic animals. *Atherosclerosis* **2018**, *268*, 207–214. [CrossRef] [PubMed]

20. Wen, L.; Xu, Y.; Wei, Q.; Chen, W.; Chen, G. Modeling and optimum extraction of multiple bioactive exopolysaccharide from an endophytic fungus of *Crocus sativus* L. *Pharmacogn. Mag.* **2018**, *14*, 36–43.

21. Zhao, M.; Yang, N.; Yang, B.; Jiang, Y.; Zhang, G. Structural characterization of water-soluble polysaccharides from *Opuntia monacantha* cladodes in relation to their anti-glycated activities. *Food Chem.* **2007**, *105*, 1480–1486. [CrossRef]

22. Vijayabaskar, P.; Babinastarlin, S.; Shankar, T.; Sivakumar, T.; Anandapandian, K.T.K. Quantification and characterization of exopolysaccharides from *Bacillus subtilis* (MTCC 121). *Adv. Biol. Res.* **2011**, *5*, 71–76.

23. Lim, J.M.; Joo, J.H.; Kim, H.O.; Kim, H.M.; Kim, S.W.; Hwang, H.J.; Yun, J.W. Structural analysis and molecular characterization of exopolysaccharides produced by submerged mycelial culture of *Collybia maculata* TG-1. *Carbohydr. Polym.* **2005**, *61*, 296–303. [CrossRef]

24. Zhang, Y.; Zhou, T.; Wang, H.; Cui, Z.; Cheng, F.; Wang, K.P. Structural characterization and *in vitro*, antitumor activity of an acidic polysaccharide from *Angelica sinensis*, (Oliv.) Diels. *Carbohydr. Polym.* **2016**, *147*, 401–408. [CrossRef] [PubMed]

25. Jeff, I.B.; Li, S.; Peng, X.; Kassim, R.M.; Liu, B.; Zhou, Y. Purification, structural elucidation and antitumor activity of a novel mannogalactoglucan from the fruiting bodies of *Lentinus edodes*. *Fitoterapia* **2013**, *84*, 338–346. [CrossRef] [PubMed]

26. Xia, Y.G.; Liang, J.; Yang, B.Y.; Wang, Q.H.; Kuang, H.X. Structural studies of an arabinan from the stems of *Ephedra sinica* by methylation analysis and 1D and 2D-NMR spectroscopy. *Carbohydr. Polym.* **2015**, *121*, 449–456. [CrossRef] [PubMed]

27. Sen, I.K.; Mandal, A.K.; Chakraborty, R.; Behera, B.; Yadav, K.K.; Maiti, T.K.; Islam, S.S. Structural and immunological studies of an exopolysaccharide from *Acinetobacter junii* BB1A. *Carbohydr. Polym.* **2014**, *101*, 188–195. [CrossRef]
28. Jeff, I.B.; Yuan, X.; Sun, L.; Kassim, R.M.; Foday, A.D.; Zhou, Y. Purification and in vitro anti-proliferative effect of novel neutral polysaccharides from *Lentinus edodes*. *Int. J. Biol. Macromol.* **2013**, *52*, 99–106. [CrossRef] [PubMed]
29. Gong, Y.J.; Zhang, J.; Gao, F.; Zhou, J.W.; Xiang, Z.N.; Zhou, C.G.; Wan, L.S.; Chen, J.C. Structure features and in vitro hypoglycemic activities of polysaccharides from different species of Maidong. *Carbohydr. Polym.* **2017**, *173*, 215–222. [CrossRef] [PubMed]
30. Fang, T.T.; Zirrolli, J.; Bendiak, B. Differentiation of the anomeric configuration and ring form of glucosyl-glycolaldehyde anions in the gas phase by mass spectrometry: isomeric discrimination between *m/z* 221 anions derived from disaccharides and chemical synthesis of *m/z* 221 standards. *Carbohydr Res.* **2007**, *342*, 217–235. [CrossRef] [PubMed]
31. Coffin, A.B.; Rubel, E.W.; Raible, D.W. Bax, Bcl2, and P53 differentially regulate neomycin- and gentamicin-induced hair cell death in the zebrafish lateral line. *J. Assoc. Res. Oto.* **2013**, *14*, 645–659. [CrossRef]
32. Wilkinson, J.F.; Dudman, W.F.; Aspinall, G.O. The extracellular polysaccharide of *Aerobacter aerogenes* A3 (S1) (*Klebsiella* type 54). *Biochem. J.* **1955**, *59*, 446–451. [CrossRef]
33. Dubois, M.; Gilles, K.A.; Rebers, P.A.; Smith, F. Colorimetric method for determination of sugar and related substances. *Anal. Chem.* **1956**, *28*, 350–356. [CrossRef]
34. Chen, Y.; Mao, W.; Wang, B.; Zhou, L.; Gu, Q.; Chen, Y.; Zhao, C.; Li, N.; Wang, C.; Shan, J.; et al. Preparation and characterization of an extracellular polysaccharide produced by the deep-sea fungus *Penicillium griseofulvum*. *Bioresource Technol.* **2013**, *132*, 178–181. [CrossRef] [PubMed]
35. Seedevi, P.; Moovendhan, M.; Viramani, S.; Shanmugam, A. Bioactive potential and structural chracterization of sulfated polysaccharide from seaweed (*Gracilaria corticata*). *Carbohydr. Polym.* **2017**, *155*, 516–524. [CrossRef] [PubMed]
36. Ou, H.C.; Santos, F.; Raible, D.W.; Simon, J.A.; Rubel, E.W. Drug screening for hearing loss: using the zebrafish lateral line to screen for drugs that prevent and cause hearing loss. *Drug Discov. Today* **2010**, *15*, 265–271. [CrossRef] [PubMed]
37. Kalinec, G.M.; Webster, P.; Lim, D.J.; Kalinec, F. A cochlear cell line as an in vitro system for drug ototoxicity screening. *Audiol. Neurootol.* **2003**, *8*, 177–189. [CrossRef] [PubMed]
38. Song, J.J.; Lee, J.D.; Lee, B.D.; Chae, S.W.; Park, M.K. Effect of diesel exhaust particles on human middle ear epithelial cells. *J. Assoc. Res. Oto.* **2012**, *76*, 334–338. [CrossRef]
39. Rah, Y.C.; Choi, J.; Yoo, M.H.; Yum, G.; Park, S.; Oh, K.H.; Lee, S.H.; Kwon, S.Y.; Cho, S.H.; Kims, S.; et al. Ecabet sodium alleviates neomycin-induced hair cell damage. *Free Radical Bio. Med.* **2015**, *89*, 1176–1183. [CrossRef]
40. Li, H.; Song, Y.; He, Z.; Chen, X.; Wu, X.; Li, X.; Bai, X.; Liu, W.; Li, B.; Wang, S.; et al. Meclofenamic acid reduces reactive oxygen species accumulation and apoptosis, inhibits excessive autophagy, and protects hair cell-like HEI-OC1 cells from cisplatin-induced damage. *Front Cell Neurosci.* **2018**, *12*, 139. [CrossRef]

Sample Availability: Samples of the compounds are not available from the authors.

molecules

MDPI

Article

Rare Earth Hydroxide as a Precursor for Controlled Fabrication of Uniform β-NaYF₄ Nanoparticles: A Novel, Low Cost, and Facile Method

Lili Xu [1], Man Wang [1], Qing Chen [1], Jiajia Yang [1], Wubin Zheng [1], Guanglei Lv [1,*], Zewei Quan [2,*] and Chunxia Li [1,*]

[1] Key Laboratory of the Ministry of Education for Advanced Catalysis Materials, Zhejiang Normal University, Jinhua 321004, China; 18329016701@163.com (L.X.); 15500100396@zjnu.edu.cn (M.W.); 1369976191@zjnu.edu.cn (Q.C.); 1547533672@zjnu.edu.cn (J.Y.); 13636130@zjnu.edu.cn (W.Z.)
[2] Department of Chemistry, Southern University of Science and Technology, Shenzhen 518055, China
* Correspondence: guanglei@zjnu.edu.cn (G.L.); quanzw@sustc.edu.cn (Z.Q.); cxli@zjnu.edu.cn (C.L.); Tel.: +86-0579-82282269 (G.L. & C.L.); +86-0755-88018399 (Z.Q.)

Received: 31 December 2018; Accepted: 17 January 2019; Published: 19 January 2019

Abstract: In recent years, rare earth doped upconversion nanocrystals have been widely used in different fields owing to their unique merits. Although rare earth chlorides and trifluoroacetates are commonly used precursors for the synthesis of nanocrystals, they have certain disadvantages. For example, rare earth chlorides are expensive and rare earth trifluoroacetates produce toxic gases during the reaction. To overcome these drawbacks, we use the less expensive rare earth hydroxide as a precursor to synthesize β-NaYF₄ nanoparticles with multiform shapes and sizes. Small-sized nanocrystals (15 nm) can be obtained by precisely controlling the synthesis conditions. Compared with the previous methods, the current method is more facile and has lower cost. In addition, the defects of the nanocrystal surface are reduced through constructing core–shell structures, resulting in enhanced upconversion luminescence intensity.

Keywords: β-NaYF₄; rare earth upconversion nanoparticles; core–shell structure

---

1. Introduction

In recent years, Lanthanide (Ln^{3+})-doped upconversion nanoparticles (UCNPs) that convert low energy photons into high energy photons through a two- or multi-photon absorption mechanism have extensively attracted researchers' attention due to their potential applications in a variety of fields, such as bioimaging [1–8], biosensing [9,10], drug delivery [11,12], and cancer therapy [13–16]. Compared with traditional fluorescent probes, such as organic fluorescent dyes and semiconductor quantum dots, UCNPs possess some unique advantages, including weak background fluorescence, large anti-Stokes shift, high photochemical stability, narrow emission bandwidth, long luminescent lifetime, high penetration depth, and low toxicity, among others [17–24]. Among reported UCNPs, hexagonal phase (β-) sodium yttrium fluoride has been shown to be one of most efficient host materials owing to its low photon cutoff energy (~350 cm⁻¹) and high chemical stability, which are able to effectively reduce non-radiative energy losses at the intermediate states of lanthanide ions [25]. So far, several methods have been reported to synthesize lanthanide-doped β-NaYF₄ nanoparticles with controlled crystalline phase, shape, and size. Solvothermal method and thermal decomposition methods are two of the most frequently used techniques to synthesize monodisperse lanthanide-doped β-NaYF₄ nanoparticles. For example, Haase reported the synthesis of β-NaYF₄ by using expensive rare earth chlorides as precursors [26]. Capobianco and co-workers synthesized β-NaYF₄ nanocrystals co-doped with Yb³⁺/Er³⁺ or Yb³⁺/Tm³⁺ via the thermal decomposition of rare earth trifluoroacetate precursors where octadecene (ODE) and oleic acid (OA) were chosen as a solvent and ligand, respectively [27]. However, there are disadvantages in this method, such as the high synthetic temperature, the complicated decomposition process, and the uncontrollable experimental conditions. More importantly, the heating of trifluoroacetate would produce toxic fluorinated and oxyfluorinated carbon gases.

In this paper, we developed a novel method by using cheap rare earth hydroxide as a precursor to synthesize monodisperse hexagonal $NaYF_4$:Yb^{3+}/Ln^{3+} core and $NaYF_4$:Yb^{3+}/Ln^{3+}@$NaGdF_4$ (Ln = Er, Tm, and Ho) core–shell nanoparticles with well-defined shapes. Compared with the previous methods, this method is low-cost and more facile. Moreover, during the reaction process no toxic gases are produced. In addition, the size of nanocrystals can be tuned by controlling the reaction conditions, such as the molar ratio of Na^+/Ln^{3+}/F^-, the volume ratio of OA and ODE, and the amount of sodium oleate (NaOA). Under 980 nm laser excitation, these core–shell nanoparticles showed intense upconversion emissions relative to $NaYF_4$:Yb/Ln (Ln = Er, Tm, and Ho) core nanoparticles.

2. Results and Discussion

It is well known that for the synthesis of nanomaterials, sodium sources and fluorine sources are two important factors in affecting the morphology and size of nanocrystals. In our experiments, we explored the effects of sodium hydroxide (NaOH), sodium oleate (NaOA), sodium trifluoroacetate (CF_3COONa), and sodium fluoride (NaF) on the morphology and size of nanocrystals. It can be easily seen that the particle size was uneven when CF_3COONa (Figure 1A) and NaF (Figure 1B) were used as sodium sources. Moreover, their powder X-ray diffraction (XRD) patterns were in line with the standard cubic one of JCPDS 06-0342 (Figure 1E). The broader width (red pentagram) of the XRD was ascribed to the peak of silica from the slide. However, $NaYF_4$:Yb^{3+}/Er^{3+} nanoparticles obtained with NaOH (Figure 1C) or NaOA (Figure 1D) as sodium sources had a uniform size and a crystal phase that matched well the standard JCPDS 16-0334 of β-$NaYF_4$ (Figure 1E). In addition, the size of the upconverting nanoparticles synthesized with NaOA was smaller than that of NaOH. This means NaOA can effectively inhibit the growth of $NaYF_4$ because of extra OA–ligands from NaOA. Taken together, the sodium source has a great influence on the size and crystal phase of the nanocrystals.

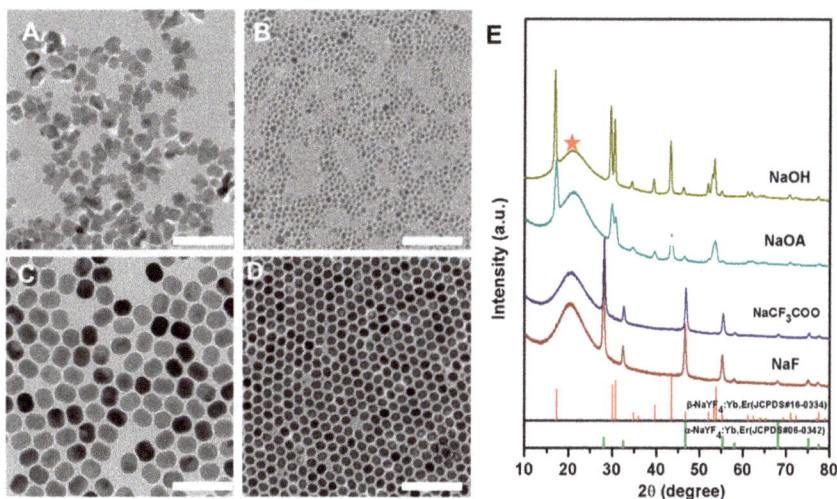

Figure 1. TEM images of $NaYF_4$:Yb^{3+}/Er^{3+} nanocrystals synthesized with NaOH (**A**), NaOA (**B**), CF_3COONa (**C**), and NaF (**D**) as well as the corresponding XRD patterns (**E**). The red pentagram represents the peak of silica from the slide. The standard diffraction patterns of the α-$NaYF_4$ (JCPDS 06-0342) and the β-$NaYF_4$ (JCPDS 16-0334) are displayed at the bottom for reference. Scale bars, 100 nm.

Based on the above results, we firstly used NaOH as sodium sources to explore the effect of Na$^+$/Ln^{3+}/F$^-$ on the size and morphology of nanocrystals. A series of NaYF$_4$:Yb^{3+}/Er^{3+} nanocrystals were synthesized with different ratios of Na$^+$/Ln^{3+}/F$^-$ when the volume ratio of OA/ODE was fixed. Figure 2 presents the TEM images and size distributions of the NaYF$_4$:Yb^{3+}/Er^{3+} nanocrystals. It can be clearly seen that when the molar ratio Na$^+$/Ln^{3+}/F$^-$ increased, the particle size of the products decreased accompanied by the shape evolution from rod to sphere. When the molar ratio of Na$^+$/Ln^{3+}/F$^-$ was 2.5:1:4, the nanoparticles are regular rods with good monodispersity. Their lengths and widths are 32 nm and 27 nm (Figure 2A,B and Figure S1, Supporting Information), which are obtained by randomly measuring more than 150 particles. When the ratio of Na$^+$/Ln^{3+}/F$^-$ was increased from 2.5:1:4 to 4:1:4, the size of the nanocrystals was reduced from 32.7 to 27.7 nm. It has been reported that the simultaneous addition of Na$^+$ and F$^-$ in the solution can produce small β-phase seeds, thereby the final growth of small-sized nanocrystals can be controlled easily [28]. Moreover, the size of the nanocrystals was decreased from 25 to 23 nm when the molar ratio of Na$^+$/Ln^{3+}/F$^-$ was increased to be 6:1:6 (Figure 2G,H). When the molar ratio was further increased to 8:1:8 (Figure 2I,J), the size of the nanocrystals was reduced to 22 nm.

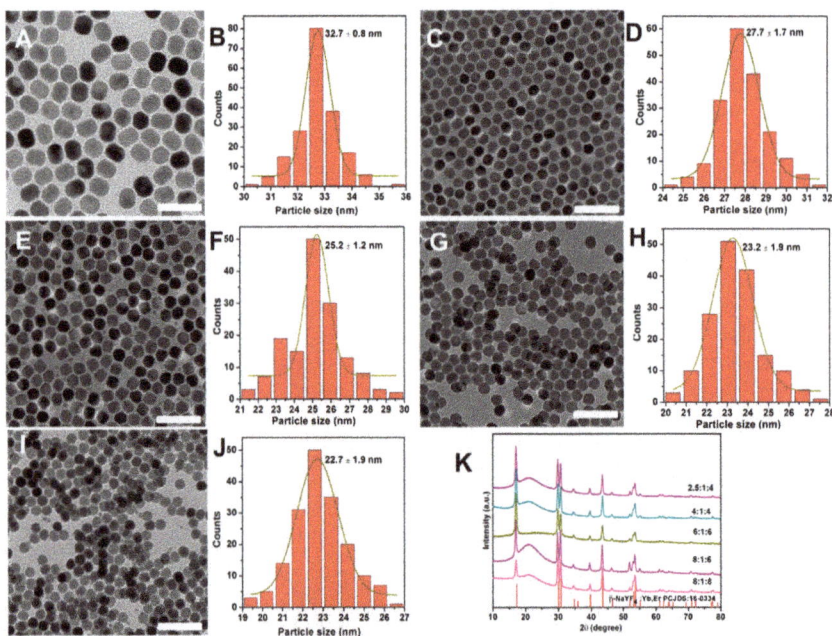

Figure 2. TEM images and size histograms of NaYF$_4$:Yb^{3+}/Er^{3+} nanocrystals synthesized with NaOH by using Na$^+$/Ln^{3+}/F$^-$ with a molar ratio of 2.5:1:4 (**A,B**), 4:1:4 (**C,D**), 6:1:6 (**E,F**), 8:1:6 (**G,H**), and 8:1:8 (**I,J**), respectively, as well as the corresponding XRD patterns (**K**); the standard diffraction pattern of the β-NaYF$_4$ (JCPDS 16-0334) is depicted at the bottom for reference. Scale bars, 100 nm.

To clarify the role of NaOA, we further studied the effect of its amount on the synthesis of NaYF$_4$:Yb^{3+}/Er^{3+} nanocrystals. Herein, we used NaOA substitute for NaOH to synthesize diverse nanoparticles with different molar ratios of Na$^+$/Ln^{3+}/F$^-$. Figure 3 shows the TEM images of as-synthesized NaYF$_4$:Yb^{3+}/Er^{3+} nanocrystals, which exhibited the morphology and size evolution. As expected, the morphology of the nanocrystals changed from rods to spheres. Meanwhile, the size was reduced from about 24 to 16 nm. As the amount of NaOA increased from 1.5 to 2.5 mmol, the size of the nanocrystals decreased to 20 nm (Figure 3A–D). Then the size of NaYF$_4$:Yb^{3+}/Er^{3+} nanocrystals was further decreased to about 17 nm when the amount of NaOA increased to 6 mmol (Figure 3E,F) and 8 mmol (Figure 3G,H), respectively. It was possible that oxygen moiety in the OA–ligands had a much stronger binding affinity to Y^{3+} ions when there were adequate OA–ligands with the increased of NaOA.

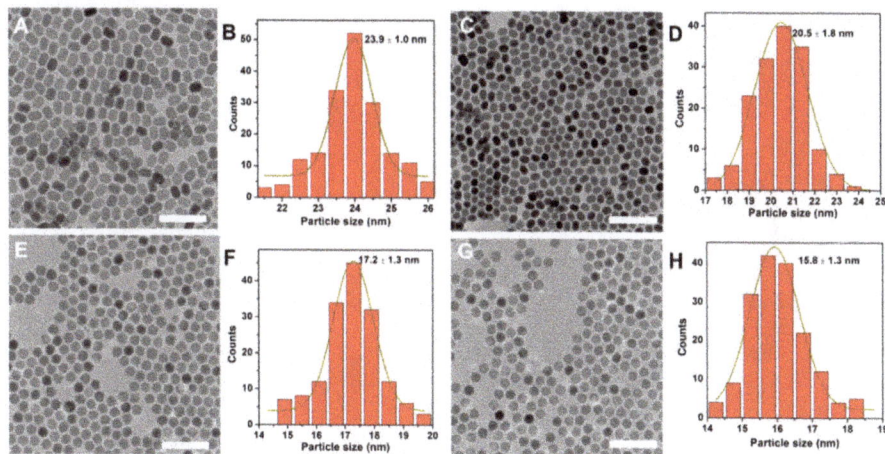

Figure 3. TEM images and size histograms of $NaYF_4$:Yb^{3+}/Er^{3+} nanocrystals synthesized with NaOA by using Na^+/Ln^{3+}/F^- withmolar ratios of 1.5:1:4 (**A,B**), 2.5:1:4 (**C,D**), 6:1:4 (**E,F**), and 8:1:4 (**G,H**), respectively. Scale bars, 100 nm.

The OA–ligands can effectively induce the orderly arrangement of Y^{3+} ions during the formation process of $NaYF_4$ [29]. Additionally, when the amount of OA–ligands was adequate, they would cover the surface of the nanoparticles to inhibit the growth of nanocrystals, which is an effective method to obtain the nanocrystals with a smaller size. Figure S2 (Supporting Information) shows the XRD patterns of the as-prepared $NaYF_4$:Yb^{3+}/Er^{3+} core nanoparticles. All the XRD patterns of samples could be matched with the pure hexagonal-phases $NaYF_4$ (JCPDS 16-0334), and no trace of other phases or impurities were observed, which clearly suggests the high crystallinities of these as-prepared nanoparticles. It has been reported that the presence of oleic acid in the solvent plays an important role in tuning the size and morphology of $NaYF_4$:Yb^{3+}/Er^{3+} nanocrystals [30]. Figure 4 shows the TEM images and size distributions of $NaYF_4$:Yb^{3+}/Er^{3+} nanoparticles prepared at different ratios of OA/ODE of 4/15 (Figure 4A,B), 8/15 (Figure 4C,D), 10/15 (Figure 4E,F), and 15/15 (Figure 4H,I). As can be seen, the resulting $NaYF_4$:Yb^{3+}/Er^{3+} nanoparticles with the ratio of OA/ODE (4/15) were hexagonal in shape with an average diameter of about 26 nm. With the increased volume ratio of OA/ODE, the particle size of $NaYF_4$:Yb^{3+}/Er^{3+} nanoparticles gradually increased, and the average diameter of nanoparticles was found to be approximately 40 nm at 8/15 of OA/ODE. The XRD of all as-prepared samples are shown in Figure S3 (Supporting Information). The same procedure was used to further synthesize $NaYF_4$ nanoparticles doped with other lanthanide elements such as 40% Yb^{3+}/0.5%Tm^{3+} (Figure S3A,B) and 18% Yb^{3+}/2% Ho^{3+} (Figure S3C,D, Supporting Information), respectively. It was observed that the size and morphology of the nanoparticles closely resembled those of the Yb^{3+}/Er^{3+} co-doped $NaYF_4$ counterpart (Figure S1, Supporting Information). These results indicate that the change of dopant ions in low doping concentrations does not alter the particle growth process [31].

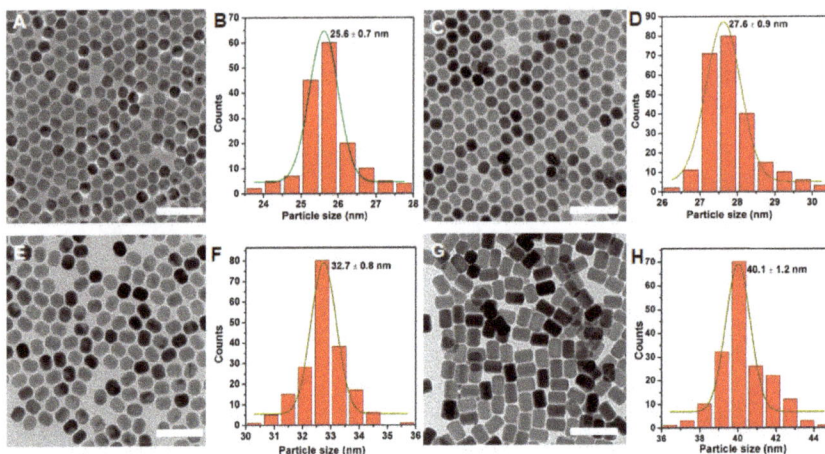

Figure 4. TEM images and size histograms of NaYF$_4$:Yb^{3+}/Er^{3+} nanocrystals synthesized at varied amounts of oleic acid (OA). The volume ratios of OA and octadecene (ODE) are 4:15 (**A,B**), 8:15 (**C,D**), 10:15 (**E,F**), and 15:15 (**G,H**), respectively. Scale bars, 100 nm.

More importantly, the emission intensity of Ln^{3+} doped UCNPs depends on dopant–host combination, particle size, shape, and phase [32–38]. The emission intensity of NaYF$_4$ core was relatively weak due to the surface defects. However, in recent years, the construction of core–shell structure has been one of the effective ways to improve the efficiency of upconversion luminescence [26,39–42]. This is because the inert shell coating can protect the luminescent activators in the core nanoparticles from the surface quenching of excitation energy [42]. In this paper, we prepared NaYF$_4$:Yb^{3+}/Ln^{3+}@NaGdF$_4$ and studied their upconversion optical properties. Figure 5A–F displays the TEM images and the corresponding size distribution of NaYF$_4$:Yb^{3+}/Ln^{3+}@NaGdF$_4$ core–shell nanoparticles. The NaYF$_4$:Yb^{3+}/Er^{3+} core-only nanoparticles with an average diameter of about 27 nm are shown in Figure S1. After coating an inert shell layer of NaGdF$_4$, the average diameter of NaYF$_4$:Yb^{3+}/Er^{3+}@NaGdF$_4$ core–shell was determined to be about 42 nm, which means a thick shell layer (thickness ~7.5 nm) was coated around the NaYF$_4$:Yb^{3+}/Er^{3+} core nanoparticles (Figure 5A,B). Similar TEM image and size distribution were obtained for NaYF$_4$:Yb^{3+}/Tm^{3+}@NaGdF$_4$ (Figure 5C,D) and NaYF$_4$:Yb^{3+}/Ho^{3+}@NaGdF$_4$ (Figure 5E,F) core–shell nanoparticles. The corresponding shell thickness was determined to be about 11 nm and 9 nm, respectively. Figure 5G depicts the upconversion emission spectra for the NaYF$_4$:Yb^{3+}/Er^{3+} core nanoparticles and the NaYF$_4$:Yb^{3+}/Er^{3+}@NaGdF$_4$ core–shell nanoparticles. As can be seen, there were two green emission bands centered at 520 nm and 540 nm, which were attributed to the electronic transition of $^2H_{11/2} \rightarrow {}^4I_{15/2}$, $^4S_{3/2} \rightarrow {}^4I_{15/2}$ of Er^{3+}. In addition, there was a red emission at 654 nm, which corresponded to the $^4F_{9/2} \rightarrow {}^4I_{15/2}$ transition of Er^{3+}. For the NaYF$_4$:Yb^{3+}/Er^{3+} core, after coating 7.5 nm NaGdF$_4$ shell, the fluorescence intensity of NaYF$_4$:Yb^{3+}/Er^{3+}@NaGdF$_4$ core–shell nanoparticles was about 100 times higher than that of NaYF$_4$:Yb^{3+}/Er^{3+} core nanoparticles at 540 nm. In contrast, for the NaYF$_4$:Yb^{3+}/Tm^{3+} core nanoparticles, the blue emissions, the ultraviolet emission, and red emission corresponded to the ($^1I_6 \rightarrow {}^3F_4$, 342 nm, $^1D_2 \rightarrow {}^3H_6$, 360 nm), ($^1D_2 \rightarrow {}^3F_4$, 450 nm, $^1G_4 \rightarrow {}^3H_6$, 475 nm), ($^1G_4 \rightarrow {}^3F_4$, 647 nm) transitions of Tm^{3+} ions, respectively. Due to the NaGdF$_4$ (~11 nm) shell coating, the emission intensity was remarkably enhanced to about 120 times at 475 nm (Figure 5G). Figure 5H shows the upconversion emission spectra of the NaYF$_4$:Yb^{3+}/Ho^{3+} core nanoparticles and the NaYF$_4$:Yb^{3+}/Ho^{3+}@NaGdF$_4$ core–shell nanoparticles. When excited at 980 nm, three upconversion fluorescence bands with maxima at green (540 nm), red (645 nm), and 750 nm regions could be observed from optimized NaYF$_4$:Yb^{3+}/Ho^{3+} nanophosphors, which corresponded to $^5S_2 \rightarrow {}^5I_8$, $^5F_3 \rightarrow {}^5I_8$, and $^5S_2 \rightarrow {}^5I_7$ transitions of Ho^{3+} ions. Similarly, when the ~9 nm NaGdF$_4$ shell was covered on the surface of the NaYF$_4$:Yb^{3+}/Ho^{3+} core, the luminescence intensity was also significantly improved. The emission intensity of NaYF$_4$:Yb^{3+}/Ho^{3+}@NaGdF$_4$ core–shell nanoparticles was determined to be about 80 times as strong as that of NaYF$_4$:Yb^{3+}/Ho^{3+} core nanoparticles.

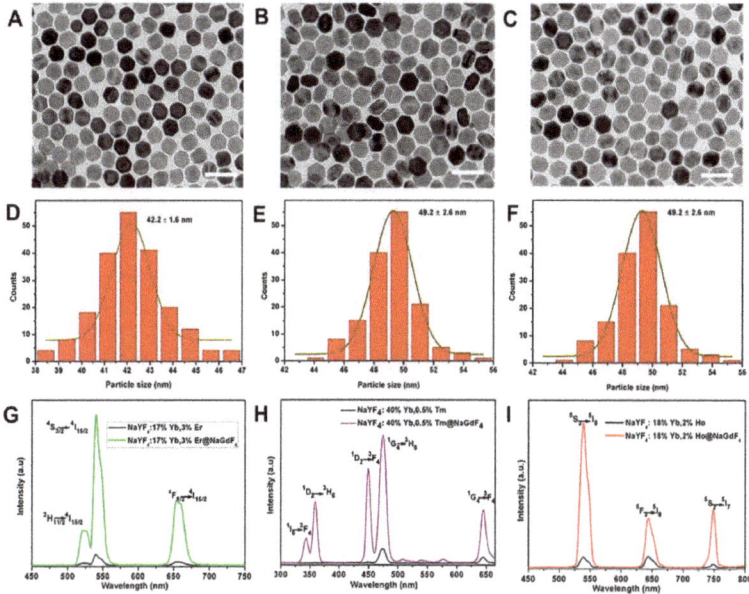

Figure 5. TEM images, size histograms, and the corresponding upconversion spectra of core–shell structured NaYF$_4$:Yb^{3+}/Er^{3+}@NaGdF$_4$ (**A,D,G**), NaYF$_4$: Yb^{3+}/Tm^{3+}@NaGdF$_4$ (**B,E,H**), and NaYF$_4$:Yb^{3+}/Ho^{3+}@NaGdF$_4$ (**C,F,I**), respectively. Scale bars, 100 nm.

3. Materials and Methods

3.1. Materials

Materials Y$_2$O$_3$ (99.99%), Yb$_2$O$_3$ (99.99%), Er$_2$O$_3$ (99.99%), Tm$_2$O$_3$ (99.99%), Ho$_2$O$_3$ (99.99%), Y$_2$O$_3$ (99.99%), GdCl$_3$·6H$_2$O (99.99%), NH$_4$F (98%), NaF (98%), and CF$_3$COONa (97%) were purchased from Aladdin (Shanghai, China). Oleic acid (OA, 90%), 1-octadecene (ODE, 90%), and sodium oleate (Na-OA, >97%) were purchased from Sigma-Aldrich (Darmstadt, Germany). Other chemical reagents, such as NaOH (90%), ethyl alcohol (99.7%), methanol (99.5%), and *n*-hexane (97%), were obtained from Shanghai Lingfeng Chemical Reagent Co., Ltd. All chemicals were used as received without further purification.

Rare earth chloride (LnCl$_3$) stock solutions of 1 M (Ln = Y, Yb) and 0.1 M (Ln = Er, Tm, and Ho) were prepared by dissolving the corresponding metal oxides in hydrochloric acid at elevated temperature. The final solutions were adjusted to pH ~6.

3.2. Synthesis of β-NaYF$_4$:Yb^{3+}/Ln^{3+} (Ln = Er, Tm and Ho) Core Nanoparticles

First, Ln(OH)$_3$ (Ln = Y^{3+}, Yb^{3+}, and Er^{3+}/Tm^{3+}/Ho^{3+}) complexes were prepared by adding NaOH of 2 M to the rare earth chloride solution, and then the obtained product was washed twice. Rare earth hydroxide was added to a 100 mL flask containing 10 mL of OA and 15 mL of ODE. The mixture was heated at 140 °C for 1 h under stirring in order to form the lanthanide–oleate complexes. After cooling down to 40 °C naturally, 8 mL of methanol solution containing NH$_4$F (4 mmol) and NaOH (2.5 mmol) was added. Subsequently, the resulting mixture solution was heated at 70 °C for 10 min to evaporate the methanol under magnetic stirring. After the temperature was raised up to 110 °C under vacuum for 30 min, the reaction mixture was heated to 300 °C under argon for 1 h and then cooled down to room temperature. The resulting nanoparticles were precipitated by adding excess ethanol, separated by centrifugation at 4000 rpm for 4 min, washed with a mixture of *n*-hexane and ethanol several times, and finally dispersed in *n*-hexane for further use.

3.3. Synthesis of β-NaYF₄:Yb/Ln@NaGdF₄ (Ln = Er, Tm, and Ho) Core–Shell Nanoparticles

In a typical procedure for the synthesis of $NaYF_4$:Yb^{3+}/Er^{3+} (17/3%)@$NaGdF_4$ nanoparticles, the $NaGdF_4$ inert shell precursor was prepared by mixing 0.5 mmol of $GdCl_3$ with 6 mL of OA and 15 mL of ODE in a 100 mL flask followed by heating at 150 °C for 40 min. Then, the obtained $NaGdF_4$ inert shell precursor was cooled down to room temperature. Subsequently, the $NaYF_4$:Yb^{3+}/Er^{3+} (17/3%) core-only nanoparticles dispersed in 3 mL of cyclohexane along with 8 mL of methanol solution of NH_4F (2 mmol) and NaOH (1.25 mmol) was added. Subsequently, the resulting mixture solution was heated at 70 °C for 10 min under magnetic stirring to evaporate the methanol. After the resulting solution for 30 min at 110 °C under vacuum, the reaction mixture was heated to 310 °C under argon for 1.5 h and then cooled down to room temperature. Finally, the obtained core–shell nanoparticle products were precipitated by the addition of ethanol, collected by centrifugation at 4000 rpm for 4 min, washed with a mixture of *n*-hexane and ethanol several times, and finally dispersed in *n*-hexane.

3.4. Instrumentation

Transmission electron microscopy (TEM) images (Hitachi, Tokyo, Japan) were acquired on a HT7700 transmission electron microscopy at an acceleration voltage of 100 kV. X-ray diffraction (XRD) patterns were recorded on an X-ray diffraction (RigakuSmartLab, Tokyo, Japan) with Cu Kαradiation (λ = 0.15418 nm) at a voltage of 45 kV and a current of 20 mA. Upconversion luminescence spectra were detected by a spectrophotometer (FluoroMax-4, Shanghai, China) equipped with a 980 nm diode laser.

4. Conclusions

In summary, we developed a novel, facile, and low-cost method to synthesize Yb^{3+}/Ln^{3+} (Ln = Er, Tm, and Ho) co-doped β-$NaYF_4$ nanocrystals. The size and morphology of the products were manipulated through the precise tuning of the ratio of Na^+/Ln^{3+}/F^-, the ratio of OA/ODE, and the quantity of the NaOA. As the ratio of Na^+/Ln^{3+}/F^- and the amount of NaOA increased, the size of the nanocrystals gradually decreased and the corresponding morphologies evolved from nanospheres to nanorods. Finally, this method can also apply to the fabrication of core–shell structured nanomaterials. The uniform $NaYF_4$:Yb^{3+}/Ln^{3+}@$NaGdF_4$ core–shell nanoparticles were prepared successfully, and their emission intensity was remarkably enhanced, which could provide prospect applications in the biomedical fields. Particularly, some photosensitizers or chemotherapy drugs could be conjugated with these UCNPs to achieve NIR-triggered drug delivery and controlled release to ameliorate the therapy efficiency of tumors.

Supplementary Materials: The following are available online, Figure S1: TEM image (A) and the corresponding size histogram (B) of NaYF₄:Yb³⁺/Er³⁺ nanocrystals synthesized with NaOH as sodium source, Figure S2: XRD patterns of the as-synthesized NaYF₄:Yb³⁺/Er³⁺ nanocrystals with NaOA at varied molar ratios of Na⁺/Ln³⁺/F⁻. The standard diffraction patterns of the β-NaYF₄ (JCPDS 16-0334) depicted at the bottom for reference, Figure S3: XRD patterns of the as-synthesized NaYF₄:Yb³⁺/Er³⁺ nanocrystals at varied amounts of oleic acid. The volume ratios of oleic acid and octadecene are 15:15, 10:15, 8:15, and 4:15, respectively. The diffraction pattern at the bottom is the literature reference for hexagonal NaYF₄ nanocrystal (JCPDS 16-0334), Figure S4: TEM images and size histograms of the NaYF₄:Yb³⁺/Er³⁺ (A, B), and NaYF₄: Yb³⁺/Ho³⁺ (C, D), respectively.

Author Contributions: Conceptualization, L.X. and G.L.; investigation and data curation, M.W. and Q.C.; writing and original draft preparation, L.X. and J.Y.; writing, reviewing, and editing, W.Z., C.L. and G.L.; and supervision, G.L., Z.Q. and C.L.

Molecules **2019**, *24*, 357

Funding: This research was funded by the National Natural Science Foundation of China (51572258, 51872263 and 51772142), Zhejiang Provincial Natural Science Foundation of China (LZ19E020001 and LQ19B050003), the Key Construction Project (2017XM022) of Zhejiang Normal University and Open Research Fund of Key Laboratory of the Ministry of Education for Advanced Catalysis Materials, and Zhejiang Key Laboratory for Reactive Chemistry on Solid Surfaces, Zhejiang Normal University.

Conflicts of Interest: The authors declare no conflict of interest.

References

1. Park, I.; Lee, K.T.; Suh, Y.D.; Hyeon, T. Upconverting nanoparticles: A versatile platform for wide-field two-photon microscopy and multi-modal in vivo imaging. *Chem. Soc. Rev.* **2015**, *44*, 1302–1317. [CrossRef] [PubMed]

2. Dong, H.; Sun, L.D.; Wang, Y.F.; Ke, J.; Si, R.; Xiao, J.W.; Lyu, G.M.; Shi, S.; Yan, C.H. Efficient Tailoring of Upconversion Selectivity by Engineering Local Structure of Lanthanides in Na_xREF_{3+x} Nanocrystals. *J. Am. Chem. Soc.* **2015**, *137*, 6569–6576. [CrossRef] [PubMed]

3. Wolfbeis, O.S. An overview of nanoparticles commonly used in fluorescent bioimaging. *Chem. Soc. Rev.* **2015**, *44*, 4743–4768. [CrossRef]

4. Wei, Z.W.; Sun, L.N.; Liu, J.L.; Zhang, J.Z.; Yang, H.R.; Yang, Y.; Shi, L.Y. Cysteine modified rare-earth up-converting nanoparticles for in vitro and in vivo bioimaging. *Biomaterials* **2014**, *35*, 387–392. [CrossRef] [PubMed]

5. Zhang, H.; Wu, Y.; Wang, J.; Tang, Z.M.; Ren, Y.; Ni, D.L.; Gao, H.B.; Song, R.X.; Jin, T.; Li, Q.; et al. In Vivo MR Imaging of Glioma Recruitment of Adoptive T-Cells Labeled with NaGdF4-TAT Nanoprobes. *Small* **2017**, *14*, 1702951–1702959. [CrossRef] [PubMed]

6. Deng, M.L.; Wang, L.Y. Unexpected luminescence enhancement of upconverting nanocrystals by cation exchange with well retained small particle size. *Nano Res.* **2014**, *7*, 782–793. [CrossRef]

7. Xu, S.; Yu, Y.; Gao, Y.F.; Zhang, Y.Q.; Li, X.P.; Zhang, J.S.; Wang, Y.F.; Chen, B.J. Mesoporous silica coating $NaYF_4:Yb,Er@NaYF_4$ upconversion nanoparticles loaded with ruthenium(II) complex nanoparticles: Fluorometric sensing and cellular imaging of temperature by upconversion and of oxygen by downconversion. *Microchim. Acta* **2018**, *185*, 454–463. [CrossRef] [PubMed]

8. Lei, X.L.; Li, R.F.; Tu, D.T.; Shang, X.Y.; Liu, Y.; You, W.W.; Sun, C.X.; Zhang, F.; Chen, X.Y. Intense near-infrared-II luminescence from $NaCeF_4:Er/Yb$ nanoprobes for in vitro bioassay and in vivo bioimaging. *Chem. Sci.* **2018**, *9*, 4682–4988. [CrossRef] [PubMed]

9. Liu, J.; Liu, Y.; Bu, W.; Bu, J.; Sun, Y.; Du, J.; Shi, J. Ultrasensitive Nanosensors Based on Upconversion Nanoparticles for Selective Hypoxia Imaging in vivo upon Near-Infrared Excitation. *J. Am. Chem. Soc.* **2014**, *136*, 9701–9709. [CrossRef] [PubMed]

10. Zheng, W.; Tu, D.T.; Huang, P.; Zhou, S.Y.; Chen, Z.; Chen, X.Y. Time-resolved luminescent biosensing based on inorganic lanthanide-doped nanoprobes. *Chem. Commun.* **2015**, *51*, 4129–4143. [CrossRef] [PubMed]

11. Jalani, G.; Tam, V.; Vetrone, F.; Cerruti, M. Seeing, Targeting and Delivering with Upconverting Nanoparticles. *J. Am. Chem. Soc.* **2018**, *140*, 10923–10931. [CrossRef] [PubMed]

12. Zhang, Y.; Yu, Z.Z.; Li, J.Q.; Ao, Y.X.; Xue, J.W.; Zeng, Z.P.; Yang, X.L.; Tan, T.T.Y. Ultrasmall-Superbright Neodymium-Upconversion Nanoparticles via Energy Migration Manipulation and Lattice Modification: 808 nm-Activated Drug Release. *ACS Nano* **2017**, *11*, 2846–2857. [CrossRef] [PubMed]

13. Zhang, C.; Chen, W.H.; Liu, L.H.; Qiu, W.X.; Yu, W.Y.; Zhang, X.Z. An O_2 Self-Supplementing and Reactive-Oxygen-Species-Circulating Amplifed Nanoplatform via H_2O/H_2O_2 Splitting for Tumor Imaging and Photodynamic Therapy. *Adv. Funct. Mater.* **2017**, *27*, 1700626–1700639. [CrossRef]

14. Lu, S.; Tu, D.T.; Hu, P.; Xu, J.; Li, R.F.; Wang, M.; Chen, Z.; Huang, M.D.; Chen, X.Y. Multifunctional Nano-Bioprobes Based on Rattle-Structured Upconverting Luminescent Nanoparticles. *Angew. Chem. Int. Ed.* **2015**, *54*, 7915–7919. [CrossRef] [PubMed]

15. Zhu, X.J.; Li, J.C.; Qiu, X.C.; Liu, Y.; Wang, F.; Li, F.Y. Upconversion nanocomposite for programming combination cancer therapy by precise control of microscopic temperature. *Nat. Commun.* **2018**, *9*, 2176–2186. [CrossRef] [PubMed]

16. Zhang, C.; Zhao, K.L.; Bu, W.B.; Ni, D.L.; Liu, Y.Y.; Feng, J.W.; Shi, J.L. Marriage of Scintillator and Semiconductor for Synchronous Radiotherapy and Deep Photodynamic Therapy with Diminished Oxygen Dependence. *Angew. Chem. Int. Ed.* **2015**, *54*, 1770–1774. [CrossRef]

17. Yang, D.M.; Ma, P.A.; Hou, Z.Y.; Cheng, Z.Y.; Li, C.X.; Lin, J. Current advances in lanthanide ion (Ln^{3+})-based upconversion nanomaterials for drug delivery. *Chem. Soc. Rev.* **2015**, *44*, 1416–1448. [CrossRef]

18. Liu, Y.S.; Tu, D.T.; Zhu, H.M.; Chen, X.Y. Photon upconversion nanomaterials. *Chem. Soc. Rev.* **2015**, *44*, 1299–1301. [CrossRef]

19. Zeng, S.J.; Wang, H.B.; Lu, W.; Yi, Z.G.; Rao, L.; Liu, H.R.; Hao, J.H. Dual-modal upconversion fluorescent/X-ray imaging using ligand-free hexagonal phase $NaLuF_4$:Gd/Yb/Er nanorods for blood vessel visualization. *Biomaterials* **2014**, *35*, 2934–2941. [CrossRef]

20. Wang, C.; Yao, W.X.; Wang, P.Y.; Zhao, M.Y.; Li, X.M.; Zhang, F. A catalase-loaded hierarchical zeolite as an implantable nanocapsule for ultrasound-guided oxygen self-sufficient photodynamic therapy against pancreatic cancer. *Adv. Mater.* **2018**, *30*, 1704833–1704840.

21. Chen, D.Q.; Chen, Y.; Lu, H.W.; Ji, Z.G. A Bifunctional Cr/Yb/Tm:$Ca_3Ga_2Ge_3O_{12}$ Phosphor with Near-Infrared Long-Lasting Phosphorescence and Upconversion Luminescence. *Inorg. Chem.* **2014**, *53*, 8638–8645. [CrossRef] [PubMed]

22. Shi, Z.L.; Duan, Y.; Zhu, X.J.; Wang, Q.W.; Li, D.D.; Hu, K.; Feng, W.; Li, F.Y.; Xu, C.X. Dual functional $NaYF_4$:Yb^{3+},Er^{3+}@$NaYF_4$:Yb^{3+}, Nd^{3+} core–shell nanoparticles for cell temperature sensing and imaging. *Nanotechnology* **2018**, *29*, 094001–094009. [CrossRef] [PubMed]

23. Dai, Y.L.; Xiao, H.H.; Liu, J.H.; Yuan, Q.H.; Ma, P.A.; Yang, D.M.; Li, C.X.; Cheng, Z.Y.; Hou, Z.Y.; Yang, P.P.; et al. In vivo multimodality imaging and cancer therapy by near-infrared light-triggered trans-platinum pro-drug-conjugated upconverison nanoparticles. *J. Am. Chem. Soc.* **2013**, *135*, 18920–18929. [CrossRef] [PubMed]

24. Yu, Z.S.; Xia, Y.Z.; Xing, J.; Li, Z.H.; Zhen, J.J.; Jin, Y.H.; Tian, Y.C.; Liu, C.; Jiang, Z.Q.; Li, J.; Wu, A.G. Y1-receptor–ligand-functionalized ultrasmall upconversion nanoparticles for tumortargeted trimodality imaging and photodynamic therapy with low toxicity. *Nanoscale* **2018**, *10*, 17038–17052. [CrossRef] [PubMed]

25. Wang, F.; Liu, X.G. Recent advances in the chemistry of lanthanide-doped upconversion nanocrystals. *Chem. Soc. Rev.* **2009**, *38*, 976–989. [CrossRef] [PubMed]

26. Homann, C.; Krukewitt, L.; Frenzel, F.; Grauel, B.; Wgrth, C.; Resch-Genger, U.; Haase, M. NaYF4:Yb,Er/NaYF4 Core/Shell Nanocrystals with High Upconversion Luminescence Quantum Yield. *Angew. Chem. Int. Ed.* **2018**, *57*, 8765–8769. [CrossRef]

27. Boyer, J.C.; Vetrone, F.; Cuccia, L.A.; Capobianco, J.A. Synthesis of colloidal upconverting $NaYF_4$ nanocrystals doped with Er^{3+}, Yb^{3+} and Tm^{3+}, Yb^{3+} via thermal decomposition of lanthanide trifluoracetate precursors. *J. Am. Chem. Soc.* **2006**, *128*, 7444–7445. [CrossRef]

28. Li, H.; Xu, L.; Chen, G.Y. Controlled Synthesis of Monodisperse Hexagonal $NaYF_4$:Yb/Er Nanocrystals with Ultrasmall Size and Enhanced Upconversion Luminescence. *Molecules* **2017**, *22*, 2113. [CrossRef]

29. Liu, D.M.; Xu, X.X.; Du, Y.; Qin, X.; Zhang, Y.H.; Ma, C.S.; Wen, S.H.; Ren, W.; Goldys, E.M.; Piper, J.A.; et al. Three-dimensional controlled growth of monodisperse sub-50 nm heterogeneous nanocrystals. *Nat. Commun.* **2016**, *7*, 10254–10261. [CrossRef]

30. Huang, X.Y. Synthesis, multicolour tuning, and emission enhancement of ultrasmall LaF_3:Yb^{3+}/Ln^{3+} (Ln = Er, Tm, and Ho) upconversion nanoparticles. *J. Mate. Sci.* **2016**, *51*, 3490–3499. [CrossRef]

31. Chen, G.Y.; Ågren, H.; Ohulchanskyy, T.Y.; Prasad, P.N. Light upconverting core–shell nanostructures: Nanophotonic control for emerging applications. *Chem. Soc. Rev.* **2015**, *44*, 1680–1713. [CrossRef] [PubMed]

32. Han, S.Y.; Deng, R.R.; Xie, X.J.; Liu, X.G. Enhancing Luminescence in Lanthanide-Doped Upconversion Nanoparticles. *Angew. Chem. Int. Ed.* **2014**, *53*, 11702–11715. [CrossRef] [PubMed]

33. Wang, F.; Han, Y.; Lim, C.S.; Lu, Y.H.; Wang, J.; Xu, J.; Chen, H.Y.; Zhang, C.; Hong, M.H.; Liu, X.G. Simultaneous phase and size control of upconversion nanocrystals through lanthanide doping. *Nature* **2010**, *463*, 1061–1065. [CrossRef] [PubMed]

34. Chen, X.; Jin, L.M.; Kong, W.; Sun, T.Y.; Zhang, W.F.; Liu, X.H.; Fan, J.; Yu, S.F.; Wang, F. Confining energy migration in upconversion nanoparticles towards deep ultraviolet lasing. *Nat. Commun.* **2016**, *7*, 10304–10309. [CrossRef] [PubMed]

35. Das, A.; Mao, C.; Cho, S.; Kim, K.; Park, W. Over 1000-fold enhancement of upconversion luminescence using water-dispersible metalinsulator-metal nanostructures. *Nat. Commun.* **2018**, *9*, 4828–4839. [CrossRef] [PubMed]

36. Cheng, T.; Marin, R.; Skripka, A.; Vetrone, F. Small and Bright Lithium-Based Upconverting Nanoparticles. *J. Am. Chem. Soc.* **2018**, *140*, 12890–12899. [CrossRef]

37. Moon, B.S.; Kim, H.E.; Kim, D.H. Ultrafast Single-Band Upconversion Luminescence in a Liquid-Quenched Amorphous Matrix. *Adv. Mater.* **2018**, *30*, 1800008. [CrossRef]
38. Cheng, X.W.; Pan, Y.; Yuan, Z.; Wang, X.W.; Su, W.H.; Yin, L.S.; Xie, X.J.; Huang, L. Er^{3+} Sensitized Photon Upconversion Nanocrystals. *Adv. Funct. Mater.* **2018**, *28*, 1800208–1800213. [CrossRef]
39. Xia, M.; Zhou, D.C.; Yang, Y.; Yang, Z.W.; Qiu, J.B. Synthesis of Ultrasmall Hexagonal $NaGdF_4$: $Yb^{3+}Er^{3+}@NaGdF_4$:$Yb^{3+}@NaGdF_4$: Nd^{3+} Active-Core/Active-Shell/Active-Shell Nanoparticles with Enhanced Upconversion Luminescence. *ECS J. Solid State Sci. Technol.* **2017**, *6*, R41–R46. [CrossRef]
40. Yi, G.S.; Chow, G.M. Water-soluble $NaYF_4$: Yb, Er(Tm)/$NaYF_4$/polymer core/shell/shell Nanoparticles with Significant Enhancement of Upconversion Fluorescence. *Chem. Mater.* **2007**, *19*, 341–343. [CrossRef]
41. Shi, R.K.; Ling, X.C.; Li, X.N.; Zhang, L.; Lu, M.; Xie, X.J.; Huang, L.; Huang, W. Tuning hexagonal $NaYbF_4$ nanocrystals down to sub-10 nm for enhanced photon upconversion. *Nanoscale* **2017**, *9*, 13739–13746. [CrossRef] [PubMed]
42. Liu, Y.; Tu, D.; Zhu, H.; Chen, X. Lanthanide-doped luminescent nanoprobes: Controlled synthesis, optical spectroscopy, and bioapplications. *Chem. Soc. Rev.* **2013**, *42*, 6924–6958. [CrossRef] [PubMed]

Sample Availability: Not available.

![molecules logo] *molecules*

MDPI

Article

Fluoropolymer-Containing Opals and Inverse Opals by Melt-Shear Organization

Julia Kredel [1], Christian Dietz [2] and Markus Gallei [1,*

[1] Ernst-Berl Institute of Technical and Macromolecular Chemistry, Technische Universität Darmstadt, Alarich-Weiss-Straße 4, 64287 Darmstadt, Germany; j.kredel@mc.tu-darmstadt.de
[2] Institute of Materials Science, Physics of Surfaces, Technische Universität Darmstadt, Alarich-Weiss-Str. 2, D-64287 Darmstadt, Germany; dietz@pos.tu-darmstadt.de
* Correspondence: m.gallei@mc.tu-darmstadt.de

Received: 22 December 2018; Accepted: 16 January 2019; Published: 17 January 2019

Abstract: The preparation of highly ordered colloidal architectures has attracted significant attention and is a rapidly growing field for various applications, e.g., sensors, absorbers, and membranes. A promising technique for the preparation of elastomeric inverse opal films relies on tailored core/shell particle architectures and application of the so-called melt-shear organization technique. Within the present work, a convenient route for the preparation of core/shell particles featuring highly fluorinated shell materials as building blocks is described. As particle core materials, both organic or inorganic (SiO_2) particles can be used as a template, followed by a semi-continuous stepwise emulsion polymerization for the synthesis of the soft fluoropolymer shell material. The use of functional monomers as shell-material offers the possibility to create opal and inverse opal films with striking optical properties according to Bragg's law of diffraction. Due to the presence of fluorinated moieties, the chemical resistance of the final opals and inverse opals is increased. The herein developed fluorine-containing particle-based films feature a low surface energy for the matrix material leading to good hydrophobic properties. Moreover, the low refractive index of the fluoropolymer shell compared to the core (or voids) led to excellent optical properties based on structural colors. The herein described fluoropolymer opals and inverse opals are expected to pave the way toward novel functional materials for application in fields of coatings and optical sensors.

Keywords: fluoropolymers; melt-shear organization; chemical resistance; solvent responsiveness; hydrophobicity; core/shell particles; emulsion polymerization; particle processing

1. Introduction

Fluor-containing polymers represent a unique class of functional materials combining different interesting properties, and such polymers have attracted significant attention in the recent past. Some of these properties are their remarkable resistance toward chemicals, high thermal stability, wetting behavior, and repellent capabilities, as well as their low refractive indices compared to other polymer materials [1–3]. Therefore, the field of applications for fluoropolymers is widespread and range over coatings, membranes, optical applications, and high performance elastomers [4–8]. Designing novel fluoropolymer and hierarchical architectures is relevant for the development of improved and new applications, for instance, in the field of three-dimensionally ordered porous coatings and photonic materials. In general, the control and understanding of surface properties is crucial for the development of advanced materials with well-defined wetting properties. Such so-called *smart surfaces* have already been used in applications such as self-cleaning surfaces, tunable optical lenses, lab-on-chip systems, microfluidic devices, and many different textile applications [9–14]. Considerable effort has been carried out on understanding the influence of designed, rough surfaces on the wetting properties, initiated by the pioneering works of Wenzel, Cassie, and Baxter long ago [15,16]. Hierarchical particle-based architectures, such as colloidal crystals and inverse opals with adjustable dimensions, have gained considerable attention due to their tremendous potential for various applications in catalysis, separation, sensors, optics, and biomedicine [17–27]. In the case of porous materials, different templating approaches have been used for the design of the final materials after removal of the template material [28–33]. Colloidal crystals can be prepared by various techniques such as particle deposition or spin coating of respective particle dispersions [21,34]. For example, Kim et al. developed omniphobic inverse opals for the preparation of omniphobic porous materials some years ago [35]. Vogel et al. reported on the infiltration of inverse opals with lubricants for gaining access to highly repellent surfaces toward many different liquids [36]. By this elegant approach, the adsorption of liquid-borne contaminants could be prevented, and reduction of ice adhesion could be accomplished. Single and dual inverse opal structures of fluoropolymer-containing materials have been developed by Wu and co-workers [37]. All these materials were obtained by the vertical deposition method or via spin-coating of particles. Another technique for the precise arrangement of polymer-based particles can be accomplished in flow fields by, e.g., combinations of melting and shear-ordering methods leading to so-called polymer opal films [27,38,39]. This so-called *melt-shear organization* technique requires core-shell particles and features the major advantage of fully solvent- and dispersion-free material processing. The hard core/soft shell particles are compressed between the plates of a moderately hot press, and the hard core particles can merge into the colloidal crystal structure yielding free-standing polymer opal films in one single step. Only recently, the feasibility of this technique was shown for inorganic core particles featuring a polymer or hybrid soft and meltable shell was reported [40–44]. This melt-shear organization technique allows for the facile preparation of almost perfectly ordered core/shell particles embedded in an elastomeric polymer matrix, and it can be applied on industrially relevant length scales [45]. The combination of fluoropolymers with this technique has not been reported for the preparation of functional opal films or inverse opals films yet. Within the present study, we report for the first time the incorporation of fluoropolymers into core/shell particle architectures, which can be advantageously used for the melt-shear organization technique. Both organic particle cores and inorganic silica core particles featuring a comparably soft fluoropolymer shell are prepared. Application of the melt-shear organization technique yield free-standing fluoropolymer opal and inverse films with remarkably distinct reflection colors and hydrophobic properties. Moreover, these novel opal and inverse opal films were investigated with respect to their optical properties, swelling capability in water, and chemical resistance toward acids and bases.

2. Results and Discussion

The following chapter is divided into six sections, starting with the design of fluorine-containing core/shell particles, followed by the preparation of opal and inverse opal films based on these particles by application of the melt-shear organization technique. Finally, the feasibility of herein investigated fluorine-containing opals and inverse opals will be elucidated with respect to their optical properties, chemical resistance, and solvent-induced structural color changes.

2.1. Bottom-Up Fabrication of Fluorine-Containing Opal and Inverse Opal Films

For the preparation of fluorine-containing opal films as well as inverse opal films the tailored design of monodisperse core interlayer shell particles is a basic prerequisite. For this purpose, the complex particle architecture was developed by a step-wise emulsion polymerization as given in Figure 1. In the case of filled opal films (Figure 1a), the hard organic core particles were generated by the polymerization of styrene, combined with the cross-linker butandiol diacrylate (BDDA), followed by a starved feed addition of an emulsion consisting of styrene and allylmethacrylate (ALMA). The particle sizes were analyzed by means of dynamic light scattering (DLS) and transmission electron microscopy (TEM), as described in detail in the following. For the next synthesis step, an additional cross-linked interlayer consisting of *n*-butylacrylate (*n*BuA) and ALMA was introduced. Herein, the residual allyl-moieties of ALMA were capable of acting as grafting anchors for the soft shell material. As already mentioned in the introduction, this step is important to ensure preservation of the spherical shape of the particle cores during processing and during the application of mechanical stress during the melt-shear-organization [27,46,47]. The major part of the outer soft shell consisted of poly(1*H*,1*H*,2*H*,2*H*-nonafluorohexylmethacrylate) (PNFHMA) (60 wt%) as a fluorine-containing polymer. For maintaining a soft and processable polymer mass for the intended extrusion step, a softer comonomer was additionally introduced as shell material. For this purpose, *n*BuA (20 wt%) and trifluoroethylacrylate (TFEA) (20 wt%) as comonomers were copolymerized with an NFHMA monomer. The intermediate glass transition temperature (T_g) decreased up to 14 °C for the copolymer, compared to the respective T_g of the respective PNFHMA homopolymers (T_g = 25 °C) (Figure S1 of the Supporting Information) [48]. These three monomers were used for the formation of the—compared to the hard core material—soft and functional fluorine-containing particle shell material. The successful formation of the core/interlayer/shell-particles was followed by DLS measurements (Figure S2a of the Supporting Information), proving an increase in the diameter after every synthesis step. In detail, the average hydrodynamic diameter of the particles were 206 ± 2 nm for the PS core particles, 211 ± 2 nm for the core/interlayer, and 246 ± 1 nm for the final fluorine-containing core/interlayer/shell particles. It can be concluded from these results that rather monodisperse particles were accessible, which will be important for the optical properties of the opal films (see next sections). The fluorine-containing particles were additionally investigated by using transmission electron microscopy (TEM) (Figure S2b–d), confirming the monodisperse and spherical character of the fluoropolymer-containing core/shell particles. These findings were essential for the fabrication of soft colloid crystal films as described in the next section. Both the DLS and TEM measurements revealed the successful formation of the core/interlayer/shell architecture and sizes of each particles were found to be in excellent agreement with expectations based on monomer consumption and the recipe for emulsion polymerization (*cf.* Experimental Section).

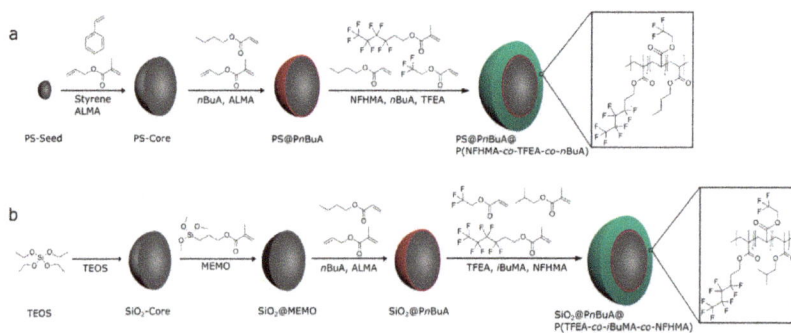

Figure 1. (**a**) Stepwise synthesis of organic fluorine-containing PS@P(NFHMA-*co*-TFEA-*co*-*n*BuA) core/interlayer/shell particles; (**b**) stepwise synthesis of inorganic/organic fluorine-containing SiO2@P(TFEA-*co*-NFHMA-*co*-*i*BuMa) core/interlayer/shell particles. Abbreviations: polystyrene (PS), allylmethacrylate (ALMA), *n*-butylacrylate (*n*BuA), 1*H*,1*H*,2*H*,2*H*-nonafluorohexylmethacrylate (NFHMA), trifluoroethylacrylate (TFEA), tetraethoxysilane (TEOS), 3-methacryloxypropyltrimethoxysilane (MEMO), and *i*-butylmethacrylate (*i*BuMA).

Core/interlayer/shell particles were also synthesized for the preparation of the corresponding inverse opal films. For this purpose, silica (SiO$_2$) particles were used as hard core particles. Pristine SiO$_2$-particles were synthesized according to the literature by van Blaaderen et al. [49], applying a sol-gel process (Stöber process). For the preparation of monodisperse SiO$_2$ particles, tetraethoxysilane (TEOS) in ethanol was used as a precursor, followed by a functionalization of the particle surface with 3-methacryloxypropyltrimethoxysilane (MEMO) [39]. For transferring the particles into the emulsion polymerization, ethanol was substituted by deionized water by azeotropic distillation. In the next step, the MEMO-functionalized SiO_2-particles were used for the preparation of the interlayer, consisting of poly(nBuA-co-ALMA). Finally, the polymerization of the outer shell was performed by copolymerizing TFEA (60 wt%), NFHMA (20 wt%), and *iso*-butylmethacrylate (*i*BMA) (20 wt%) as monomers. These inorganic core particles featuring an organic interlayer/shell polymer were examined with respect to morphology and average size by DLS measurements and TEM (Figure 2). All data on the different particles are compiled in Table 1.

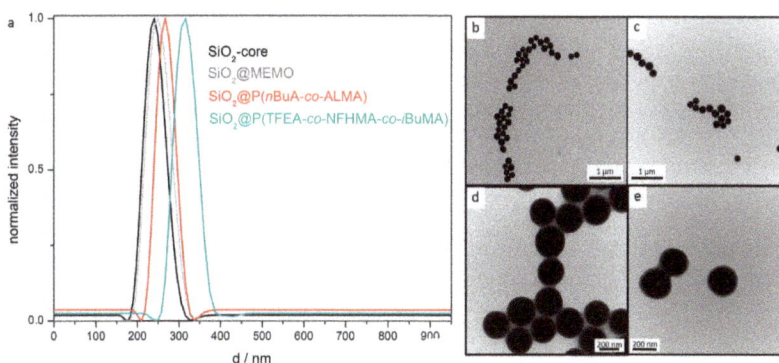

Figure 2. (**a**) DLS measurements of SiO_2 core particles, MEMO-functionalized SiO_2 particles, SiO_2 core/P(nBuA-co-AMLA)-interlayer particles, and SiO_2 core/P(nBuA-co-ALMA)interlayer/ P(TFEA-co-NFHMA-co-*i*BuMA)-shell particles; (**b**) TEM image of SiO_2-core particles; (**c**) TEM image of MEMO-functionalized SiO_2 particles; (**d**) TEM image of SiO_2 core/interlayer particles; (**e**) TEM image of SiO_2 core interlayer shell particles.

Table 1. Average hydrodynamic diameter of particles measured by means of DLS and average diameters of the dried particle, as determined by TEM. In the case of TEM analysis, 50 particles were measured with respect to their size.

Particle	DLS (d/nm)	TEM (d/nm)
PS	206 ± 2	192 ± 10
PS@P(nBuA-co-ALMA)	211 ± 2	202 ± 6
PS@P(NFHMA-co-TFEA-co-nBuA)	246 ± 1	238 ± 11
SiO_2	240 ± 1	232 ± 6
SiO_2@MEMO	249 ± 4	234 ± 11
SiO_2@P(nBuA-co-ALMA)	271 ± 1	261 ± 8
SiO_2@P(TFEA-co-NFHMA-co-iBuMa)	315 ± 9	303 ± 4

In summary, the DLS and TEM measurements again evidenced the successful preparation of well-defined spherical particles and a continuously increasing particle size for each synthesis step. While the particles in Figure 2b,c were still clearly separated from each other, the particles in Figure 2d,e are obviously connected by a soft polymer shell. DLS measurements in general give a larger diameter for particle systems compared to the TEM images, because the hydrodynamic diameters of the particles is determined, which is larger compared to particles in the dried state. For gaining insights into the composition and the thermal properties of the designed functional particles comprising the PS core, the PS@P(NFHMA-co-TFEA-co-nBuA) as well as the SiO_2@P(TFEA-co-NFHMA-co-iBuMA) core/interlayer/shell particles, differential scanning calorimetry (DSC) measurements were performed. The measured glass transition temperature of the (NFHMA-co-TFEA-co-nBuA)-polymer shell was found to be 14 °C, which was significantly lower than the T_g value found for the PS core, i.e., 100 °C (Figure S1). Moreover, the T_g value of the shell was found to be in between the T_gs of the pure components, i.e., 25 °C for PNFHMA [48], −54 °C for P*n*BuA [50], and −10 °C for PTFEA, confirming the successful copolymerization of the corresponding monomers.

Compared to this organic particle system, the glass transition temperature for the SiO$_2$@P(TFEA-*co*-NFHMA-*co*-*i*BuMA) core/shell particles was found to be 27 °C (Figure S3) due to the content of P*i*BuMA, which featured a glass transition temperature of 53 °C for the homopolymers [50]. Hence, successful copolymerization of the respective monomers was evidenced by the presence of the intermediate glass transition temperature. Moreover, these moderate values for the glass transition temperatures should enable processing by means of melt-shear organization of the core/shell particles, which will be described in the following.

For preparation of the elastomeric opal films using the melt-shear organization technique, the core/interlayer/shell particles were precipitated from their dispersion followed by drying at 40 °C. For homogenization of the obtained particle mass as well as for the addition of UV-cross-linking reagents (Irgacure 184, benzophenone, and 1,4-butanedioldiacrylate (BDDA)), the sticky particle mass was mixed using a microextruder at 90 °C (see Experimental Section). During this step, the addition of cross-linking reagents is important for subsequent UV-induced cross-linking reaction of the opal film. For this reason, cross-linking reagents that do not initiate chemical reactions during the melt-shear organization process but can initiate a posteriori are necessary. A cross-linked network, generated by the UV irradiation of the opal films, enhances the mechanical—and therefore the optical—properties of the opal films [46,47,51]. In order to additionally enhance the reflection colors of the opal films, 0.05 wt% carbon black (special black 4, Degussa) powder as an absorber was added during the extrusion. In general, carbon black powder has been found to dramatically enhance the perceived reflection color due to spectrally resonant scattering inside the opal structure without affecting the lattice quality [52]. Moreover, because of the small amount of added carbon black, it is not expected that the refractive index of the fluoropolymer-containing matrix material will be significantly increased. In the next step, the extruded polymer strands were subjected to the melt-shear organization process (Figure 3), allowing the core/shell particles to merge into a colloid crystal structure.

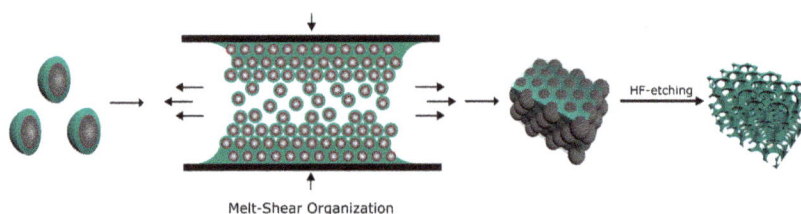

Melt-Shear Organization

Figure 3. Fabrication of opal films using the melt-shear organization technique and preparation of an inverse opal by HF etching of the SiO$_2$ core particles.

For this purpose, the particle mass was placed between the plates of a press followed by increasing the temperature and applying a pressure up to 100 bar (*cf.* Experimental Section). Within this film formation step, the soft shells generated a continuous matrix embedding the hexagonally arranged hard core particles. The latter formed the final colloidal crystal structure. In order to gain elastomeric properties and to maintain the stability upon etching for the preparation of an inverse opal film, the opal film was irradiated with UV-light to initiate the cross-linking reaction inside the polymer matrix. In this way, an elastomeric fluorine-containing opal film with iridescent reflection colors was prepared as studied in detail in the next section.

2.2. Optical Properties and Morphology of the PS@P(NFHMA-co-TFEA-co-nBuA) Opal Film

For the preparation of the elastomeric fluorine-containing opal films, some basic requirements must be fulfilled: (i) a periodic close-packed arrangement of the hard spheres, (ii) the monodispersity of the core particles, and (iii) a refractive index contrast between the hard core, respectively the voids, and the matrix material.

According to Bragg´s law of diffraction combined with Snell´s law (Equation (1)), the perceived wavelength of reflection (λ_{111}) is influenced by the average refractive index n_{eff}, which can be calculated by considering the refractive index n_i and the volume fractions ϕ_i for all ingredients of the opal (Equation (2)). The wavelength of reflection depends also on the periodicity α_{111} and the angle of incident light θ [24].

$$\lambda_{111} = 2\alpha_{111}\left(n_{eff}^2 - \sin^2\theta\right)^{1/2} \tag{1}$$

$$n_{eff} = \sum n_i \phi_i \tag{2}$$

For this purpose, an organic core-particle consisting of polystyrene ($n_{eff} \cong 1.59$) [53] and a shell material containing a high content of poly(NFHMA) ($n_{eff} \cong 1.35$) [54] was used. Compared to previously reported organic core/shell opal films, this combination featured a higher refractive index contrast ($\Delta n_{eff} \leq 0.24$). From the photographs obtained by scanning electron microscopy (SEM) (Figure 4a), it can be concluded that the particle-based film consists of closely packed and hexagonally aligned layers of particles. Additional to these findings, transmission electron microscopy (TEM) images obtained for ultra-thin slices prepared from the opal film are shown in Figure 4b, revealing the individual PS particles embedded in a lighter appearing matrix. The lighter matrix corresponded to the fluoropolymer shell material, having less electron contrast compared to the aromatic PS core particles. However, in the case of the TEM images, the core particles appeared to be more distorted. This can be explained by the fact that the spherical domains having a size of 192 ± 10 nm are much larger than the microtomed thin slices, which are approximately 50–70 nm in thickness. During ultramicrotoming of the bulk polymer films with spherical domains inside a soft matrix, the probability for perfect cutting of spherical objects is reduced, which is a known problem for spherical domains during sample preparation by using ultramicrotomy [55].

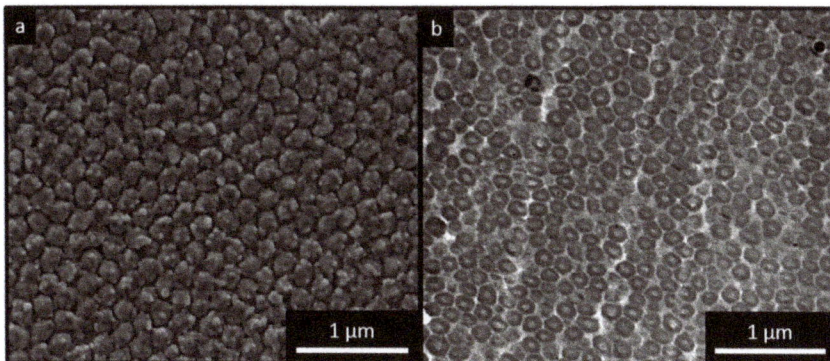

Figure 4. (**a**) SEM image: surface of PS@P(NFHMA-*co*-TFEA-*co*-*n*BuA) opal film; (**b**) TEM image: ultra-thin cut of the cross-section of a PS@P(NFHMA-*co*-TFEA-*co*-*n*BuA) opal film.

To determine the position of the Bragg peak and evidencing a good optical performance based on the ordered particles, angle-dependent UV/Vis measurements were performed (Figure 5). The measurements were recorded for angles of incident between 90° and 60°. Within the corresponding spectra, a distinct Bragg peak at a wavelength of 558 nm was observed. At lower angles of incidence, the value for the reflected light, i.e., the Bragg peak was blue-shifted from 558 nm at 90° to 486 nm at 60° according to Bragg´s law of diffraction. This finding proved the existence of a structural color for the herein designed fluorine-containing opal films. Moreover, it can be concluded that the melt-shear organization of the PS@P(NFHMA-*co*-TFEA-*co*-*n*BuA) particles leads to a regularly ordered particle-based film with brilliant reflection colors.

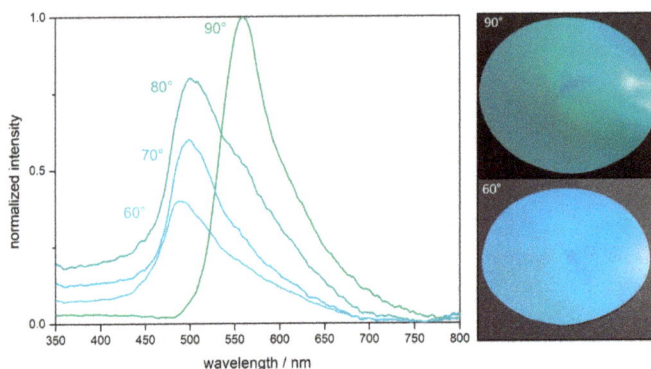

Figure 5. Angle-dependent UV/Vis reflection spectra of the opal film prepared from PS@P(NFHMA-*co*-TFEA-*co*-*n*BuA) core/interlayer/shell particles and corresponding photographs of the opal film at an angle of view of 90° and 60°. The final opal discs featured a diameter of 8 cm.

2.3. Chemical Resistance of the PS@P(NFHMA-co-TFEA-co-nBuA) Opal Film

In general, polymers having a high content of fluorine-containing monomers feature a good chemical resistance toward acids or bases [56]. In order to investigate the chemical resistance, the here prepared opal films were exposed to a strong acid (hydrochloric acid pH = 1) and base (potassium hydroxide pH = 13), followed by subsequent UV/Vis measurements of the samples. In Figure S4, UV/Vis spectroscopy measurements of the opal films treated with potassium hydroxide solution and hydrochloric acid are shown. The untreated film featured a reflection color maximum at a wavelength of approximately 525 nm. When the fluorine-containing opal film was treated with deionized water, the reflection peak of the opal film slightly shifted to 544 nm, which was due to the swelling capability of the matrix material. However, treatment of the swollen opal films with concentrated acid or concentrated base did not lead to a significant change of the optical properties. In more detail, the reflection color of the opal film treated for 20 min with potassium hydroxide was shifted only toward 6 nm and Bragg peak maximum was finally located at a wavelength of 550 nm at an angle of view of 90°. When the opal film was treated for 20 min with hydrochloric acid, the peak was located at a wavelength of 540 nm, i.e., only with a shift of 4 nm. Therefore, UV/Vis measurements proved the excellent order and resistance of the opal films under harsh conditions.

2.4. Hydrophobicity and Solvent Response of the PS@P(NFHMA-co-TFEA-co-nBuA) Opal Film

The hydrophobic character of the fluorine-containing opals was determined by contact angle measurements. A contact angle between the opal film surface and a sessile drop of deionized water of $106 \pm 3°$ was obtained, which categorized the surface hydrophobic (see Figure S5a). As described in the introduction, fluoropolymers have already found an extensive application as a low surface energy material for water repellency applications [37,57]. In general, the wettability of surfaces with liquid depends on two factors: (i) the chemical factor based on the low surface energy and (ii) the geometrical factors mainly given by roughness and tailored architecture of the surface. As can be concluded from the SEM image in Figure 4a, the surface of the opal film is not exceptional rough or structured, and the water repellency effect is therefore considered to stem from the chemical factor, i.e., the used fluorine-containing polymers. Despite the high chemical resistance and water repellency, the domain sizes of the opal film could be influenced upon treatment with various organic solvents. When the distance between the spheres forming the colloidal crystal stack is in the range of half the wavelength of visible light, structural colors can be observed, as described by Bragg´s law of diffraction. Solvent treatment caused a swelling-induced volume change of the matrix polymer, resulting in the increase of surface plane spacing of the closely packed PS spheres embedded in the cross-linked opal film. Moreover, the Bragg conditions for the respective colors was also influenced by the refractive index of the solvent used for matrix swelling, which has been described in earlier works [23,24,58]. Typically, the swelling behavior of the matrix materials leads to a shift of the reflection peak maximum toward a higher wavelength. In other words, the volume expansion of the polymer matrix induced by solvent treatment leads to a red-shift of the reflection peak maxima. Exemplarily, the peak maximum shifted from 525 nm for the untreated opal film to 808 nm after treatment with THF (Figure 6). Depending on the swelling capability, the used media (water, ethanol, acetonitrile, acetone, ethyl methyl ketone (EMK), and THF) will lead to a more pronounced reflection peak maximum shift within this order. Noteworthy, after complete evaporation of the used solvents, the fluorine-containing opals reached the original peak maximum value after repeated solvent treatment-evaporation cycles, at least three times.

In summary, the convenient preparation of fluoropolymer-containing opal films featuring angle-dependent reflection colors and solvent-responsive behavior were prepared. Moreover, the reflection peak position of the fluorine-containing opal films were not altered by treatment with concentrated acids and bases.

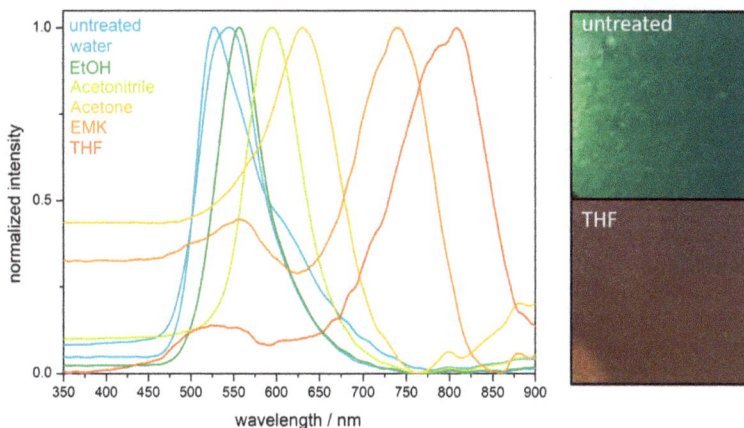

Figure 6. UV/Vis reflection spectra of a solvent-treated opal film based on PS@P(NFHMA-*co*-TFEA-*co-n*BuA) core/interlayer/shell particles (**left**) and exemplary photographs of the opal film prior to and after treatment with THF (**right**). Abbreviations: ethanol (EtOH), ethyl methyl-ketone (EMK), and tetrahydrofurane (THF).

2.5. Optical and Structural Properties of the SiO_2@P(TFEA-co-NFHMA-co-iBuMA) Inverse Opal Film

For the preparation of fluoropolymer inverse opal films, SiO_2 core particles with a comparably soft fluorine-copolymer shell were subjected to the melt-shear-organization technique. The silica-core particles were etched by HF treatment in a subsequent step in order to gain access to the inverted opal structure (see Experimental Section). The opal films prior to the etching process featured no bright reflection color, since the refractive index contrast Δn_{eff} between the silica core particles ($n_{eff} = 1.43$) [59] and the fluoropolymer matrix ($n_{eff} \leq 1.41$) was rather low [57]. However, after removal of the silica core particles, the ordered pores inside the fluoropolymer matrix led to a more pronounced refractive index contrast and therefore to the appearance of iridescent structural colors. Figure 7a shows photographs of the opal film filled with SiO_2 cores featuring no reflection colors, while in Figure 7b the inverted opal film is given showing structural colors.

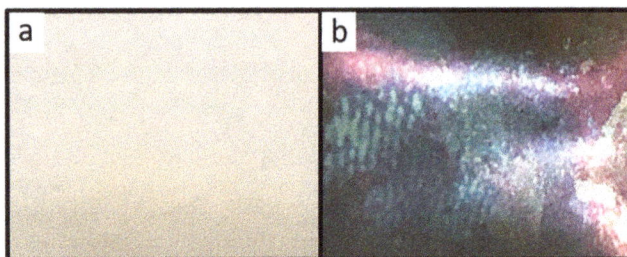

Figure 7. (**a**) Photographic picture of the filled opal film based on SiO$_2$@P(TFEA-*co*-NFHMA-*co*-*i*BuMA) core/interlayer/shell particles, without reflection colors; (**b**) photographic picture of the inverse opal film based on SiO$_2$@P(TFEA-*co*-NFHMA-*co*-*i*BuMA) core/interlayer/shell particles, with reflection colors. The cutout of the inverse opal discs featured a diameter of 2 cm.

In order to obtain more intensive reflection colors, a higher content of pores was of utmost importance. Therefore, the film was treated with hydrofluoric acid for full removal of the SiO$_2$ cores after several days. Corresponding SEM studies revealed that the surface of the inverse fluoropolymer opal film is open-porous with a uniform diameter of the pores of 175 ± 5 nm (Figure 8a). The corresponding cross-section (Figure 8b) revealed that the SiO$_2$ core particles were also removed in the interior of the opal film. It has to be mentioned that the pore order and pore size distribution seemed to be influenced with respect to their spherical shape within the SEM photographs for the cross-section compared to the film topography. This might be caused by sample preparation using freeze-fracturing for the comparably soft inverse opal films. Moreover, for the preparation of the fluoropolymer inverse opal films, a slightly different polymer composition was chosen. In comparison to the opal films described in Section 2.2, a lower fraction of NFHMA (20 wt%) and a higher amount of TFEA (60 wt%) was used. The reason for changing the composition and for additionally introducing *i*BuMA (20 wt%) instead of *n*BuA was the softness of the final inverse opal film leading to a pore collapse after the etching process. While using a fluoropolymer copolymer with P*i*BMA, the glass transition temperature could be increased to 25 °C, suitable for convenient core/shell particle processing and for the preparation of a free-standing inverse opal film. The T$_g$ was determined to be 25 °C, as studied by DSC measurements (Figure S3).

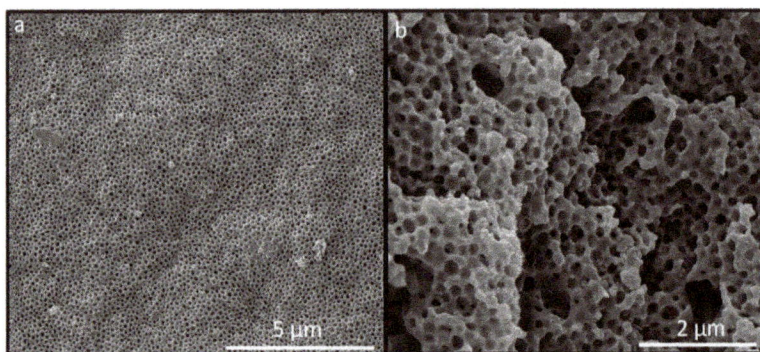

Figure 8. SEM photograph of the topography (**a**) and corresponding cross-section SEM photograph (**b**) of an inverse opal film based on SiO$_2$@P(TFEA-*co*-NFHMA-*co*-*i*BuMA) core/interlayer/shell particles after melt-shear organization and subsequent HF etching.

To further investigate the pore order and to visualize influence on the morphology upon water treatment, atomic force microscopy (AFM) studies were additionally carried out. Corresponding AFM measurements for the dried film and for the inverse opal film in water are given in Figure 9. In good accordance with the SEM images in Figure 8, AFM studies confirmed the presence of well-ordered hexagonally aligned pores. The average pore size in the dried state was determined to be 179 ± 8 nm, which was in good agreement with the pore size determined in the corresponding SEM image (175 ± 5 nm). Figure 9b give the results for AFM measurement of the same inverse opal film in water. As a finding, the average pore size was found to 153 ± 10 nm in diameter, which reflects a slight swelling capability of the water-treated inverse opal matrix material.

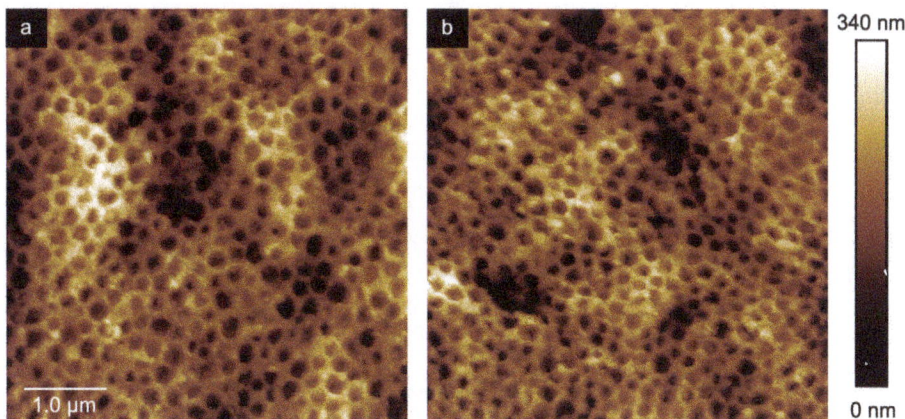

Figure 9. AFM image of the dried inverse opal film obtained after etching of the SiO$_2$@P(TFEA-*co*-NFHMA-*co-i*BuMA)-based particle films (**a**) and the same film measured in water (**b**) Scale bars correspond to 1 μm (see text).

2.6. Chemical Resistance and Solvent-Responsivness of the Inverse Opal Film

As already mentioned in the previous section, the hydrophobic character of the fluoropolymer-based opal materials was proven by contact angle measurements. Here, contact angle measurements for the inverse opal film derived after etching of SiO$_2$@P(TFEA-*co*-NFHMA-*co-i*BuMA)-based particle films, resulted in a contact angle of 102 ± 2° (Figure S5b). Along with the results obtained from the AFM measurements, a slight swelling of the inverse opal film upon water treatment was observed, which also led to a shift of the maximum peak position during UV/Vis spectroscopy measurements (Figure 10). In detail, the dried inverse opal film featured a reflection color peak at a wavelength of 518 nm, while the peak shifted to a wavelength of 570 nm upon water treatment. This was found to be a remarkable difference compared to the results obtained for the filled opals as shown in Figure 6, revealing a peak maximum shift of only 16 nm. This finding underpins the more pronounced sensing capability for water (and other media) for the inverse opal structures compared to the filled opal films. This effect is even more pronounced while using organic solvents (Figure 10). The peak maxima shifted in polar media, such as water (peak maximum at 570 nm) or ethanol (612 nm), tetrahydrofuran (704 nm), and acetonitrile (836 nm), respectively. This observation can be explained by the swelling capability induced by the different solvents for the fluorine-containing matrix. In contrast, the appearance of the structural colors is only slightly affected by the change of the refractive index contrast, since the different refractive indices of the used solvents are similar, i.e., 1.33 for water, 1.36 for ethanol, 1.37 for ethyl methyl ketone, 1.40 for THF, and 1.34 for acetonitrile [60,61].

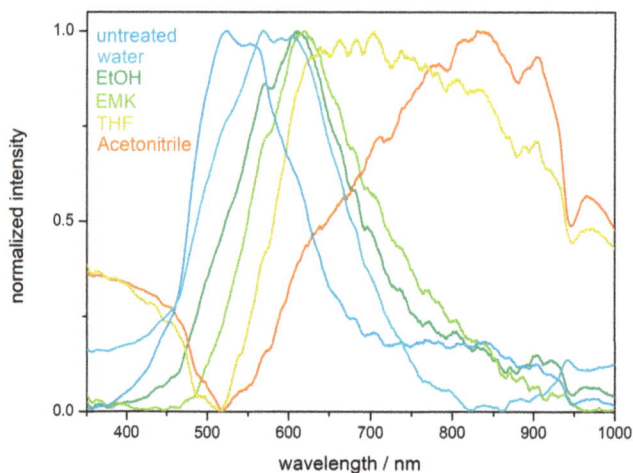

Figure 10. UV/Vis reflection spectra of solvent-treated inverse opal film based on P(TFEA-*co*-NFHMA-*co-i*BuMA). Abbreviations: ethanol (EtOH), ethyl methyl ketone (EMK), and tetrahydrofuran (THF).

Finally, the chemical resistance of the inverse opal film caused by fluorine-containing matrix was also examined by treatment with a strong acid (hydrochloric acid pH = 1) and base (potassium hydroxide pH = 13) by UV/Vis spectroscopy measurements (Figure S6). Despite the slightly reduced fluorine-content inside the inverse opal matrix, only a small shift of the reflection peak maximum of approximately 10 nm was obtained from the corresponding spectra. This finding proved the high chemical resistance also for the inverse opal films upon potassium hydroxide and hydrochloric acid treatment.

3. Experimental

3.1. Materials and Methods

1*H*,1*H*,2*H*,2*H*-Nonafluorohxylmethacrylate (NFHMA, 95%) was obtained from ABCR (Karlsruhe, Germany), trifluoroethylacrylat (TFEA, >98%) from TCI (Eschborn, Germany), Butandioldiacrylate (BDDA), and Irgacure 184 from BASF (Ludwigshafen, Germany). Dowfax 2A1, a surfactant having a dual polar head group and a non-polar alkyl chain was obtained from Dow Chemicals (Midland, MI, USA), carbon black (Special Black 4) by Degussa GmbH (Essen, Germany), and benzophenon, the UV-initiator, was donated by Merck Chemicals (Darmstadt, Germany). All other chemicals were purchased from VWR (Radnor, PA, USA), Acros Organics, Fisher Scientific (Schwerte, Germany), and Sigma Aldrich (St. Louis, MO, USA) and used as received, if not otherwise mentioned. Prior to use for the polymerization, inhibitors were removed from the monomers *n*BuA, *i*BuMA, and styrene by passing through an alumina column (basic, 50–200 µm, Acros Organics).

Dynamic light scattering (DLS) measurements of the particle dispersions were performed with a Zetasizer ZS90 (Malvern Instruments, Malvern, UK). The measurements were carried out at 25 °C at an angle of 90°. For the evaluations, the z-weight average hydrodynamic diameter was used. Transmission electron microscopy (TEM) was realized on a Zeiss EM 10 electron microscope (Oberkochen, Germany) operating at 60 kV. The images were recorded in bright field mode with a slow scan CCD camera obtained from TRS Tröndle (Morrenweis, Germany). The control of the camera was computer-assisted using ImageSP from TRS. Scanning electron microscopy (SEM) was performed on a Philips XL30 FEG (Amsterdam, Netherlands) with an operating voltage of 5–10 kV. The samples were previously coated for 100 s at 30 mA with a thin gold layer, using a Quorum Q300T D sputter coater (Lewes,

UK). Angle dependent reflection measurements were performed using a custom built goniometer setup measured in steps of 10° of scattering angle. All other reflection spectra were recorded using a Vis/-NIR fiber spectrometer USB 2000, Ocean Optics (Ostfildern, Germany). For the measurements, a deuterium/tungsten halogen lamp DT mini 2, Ocean Optics was used. Measurements in water or solvents were carried out at normal light incidence (90°). Thermal properties of the core interlayer shell particles were characterized using a differential scanning calorimeter (DSC) from Mettler Toledo (Columbus, OH, USA) DSC-1 in the temperature range from −50 °C to 150 °C with a heating rate of 20 K min^{-1} in a nitrogen atmosphere. Atomic force microscopy measurements were accomplished in the PeakForce Tapping mode with an Icon Dimension Bruker AXS (Santa Barbara, CA, USA). Images with dimensions of 5 × 5 μm^2 (512 × 512 pixel) were taken in air and deionized water at a scan rate of 1 Hz using a maximum force of 2 nN. The inverse optical sensitivity of the laser detection system was calibrated by pushing the cantilever with the tip onto a stiff substrate (sapphire) and relating the deflection of the laser spot on the photosensitive diode with the movement of the z-piezo. The cantilever spring constant was 0.8 nN/nm measured by the thermal noise method [62]. The contact angle (CA) was measured using the sessile-drop-method with a Contact angle system DATAPhysics OCA 15 EC (Filderstadt, Germany) using 2 μL droplets of deionized water. The measurements were conduced in a controlled climatic chamber at a temperature of 23 ± 2 °C and a relative humidity of 40%. Contact angles were determined geometrically using the SCA20 software by aligning a tangent from the surficial contact point along the droplet profile.

3.2. Synthesis of PS@PBuA@P(NFHMA-co-TFEA-co-nBuA) Core/Interlayer/Shell-Particles

The stepwise generation of fluorine-containing polymer based particles is illustrated in Figure 1a. The corresponding monodisperse PS particles were synthesized according to starved feed emulsion polymerization protocols. A 1 L reactor under an argon stream, equipped with a reflux condenser and a stirrer, was heated up to 80 °C and filled with a dispersion of 3.6 g of styrene, 0.4 g of BDDA, 0.1 g of SDS, and 280 g of degassed water. The polymerization was initiated by adding solutions of 70 mg of sodiummetabisulfite (SBS) in 5 g of water and 500 mg of sodiumperoxodisulfate (SPS) in 5 g of water. After 20 min, a monomer emulsion containing 3.5 g of ALMA, 31.5 g of styrene, 180 mg of KOH, 164 mg of SDS 105 mg of Igepal, and 40.9 g of water was added continuously with a flow rate of 0.5 mL/min. After one hour of reaction time, the polystyrene particles with an average diameter of 206 nm were characterized and stored for further use.

The core-shell particle featuring a copolymer of PnBuA, PNFHMA, and PTFEA as a soft shell was synthesized in a 100 mL double-wall reactor equipped with a stirrer and reflux condenser in an argon atmosphere at 80 °C. For this purpose, 60 g of the PS-particle dispersion with a solid content of 8.4 wt% was filled into the reactor, and the emulsion polymerization was initiated by the addition of 14 mg of SBS and 77 mg of SPS dissolved in 2 mL of deionized water. After 20 min reaction time, a monomer emulsion (ME1) consisting of 75 mg of ALMA, 425 mg of nBuA, 10 mg of Dowfax 2A1, 5 mg of SDS, and 2.5 g of deionized water was added with a flow rate of 0.2 mL min^{-1} using a rotary piston pump. After the complete addition of ME1, a solution of 12 mg of SPS in 2 g of water was added and a second monomer emulsion (ME2) consisting of 2.7 g of NFHMA, 0.9 g of TFEA, 0.9 g of nBuA, 45 mg of KOH, 16 mg of SDS, 13 mg of Dowfax 2A1, and 5.22 g of water were continuously added with a flow rate of 0.2 mL min^{-1}. After complete addition of the ME2, the temperature of polymerization was maintained for an additional hour prior to cooling the vessel to room temperature. The average particle diameter of the core shell particle was determined to be 246 ± 1 nm (DLS).

3.3. Synthesis of SiO$_2$@PBuA@P(TFEA-co-NFHMA-co-iBuMA) Core/Interlayer/Shell-Particles

Silica particle dispersion with a solid content of 2.5 wt% in ethanol were prepared using a sol-gel process (Stöber process) according to the protocol described by van Blaaderen [49]. A mixture of 1.3 L of ethanolic silica dispersion with 1.6 mL of 3-methacryloxypropyltrimethoxysilane (MEMO) was heated to 60 °C and stirred for 2 h. After functionalization, ammonia was carefully removed by azeotropic distillation at 60 °C under reduced pressure, while the volume was kept constant by tcontinuous addition of ethanol. When the dispersion was free of ammonia, the volume was reduced to 300 mL at 60 °C. For transferring the silica particle dispersion into an aqueous medium for the intended emulsion polymerization, a solution of 50 mg of SDS in 100 mL of deionized water was added, and ethanol was removed by azeotropic distillation. The volume was kept constant by continuous addition of water. The final MEMO-functionalized aqueous dispersion featured a silica solid content of 8.83 wt%. The average particle diameter of the particles after functionalization was 249 ± 4 nm, as determined by DLS measurements.

For synthesis of the hybrid core-shell particles, 70 g of the pristine SiO_2 particle dispersion was filled in a 100 mL double wall reactor. The emulsion polymerization was initiated at 80 °C by the addition of a solution of 16 mg of SBS and 94 mg of SPS in 2 mL of deionized water. After 20 min, a monomer emulsion (ME1) consisting of 83 mg of ALMA, 0.47 g of *n*BuA, 11 mg of Dowfax 2A1, 6 mg of SDS, and 2.77 g of water were added by a rotary piston pump with a flow rate of 0.2 mL min^{-1}. After a 15 min reaction time, 15 mg of SPS in 2 mL of water was added and another monomer emulsion (ME2) composed of 3.55 g of TFEA, 0.9 g of NFHMA, 0,9 g of *i*BuMA, 7 mg of KOH, 16 mg of SDS, 12 mg of Dowfax 2A1, and 5.2 g of water was continuously added with a flow rate of 0.2 mL min^{-1}. After an additional hour, the product was cooled to room temperature. The diameter of the core/shell particle was 315 ± 9 nm, as determined by DLS measurements.

3.4. Particle Processing and Preparation of Opal and Inverse Opal Films

For the preparation of elastomeric opal films, the obtained PS@P*n*BuA@P(NFHMA-*co*-TFEA-*co*-*n*BuA) particles were precipitated in methanol containing 20% of saturated sodium chloride solution. The precipitate was filtered, washed with water, and dried under vacuum at 40 °C. For homogenization, the precursor powder was mixed with 7.25 wt% BDDA, 1 wt% benzophenone, 1 wt% Irgacure 184, and 0.05 wt% carbon black (special black 4, Degussa) in a microextruder (HAAKE Minilab II350, Thermo Scientific, Waltham, MA, USA) at 90 °C. For opal film formation, a 2 g portion of the polymeric mass was covered by two PET foils and heated between two steel-plates in a Collin laboratory press (Dr. Collin GmbH, Ebersberg, Germany). The particle mass was transduced into an opal disc film of approximately 8 cm in size by using the melt-shear organization technique at 90 °C and 100 bar. Subsequently, the opal films were irradiated with a mercury lamb (UVA Cube 2000, Dr Hoenle AG, Gräfelfing, Germany) with an output power of 1000 W at a distance of 10 cm for 3 min at both sides, for cross-linking reactions. For conversion of the opal film containing SiO_2 core particles into an inverse opal film, the cores were removed by etching with hydrofluoric acid (HF, 10 wt% in water) for 4 days. The films were washed with plenty of deionized water several times.

4. Conclusions

In conclusion, the current work demonstrated an efficient protocol for preparation of fluorine-containing core/interlayer/shell particles. For the hard core particle preparation, inorganic SiO_2 or PS particles were used, whereas the soft shell was formed by a stepwise emulsion polymerization of highly fluorinated monomers, i.e., 1*H*,1*H*,2*H*,2*H*-nonafluorohexylmethacrylate (NFHMA) and trifluoroethylacrylate (TFEA), leading to well-defined fluorine-containing core/shell particles. The melt-shear organization technique was applied for the preparation of easily scalable fluoropolymer opal disc films. The opal film based on SiO_2 core particles were subjected to an etching protocol using hydrofluoric acid to gain access to fluoropolymer inverse opal structures. Characterization of the particles, opal films, and inverse opal films was carried out using DLS, TEM, SEM, AFM, DSC, UV/Vis spectroscopy, and contact angle measurements, evidencing the size, uniformity, and order after the melt-shear organization and subsequent etching. The feasibility of fluorine-containing polymers as functional matrix materials, leading to opal films featuring brilliant reflection colors, was shown by the good refractive index contrast compared to the PS core material (or air voids in the case of the inverse opals). Finally, swelling capability and stimuli-responsiveness

Molecules **2019**, *24*, 333

upon treatment with different solvents were investigated, and the reversible switching behavior of the optical properties was shown. Moreover, the inverse opals revealed good hydrophobic properties and excellent chemical resistance toward strong acids and bases. Along with the good structural colors, we envisage the herein investigated fluorine-containing opal and inverse opal films as promising candidates in the field of robust coatings and switchable optical sensing devices.

Supplementary Materials: The following information are available online, Figure S1: DSC thermogram of PS based core/shell particles; Figure S2: DLS measurements and TEM images of PS based core/shell particles; Figure S3: DSC thermogram of SiO_2 based core/shell particles; Figure S4: UV/Vis spectra of an opal film; Figure S5: Photograph of drops of water on opal and invers opal films; Figure S6: UV/Vis spectra of an inverse opal film.

Author Contributions: Conceptualization: J.K. and M.G.; methodology: J.K. and M.G.; validation: J.K. and M.G.; AFM analysis: C.D.; investigation: J.K.; writing—original draft preparation: M.G. and J.K.; writing—review and editing: J.K. and M.G.; supervision: M.G.; project administration: M.G.; funding acquisition: M.G.

Funding: This research received no external funding.

Acknowledgments: The authors thank the company Merck KGaA for the scientific collaboration.

Conflicts of Interest: The authors declare no conflict of interest.

References

1. Drobny, J.G. *Technology of Fluoropolymers*, 2nd ed.; CRC Press: New York, NY, USA, 2009.
2. Munekata, S. Fluoropolymers as coating material. *Progr. Org. Coat.* **1988**, *16*, 113–134. [CrossRef]
3. Smith, D.W.; Iacono, S.T.; Iyer, S.S. *Handbook of Fluoropolymer Science and Technology*, 1st ed.; Wiley: Hoboken, NJ, USA, 2014.
4. Cui, Z.; Drioli, E.; Lee, Y.M. Recent progress in fluoropolymers for membranes. *Progr. Polym. Sci.* **2014**, *39*, 164–198. [CrossRef]
5. Ameduri, B.; Boutevin, B. *Well-Architectured Fluoropolymers: Synthesis, Properties and Applications*; Elsevier: Amsterdam, The Netherlands, 2004.
6. Gardiner, J. Fluoropolymers: Origin, Production, and Industrial and Commercial Applications. *Aust. J. Chem.* **2015**, *68*, 13–22. [CrossRef]
7. Moore, A.L. *Fluoroelastomers Handbook: The Definitive User's Guide and Databook*; William Andrew Publishing: Norwich, NY, USA, 2005.
8. Dams, R.; Hintzer, K. Chapter 1 Industrial Aspects of Fluorinated Oligomers and Polymers. In *Fluorinated Polymers: Volume 2: Applications*; The Royal Society of Chemistry: London, UK, 2017; Volume 2, pp. 1–31.
9. Sun, W.; Zhou, S.; You, B.; Wu, L. A facile method for the fabrication of superhydrophobic films with multiresponsive and reversibly tunable wettability. *J. Mater. Chem. A* **2013**, *1*, 3146–3154. [CrossRef]
10. Shi, F.; Song, Y.; Niu, J.; Xia, X.; Wang, Z.; Zhang, X. Facile Method to Fabricate a Large-Scale Superhydrophobic Surface by Galvanic Cell Reaction. *Chem. Mater.* **2006**, *18*, 1365–1368. [CrossRef]
11. Feng, C.L.; Zhang, Y.J.; Jin, J.; Song, Y.L.; Xie, L.Y.; Qu, G.R.; Jiang, L.; Zhu, D.B. Reversible Wettability of Photoresponsive Fluorine-Containing Azobenzene Polymer in Langmuir-Blodgett Films. *Langmuir* **2001**, *17*, 4593–4597. [CrossRef]
12. Pei, Y.; Travas-Sejdic, J.; Williams, D.E. Reversible electrochemical switching of polymer brushes grafted onto conducting polymer films. *Langmuir* **2012**, *28*, 8072–8083. [CrossRef]
13. Hu, J.; Meng, H.; Li, G.; Ibekwe, S.I. A review of stimuli-responsive polymers for smart textile applications. *Smart Mater. Struct.* **2012**, *21*, 053001. [CrossRef]
14. Drelich, J.; Chibowski, E.; Meng, D.D.; Terpilowski, K. Hydrophilic and superhydrophilic surfaces and materials. *Soft Matter* **2011**, *7*, 9804–9828. [CrossRef]
15. Wenzel, R.N. Resistance of Solid Surfaces to Wetting by Water. *Ind. Eng. Chem. Res.* **1936**, *28*, 988–994. [CrossRef]
16. Cassie, A.B.D.; Baxter, S. Wettability of porous surfaces. *Trans. Faraday Soc.* **1944**, *40*, 546–550. [CrossRef]
17. Whitesides, G.M. Nanoscience, nanotechnology, and chemistry. *Small* **2005**, *1*, 172–179. [CrossRef] [PubMed]
18. Piao, Y.; Burns, A.; Kim, J.; Wiesner, U.; Hyeon, T. Designed Fabrication of Silica-Based Nanostructured Particle Systems for Nanomedicine Applications. *Adv. Funct. Mater.* **2008**, *18*, 3745–3758. [CrossRef]

19. Phillips, K.R.; Vogel, N.; Hu, Y.; Kolle, M.; Perry, C.C.; Aizenberg, J. Tunable Anisotropy in Inverse Opals and Emerging Optical Properties. *Chem. Mater.* **2014**, *26*, 1622–1628. [CrossRef]

20. Huang, X.; Chen, J.; Lu, Z.; Yu, H.; Yan, Q.; Hng, H.H. Carbon inverse opal entrapped with electrode active nanoparticles as high-performance anode for lithium-ion batteries. *Sci. Rep.* **2013**, *3*, 2317. [CrossRef] [PubMed]

21. Schäfer, C.G.; Vowinkel, S.; Hellmann, G.P.; Herdt, T.; Contiu, C.; Schneider, J.J.; Gallei, M. A polymer based and template-directed approach towards functional multidimensional microstructured organic/inorganic hybrid materials. *J. Mater. Chem. C* **2014**, *2*, 7960–7975. [CrossRef]

22. Schäfer, C.G.; Gallei, M.; Zahn, J.T.; Engelhardt, J.; Hellmann, G.P.; Rehahn, M. Reversible Light-, Thermo-, and Mechano-Responsive Elastomeric Polymer Opal Films. *Chem. Mater.* **2013**, *25*, 2309–2318. [CrossRef]

23. Schäfer, C.G.; Biesalski, M.; Hellmann, G.P.; Rehahn, M.; Gallei, M. Paper-supported elastomeric opal films for enhanced and reversible solvatochromic response. *J. Nanophotonics* **2013**, *7*. [CrossRef]

24. Schäfer, C.G.; Winter, T.; Heidt, S.; Dietz, C.; Ding, T.; Baumberg, J.J.; Gallei, M. Smart polymer inverse-opal photonic crystal films by melt-shear organization for hybrid core–shell architectures. *J. Mater. Chem. C* **2015**, *3*, 2204–2214. [CrossRef]

25. Schaffner, M.; England, G.; Kolle, M.; Aizenberg, J.; Vogel, N. Combining Bottom-Up Self-Assembly with Top-Down Microfabrication to Create Hierarchical Inverse Opals with High Structural Order. *Small* **2015**, *11*, 4334–4340. [CrossRef]

26. Phillips, K.R.; England, G.T.; Sunny, S.; Shirman, E.; Shirman, T.; Vogel, N.; Aizenberg, J. A colloidoscope of colloid-based porous materials and their uses. *Chem. Soc. Rev.* **2016**, *45*, 281–322. [CrossRef]

27. Gallei, M. Functional Polymer Opals and Porous Materials by Shear-Induced Assembly of Tailor-Made Particles. *Macromol. Rapid Commun.* **2018**, *39*, 1700648. [CrossRef] [PubMed]

28. Schüth, F.; Schmidt, W. Microporous and Mesoporous Materials. *Adv. Mater.* **2002**, *14*, 629–638. [CrossRef]

29. Stein, A. Advances in Microporous and Mesoporous Solids—Highlights of Recent Progress. *Adv. Mater.* **2003**, *15*, 763–775. [CrossRef]

30. Thomas, A.; Goettmann, F.; Antonietti, M. Hard Templates for Soft Materials: Creating Nanostructured Organic Materials. *Chem. Mater.* **2008**, *20*, 738–755. [CrossRef]

31. Llusar, M.; Sanchez, C. Inorganic and Hybrid Nanofibrous Materials Templated with Organogelators. *Chem. Mater.* **2008**, *20*, 782–820. [CrossRef]

32. Joshi, R.K.; Schneider, J.J. Assembly of one dimensional inorganic nanostructures into functional 2D and 3D architectures. Synthesis, arrangement and functionality. *Chem. Soc. Rev.* **2012**, *41*, 5285–5312. [CrossRef] [PubMed]

33. Scheid, D.; Cherkashinin, G.; Ionescu, E.; Gallei, M. Single-source magnetic nanorattles by using convenient emulsion polymerization protocols. *Langmuir* **2014**, *30*, 1204–1209. [CrossRef]

34. Galisteo-López, J.F.; Ibisate, M.; Sapienza, R.; Froufe-Pérez, L.S.; Blanco, Á.; López, C. Self-assembled photonic structures. *Adv. Mater.* **2011**, *23*, 30–69. [CrossRef]

35. Kang, H.; Lee, J.S.; Chang, W.S.; Kim, S.H. Liquid-impermeable inverse opals with invariant photonic bandgap. *Adv. Mater.* **2015**, *27*, 1282–1287. [CrossRef]

36. Vogel, N.; Belisle, R.A.; Hatton, B.; Wong, T.S.; Aizenberg, J. Transparency and damage tolerance of patternable omniphobic lubricated surfaces based on inverse colloidal monolayers. *Nat. Commun.* **2013**, *4*, 2167. [CrossRef]

37. Wu, Y.; Zhou, S.; Wu, L. Fabrication of Robust Hydrophobic and Super-Hydrophobic Polymer Films with Onefold or Dual Inverse Opal Structures. *Macromol. Mater. Eng.* **2016**, *301*, 1430–1436. [CrossRef]

38. Ruhl, T.; Spahn, P.; Hellmann, G.P. Artificial opals prepared by melt compression. *Polymer* **2003**, *44*, 7625–7634. [CrossRef]

39. Finlayson, C.E.; Baumberg, J.J. Generating Bulk-Scale Ordered Optical Materials Using Shear-Assembly in Viscoelastic Media. *Materials* **2017**, *10*, 688. [CrossRef] [PubMed]

40. Scheid, D.; Stock, D.; Winter, T.; Gutmann, T.; Dietz, C.; Gallei, M. The pivotal step of nanoparticle functionalization for the preparation of functional and magnetic hybrid opal films. *J. Mater. Chem. C* **2016**, *4*, 2187–2196. [CrossRef]

41. Winter, T.; Su, X.; Hatton, T.A.; Gallei, M. Ferrocene-Containing Inverse Opals by Melt-Shear Organization of Core/Shell Particles. *Macromol. Rapid Commun.* **2018**, *39*. [CrossRef]

42. Vowinkel, S.; Schäfer, C.G.; Cherkashinin, G.; Fasel, C.; Roth, F.; Liu, N.; Dietz, C.; Ionescu, E.; Gallei, M. 3D-ordered carbon materials by melt-shear organization for tailor-made hybrid core–shell polymer particle architectures. *J. Mater. Chem. C* **2016**, *4*, 3976–3986. [CrossRef]

43. Vowinkel, S.; Malz, F.; Rode, K.; Gallei, M. Single-source macroporous hybrid materials by melt-shear organization of core-shell particles. *J. Mater. Sci.* **2017**, *52*, 11179–11190. [CrossRef]

44. Vowinkel, S.; Boehm, A.; Schäfer, T.; Gutmann, T.; Ionescu, E.; Gallei, M. Preceramic Core-Shell Particles for the Preparation of Hybrid Colloidal Crystal Films by Melt-Shear Organization and Conversion into Porous Ceramics. *Mater. Des.* **2018**, *160*, 926–935. [CrossRef]

45. Zhao, Q.; Finlayson, C.E.; Snoswell, D.R.E.; Haines, A.; Schäfer, C.; Spahn, P.; Hellmann, G.P.; Petukhov, A.V.; Herrmann, L.; Burdet, P.; et al. Large-scale ordering of nanoparticles using viscoelastic shear processing. *Nat. Commun.* **2016**, *7*, 11661. [CrossRef]

46. Schäfer, C.G.; Smolin, D.A.; Hellmann, G.P.; Gallei, M. Fully Reversible Shape Transition of Soft Spheres in Elastomeric Polymer Opal Films. *Langmuir* **2013**, *29*, 11275–11283. [CrossRef] [PubMed]

47. Schäfer, C.G.; Viel, B.; Hellmann, G.P.; Rehahn, M.; Gallei, M. Thermo-cross-linked Elastomeric Opal Films. *ACS Appl. Mater. Interfaces* **2013**, *5*, 10623–10632. [CrossRef] [PubMed]

48. Jiang, J.; Zhang, G.; Wang, Q.; Zhang, Q.; Zhan, X.; Chen, F. Novel Fluorinated Polymers Containing Short Perfluorobutyl Side Chains and Their Super Wetting Performance on Diverse Substrates. *ACS Appl. Mater. Interfaces* **2016**, *8*, 10513–10523. [CrossRef] [PubMed]

49. Graf, C.; van Blaaderen, A. Metallodielectric Colloidal Core-Shell Particles for Photonic Applications. *Langmuir* **2002**, *18*, 524–534. [CrossRef]

50. Schneider, H.A. Polymer class specificity of the glass temperature. *Polymer* **2005**, *46*, 2230–2237. [CrossRef]

51. Viel, B.; Ruhl, T.; Hellmann, G.P. Reversible Deformation of Opal Elastomers. *Chem. Mater.* **2007**, *19*, 5673–5679. [CrossRef]

52. Pursiainen, O.L.J.; Baumberg, J.J.; Winkler, H.; Viel, B.; Spahn, P.; Ruhl, T. Nanoparticle-tuned structural color from polymer opals. *Opt. Express* **2007**, *15*, 9553–9561. [CrossRef]

53. Katritzky, A.R.; Sild, S.; Karelson, M. Correlation and Prediction of the Refractive Indices of Polymers by QSPR. *J. Chem. Inf. Comp. Sci.* **1998**, *38*, 1171–1176. [CrossRef]

54. Yao, W.; Li, Y.; Huang, X. Fluorinated poly(meth)acrylate: Synthesis and properties. *Polymer* **2014**, *55*, 6197–6211. [CrossRef]

55. Gleinser, W.; Maier, D.; Schneider, M.; Weese, J.; Friedrich, C.; Honerkamp, J. Estimation of sphere-size distributions in two-phase polymeric materials from transmission electron microscopy data. *J. Appl. Polym. Sci.* **1994**, *53*, 39–50. [CrossRef]

56. De Francisco, R.; Tiemblo, P.; Hoyos, M.; González-Arellano, C.; García, N.; Berglund, L.; Synytska, A. Multipurpose Ultra and Superhydrophobic Surfaces Based on Oligodimethylsiloxane-Modified Nanosilica. *ACS Appl. Mater. Interfaces* **2014**, *6*, 18998–19010. [CrossRef]

57. García-Domenech, R.; de Julián-Ortiz, J.V. Prediction of Indices of Refraction and Glass Transition Temperatures of Linear Polymers by Using Graph Theoretical Indices. *J. Phys. Chem. B* **2002**, *106*, 1501–1507. [CrossRef]

58. Schäfer, C.G.; Lederle, C.; Zentel, K.; Stuhn, B.; Gallei, M. Utilizing stretch-tunable thermochromic elastomeric opal films as novel reversible switchable photonic materials. *Macromol. Rapid Commun.* **2014**, *35*, 1852–1860. [CrossRef] [PubMed]

59. Hart, S.J.; Terray, A.V. Refractive-index-driven separation of colloidal polymer particles using optical chromatography. *Appl. Phys. Lett.* **2003**, *83*, 5316–5318. [CrossRef]

60. Aralaguppi, M.I.; Jadar, C.V.; Aminabhavi, T.M. Density, Viscosity, Refractive Index, and Speed of Sound in Binary Mixtures of Acrylonitrile with Methanol, Ethanol, Propan-1-ol, Butan-1-ol, Pentan-1-ol, Hexan-1-ol, Heptan-1-ol, and Butan-2-ol. *J. Chem. Eng. Data* **1999**, *44*, 216–221. [CrossRef]

61. Awwad, A.M.; Al-Dujaili, A.H. Density, Refractive Index, Permittivity, and Related Properties for N-Formylmorpholine + Ethyl Acetate and + Butanone at 298.15 K. *J. Chem. Eng. Data* **2001**, *46*, 1349–1350. [CrossRef]

62. Butt, H.J.; Jaschke, M. Calculation of thermal noise in atomic force microscopy. *Nanotechnology* **1995**, *6*, 1. [CrossRef]

Sample Availability: Samples of the compounds (particles, opals and inverse opals) are available from the authors.

molecules

MDPI

Article

VE-Albumin Core-Shell Nanoparticles for Paclitaxel Delivery to Treat MDR Breast Cancer

Bo Tang [1], Yu Qian [1], Yi Gou [1], Gang Cheng [2] and Guihua Fang [1,*]

[1]   School of Pharmacy, Nantong University, 19 Qixiu Road, Nantong 226001, Jiangsu, China;
      tangbo@ntu.edu.cn (B.T.); syfsxyhh18@163.com (Y.Q.); gouyi@ntu.edu.cn (Y.G.)
[2]   School of Pharmacy, Shenyang Pharmaceutical University, 103 Wenhua Road, Shenhe District,
      Shenyang 110016, Liaoning, China; chenggang63@hotmail.com
*   Correspondence: fangguihua@ntu.edu.cn; Tel.: +86-153-6559-3488

Academic Editors: Jianxun Ding, Yang Li and Mingqiang Li
Received: 19 September 2018; Accepted: 20 October 2018; Published: 25 October 2018

Abstract: Multi-drug resistance (MDR) presents a serious problem in cancer chemotherapy. In this study, Vitamin E (VE)-Albumin core-shell nanoparticles were developed for paclitaxel (PTX) delivery to improve the chemotherapy efficacy in an MDR breast cancer model. The PTX-loaded VE-Albumin core-shell nanoparticles (PTX-VE NPs) had small particle sizes (about 100 nm), high drug entrapment efficiency (95.7%) and loading capacity (12.5%), and showed sustained release profiles, in vitro. Docking studies indicated that the hydrophobic interaction and hydrogen bonds play a significant role in the formation of the PTX-VE NPs. The results of confocal laser scanning microscopy analysis demonstrated that the cell uptake of PTX was significantly increased by the PTX-VE NPs, compared with the NPs without VE (PTX NPs). The PTX-VE NPs also exhibited stronger cytotoxicity, compared with PTX NPs with an increased accumulation of PTX in the MCF-7/ADR cells. Importantly, the PTX-VE NPs showed a higher anti-cancer efficacy in MCF-7/ADR tumor xenograft model than the PTX NPs and the PTX solutions. Overall, the VE-Albumin core-shell nanoparticles could be a promising nanocarrier for PTX delivery to improve the chemotherapeutic efficacy of MDR cancer.

Keywords: Vitamin E; albumin; core-shell nanoparticles; paclitaxel; multi-drug resistance; breast cancer

---

1. Introduction

Multi-drug resistance (MDR) is a large obstacle to the success of cancer chemotherapy and is crucial to cancer metastasis and recovery [1]. The well-known P-glucoprotein (P-gp), an ATP-binding cassette transporter, which is over expressed on the surface membrane of cancer cells, is one of the major reasons for the cancer MDR [2]. Many P-gp substrates, such as paclitaxel, were expelled out of the cancer cells, resulting in the reduction of intracellular drug accumulation, thereby leading to the treatment failure [3]. Therefore, it is urgent to explore a more effective strategy for overcoming the cancer MDR.

Paclitaxel (PTX), a water-insoluble compound, is used widely as a fist-line drug in clinical treatment against variety of cancers [4]. PTX is commonly formulated as Taxol®, which uses Cremophor EL and dehydrated ethanol (50:50, V/V) as delivery vehicles to enhance its solubility. But this formulation often causes side effects, such as hypersensitivity, neuropathy, and neurotoxicity, which associate highly with Cremophor EL [5]. To mitigate these side effects, Abraxane®, a new formulation, was developed by using the high affinity between paclitaxel and serum albumin to prepare paclitaxel/albumin nanocomplex [5]. This formulation was approved by Food and Drug Administration in 2005. However, its anticancer effect is still greatly affected by cancer MDR.

P-gp inhibitors have been studied for over P-gp mediated drug efflux, such as verapamil, dexverapamil, and tariquidar [6]. These P-gp inhibitors have been evaluated in clinic, but have not exhibited a good improvement in the therapeutic efficiency. These failures were mainly ascribed to the undesired toxicities, which have urged us to seek for new, more effective compounds with low toxicity and fewer side effects. Vitamin E (VE) is a lipid-soluble antioxidant, which protects lipids and membranes from oxidative damage [7]. It was reported that VE were not only able to overcome MDR by inhibition of ATPase activity, but also did not consider the toxicity [8,9]. In addition, water insoluble anticancer drugs, like PTX, can be loaded well in the VE-based emulsion for parental delivery [10]. Therefore, the use of VE as a P-gp inhibitor will be an attractive candidate to overcome the cancer MDR. However, VE is water-insoluble, which affects its administration in clinic.

Nano-drug delivery systems have been extensively investigated for anti-cancer drug delivery [11–21]. The nanocarriers could not only increase the water solubility of drugs but circumvent the P-gp efflux pump with entering the cancer cells by an endocytosis process [22]. Moreover, the nanoparticles provided a promising strategy for co-delivery of multiple drugs, in a single carrier, to improve the therapeutic efficiency of cancers [23–28]. Various nanocarriers have been developed for the co-delivery of anti-cancer drugs and P-gp inhibitors, such as the co-delivery of paclitaxel and borneol in lipid-albumin nanocomplex [29–31], docetaxel and verapamil in polymeric micelles [32], and paclitaxel and curcumin in lipid-albumin hybrid nanoparticles [33]. Additionally, VE is a lipid-soluble oil, which could be well encapsulated by the albumin to fabricate the VE-Albumin core-shell nanoparticles, through a hydrophobic interaction; VE and albumin could also make interactions with PTX. These interactions are beneficial for increasing the PTX-loading efficiency. Therefore, we speculate that co-delivery of the PTX and the VE with the VE-Albumin core-shell nanoparticles could improve the therapeutic efficiency of PTX against MDR cancers.

In this study, bovine serum albumin was used as a carrier to fabricate the VE-albumin core-shell nanoparticles co-delivery of the PTX and the VE. VE as the oil core of nanoparticles, not only increases the PTX-loading efficiency but also overcomes the P-gp-mediated drug efflux. The physicochemical properties and in vitro release were characterized. The cytotoxicity and cellular uptake were also investigated. Moreover, the anti-cancer effect was evaluated in breast cancer xenografts, in mice. It was speculated that the VE-albumin core-shell nanoparticles would be a suitable drug delivery system for anticancer drug delivery to over MDR, in cancer.

2. Materials and Methods

2.1. Materials and Animals

Paclitaxel was obtained from the Tianfeng Bioengineering Technology Co., Ltd. (Shenyang, China). Vitamin E and bovine serum albumin was obtained from Sigma-Aldrich (St. Louis, MO, USA). Cremophor EL was obtained from BASF Corporation (Ludwigshafen, Germany). 3-(4,5-Dimethyl-thiazol-2-yl)-2,5-diphenyl-tetrazolium bromide (MTT) was obtained from Sigma (St. Louis, MO, USA). RPMI-1640 and fetal bovine serum were obtained from Gibco (BRL, Gaithersburg, MD, USA). All other chemicals and solvents were of analytical or chromatographic grade and were used without further purification.

The PTX solution was prepared according to the clinical formulation. In brief, PTX (0.012 g) was dissolved in anhydrous ethanol (1 mL) and Cremophor EL (1 mL), under magnetic stirring. PTX solution was diluted with saline, before the test.

Balb/c mice (16–18 g) were obtained from the Experimental Animal Center (Nantong University, China). All animal experiments were approved by Nantong University Ethics Committee (20180512-001) and conformed to the Guidelines for the Use of Laboratory Animals.

2.2. Preparation of PTX NPs and PTX-VE NPs

The PTX NPs and PTX-VE NPs were fabricated by a desolvation-ultrasonication technique, as described in our previous report, with some modifications [29]. In brief, BSA (0.20 g) was dissolved in 5 mL deionized water, with magnetic stirring. VE (0.01 g) and PTX (0.01 g) were dissolved in

0.3 mL anhydrous ethanol. Then, the VE and PTX mixed solution were added dropwise to the BSA solution, with magnetic stirring. The mixtures were dispersed by probe ultrasonication (JY92-II, Ningbo Scientz Biotechnology Co., Ltd., Ningbo, China) at 400 W, for 4 min, in an ice bath with a 3 s pulse-on period and a 1 s pulse-off period. After sonication, anhydrous ethanol was evaporated by a rotator RE-2000 (Ya Rong Biochemical Instrument Factory, Shanghai, China). Subsequently, the samples were centrifuged at 3000 rpm, for 10 min, to remove the unloaded drug and impurities, and passed through a 0.45 μm filter membrane for removing the larger particles. The obtained suspensions were kept at 4 °C.

2.3. Characterization of PTX-VE NPs

2.3.1. Size Distribution and Morphology

The particle size and polydispersity index (P.I.) of the NPs was determined by dynamic light scattering (PSS NICOMP 380, Santa Barbara, CA, USA). The morphology of NPs was evaluated by transmission electron microscopy (TEM) (JEOL, Tokyo, Japan). In brief, the samples were diluted with distilled water and dropped onto a copper grid. The excess sample was removed with filter paper. Then, 2% phosphotungstic acid staining solution was dropped onto the grid. Finally, the sample was air-dried and assessed with TEM.

2.3.2. Determination of Entrapment Efficiency (EE) and Loading Capacity (LC)

The EE and LC were determined by the method described in our previous research [29]. To separate PTX from the PTX-VE NPs, acetonitrile was added to precipitate BSA, via sonication, for 5 min. After centrifugation at 12,000 rpm for 10 min, the PTX concentration in the supernatant was determined by HPLC. The drug encapsulation efficiency (EE) and loading capacity (LC) were calculated as follows:

$$EE(\%) = \frac{W_{drug\ in\ NPs}}{W_{total\ drug}} \times 100 \tag{1}$$

$$LC(\%) = \frac{W_{drug\ in\ NPs}}{W_{excipients\ and\ drug}} \times 100 \tag{2}$$

2.4. Docking Studies

Geometry of VE was optimized in gas phase. The calculation was conducted using the GAMESS suit of codes with the hybrid functional B3LYP. The 6-31G (dp) basis set was used for all the elements.

Docking studies were executed using Autodock Vina.9 [27]. BSA crystalline protein structure was obtained from protein databank (Bovine serum albumin: 4OR0, http://www.rcsb.org). The structure of VE used for docking studies was optimized with DFT calculations. Protein structures were altered to include polar hydrogen atoms. During docking studies, the protein structure was kept rigid. Rotation in the VE complex and the PTX complex was permitted for all single bonds.

2.5. In Vitro Drug Release

The release behavior of PTX from the PTX-VE NPs was evaluated by a dialysis method. The phosphate buffered saline (PBS, pH 7.4) containing 0.5% w/v Tween 80 was used as the dissolution medium. In brief, the samples were suspended in a flask and then immersed in the dissolution medium at 37 °C under at 120 rpm. The amount of PTX released was determined by HPLC.

2.6. Cytotoxicity of PTX-VE NPs

The MCF-7 and MCF-7/ADR cell line were purchased from Nanjing Kaiji Biotech. Ltd. Co. (Nanjing, China). The cells were cultured in RPMI medium, supplemented with 10% FBS and 1% penicillin-streptomycin, at 37 °C, with 5% CO_2 and 95% relative humidity. The Cells were seeded in 96-well plates, at a density of 5.0×10^3 cells/well. After 48 h incubation, the growth medium was replaced with 200 µL medium containing VE, PTX solution, and PTX-VE NPs with different concentrations, respectively. Then, each well was added with 20 µL MTT (5 mg/mL) solution and was incubated for an additional 4 h. The culture medium was removed and 200 µL DMSO was added to each well to dissolve the formazan. The absorbance at 492 nm was measured in a microplate reader (Model 500, San Francisco, CA, USA). The results were expressed as % cell viability (OD of treated group/OD of control group $\times 100$).

2.7. Rhodamine 6G Accumulation in MCF-7/ADR Cells

MCF-7/Adr cells were seeded onto the cover glasses at 1×10^5 cells in 6-well plate, for 24 h. The cells were washed and incubated with Rhodamine 6G solution (Rho solution), Rho NPs, and Rho-VE NPs, at 37 °C, for 2 h. Then the cells were washed twice with 4 °C PBS and fixed with 4% paraformaldehyde, for 20 min. The nuclei were counterstained by 4',6-diamidino-2-phenylindole (DAPI). The Rho 6G fluorescence was visualized by confocal laser scanning microscope (CLSM, TCS SP2/AOBS, LEICA, Bensheim, Germany).

2.8. Therapeutic Efficacy in Resistant Breast Cancer Xenografts Mice

The anti-tumor effect was evaluated in MCF-7/ADR tumor bearing Balb/c mice model. MCF-7/ADR cells were injected subcutaneously, at 2×10^7 cells, in the armpit of the right anterior limb. When the tumor sizes reached about 100 mm^3, the mice were randomly divided into four groups (n = 6), and the formulations of PTX solution, PTX NPs, and PTX-VE NPs were intravenously administrated at a dose of 10 mg/kg, at two-day intervals, for five times, with physiological saline as control. Tumor volumes and Body weights were measured with a caliper every other day. Tumor volume (V) was determined by the following formula:

$$V = \frac{\pi a b^2}{6} \tag{3}$$

where a and b represent the long and short axis of tumor, respectively.

2.9. Statistical Analysis

Analysis of statistical significance was performed with the SPSS statistics software 16.0. The data are presented as the mean $\pm$ SD. Student's *t*-test was used to analyze the differences. The differences were considered significant at $p < 0.05$.

3. Results and Discussion

3.1. Preparation and Characterization of PTX-VE NPs

The PTX-VE NPs were fabricated by the desolvation-ultrasonication method. The PTX and VE mixed ethanol solution was added to the BSA solution, and the mixture was subjected to probe sonication. Subsequently, the PTX-VE NPs was formed by the interaction among the PTX, VE, and BSA. Both PTX and VE have high affinity with albumin [34,35], the PTX and VE can bound tightly to BSA via hydrophobic interactions and hydrogen bond, which were in favor of the formation of PTX-VE NPs.

The average size of the PTX NPs and PTX-VE NPs was approximately 100 nm (Table 1), which was considered effective for the accumulation of nanoparticles, in tumor tissue, via passive targeting. The PTX encapsulation in the PTX-VE NPs was increased significantly with a loading capacity > 12%, which was five-fold more than that of PTX NPs, owing mainly to the solubility of PTX in the VE oil. The NPs also have a higher drug encapsulation efficiency (>90%). The morphology of the NPs was evaluated by TEM, which showed a uniform and spherical shape (Figure 1).

Table 1. Physicochemical characteristics of PTX NPs and PTX-VE NPs (n = 3).

Formulation	Size (nm)	P.I.	Zeta Potential (mV)	EE (%)	LC (%)
PTX NPs	101.2 ± 2.8	0.167 ± 0.03	−2.15 ± 0.6	91.2 ± 3.0	2.5 ± 0.08
PTX-VE NPs	106.9 ± 3.2	0.172 ± 0.02	−20.64 ± 0.8	95.7 ± 2.1	12.5 ± 0.15

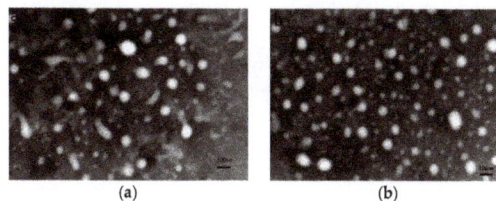

(a)   (b)

Figure 1. TEM image of PTX NPs (a) and PTX-VE NPs (b).

3.2. Docking Studies

Docking studies were performed to investigate the interaction between PTX and BSA, VE and BSA, respectively (Figure 2). The crystalline structure of BSA was obtained from the protein databank. In thermodynamics, a negative binding affinity demonstrates a favorable interaction system. The calculated binding affinity of PTX and VE with BSA were −8.3 kcal/mol and −7.5 kcal/mol, respectively, indicating that PTX and VE have strong interaction with BSA. As shown in Figure 2, there were hydrogen bonds between the Glu 424, Ser 109, Lys 114 of BSA and the nitrogen atom, oxygen atom, oxygen atom of PTX, respectively. There also existed hydrogen bonds between the Arg144, Asp108 and the hydroxy of the VE, respectively. Moreover, hydrophobic interactions were another important interaction between the PTX/VE and the BSA. The hydrophobic interactions were established between the Ile 522, Leu 189, Val 423, Ala 193, and PTX; these interactions also occurred between the Ile 455, Val 425, Leu189, Ala 193, and VE. In addition, the hydrophobic interactions and hydrogen bonds between the VE and the PTX cannot be ignored. In summary, the intermolecular force, such as hydrophobic interactions and hydrogen bonds played significantly important role in the formation of the PTX-VE NPs.

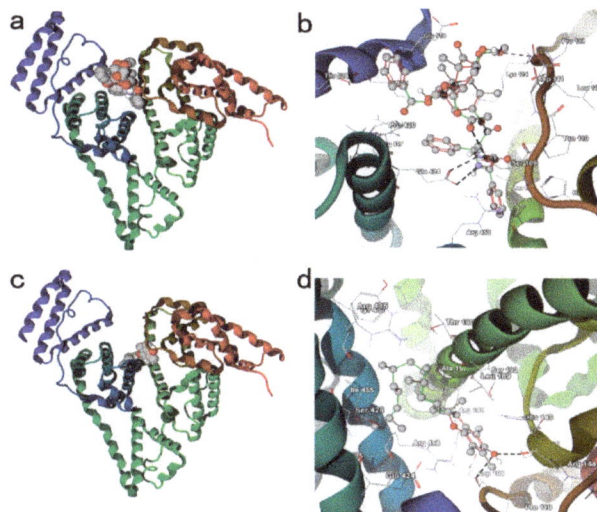

Figure 2. The molecular docking BSA with PTX (a,b) and VE (c,d).

3.3. In Vitro Drug Release

The profiles of the PTX release from the NPs are showed in Figure 3. Nearly 90% of the PTX was released from the PTX solution, in 8 h. Both, PTX NPs and PTX-VE NPs exhibited a sustained release profile, as compared with the PTX solution. After 24 h, the PTX release was about 76% and 60% for the PTX NPs and PTX-VE NPs, respectively. The PTX release from the PTX-VE NPs was slower than that of the PTX NPs. The release of PTX from the NPs might be affected by interactions among PTX, VE, and albumin, such as Hydrogen bonds, π-π stacking, and hydrophobic interactions. In addition, the relative rapid release of PTX from NPs, in the early stage, was likely due to the diffusion of drugs absorbed at the outer shell, while the sustained release in the late stage was probably related to the gradual diffusion of drugs in the inner core of NPs.

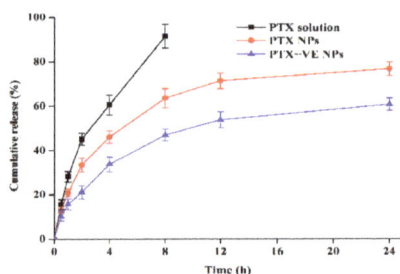

Figure 3. In vitro release profile of PTX from PTX solution, PTX NPs, and PTX-VE NPs, in phosphate buffered saline (0.5% of Tween 80 in PBS, pH 7.4), at 37 ± 0.5 °C (n = 3).

3.4. Cytotoxicity of PTX-VE NPs

The cytotoxicity study of the PTX solution, PTX NPs, and the PTX-VE NPs was evaluated using the MTT assay and the results are show in Figure 4. It is clear that cell viabilities for all the formulations in MCF-7 cells were lower than that at concentration 1–100 µg/mL in MCF-7/ADR cells, due to P-gp-mediated efflux, which could reduce the drug accumulation in the cells. The PTX-VE NPs exhibited better cytotoxic effect than the PTX NPs, indicating that the PTX-VE NPs could exert stronger MDR-overcoming effects on the MCF-7/ADR cells. In addition, the cytotoxicity of the PTX-VE NPs showed no significant difference, as compared with the PTX solution. The reason was deduced that the cytotoxicity of the PTX solution was partially attributed to the use of Cremophor EL/ethanol mixture, which had 42.1% cytotoxicity on the MCF-7 cells [36]. However, in the case of the PTX-VE NPs, the nanocarriers are biocompatible, and the cytotoxic effect was induced mainly by the PTX incorporated in the nanoparticles [37].

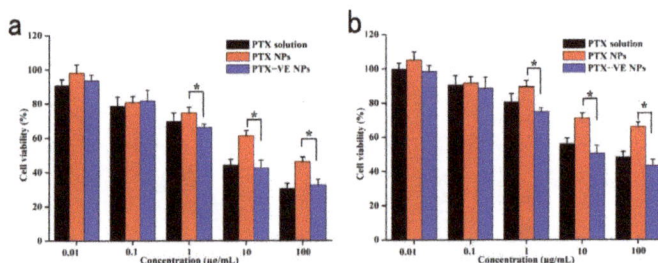

Figure 4. In vitro cytotoxicity of PTX solution, PTX NPs, and PTX-VE NPs, against MCF cells (**a**) and MCF-7/ADR cells (**b**). Data represented the mean $\pm$ S.D. (n = 3). * $p < 0.05$, significant difference.

3.5. Rhodamine 6G Accumulation in MCF-7/ADR Cells

As a substrate of the P-gp efflux pump, the Rhodamine 6G (Rho) was often used as a fluorescent probe to evaluate the whether the improved cytotoxicity of the PTX-VE NPs was due to the increased intracellular drug delivery in MCF-7/ADR cells. Figure 5 shows the fluorescence images in the MCF-7/ADR cells, after 1 h, following the different Rho-labeled formulation treatments. The cells treated with Rho solution exhibited the least amounts of fluorescence signals, which could be attributed to the uptake inhibition of Rho by the P-gp efflux transporter, overexpressed in the MCF-7/ADR cells. Cells treated with the Rho NPs showed increased fluorescence signals, in comparison with the Rho solution. This was consistent with the research that the nanocarriers were able to circumvent the P-gp efflux pump by entering the cells through an endocytic process. However, the increased cell uptake of the Rho NPs was still restricted. The reason inferred was that the rapid release of encapsulated drugs in the cytoplasm was still probably shuttled out of cells, by the P-gp, which is overexpressed on the MCF-7/ADR cell membranes. Furthermore, the strongest fluorescence signals were observed in the cells treated with the Rho-VE NPs. This could be ascribed to the synergic combination of the nanocarrier and the VE. The nanocarrier increased the intracellular drug delivery by circumventing the P-gp efflux pump and the P-gp ATPase activity was probably inhibited by the released VE from the nanoparticles in cells.

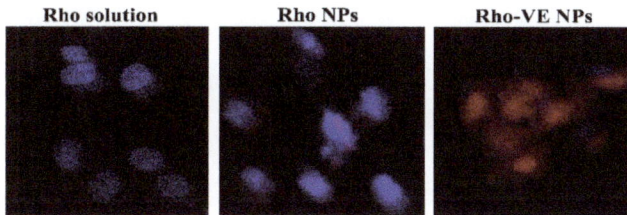

Rho solution **Rho NPs** **Rho-VE NPs**

Figure 5. Confocal laser scanning microscope images of various Rhodamine 6G (Rho) labeled formulations in the MCF-7/ADR cells. Original magnification: 10 × 20.

3.6. Anti-Tumor Effect and Safety in the Xenograft Model

The anti-tumor effect of the PTX-VE NPs was studied in the Balb/c mice bearing the MCF-7/ADR tumor xenograft model. The changes of tumor value and body weight were shown in Figure 6a, the tumor growth rates of different formulations, in ascending order, was PTX-VE NPs, PTX NPs, PTX solution and Saline, respectively. The results indicated that the PTX-VE NPs were the most effective formulation to treat MDR breast cancer. During the treatment, the relative body weight change of mice was also measured, as shown in Figure 6b. No significant body weight loss was observed for the PTX-VE NPs, PTX NPs, PTX solution and Saline, throughout the experiment, demonstrating good safety and low systemic toxicity of these formulations.

Accordingly, PTX-VE NPs was a safe and efficient drug delivery system to overcome MDR. The reasons could be deduced to the following aspects, as shown in Figure 7. First, the PTX-VE NPs with particle size less than 200nm, which was able to avoid the rapid reticuloendothelial system uptake and achieved a high accumulation in the tumor site by an enhanced permeability and retention (EPR) effect [35]. Second, the NPs could utilize the albumin receptor (gp60) mediated endocytosis through the blood vessel endothelial cells in the angiogenic tumor vasculature and an increased intratumoral accumulation [10,26]. Finally, the combination of the PTX and the VE in the NPs, would significantly enhance the intracellular drug uptake efficiency in MDR tumor cells and improve the therapeutic efficiency of resistant cancers.

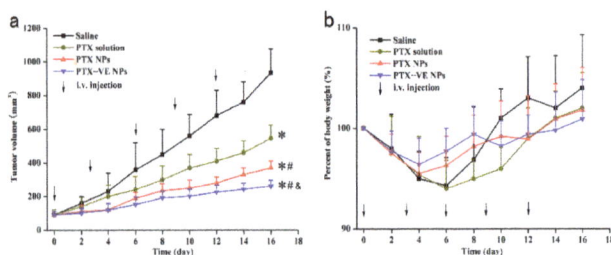

Figure 6. Anti-tumor effect of PTX formulations on the Balb/c mice implanted with MCF-7/ADR cells. Mice were injected through tail vein with Saline, PTX solution, PTX NPs and PTX-VE NPs, at a dosage of 10 mg/kg. The changes of tumor volume (**a**) and body weight (**b**) during 16-day treatments. (n = 6). * $p < 0.05$, significant difference when compared with Saline, # $p < 0.05$, significant difference when compared with PTX solution. & $p < 0.05$, significant difference when compared with PTX NPs.

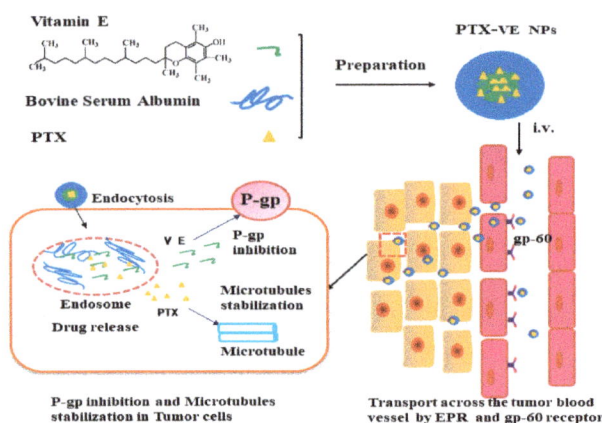

Figure 7. Schematic illustration of the PTX-VE NPs to reversal of MDR.

4. Conclusions

The present study fabricated the PTX loaded VE-Albumin core-shell nanoparticles to treat MDR breast cancer. The fabricated nanoparticles showed satisfying particle size, drug loading, and sustained-release profile, in vitro. Compared with PTX NPs, PTX-VE NPs exhibited higher cellular uptake and stronger cytotoxic effect in MDR breast cancer cells. Therefore, the PTX loaded VE-Albumin core-shell nanoparticles could be a potential strategy for the treatment of MDR breast cancer.

Author Contributions: Conceptualization, B.T. and G.C.; methodology, G.H.F.; software, Y.G.; validation, G.H.F.; formal analysis, GH.F. and B.T.; investigation, G.H.F. and Y.Q.; data curation, Y.Q.; writing—original draft preparation, B.T.; writing—review and editing, G.H.F.; funding acquisition, B.T.

Funding: This work was supported by the Natural Science Fund for Colleges and Universities in Jiangsu Province, No. 17KJB350009 and Natural Science Foundation of Jiangsu Province, No. BK20170445.

Conflicts of Interest: The authors declare no conflict of interest.

References

1. Panchagnula, R. Pharmaceutical aspects of paclitaxel. *Int. J. Pharm.* **1998**, *172*, 1–15. [CrossRef]
2. Sofias, A.M.; Dunne, M.; Storm, G.; Allen, C. The battle of "nano" paclitaxel. *Adv. Drug Deliv. Rev.* **2017**, *122*, 20–30. [CrossRef] [PubMed]

3. Palmeira, A.; Sousa, E.; Vasconcelos, M.H.; Pinto, M.M. Three decades of P-gp inhibitors: Skimming through several generations and scaffolds. *Curr. Med. Chem.* **2012**, *19*, 1946–2025. [CrossRef] [PubMed]

4. Galli, F.; Azzi, A.; Birringer, M.; Cook-Mills, J.M.; Eggersdorfer, M.; Frank, J.; Cruciani, G.; Lorkowski, S.; Özer, N.K. Vitamin E: Emerging aspects and new directions. *Free Radic. Biol. Med.* **2017**, *102*, 16–36. [CrossRef] [PubMed]

5. Meydani, S.N.; Meydani, M.; Rall, L.C.; Morrow, F.; Blumberg, J.B. Assessment of the safety of high-dose, short-term supplementation with vitamin E in healthy older adults. *Am. J. Clin. Nutr.* **1994**, *60*, 704–709. [CrossRef] [PubMed]

6. Tang, J.; Fu, Q.; Wang, Y.; Racette, K.; Wang, D.; Liu, F. Vitamin E reverses multidrug resistance in vitro and in vivo. *Cancer Lett.* **2013**, *336*, 149–157. [CrossRef] [PubMed]

7. Constantinides, P.P.; Tustian, A.; Kessler, D.R. Tocol emulsions for drug solubilization and parenteral delivery. *Adv. Drug Deliv. Rev.* **2004**, *56*, 1243–1255. [CrossRef] [PubMed]

8. Gou, J.; Feng, S.; Liang, Y.; Fang, G.; Zhang, H.; Yin, T.; Zhang, Y.; He, H.; Wang, Y.; Tang, X. Polyester–solid lipid mixed nanoparticles with improved stability in gastro-intestinal tract facilitated oral delivery of larotaxel. *Mol. Pharm.* **2017**, *14*, 3750–3761. [CrossRef] [PubMed]

9. Gou, Y.; Zhang, Y.; Zhang, Z.; Wang, J.; Zhou, Z.; Liang, H.; Yang, F. Design of an anticancer copper(II) prodrug based on the Lys199 residue of the active targeting human serum albumin nanoparticle carrier. *Mol. Pharm.* **2017**, *14*, 1861–1873. [CrossRef] [PubMed]

10. Wang, F.; Chen, L.; Zhang, R.; Chen, Z.; Zhu, L. RGD peptide conjugated liposomal drug delivery system for enhance therapeutic efficacy in treating bone metastasis from prostate cancer. *J. Control. Release* **2014**, *196*, 222–233. [CrossRef] [PubMed]

11. Zhu, H.; Cao, J.; Cui, S.; Qian, Z.; Gu, Y. Enhanced tumor targeting and antitumor efficacy via hydroxycamptothecin-encapsulated folate-modified N-Succinyl-N'-Octyl chitosan micelles. *J. Pharm. Sci.* **2013**, *102*, 1318–1332. [CrossRef] [PubMed]

12. Zhu, H.; Zhang, S.; Ling, Y.; Meng, G.; Yang, Y.; Zhang, W. pH-responsive hybrid quantum dots for targeting hypoxic tumor siRNA delivery. *J. Control. Release* **2015**, *220*, 529–544. [CrossRef] [PubMed]

13. Han, Y.-H.; Kankala, R.; Wang, S.-B.; Chen, A.-Z. Leveraging engineering of indocyanine green-encapsulated polymeric nanocomposites for biomedical applications. *Nanomaterials* **2018**, *8*, 360. [CrossRef] [PubMed]

14. Kankala, R.K.; Chen, B.-Q.; Liu, C.-G.; Tang, H.-X.; Wang, S.-B.; Chen, A.-Z. Solution-enhanced dispersion by supercritical fluids: An ecofriendly nanonization approach for processing biomaterials and pharmaceutical compounds. *Int. J. Nanomed.* **2018**, *13*, 4227–4245. [CrossRef] [PubMed]

15. Kankala, R.K.; Zhang, Y.S.; Wang, S.B.; Lee, C.H.; Chen, A.Z. Supercritical Fluid technology: An emphasis on drug delivery and related biomedical applications. *Adv. Healthc. Mater.* **2017**, *6*. [CrossRef] [PubMed]

16. Fang, G.; Tang, B.; Chao, Y.; Xu, H.; Gou, J.; Zhang, Y.; Xu, H.; Tang, X. Cysteine-Functionalized nanostructured lipid carriers for oral delivery of docetaxel: A permeability and pharmacokinetic study. *Mol. Pharm.* **2015**, *12*, 2384–2395. [CrossRef] [PubMed]

17. Fang, G.; Tang, B.; Chao, Y.; Zhang, Y.; Xu, H.; Tang, X. Improved oral bioavailability of docetaxel by nanostructured lipid carriers: In vitro characteristics, in vivo evaluation and intestinal transport studies. *RSC Adv.* **2015**, *5*, 96437–96447. [CrossRef]

18. Fang, G.; Tang, B.; Liu, Z.; Gou, J.; Zhang, Y.; Xu, H.; Tang, X. Novel hydrophobin-coated docetaxel nanoparticles for intravenous delivery: In vitro characteristics and in vivo performance. *Eur. J. Pharm. Sci.* **2014**, *60*, 1–9. [CrossRef] [PubMed]

19. Markman, J.L.; Rekechenetskiy, A.; Holler, E.; Ljubimova, J.Y. Nanomedicine therapeutic approaches to overcome cancer drug resistance. *Adv. Drug Deliv. Rev.* **2013**, *65*, 1866–1879. [CrossRef] [PubMed]

20. Gou, Y.; Zhang, Z.; Li, D.; Zhao, L.; Cai, M.; Sun, Z.; Li, Y.; Zhang, Y.; Khan, H.; Sun, H.; et al. HSA-based multi-target combination therapy: Regulating drugs' release from HSA and overcoming single drug resistance in a breast cancer model. *Drug Deliv.* **2018**, *25*, 321–329. [CrossRef] [PubMed]

21. Khan, I.U.; Khan, R.U.; Asif, H.; Alamgeer; Khalid, S.H.; Asghar, S.; Saleem, M.; Shah, K.U.; Shah, S.U.; Rizvi, S.A.A.; et al. Co-delivery strategies to overcome multidrug resistance in ovarian cancer. *Int. J. Pharm.* **2017**, *533*, 111–124. [CrossRef] [PubMed]

22. Qi, S.-S.; Sun, J.-H.; Yu, H.-H.; Yu, S.-Q. Co-delivery nanoparticles of anti-cancer drugs for improving chemotherapy efficacy. *Drug Deliv.* **2017**, *24*, 1909–1926. [CrossRef] [PubMed]

23. Kankala, R.K.; Tsai, P.-Y.; Kuthati, Y.; Wei, P.-R.; Liu, C.-L.; Lee, C.-H. Overcoming multidrug resistance through co-delivery of ROS-generating nano-machinery in cancer therapeutics. *J. Mater. Chem. B* **2017**, *5*, 1507–1517. [CrossRef]

24. Xu, P.-Y.; Kankala, R.K.; Pan, Y.-J.; Yuan, H.; Wang, S.-B.; Chen, A.-Z. Overcoming multidrug resistance through inhalable siRNA nanoparticles-decorated porous microparticles based on supercritical fluid technology. *Int. J. Nanomed.* **2018**, *13*, 4685–4698. [CrossRef] [PubMed]

25. Kankala, R.K.; Liu, C.-G.; Chen, A.-Z.; Wang, S.-B.; Xu, P.-Y.; Mende, L.K.; Liu, C.-L.; Lee, C.-H.; Hu, Y.-F. Overcoming multidrug resistance through the synergistic effects of hierarchical pH-sensitive, ROS-generating nanoreactors. *ACS Biomater. Sci. Eng.* **2017**, *3*, 2431–2442. [CrossRef]

26. Tang, B.; Fang, G.; Gao, Y.; Liu, Y.; Liu, J.; Zou, M.; Cheng, G. Liprosomes loading paclitaxel for brain-targeting delivery by intravenous administration: In vitro characterization and in vivo evaluation. *Int. J. Pharm.* **2014**, *475*, 416–427. [CrossRef] [PubMed]

27. Tang, B.; Fang, G.; Gao, Y.; Liu, Y.; Liu, J.; Zou, M.; Cheng, G. Co-encapsulation of borneol and paclitaxel by liprosomes improved anti-tumor effect in a xenografted glioma model. *RSC Adv.* **2015**, *5*, 106613–106620. [CrossRef]

28. Tang, B.; Fang, G.; Gao, Y.; Liu, Y.; Liu, J.; Zou, M.; Wang, L.; Cheng, G. Lipid-albumin nanoassemblies co-loaded with borneol and paclitaxel for intracellular drug delivery to C6 glioma cells with P-gp inhibition and its tumor targeting. *Asian J. Pharm. Sci.* **2015**, *10*, 363–371. [CrossRef]

29. Guo, Y.; He, W.; Yang, S.; Zhao, D.; Li, Z.; Luan, Y. Co-delivery of docetaxel and verapamil by reduction-sensitive PEG-PLGA-SS-DTX conjugate micelles to reverse the multi-drug resistance of breast cancer. *Colloids Surf. B Biointerfaces* **2017**, *151*, 119–127. [CrossRef] [PubMed]

30. Ruttala, H.B.; Ko, Y.T. Liposomal co-delivery of curcumin and albumin/paclitaxel nanoparticle for enhanced synergistic antitumor efficacy. *Colloids Surf. B Biointerfaces* **2015**, *128*, 419–426. [CrossRef] [PubMed]

31. Hu, L.; Chen, D.-Y. Application of headspace solid phase microextraction for study of noncovalent interaction of borneol with human serum albumin. *Acta Pharm. Sin.* **2009**, *30*, 1573–1576. [CrossRef] [PubMed]

32. Paal, K.; Müller, J.; Hegedûs, L. High affinity binding of paclitaxel to human serum albumin. *Eur. J. Biochem.* **2001**, *268*, 2187–2191. [CrossRef] [PubMed]

33. Gelderblom, H.; Verweij, J.; Nooter, K.; Sparreboom, A. Cremophor EL: The drawbacks and advantages of vehicle selection for drug formulation. *Eur. J. Cancer* **2001**, *37*, 1590–1598. [CrossRef]

34. Nosrati, H.; Abbasi, R.; Charmi, J.; Rakhshbahar, A.; Aliakbarzadeh, F.; Danafar, H.; Davaran, S. Folic acid conjugated bovine serum albumin: An efficient smart and tumor targeted biomacromolecule for inhibition folate receptor positive cancer cells. *Int. J. Biol. Macromol.* **2018**, *117*, 1125–1132. [CrossRef] [PubMed]

35. Alexis, F.; Pridgen, E.; Molnar, L.K.; Farokhzad, O.C. Factors affecting the clearance and biodistribution of polymeric nanoparticles. *Mol. Pharm.* **2008**, *5*, 505–515. [CrossRef] [PubMed]

36. Wu, Q.; Yang, Z.; Nie, Y.; Shi, Y.; Fan, D. Multi-drug resistance in cancer chemotherapeutics: Mechanisms and lab approaches. *Cancer Lett.* **2014**, *347*, 159–166. [CrossRef] [PubMed]

37. Krishna, R.; Mayer, L.D. Multidrug resistance (MDR) in cancer: Mechanisms, reversal using modulators of MDR and the role of MDR modulators in influencing the pharmacokinetics of anticancer drugs. *Eur. J. Pharm. Sci.* **2000**, *11*, 265–283. [CrossRef]

Sample Availability: Samples are not available from the authors.

![molecules logo] *molecules*

MDPI

Article

Ultrasensitive (Co)polymers Based on Poly(methacrylamide) Structure with Fining-Tunable pH Responsive Value

Haiming Fan [1], Po Li [2], Wei Li [2], Hui Li [1] and Xiaonan Huang [2,*]

[1] Shandong Provincial Key Laboratory of Oilfield Chemistry, School of Petroleum Engineering, China University of Petroleum (East China), Qingdao 266580, China; Haimingfan@126.com (H.F.); lihui0714_upc@163.com (H.L.)
[2] Department of Chemistry, Capital Normal University, 105 West 3rd Ring North Rd, Beijing 100048, China; 2150702055@cnu.edu.cn (P.L.); 2120702055@cnu.edu.cn (W.L.)
* Correspondence: huangxn@cnu.edu.cn

Academic Editors: Jianxun Ding, Yang Li and Mingqiang Li
Received: 25 June 2018; Accepted: 15 July 2018; Published: 27 July 2018

Abstract: Novel pH responsive copolymers with tertiary amine groups were prepared by free radical polymerization with 2-(dialkylamino)ethyl methacrylate monomers. These polymers were pH sensitive with the ability to be responsively fine-tuned in aqueous solution, which was proven through titration, transmittance measurements, and proton nuclear magnetic resonance spectroscopy. The polymers were soluble in water at low pH values, induced by electrostatic repulsion between amine groups, and aggregated above their pK_a value due to the hydrophobic effect of the alkyls. The pH responsive values were precisely tuned from 7.4 to 4.8 by increasing the hydrophobic monomer ratio. Our work provides a novel approach for the development of ultrasensitive pH-responsive polymers for application in biomedical materials.

Keywords: fine-tuning; pH responsive; poly(methacrylamide)s; phase transition

---

1. Introduction

The development of tumor-targeted drug delivery systems have attracted increasing interest given their potential to address therapeutic issues in clinical practice [1–7]. Many studies have focused on designing and improving specific transport and activation of intelligent materials that have stimuli responsive properties [8–13]. An interesting aspect of the physiology of tumors is their lower extra-cellular pH (6.5–7.2) compared to the surrounding normal tissues and blood, thus pH can be considered as an ideal trigger for tumor tissues and tumor cells [14]. To target the acidic bio-compartments, several smart materials with pH-responsive properties were designed to increase target efficacy. For instance, the Kobayashi group developed a small molecule system to target the tissue of the mice. This pH of this system can transition across two values of pH value, which means a 200-fold proton concentration change [15–17]. Despite these remarkable advances, selective targeting by pH-sensitive materials of different endocytic components is challenging owing to the small pH differences in these compartments, such as early endosomes (5.9–6.2) versus lysosomes (5.0–5.5) [18–20].

Various pH-responsive polymer systems have been studied, such as liposomes [21], micelles [22,23], nanoparticles [24], and nanogels [25–27]. The pH-responsive polymers can undergo phase transition at a specific pH values and direct spontaneous assemblies [28–31]. Polyelectrolytes, which contain weak acidic or basic groups, have been exploited as pH responsive polymers, inspiring advances in polymer research. Polybases, like poly(2-(diethylamino)ethyl methacrylate) (PDEAEMA), poly(4 or 2-vinylpyridine) (PVP), and poly(vinyl imidazole), were developed for biomedical applications in gene transfer, bio-imaging, and controlled drug release, due to their excellent physicochemical properties [32–34]. The amine group, pyridine, and imidazole group in polymer chains can accept protons under acidic condition and release them in basic environments to induce phase transition [35,36]. For instance, PDEAEMA dissolves in acidic aqueous solution due to the electrostatic repulsions between the protonated amine groups, but undergoes a precipitation above pH 7.5, caused by the hydrophobic effect of substituted ethyl group. However, a challenging issue faced by these widely used polybases is that their phase transition values from the solution state to the precipitation state are usually fixed. Therefore, they are less accurate when targeting cell organelles with different pH microenvironments. The broad pH transition range of these polybases also limits the increase in their responsive property. To develop responsive polymers with narrow transition pH ranges, Armes et al. [37] reported a novel polybase system, poly(2-(diisopropylamino)ethyl methacrylate) (PDPAEMA), with a phase transition at pH 6.3 and a sharp pH transition, providing a method by which to prepare ultra-sensitive pH-responsive polymers.

In this work, we developed a series of novel reversible pH-responsive (co)polymers with a fine-tunable pH transition value. These polymers were synthesized using a free radical polymerization method with tertiary amine-based monomers, 2-(dibutylamino)ethyl methacrylate and 2-(dimethylamino)ethyl methacrylate. At low pH, the (co)polymers were in the unimer state due to the electrostatic interaction and hydrophilicity of the protonated tertiary amine units. At high pH, the stronger hydrophobicity of the alkyl groups on the deprotonated amines led to (co)polymer aggregation. All the (co)polymers showed rapid phase separation at their pK_a, and the pH transition range can be controlled within a very narrowly defined value.

2. Results

2.1. Synthesis of the (Co)polymers

A series of random copolymers poly(2-(dimethylamino)ethyl methacrylate)-*co*-2-(dibutylamino) ethyl methacrylate (P(DMAEMA-*co*-DBAEMA)) were obtained by conventional free radical polymerization with varying feed ratios of the monomers containing different tertiary amine groups. Comparing the signal integrals of the proton peaks at 2.3 ppm and at 2.5 ppm (Figure 1, Figures S1–S4), the compositions of the copolymers could be determined, as summarized in Table 1. The molar ratios of the monomer units in the copolymers were relatively close to the feed ratios. Due to the methacrylic acid ester structure of the two monomers, random sequences of two monomer units were expected in the copolymer chains.

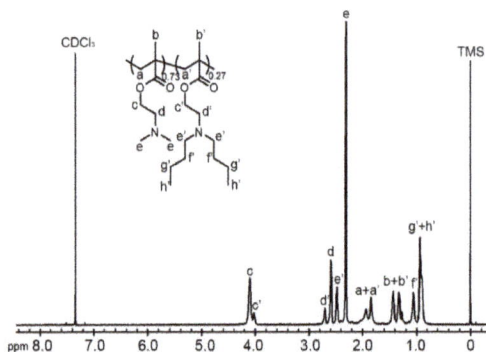

Figure 1. Proton nuclear magnetic resonance (^{1}H-NMR) spectrum of poly(2-(dimethylamino)ethyl methacrylate)-*co*-2-(dibutylamino)ethyl methacrylate (P(DMAEMA$_{0.73}$-*co*-DBAEMA$_{0.27}$)) in CDCl$_3$.

Table 1. Characterization of (co)polymers.

Copolymer	DMA Content		Yield (%)	M_n [b] ($\times 10^4$)	M_w [b] ($\times 10^4$)	PDI	pK_a [c]
	In Feed	In Polymer [a]					
PDMAEMA	100	100	73	1.21	1.85	1.53	7.4
P(DMAEMA$_{0.73}$-*co*-DBAEMA$_{0.27}$)	75	73	67	1.06	1.47	1.39	6.8
P(DMAEMA$_{0.51}$-*co*-DBAEMA$_{0.49}$)	50	51	62	1.01	1.84	1.82	6.3
P(DMAEMA$_{0.29}$-*co*-DBAEMA$_{0.71}$)	25	29	65	1.13	2.03	1.80	5.5
PDBAEMA	0	0	56	1.40	2.59	1.85	4.8

[a] Percent molar ratio determined by hydrogen nuclear magnetic resonance (^{1}H-NMR); [b] Determined by gel permeation chromatography (GPC) in tetrahydrofuran (THF) with polystyrene standards; [c] Determined by titration during the heating process.

2.2. pH Titration of Different Copolymers and Measurement of pKa

For pH-responsive polymers, the pK_a value is a critical parameter that can be evaluated through titration experiments. Figure 2A shows the potentiometric titration curves for 0.5 mg/mL (co)polymer solutions. NaOH solution was added to neutralize the protons. Potentiometric titration curves were produced for (co)polymers by plotting the solution pH against the volume of added NaOH. Starting from pH ~3, the addition of small aliquots of NaOH would neutralize the free protons and increase the solution pH. The pH value of PDMAEMA increased continuously with higher NaOH content, whereas for the other (co)polymers, especially PDBAEMA, titration curves were obtained with a plateau at the polymer buffering region. This indicates that the added NaOH was consumed during the deprotonation of the tertiary amine groups on the polymer pendent chain, which corresponds to the phase transition of the polymer. With the completion of the process, further addition of base increased the solution pH rapidly. For each sample, the pK_a values were calculated as the pH halfway between the two equivalence points in the titration curve. With increasing DBAEMA monomer amounts from 0% to 100%, as shown in Figure 2B, the average pK_a values of the polymers decreased from 7.4 to 4.8, which was attributed to the hydrophobic property of butyl on amine. An approximately linear relationship was observed between the pK_a values and DBAEMA monomer molar ratios of the (co)polymers, which indicates that the pK_a of the pH sensitive polycations can be controlled easily and precisely by adjusting the hydrophobic properties of the pendent group.

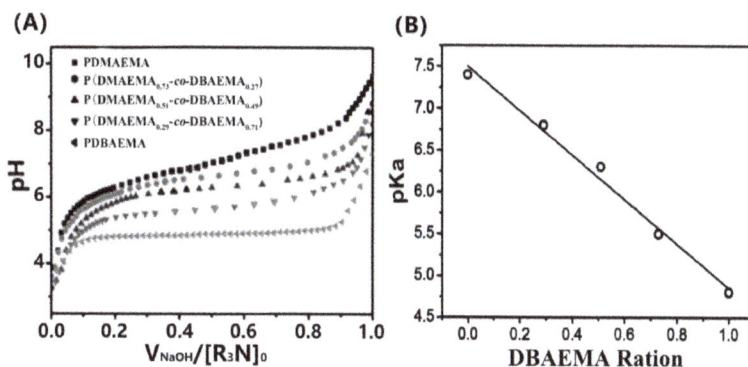

Figure 2. (**A**) pH titration curves of (co)polymers. The volumes of NaOH (V_{NaOH}) were normalized to the initial amount of amine ([R$_3$N]$_0$ in mmol); (**B**) pKa values of the tertiary amine-based polymers as a function of DBAEMA ratio.

2.3. Phase Transition Behaviors of the Fine-Tuning pH Responsive Polymers

The tenability of the (co)polymers' pH-responsive properties were evaluated by transmittance experiments. Figure 3A,B show the optical images and transmittance changes of the polymer solutions as NaOH was added from initial acid solutions, where all of the polymers were completely dissolved. Over the entire pH range in the measurements, the PDMAEMA solution remained clear with a transmittance around 100%, which was likely due to the hydrophilicity of the dimethyl amino groups, in good agreement with the titration experiment. However, with the incorporation of DBAEMA, the turbidity of the polymer solutions increased sharply at critical pH values due to the aggregation of the polymer chains. The phase transition pH range decreased with the increasing ratio of DBAEMA. Compared to conventional pH responsive materials, the DBAEMA-based polymers displayed a much sharper pH-dependent phase transition, where the transmittance from 100% to 0% was limited to 0.3 pH, indicating that only double the proton concentration could lead to complete turbidity of the copolymers in aqueous solutions. Furthermore, the reversibility of the pH-responsive tertiary amine-based polymers was confirmed by the PDBAEMA transmittance measurement. We also analyzed the polymer solution during cyclic tests between pH 4.0 and pH 7.4. As shown in Figure 3C, the dissolution and aggregation of the polymer solutions are fully reversible for at least five cycles, which indicates the novel pH sensitive materials can be used repeatedly.

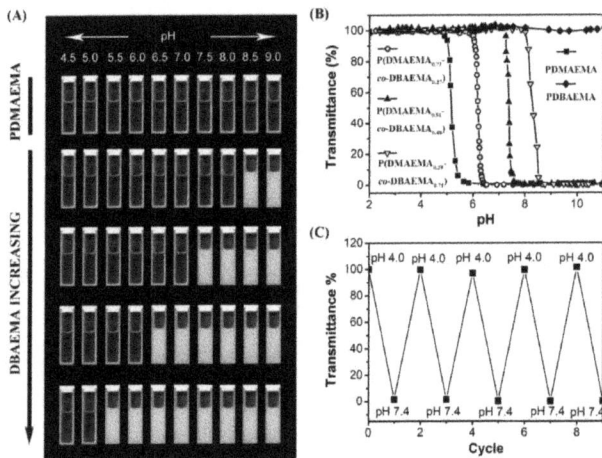

Figure 3. (**A**) Optical images of (co)polymer solutions at different pH. (**B**) The transmittance curves of (co)polymer solutions with increasing pH values. (**C**) The pH reversibility study of PDBAEMA solution with cycles between pH 4.0 and pH 7.4.

Proton nuclear magnetic resonance spectroscopy (^{1}H-NMR) was also employed to investigate the effect of pH on the phase transition behaviors of the polymers in aqueous solution. PDBAEMA and P(DMAEMA$_{0.73}$-co-DBAEMA$_{0.27}$) were investigated as a function of pH in order to compare the aggregation behavior of the polymers with different tertiary amine groups. 1,4-Dioxane, with a single peak at 3.7 ppm, was used as the internal standard, as its intensity remained almost the same before and after the polymer underwent a phase transition. The spectra of the PDBAEMA and P(DMAEMA$_{0.73}$-co-DBAEMA$_{0.27}$) were recorded in the pH range of 4.1 to 8.5, and Figure 4 shows the ^{1}H-NMR spectra of the polymers in D$_2$O with the tertiary amines in different ionization states. At pH 4.1 and 5.3, the amine groups were protonated; both PDBAEMA and P(DMAEMA$_{0.73}$-co-DBAEMA$_{0.27}$) were dissolved in the deuterated solution and proton resonance peaks for the segments of each polymer were easily visualized. However, as pH increased from 5.5 to 6.6, the signal intensities of the PDBAEMA dramatically decreased and disappeared at pH 7.8, when PDBAEMA became completely deprotonated. The suppressed resonances of the polymer were due to the limitation of the motion of the polymer chain, which can be attributed to the phase transition behavior driven by the hydrophobic effect. For P(DMAEMA$_{0.73}$-co-DBAEMA$_{0.27}$), although the signal integrations of the polymer signals decreased gradually when the pH increased above 6.3, the peaks of DMAEMA segments were still observable even at pH 8.5. This drastic difference between PDBAEMA and P(DMAEMA$_{0.73}$-co-DBAEMA$_{0.27}$) spectra can be attributed to the dehydration behaviors of the polymers upon neutralization. The hydrophobic effect involving the butyl groups on the PDBAEMA pendent segments enabled the polymer chains to pack densely in aqueous solution, thus leading to more efficient dehydration and a sharper phase transition.

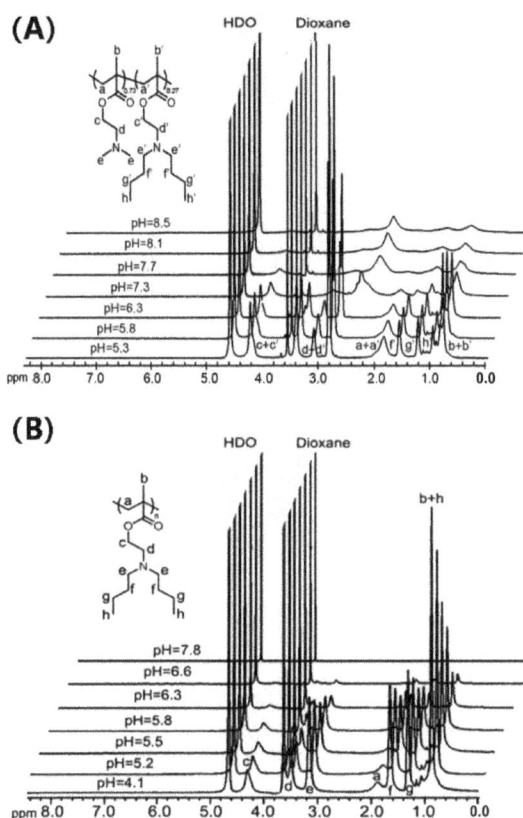

Figure 4. ^{1}H-NMR spectra of (**A**) P(DMAEMA$_{0.73}$-co-DBAEMA$_{0.27}$) and (**B**) PDBAEMA at different pH in D$_2$O.

3. Materials and Methods

3.1. Materials

2-(dibutylamino) ethanol was purchased from Alfa Aesar Company (Haverhill, MA, USA) and 2-(dimethylamino)ethyl methacrylate (DMAEMA) was purchased from Sigma-Aldrich (St. Louis, MO, USA). CDCl$_3$, D$_2$O, and 1,4-Dioxane were purchased from Acros Co. (Beijing, China) and used as received. Tetrahydrofuran (THF) was freshly purified by distillation over sodium prior to use.

2,2-Azoisobutyronitrile (AIBN) was recrystallized three times from methanol. Other solvents and reagents were purchased from Beijing Chemical Reagent Co. (Beijing, China) and used as received.

3.2. Synthesis of the DBA Monomer

The tertiary amine-based methacrylate monomer 2-(Dibutylamino)ethyl methacrylate (DBAEMA) was synthesized by following a previously reported method [38]. 2-(dibutylamino)ethyl ethanol (17.3 g, 0.1 mol), triethylamine (10.1 g, 0.1 mol), and inhibitor hydroquinone (0.11 g, 0.001 mol) were dissolved in 100 mL THF and placed in a three-neck flask. To this solution, methacryloyl chloride (10.4 g, 0.1 mol) was added dropwise with constant stirring. The resulting solution was refluxed in THF for 2 h and then filtered to remove the precipitated white triethylamine-HCl salts with completion of the reaction. After removing THF solvent by rotary evaporator, the resulting residue was distilled in vacuo (83–87 °C at 0.05 mm Hg) as a colorless liquid. The obtained DBAEMA monomer was characterized by ^{1}H-NMR. ^{1}H-NMR (TMS, CDCl$_3$, ppm): 6.09 (br, 1H, CHH=C(CH$_3$)-), 5.55 (br, 1H, CHH=C(CH$_3$)-), 4.19 (t, J = 6.3 Hz, 2H, -OCH_2CH$_2$N-), 2.73 (t, J = 6.3 Hz, 2H, -OCH$_2$CH_2N-), 2.46 (t, J = 7.6 Hz, 2H, -N(CH_2CH$_2$CH$_2$CH$_3$)$_2$), 1.93 (s, 3H, CH$_2$=C(CH_3)-), 1.41 (m, 4H, -N(CH$_2$CH_2CH$_2$CH$_3$)$_2$), 1.29 (m, 4H, -N(CH$_2$CH$_2$CH_2CH$_3$)$_2$), 0.89 (t, J = 7.3 Hz, 6H, -N(CH$_2$CH$_2$CH$_2$CH_3)$_2$), Yield: 56%.

3.3. Synthesis of the (Co)polymers

The P(DMAEMA-*co*-DBAEMA) copolymers, PDMAEMA, and PDBAEMA were prepared by free radical polymerization in THF with varying DBAEMA to DMAEMA molar feed ratios of 0:4, 1:3, 2:2, 3:1, and 4:0, respectively. Typical procedures employed for the polymerization were as follows. Monomers were dissolved in THF at a total concentration of 0.10 g/mL, to which azodiisobutyronitrile (AIBN) (1.0 mol% relative to monomers) was added as a free radical initiator. After three cycles of freeze-thaw to thoroughly remove oxygen, the tube was sealed under reduced pressure and the polymerization was performed at 60 °C for 24 h. The polymers were purified by precipitation from diethyl ether twice, collected by filtration, and dried under vacuum to obtain white powders. The molecular weights and the composition of the (co)polymers were determined by gel permeation chromatography (GPC) and ^{1}H-NMR, respectively.

3.4. Characterization of the (Co)polymers

^{1}H-NMR spectra of the monomers and (co)polymers were recorded in CDCl$_3$ on a Varian-600 MHz spectrometer with tetramethylsilane (TMS) as the internal reference. The molecular weight and molecular weight distributions of the (co)polymers were measured with gel permeation chromatography (GPC) equipment consisting of WGE3010 pump, 3010 refractive index detector, and WGE Styra gel columns. The temperature of the columns was 35 °C and THF was used as an eluent at a flow rate of 1 mL/min. A series of narrowly dispersed polystyrene samples were used as standards and Millennium 32 software was used to calculate the molecular weight and polydispersity.

3.5. pH Titration

The pK_a values of different (co)polymers were detected by pH titration. When the prepared polymer was dissolved in 0.1 N HCl to reach a final concentration of 5–10 mg/mL, the pH titration experiment was performed by adding small volumes (50–100 μL increments) of 0.1 N NaOH solution under stirring. The pH values of the solution were measured continuously using a Sartorius (Germany) pH meter with a microelectrode.

3.6. Turbidity Measurements by UV/Vis Spectroscopy

The pH-dependent phase transition of the polymers was determined by the turbidity of the polymer solutions as a function of pH. The transmittance of the 1 mg/mL (co)polymer solutions in 0.1 N HCl were measured at 500 nm through a 1 cm quartz cell on a Shimadzu 2550 ultraviolet

(UV)-vis spectrometer. To adjust the pH, 0.2 N NaOH was added, and the final volumes of the polymer solutions increased about 10% compared with initial volumes. The pH values were monitored with a digital internal pH meter. Polymer-free deionized (DI) water was used as a reference.

3.7. ^{1}H-NMR Measurements of pH-Dependent Phase Transition

P(DMAEMA$_{0.73}$-co-DBAEMA$_{0.27}$) and PDBA were first dissolved in the deuterated buffers at a concentration of 10 mg/mL, and 1,4-dioxane was used as an internal standard. To adjust the pH value, 0.1 N NaOD deuterated solution was added and the Varian Mercury Plu600 MHz NMR spectrometer was used to record the ^{1}H-NMR spectra at different pH.

4. Conclusions

A series of novel pH-responsive polymers based on tertiary amine functional groups were developed by easy free radical copolymerization. The pK_a values of the polymers were precisely tuned by adjusting the feed ratio between the two monomers, DMAEMA and DBAEMA. The phase transition of polymers occurred in narrow pH ranges, which was demonstrated by transmittance and ^{1}H-NMR detection. Below pK_a, the positive charges of the protonated amine groups maintained the solubility of the polymers in aqueous solution, whereas when pH was greater than pK_a, the hydrophobic butyl groups on neutralized PDBAEMA segments rapidly induced the aggregation and precipitation of the polymers. Compared to widely used PDMAEMA and PDEAEMA, the developed polymers with PDBAEMA segments displayed much sharper pH transition ranges. The finely tunable transition pH values mean the polymers are a promising platform for drug delivery and biomedicine applications, where the encapsulated drugs at physiological pH would be triggered to release in acidic microenvironments.

Supplementary Materials: The following are available online.

Author Contributions: H.F., P.L., and X.H. conceived and designed the experiments; P.L. and W.L. performed the experiments; H.F. and X.H. analyzed the data; H.L. contributed reagents/materials/analysis tools; X.H. wrote the paper.

Acknowledgments: This work was supported by the National Natural Science Foundation of China (Nos. 20474005, 51574267), the General Program of Science and Technology Development Project of Beijing Municipal Education Commission (No. KM201310028007), the Fundamental Research Funds for the Central Universities (18CX05013), and Scientific Research Staring Foundation for the Returned Overseas Beijing Scholars (Grant No. 009125403700) and the Program for Changjiang Scholars and Innovative Research Team in University (IRT_14R58).

Conflicts of Interest: The authors declare no conflict of interest.

References

1. Allen, T.M.; Cullis, P.R. Drug delivery systems: Entering the mainstream. *Science* **2004**, *303*, 1818–1822. [CrossRef] [PubMed]
2. Zhang, S.; Greenfield, M.A.; Mata, A.; Palmer, L.C.; Bitton, R.; Mantei, J.R.; Conrado Aparicio, C.; Cruz, M.O.; Stupp, S.I. A self-assembly pathway to aligned monodomain gels. *Nat. Mater.* **2010**, *9*, 594–601. [CrossRef] [PubMed]
3. Bellomo, E.G.; Wyrsta, M.D.; Pakstis, L.; Pochan, D.J.; Deming, T.J. Stimuli-responsive polypeptide vesicles by conformation-specific assembly. *Nat. Mater.* **2004**, *3*, 244–248. [CrossRef] [PubMed]
4. So, M.K.; Xu, C.J.; Loening, A.M.; Gambhir, S.S.; Rao, J.H. Self-illuminating quantum dot conjugates for in vivo imaging. *Nat. Biotechnol.* **2006**, *24*, 339–343. [CrossRef] [PubMed]
5. Wang, C.S.; Zhao, T.; Li, Y.; Huang, G.; White, M.A.; Gao, J.M. Investigation of Endosome and Lysosome Biology by Ultra pH-Sensitive Nanoprobes. *Adv. Drug Deliv. Rev.* **2017**, *113*, 87–96. [CrossRef] [PubMed]
6. Kopecek, J. Polymer-drug conjugates: Origins, progress to date and future directions. *Adv. Drug Deliv. Rev.* **2013**, *65*, 49–59. [CrossRef] [PubMed]

7. Zhao, T.; Huang, G.; Li, Y.; Yang, S.; Ramazani, S.; Lin, Z.; Wang, Y.; Ma, X.; Zeng, Z.; Luo, M.; et al. A transistor-like pH nanoprobe for tumour detection and image-guided surgery. *Nat. Biomed. Eng.* **2016**, *1*, 0006–0013. [CrossRef] [PubMed]

8. Engin, K.; Leeper, D.B.; Cater, J.R.; Thistlethwaite, A.J.; Tupchong, L.; Mcfarlane, J.D. Extracellular pH distribution in human tumors. *Int. J. Hyperth.* **1995**, *11*, 211–216. [CrossRef] [PubMed]

9. Van Sluis, R.; Bhujwalla, Z.M.; Raghunand, N.; Ballesteros, P.; Alvarez, J.; Cerdan, S.; Galons, J.P.; Gillies, R.J. In vivo imaging of extracellular pH using H-1 MRSI. *Magn. Reson. Med.* **1999**, *41*, 743–750. [CrossRef]

10. Yang, L.; Tang, H.L.; Sun, H. Progress in Photo-Responsive Polypeptide Derived Nano-Assemblies. *Micromachines* **2018**, *9*, 296–313. [CrossRef]

11. Dai, Y.Q.; Sun, H.; Pal, S.; Zhang, Y.L.; Park, S.; Kabb, C.; Wei, D.; Sumerlin, B. Near-IR-induced dissociation of thermally-sensitive star polymers. *Chem. Sci.* **2017**, *8*, 1815–1821. [CrossRef] [PubMed]

12. Sun, H.; Kabb, C.; Dai, Y.Q.; Hill, M.; Ghiviriga, I.; Bapat, A.; Sumerlin, B. Macromolecular metamorphosis via stimulus-induced transformations of polymer architecture. *Nat. Chem.* **2017**, *9*, 817–823. [CrossRef] [PubMed]

13. Alfurhood, J.; Sun, H.; Kabb, C.; Tucker, B.; Matthews, J.; Luesch, H.; Sumerlin, B. Poly(N-(2-hydroxypropyl)methacrylamide)—Valproic acid conjugates as block copolymer nanocarriers. *Polym. Chem.* **2017**, *8*, 4983–4987. [CrossRef] [PubMed]

14. Yan, E.Y.; Ding, Y.; Chen, C.J.; Li, R.T.; Hu, Y.; Jiang, X.Q. Polymer/silica hybrid hollow nanospheres with pH-sensitive drug release in physiological and intracellular environments. *Chem. Commun.* **2009**, *19*, 2718–2720. [CrossRef] [PubMed]

15. Ulbrich, K.; Subr, V. Polymeric anticancer drugs with pH-controlled activation. *Adv. Drug Deliv. Rev.* **2004**, *56*, 1023–1050. [CrossRef] [PubMed]

16. Bae, Y.; Fukushima, S.; Harada, A.; Kataoka, K. Design of Environment-Sensitive Supramolecular Assemblies for Intracellular Drug Delivery: Polymeric Micelles that are Responsive to Intracellular pH Change. *Angew. Chem.* **2003**, *115*, 4788–4791. [CrossRef]

17. Li, Y.; Zhao, T.; Wang, C.S.; Lin, Z.Q.; Huang, G.; Sumer, B.D.; Gao, J.M. Molecular basis of cooperativity in pH-triggered supramolecular self-assembly. *Nat. Commun.* **2016**, *7*, 13214–13222. [CrossRef] [PubMed]

18. Urano, Y.; Asanuma, D.; Hama, Y.; Koyama, Y.; Barrett, T.; Kamiya, M.; Nagano, T.; Watanabe, T.; Hasegawa, A.; Choyke, P.L.; et al. Selective molecular imaging of viable cancer cells with pH-activatable fluorescence probes. *Nat. Med.* **2009**, *15*, 104–109. [CrossRef] [PubMed]

19. Hama, Y.; Urano, Y.; Koyama, Y.; Choyke, P.L.; Kobayashi, H. D-galactose receptor–targeted in vivo spectral fluorescence imaging of peritoneal metastasis using galactosamin-conjugated serum albumin-rhodamine green. *J. Biomed. Opt.* **2007**, *12*, 051501–051509. [CrossRef] [PubMed]

20. Koyama, Y.; Hama, Y.; Urano, Y.; Nguyen, D.M.; Choyke, P.L.; Kobayashi, H. Spectral fluorescence molecular imaging of lung metastases targeting HER2/neu. *Clin. Cancer Res.* **2007**, *13*, 2936–2945. [CrossRef] [PubMed]

21. Yuba, E.; Harada, A.; Sakanishi, Y.; Kono, K. Carboxylated hyperbranched poly(glycidol)s for preparation of pH-sensitive liposomes. *J. Control. Release* **2011**, *149*, 72–80. [CrossRef] [PubMed]

22. Liu, J.; Li, H.; Jiang, X.; Zhang, C.; Ping, Q. Novel pH-sensitive chitosan-derived micelles loaded with paclitaxel. *Carbohydr. Polym.* **2010**, *82*, 432–439. [CrossRef]

23. Kim, J.H.; Li, Y.; Kim, M.S.; Kang, S.W.; Jeong, J.H.; Lee, D.S. Synthesis and evaluation of biotin-conjugated pH-responsive polymeric micelles as drug carriers. *Int. J. Pharm.* **2012**, *427*, 435–442. [CrossRef] [PubMed]

24. Prajakta, D.; Ratnesh, J.; Chandan, K.; Suresh, S.; Grace, S.; Meera, V.; Vandana, P. Curcumin Loaded pH-Sensitive Nanoparticles for the Treatment of Colon Cancer. *J. Biomed. Nanotechnol.* **2009**, *5*, 445–455. [CrossRef] [PubMed]

25. Sahu, P.; Kashaw, S.K.; Kushwah, V.; Sau, S.; Jain, S.; Iyer, A.K. PH responsive biodegradable nanogels for sustained release of bleomycin. *Bioorg. Med. Chem.* **2017**, *25*, 4595–4613. [CrossRef] [PubMed]

26. Curcio, M.; Diaz-Gomez, L.; Cirillo, G.; Concheiro, A.; Iemma, F.; Alvarez-Lorenzo, C. PH/redox dual-sensitive dextran nanogels for enhanced intracellular drug delivery. *Eur. J. Pharm. Biopharm.* **2017**, *117*, 324–332. [CrossRef] [PubMed]

27. Zhang, X.; Achazi, K.; Haag, R. Boronate Cross-linked ATP- and pH-Responsive Nanogels for Intracellular Delivery of Anticancer Drugs. *Adv. Healthc. Mater.* **2015**, *4*, 585–592. [CrossRef] [PubMed]

28. Wang, Y.G.; Wang, C.S.; Li, Y.; Huang, G.; Zhao, T.; Ma, X.P.; Wang, Z.H.; Sumer, B.D.; White, M.A.; Gao, J.M. Digitization of Endocytic pH by Hybrid Ultra-pH-Sensitive Nanoprobes at Single-Organelle Resolution. *Adv. Mater.* **2016**, *29*, 1603794–1603802. [CrossRef] [PubMed]

29. Lee, Y.; Miyata, K.; Oba, M.; Ishii, T.; Fukushima, S.; Han, M.; Koyama, H.; Nishiyama, N.; Kataoka, K. Charge-Conversion Ternary Polyplex with Endosome Disruption Moiety: A Technique for Efficient and Safe Gene Delivery. *Angew. Chem. Int. Ed.* **2008**, *120*, 5241–5244. [CrossRef]

30. Weaver, J.V.M.; Williams, R.T.; Royles, B.J.L.; Findlay, P.H.; Cooper, A.I.; Rannard, S.P. PH-responsive branched polymer nanoparticles. *Soft Matter* **2008**, *4*, 985–992. [CrossRef]

31. Bae, Y.; Nishiyama, N.; Kataoka, K. In vivo antitumor activity of the folate-conjugated pH-Sensitive polymeric micelle selectively releasing adriamycin in the intracellular acidic compartments. *Bioconj. Chem.* **2007**, *18*, 1131–1139. [CrossRef] [PubMed]

32. Pal, A.; Pal, S. Synthesis of poly (ethylene glycol)-block-poly (acrylamide)-block-poly (lactide) amphiphilic copolymer through ATRP, ROP and click chemistry: Characterization, micellization and pH-triggered sustained release behaviour. *Polymer* **2017**, *127*, 150–158. [CrossRef]

33. Gil, E.S.; Hudson, S.M. Stimuli-reponsive polymers and their bioconjugates. *Prog. Polym. Sci.* **2004**, *29*, 1173–1222. [CrossRef]

34. Ganta, S.; Devalapally, H.; Shahiwala, A.; Amiji, M. A review of stimuli-responsive nanocarriers for drug and gene delivery. *J. Control. Release* **2008**, *126*, 187–204. [CrossRef] [PubMed]

35. Shen, Y.Q.; Tang, H.D.; Zhan, Y.H.; Van, K.; Murdoch, W.J. Degradable poly (β-amino ester) nanoparticles for cancer cytoplasmic drug delivery. *Nanomedicine* **2009**, *5*, 192–201. [CrossRef] [PubMed]

36. Zhang, L.; Guo, R.; Yang, M.; Jiang, X.; Liu, B. Thermo and pH dualresponsive nanoparticles for anti-cancer drug delivery. *Adv. Mater.* **2007**, *19*, 2988–2992. [CrossRef]

37. Topham, P.D.; Howse, J.R.; Mykhaylyk, O.O.; Armes, S.P.; Jones, R.A.L.; Ryan, A.J. Synthesis and solid state properties of a poly(methyl methacrylate)-block-poly(2-(diethylamino)ethyl methacrylate)-block-poly(methyl methacrylate) triblock copolymer. *Macromolecules* **2006**, *39*, 5573–5576. [CrossRef]

38. Zhou, K.J.; Wang, Y.G.; Huang, X.N.; Luby-Phelps, K.; Sumer, B.D.; Gao, J.M. Tunable, Ultrasensitive pH-Responsive Nanoparticles Targeting Specific Endocytic Organelles in Living Cells. *Angew. Chem. Int. Ed.* **2011**, *50*, 6109–6114. [CrossRef] [PubMed]

Sample Availability: Not Available.

molecules

MDPI

Article

Effect of Hydrophobic Polypeptide Length on Performances of Thermo-Sensitive Hydrogels

Jiandong Han [1,2], Xingyu Zhao [2], Weiguo Xu [2], Wei Wang [1,*], Yuping Han [2,3,*] and Xiangru Feng [2,*]

1   Department of Chemistry, Changchun University of Science and Technology, Changchun 130022, China; jdhan@ciac.ac.cn
2   Key Laboratory of Polymer Ecomaterials, Changchun Institute of Applied Chemistry, Chinese Academy of Sciences, Changchun 130022, China; star20012002@163.com (X.Z.); wgxu@ciac.ac.cn (W.X.)
3   Department of Urology, China-Japan Union Hospital of Jilin University, Changchun 130033, China
*   Correspondence: weiwanglg@163.com (W.W.); hyp181818@126.com (Y.H.); xrfeng@ciac.ac.cn (X.F.)

Received: 12 April 2018; Accepted: 21 April 2018; Published: 25 April 2018

Abstract: Thermosensitive gels are commonly used as drug carriers in medical fields, mainly due to their convenient processing and easy functionalization. However, their overall performance has been severely affected by their unsatisfying biocompatibility and biodegradability. To this end, we synthesized poly(L-alanine) (PLAla)-based thermosensitive hydrogels with different degrees of polymerization by ring-opening polymerization. The obtained mPEG$_{45}$−PLAla copolymers showed distinct transition temperatures and degradation abilities. It was found that slight changes in the length of hydrophobic side groups had a decisive effect on the gelation behavior of the polypeptide hydrogel. Longer hydrophobic ends led to a lower gelation temperature of gel at the same concentration, which implied better gelation capability. The hydrogels showed rapid gelling, enhanced biocompatibility, and better degradability. Therefore, this thermosensitive hydrogel is a promising material for biomedical application.

Keywords: amphiphilicity; phase change; polyamino acids; degradability

---

1. Introduction

The application of hydrogels in the biomedical field has rapidly increased over the past decade, including use in three-dimensional (3D) cell culture [1–3], drug delivery [4–6], and tissue engineering [7–9]. Polymer hydrogels can trap a large amount of moisture inside for easier cell membrane penetration and drug transmission. They enjoy excellent physical properties and exhibit a controllable degradation process. On account of the incomparable convenience, the in situ gelation of biodegradable hydrogels has aroused great interest in many researchers. Among numerous biodegradable hydrogels, thermosensitive hydrogels have specific advantages. On the one hand, they are rather safe for clinical utilization because the common thermosensitive gel preparation process does not involve the use of organic solvents. On the other hand, the gelling conditions of heat-sensitive hydrogels are easy to control, making them applicable in the field of biomedicine [10–12].

Thermosensitive hydrogels can be generated from block copolymers consisting of hydrophilic poly(ethylene glycol) (PEG) and hydrophobic moieties such as poly(lactic acid) (PLA) [13,14], poly(lactic-co-glycolic acid) (PLGA) [15–17], poly(ε-caprolactone) (PCL) [18–20], polyamidoamine (PAMAM) [21], and so on. Thermosensitive hydrogels are soluble at low temperatures, which facilitates the introduction of chemotherapeutic drugs [22–24], functional proteins [25], and cells. Due to the amorphous form of the hydrogel at low temperatures, it can be seamlessly filled into injury sites [26,27]. At body temperature, the soluble hydrogels turn into a gel state and provide a depot for the encapsulated medical agents to achieve sustained release and long-term therapeutic effect [28]. For example, growth factors can be slowly and continuously released locally for the efficient repair of bone and nerve tissues [29–31]. The hydrogels can also function as 3D scaffolds and as a nutrient resource for cell growth in tissue engineering [32], aiming to help cells grow in a more uniform manner and promote cell proliferation [33].

In 2012, Qian's team prepared a bionic hydrogel composed of three components (i.e., triblock PEG−PCL−PEG copolymer (PECE), nano-hydroxyapatite (n-HA), and collagen), which showed satisfactory efficacy in skull repair [34]. In 2015, Chen's team engineered a thermosensitive PLGA−PEG−PLGA hydrogel loaded with 5-fluorouracil for the prevention of postoperative tendon adhesion, and acquired a good histological score [35]. However, despite the wide application of thermosensitive hydrogels in various diseases in clinical practice, they are not free from flaws. Some hydrogels are difficult to gelatinize, and the degradation time is uncontrollable. Moreover, hydrogels with poor biocompatibility always lead to severe tissue damages. Therefore, there is an urgent need to develop thermosensitive polypeptide polymers with good biocompatibility and biodegradability to avoid the above problems.

Towards this aim, peptide systems have been developed to avoid evoking the formation of an acidic microenvironment during degradation, and thus reduce the damages to surrounding tissues and minimize the adverse effects toward the bioactivity of the loaded protein or cells. Apart from this, the gelation performance of the polypeptide copolymer can be tuned by copolymerization with various hydrophobic or hydrophilic amino acid monomers, which widens the clinical applications. In 2011, Jeong's team reported the synthesis of poly(alanine-*co*-leucine)−poloxamer−poly(alanine-*co*-leucine) (PAL−PLX−PAL) hydrogels, which exhibited a sustained drug release pattern without producing evident inflammatory effects [22]. In 2017, methoxy poly(ethylene glycol)-*block*-poly(L-alanine-*co*-L-phenylalanine) mPEG-*b*-PLAF hydrogel was synthesized by the ring-opening polymerization (ROP) reaction of L-alanine N-carboxyanhydrides (L-Ala NCA) and L-phenylalanine NCA, with amino-terminated mPEG (mPEG-NH_2) as a macroinitiator. Encapsulated with combretastatin A4 (CA4) and doxorubicin (DOX), mPEG−*b*−PLAF hydrogel had an excellent inhibitory effect on the proliferation of B16F10 melanoma cells [36].

In this work, mPEG−poly(L-alanine) (mPEG−PLAla) polymers with different degrees of polymerization (DPs) were synthesized and named mPEG$_{45}$−PLAla$_{30}$, mPEG$_{45}$−PLAla$_{22}$, and mPEG$_{45}$−PLAla$_{14}$. The influence of different DPs on the gelation properties of hydrogels was studied. It was found that hydrogels with higher DP showed better gelation ability than those with lower DP. In addition, the gelation behavior, tissue safety, and cytotoxicity of these polymers in vitro and in vivo were also tested and evaluated. The study revealed that these hydrogels had notable potential for employment in the biomedical field.

2. Results and Discussion

2.1. Material Synthesis and Structure Characterization

Triphosgene and L-alanine were synthesized into L-Ala NCA in dried tetrahydrofuran (THF) solvent, as shown in Scheme 1. The polypeptide hydrogels were produced via the ROP of L-Ala NCA initiated by mPEG$_{45}$-NH_2. The mPEG$_{45}$−PLAla copolymers with different DPs were synthesized by altering the feed amount. The DPs of PLAla were determined by contrasting the integral of the methyl peaks of side chains (−CH_3) with the methylene peak of PEG (−CH_2CH_2O−). The DPs of mPEG$_{45}$−PLAla$_{30}$, mPEG$_{45}$−PLAla$_{22}$, and mPEG$_{45}$−PLAla$_{14}$ were 30, 22, and 14, respectively.

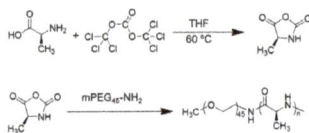

Scheme 1. Synthetic routes of L-alanine N-carboxyanhydrides (l-Ala NCA) and methoxy poly(ethylene glycol)−poly(l-alanine) (mPEG$_{45}$−PLAla).

The typical proton nuclear magnetic resonance (^{1}H NMR) spectrum of block copolymer is shown in Figure 1a, and all peaks could be accurately assigned. The peaks at 1.47, 3.40, 3.50–3.80, and 4.64 ppm stood for the protons of alanine methyl, mPEG terminal methoxy, mPEG backbone, and polymer methine, respectively, indicating the successful synthesis of the three polymers.

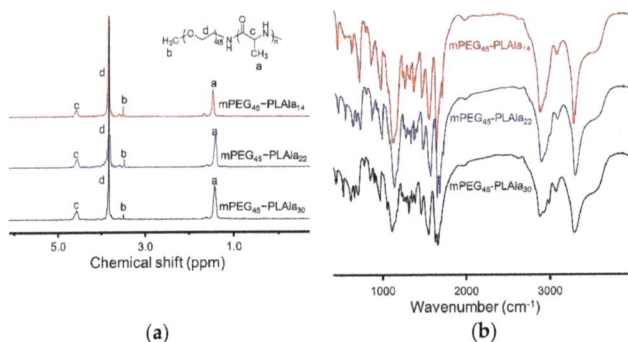

Figure 1. Structure characterization of mPEG$_{45}$−PLAla. (a) ^{1}H NMR spectra of mPEG$_{45}$−PLAla$_{30}$, mPEG$_{45}$−PLAla$_{22}$, and mPEG$_{45}$−PLAla$_{14}$, where the subscripts indicate the degree of polymerization (DP); (b) Fourier-transform infrared (FT-IR) spectra of mPEG$_{45}$−PLAla$_{30}$, mPEG$_{45}$−PLAla$_{22}$, and mPEG$_{45}$−PLAla$_{14}$.

The secondary structure of the copolymer was studied by FT-IR. In Figure 1b, characteristic peaks of amide bond at 1627 cm^{-1} and 1544 cm^{-1} were observed, indicating that all copolymers went through the main β-sheet conformation.

2.2. Gelation Ability and Internal Appearance

The tube inversion method was used to determine the transition temperatures. Polypeptides were solubilized in a phosphate-buffered saline (PBS) solution, and went through a solution–gel transition with the change in temperature. The samples were defined as gel if they stayed still when the vial was inverted for as long as 30 s. The phase diagrams of mPEG$_{45}$−PLAla$_{30}$ and mPEG$_{45}$−PLAla$_{22}$ copolymers are shown in Figure 2a,b. Only copolymers with a DP of no less than 22 could undergo solution–gel transition at the concentration of 3.0–8.0 wt. %. It was noticeable that mPEG$_{45}$−PLAla$_{22}$ showed a higher gel transition temperature than mPEG$_{45}$−PLAla$_{30}$. The critical gelation temperatures (CGTs) of mPEG$_{45}$−PLAla$_{30}$ and mPEG$_{45}$−PLAla$_{22}$ were 5 °C and 15 °C, respectively, at the concentration of 5.0 wt. %. The lower CGT of mPEG$_{45}$−PLAla$_{30}$ could be explained by longer hydrophobic end than that of mPEG$_{45}$−PLAla$_{22}$ [37].

Solution phase behaviors of the three copolymers at different temperatures were investigated. The images in Figure 2a,b showed the mPEG$_{45}$−PLAla$_{30}$ and mPEG$_{45}$−PLAla$_{22}$ solutions in PBS (5.0 wt. %) at 4 °C and 37 °C. Although all copolymers were dispersed in PBS at 4 °C, the mPEG$_{45}$−PLAla$_{14}$ formed a clear solution while mPEG$_{45}$−PLAla$_{30}$ and mPEG$_{45}$−PLAla$_{22}$ showed a turbid state. When temperature rose to 37 °C, mPEG$_{45}$−PLAla$_{30}$ and mPEG$_{45}$−PLAla$_{22}$ were observed to be stably gelled within 5 min, while mPEG$_{45}$–PLala$_{14}$ remained in a viscous flow state. This was because mPEG$_{45}$–PLala$_{14}$ had better water solubility. Further, the morphology of the mPEG$_{45}$–PLAla$_{30}$ hydrogel (5.0 wt. %) was observed by scanning electron microscope (SEM). As can be seen in Figure 2c, it was found that there were numerous tiny pores in the mPEG$_{45}$–PLAla$_{30}$ gel, which was due to the 3D network structure of the gel. These pores showed a size of about 100 μm and were evenly distributed within the hydrogel, making them suitable as a depot for therapeutic agents. These findings prove that mPEG$_{45}$–PLAla$_{30}$ and mPEG$_{45}$–PLAla$_{22}$ hydrogels feature good stability at body temperature and are suitable for biomedical applications.

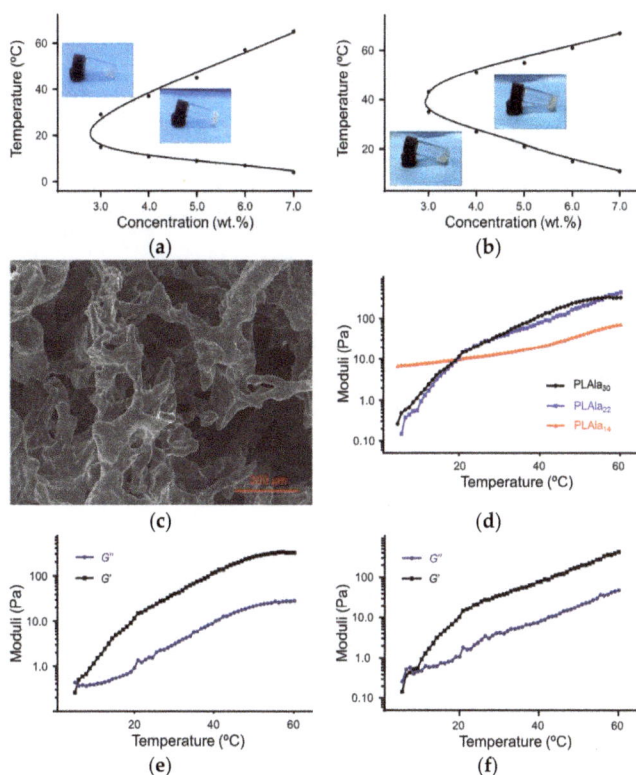

Figure 2. Gelatinization characteristics of mPEG$_{45}$–PLAla. (**a**) Solution–gel phase diagrams of the mPEG$_{45}$–PLAla$_{30}$ and (**b**) mPEG$_{45}$–PLAla$_{22}$ copolymer solutions; (**c**) Scanning electron microscope (SEM) image of mPEG$_{45}$–PLAla$_{30}$ hydrogels formed at 40 °C; (**d**) G' of the mPEG$_{45}$–PLAla$_{30}$ (PLAla$_{30}$), mPEG$_{45}$–PLAla$_{22}$ (PLAla$_{22}$), and mPEG$_{45}$–PLAla$_{14}$ (PLAla$_{14}$) at the concentration of 5.0 wt. % in phosphate-buffered saline (PBS) solution; (**e**) Changes of G' and G'' of mPEG$_{45}$–PLAla$_{30}$ and (**f**) mPEG$_{45}$–PLAla$_{22}$ in PBS solutions (5 wt. %).

2.3. Mechanical Performance Test

Thermally induced storage modulus (G') and loss modulus (G'') changes of the three copolymers were obtained by dynamic mechanical analysis. G' represents the systematic gel-like behavior of the elastic component of the complex modulus, and G'' is an index of the viscous component of the complex modulus and a measure of the sol-like behavior. The intersection of G' and G'' reflects the sol–gel transition. As shown in Figure 2d, the G' of mPEG$_{45}$–PLAla$_{30}$ and mPEG$_{45}$–PLAla$_{22}$ obviously increased along with rising temperatures. However, only a minor increment in G' was seen in mPEG$_{45}$–PLAla$_{14}$, and no significant change in G' was detected, even when the temperature rose to about 50 °C, implying that there was no sol–gel modification of mPEG$_{45}$–PLAla$_{14}$ within the experimental temperature range. The reason was that hydrophilic mPEG was longer than the hydrophobic end made up of polypeptides. Therefore, with a longer hydrophilic segment, mPEG$_{45}$–PLAla$_{14}$ showed better water solubility than mPEG$_{45}$–PLAla$_{30}$ and mPEG$_{45}$–PLAla$_{22}$, thus featuring less obvious sol–gel change. Notably, the length of the hydrophobic side chain had a remarkable effect on the gelation function of the mPEG–peptide block copolymer.

In Figure 2e,f, G' was smaller than G'' when the temperature was lower than CGTs (5 °C in mPEG$_{45}$–PLAla$_{30}$ and 12 °C in mPEG$_{45}$–PLAla$_{22}$), which reflected the viscous state of hydrogels. As the temperature increased above CGTs, G' was sharply elevated and surpassed G'', confirming the gel formation. This result was in accordance with the findings in Figure 2a,b.

2.4. Mechanism of Gelatinization

In order to study the mechanism of sol–gel transition, the nuclear magnetic peak changes, diameter changes, and conformation evolution of polypeptides in response to temperature were tested by carbon nuclear magnetic resonance (^{13}C-NMR), dynamic light scattering (DLS), and circular dichroism (CD). As shown in Figure 3a, when temperature rose from 20 to 60 °C, the characteristic peak of PEG gradually moved from 69.7 to 70.3 ppm, which revealed continuous dehydration of the PEG block during the heat-induced sol–gel transition process. Two main reasons accounted for the gelation of polymers. First, the interaction between hydrophobic blocks became stronger as temperature increased. Secondly, dehydration of PEG segments facilitated the aggregation of the hydrogels [38].

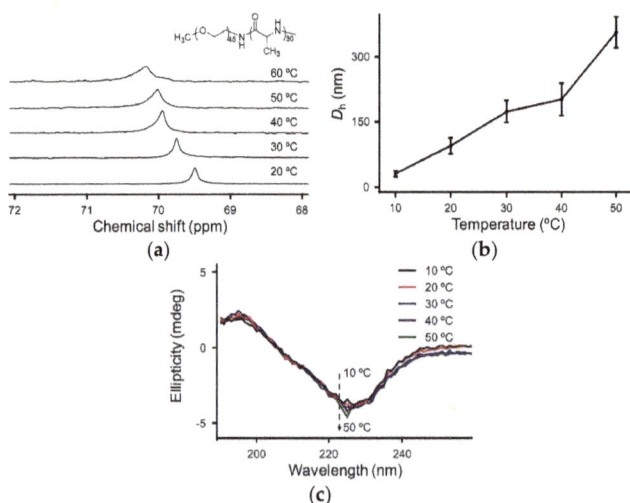

Figure 3. Gelation mechanism of the hydrogels. (**a**) ^{13}C-NMR spectra (in D$_2$O) of 5.0 wt. % mPEG$_{45}$–PLAla$_{30}$ solution as a function of temperature; (**b**) Average hydrodynamic diameter (D_h) of micelles of mPEG$_{45}$–PLAla$_{30}$ as a function of temperature in water (5.0 μg mL^{-1}); (**c**) Circular dichroism (CD) spectra of mPEG$_{45}$–PLAla$_{30}$ (0.05 mg mL^{-1}) in aqueous solution as a function of temperature.

As seen in Figure 3b, the diameter changes of mPEG$_{45}$–PLAla$_{30}$ were determined at the concentration of 5.0 μg mL^{-1}. At 10 °C, the average hydrodynamic diameter (D_h) of mPEG$_{45}$–PLAla$_{30}$ was about 31.4 nm. When the temperature rose to 20 °C and 50 °C, the number swiftly increased to 95.9 nm and 356.2 nm, respectively. The dramatic changes in particle size could be explained by the interaction between the shell of PEG and core of polypeptides caused by the dehydration processof PEG [38].

In Figure 3c, the CD spectrum illustrated the alteration of the secondary structure of the aqueous mPEG$_{45}$−PLAla$_{30}$ during the sol–gel transition process. The two typical bands corresponding to the β-sheet conformation, a positive Cotton band at 195 nm and a negative Cotton band at 226 nm, were clearly shown. The above results revealed that the synthesized thermosensitive hydrogel had good physicochemical properties and a clear solution−gel transformation mechanism.

2.5. Degradability Test and Pathological Analysis

Since hydrogels are often used in the biomedical field, biodegradability is a decisive factor for clinical application. If the degradation rate of the hydrogel is too high, it will lead to a rapid release of the payload. In contrast, if the hydrogel requires a long time to degrade, then it will probably remain at the injection site even after all drugs are released, which is inconvenient for further treatment. The in vitro degradation of mPEG$_{45}$−PLAla$_{30}$ hydrogel (5.0 wt. %) was evaluated in PBS and PBS with elastase K or α-chymotrypsin. As shown in Figure 4a, there was a mass loss of over 70% and 67% hydrogel in the elastase K and α-chymotrypsin groups on day 15, respectively—considerably higher than 20% in PBS. The results could be explained by the absence of elastase K or α-chymotrypsin, where the mass loss of gels was merely attributed to the surface erosion of the hydrogels. However, with elastase K or α-chymotrypsin, the polypeptide chains of the hydrogels were also fast degrading, which sped up the loss of gels.

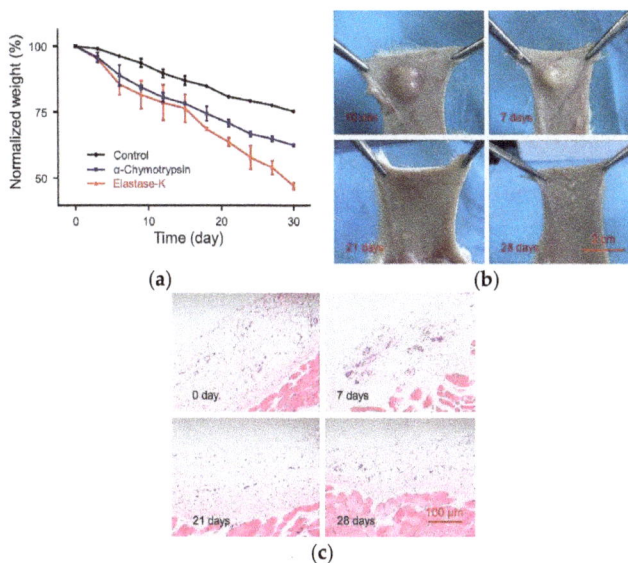

Figure 4. Degradation of hydrogel in vitro and in vivo, and histological analysis of skin. (**a**) Mass loss curves of in vitro degradation of hydrogels in PBS, and PBS with elastase-K or α-chymotrypsin (0.2 mg mL^{-1}) groups; (**b**) Images of in vivo gel maintenance at 10 min, 7, 14, 21, and 28 days after the injection of 5.0 wt. % mPEG$_{45}$−PLAla$_{30}$ hydrogels; (**c**) Hematoxylin and eosin (H&E) images of the skin tissue near the hydrogels on day 7, 14, 21, and 28, respectively.

The degradation process of hydrogels in vivo was monitored for 28 days, and images were taken at 10 min, 7, 14, 21, and 28 days after hydrogel treatment. At 4 °C, 500.0 µL of mPEG$_{45}$−PLAla$_{30}$ solution (5.0 wt. %) was subcutaneously injected into Sprague−Dawley (SD) rats through 21-gauge syringe needles. In Figure 4b, the solution rapidly turned into gel in 10 min after injection. Seven days later, only a small portion of the hydrogel was degraded. On day 14, the size of the hydrogel shrank to less than 50%, while on day 28, the hydrogel was completely degraded. This degradation rate of hydrogel was suitable for application in biomedical fields requiring long-term treatment, such as tissue repair. As shown in Figure 4c, at different time intervals, the tissue surrounding the gel was surgically separated and processed by H&E staining to examine the condition of tissue damage. Basically, no histological damage or immune response was found at any time point.

2.6. Safety Evaluation

The cytotoxicity of hydrogel mPEG$_{45}$−PLAla$_{30}$ in vitro was evaluated by methyl thiazolyl tetrazolium (MTT) assay and hemolysis test. In Figure 5a, L929 cells treated with the highest concentration of mPEG$_{45}$−PLAla$_{30}$ (100.0 ìg mL^{-1}) for 24 h retained almost 100% viability, confirming its excellent biocompatibility. Moreover, the effect of different concentrations of gel on hemolysis was also investigated. In this part, hemolysis was measured spectroscopically according to previously reported methods [39]. As shown in Figure 5b, no hemolysis was detected in blood samples treated with all test concentrations of mPEG$_{45}$−PLAla$_{30}$, which further confirmed the outstanding biocompatibility. Overall, the findings above revealed that the hydrogel showed little toxicity, both in vitro and in vivo. Therefore, it is a promising material for usage in clinical practice.

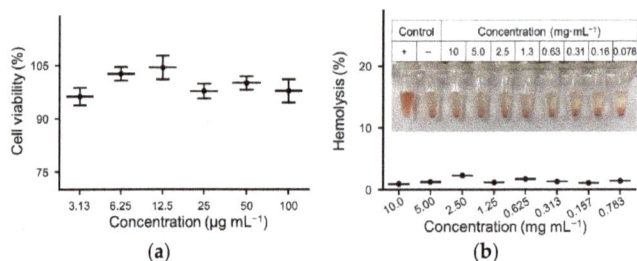

Figure 5. Biological compatibility of mPEG$_{45}$−PLAla$_{30}$. (**a**) In vitro cytotoxicity of the mPEG$_{45}$−PLAla$_{30}$ toward L929 cells. Data were presented as mean ± SD (n = 5); (**b**) Hemolysis experiments of mPEG$_{45}$−PLAla$_{30}$. Data are presented as mean ± SD (n = 3).

3. Materials and Methods

3.1. Materials

mPEG−OH, number-average molecular weight (Mn) = 2000 g mol^{-1}, was purchased from Sigma-Aldrich (St. Louis, MO, USA). The mPEG$_{45}$−NH$_2$ was synthesized conforming to the previously reported protocol in our work [40]. THF and toluene were refluxed with sodium and distilled under nitrogen before usage. *N,N*-Dimethylformamide (DMF) was stored over calcium hydride (CaH$_2$) and purified by vacuum distillation. All the other reagents and solvents were bought from Sinopharm Chemical Reagent Co. Ltd., Beijing, China, and used as obtained.

3.2. Phase Diagram

The sol–gel transition behavior of the copolymers in PBS (pH 7.4) was determined by inverting test method with a temperature increment of 2 °C per step. Samples with concentrations ranging from 3.0–7.0 wt. % were dissolved in PBS and stirred at 0 °C for 12 h. The copolymer solution (0.2 mL) was introduced into the test tube with an inner diameter of 10.0 mm. The sol–gel transition temperature was recorded if no flow was observed within 30 s after inverting the test tube. Each data point was the average of three measurements.

3.3. Dynamic Mechanical Analysis

Rheological experiments were performed on a US 302 Rheometer (Anton Paar, Graz, Austria). The copolymer solution was placed between parallel plates of 25.0 mm in diameter with a gap of 0.5 mm. To prevent the evaporation of water, the outer edge of the sandwiched sample was sealed by a thin layer of silicon oil. The data were collected under a controlled strain γ of 1% and a frequency of 1 rad s^{-1}. The heating rate was 1 °C min^{-1}.

3.4. In Vitro Gel Degradation

For this, 0.5 mL of mPEG$_{45}$−PLAla$_{30}$ hydrogel was incubated in PBS (5.0 wt. %) in vials (diameter = 16.0 mm) at 37 °C for 10 min. PBS solutions (pH 7.4) containing 0.2 mg mL^{-1} elastase K or 0.2 mg mL^{-1} α-chymotrypsin were used as degradation media, and hydrogels incubated in PBS were only used as a control. Different solutions (2.0 mL) were added to the top of the gels at 37 °C and the entire medium was changed daily. The weight of the remaining gel was measured daily.

3.5. In Vivo Gel Degradation

SD rats (about 180.0 g, provided by Beijing Vital River Laboratory Animal Technology Co., Ltd., Beijing, China) were used for the detection of gel degradation in vivo. Rats were anesthetized by inhalation of ether before 0.5 mL of mPEG$_{45}$−PLAla$_{30}$ PBS solutions (5.0 wt. %) were injected into the dorsal subcutaneous area of the rats using a 21-gauge needle. The rats were sacrificed on day 7, day 14, day 21, and day 28, respectively, to monitor the degradation behavior of the gel. The experiments on animals were carried out according to the guide for the care and use of laboratory animals, provided by Jilin University, Changchun, China, and the procedure was approved by the local Animal Ethics Committee.

3.6. Cytotoxicity Measurement

The relative cytotoxicity was assessed by MTT viability assay against L929 mouse fibroblasts cells. L929 cells were cultured in complete Dulbecco's modified Eagle's medium (DMEM) supplemented with 10.0% (v/v) FBS, penicillin (50.0 IU mL^{-1}), and streptomycin (50.0 IU mL^{-1}) at 37 °C in a 5.0% (v/v) carbon dioxide atmosphere. L929 cells with a density of 6000 cells per well were planted in 96-well plates in 180.0 µL of DMEM. After incubation for 24 h, 20.0 µL of copolymer solutions at different concentrations (31.3–1000.0 µg mL^{-1}) were added. L929 cells were incubated with copolymers for another 24 h before 20.0 ìL of PBS solution containing MTT (0.05 mg mL^{-1}) was added and incubated for a further 4 h. Then, the media was replaced with 160.0 µL of dimethyl sulfoxide (DMSO). The absorbance of the solution was measured on a Bio-Rad 680 microplate reader (Hercules, CA, USA) at 490 nm. Cell viability (%) was calculated according to the following Equation (1).

Measurements were done in five replicates.

$$\text{Cell viability } (\%) = A_{\text{sample}} / A_{\text{control}} \times 100, \tag{1}$$

where A_{sample} and A_{control} denote the absorbances of sample and control, respectively.

4. Conclusions

In this work, thermosensitive hydrogels mPEG$_{45}$−PLAla with three different DP (14, 22, and 30) were synthesized by ROP of L-Ala NCA monomer initiated by mPEG$_{45}$−NH$_2$. The effect of different DP on the solution−gel transition was investigated. It is worth noting that mPEG$_{45}$−PLAla$_{30}$ had a lower sol–gel transition temperature than mPEG$_{45}$−PLAla$_{22}$, attributed to the longer hydrophobic segment. In addition, mPEG−PLAla hydrogel had good stability and high mechanical strength after gelation. Moreover, mPEG−PLAla hydrogel did not cause any tissue damage, inflammatory reaction, or hemolysis reaction during degradation, confirming its good biocompatibility. In summary, the mPEG−PLAla hydrogel designed in our study had improved biocompatibility, appropriate degradation, and gelation ability. Therefore, the polypeptide hydrogels showed promise for application in tissue repair and regeneration, cell 3D culture, and treatment of cancer. Moreover, by loading functional drugs, the hydrogels can also be used for postoperative recovery and anti-infection. Based on the above findings, polypeptide thermosensitive hydrogels have broad prospects in the biomedical field.

Supplementary Materials: The following are available online.

Author Contributions: W.X. and W.W. conceived and designed the experiments; J.H. performed the experiments; J.H., X.Z. and X.F. analyzed the data; and Y.H. contributed reagents and materials. J.H. initiated and wrote this article; X.F. helped to write and corrected the manuscript; and Y.H. discussed and suggested ideas for improvement of this article.

Acknowledgments: This research was supported by the Key Technology Research and Development Project of Jilin Department of Science and Technology (No. 20180201083GX) and Natural Science Foundation of Jilin Department of Science and Technology (20170101107JC).

Conflicts of Interest: The authors declare no conflict of interest.

References

1. Ravi, M.; Paramesh, V.; Kaviya, S.R.; Anuradha, E.; Solomon, F.D. 3D cell culture systems: Advantages and applications. *J. Cell Physiol.* **2015**, *230*, 16–26. [CrossRef] [PubMed]
2. Tibbitt, M.W.; Anseth, K.S. Hydrogels as extracellular matrix mimics for 3D cell culture. *Biotechnol. Bioeng.* **2009**, *103*, 655–663. [CrossRef] [PubMed]
3. Zhao, T.; Sellers, D.L.; Cheng, Y.; Horner, P.J.; Pun, S.H. Tunable, injectable hydrogels based on peptide-cross-linked, cyclized polymer nanoparticles for neural progenitor cell delivery. *Biomacromolecules* **2017**, *18*, 2723–2731. [CrossRef] [PubMed]
4. Lock, L.L.; Lo, Y.; Mao, X.; Chen, H.; Staedtke, V.; Bai, R.; Ma, W.; Lin, R.; Li, Y.; Liu, G.; et al. One-component supramolecular filament hydrogels as theranostic label-free magnetic resonance imaging agents. *ACS Nano* **2017**, *11*, 797–805. [CrossRef] [PubMed]
5. Deng, Y.; Yang, F.; Cocco, E.; Song, E.; Zhang, J.W.; Cui, J.J.; Mohideen, M.; Bellone, S.; Santin, A.D.; Saltzman, W.M. Improved i.P. Drug delivery with bioadhesive nanoparticles. *Proc. Natl. Acad. Sci. USA* **2016**, *113*, 11453–11458. [CrossRef] [PubMed]
6. Zhu, M.; Wei, K.; Lin, S.; Chen, X.; Wu, C.-C.; Li, G.; Bian, L. Bioadhesive polymersome for localized and sustained drug delivery at pathological sites with harsh enzymatic and fluidic environment via supramolecular host-guest complexation. *Small* **2018**, *14*. [CrossRef] [PubMed]
7. Qu, Y.; Wang, B.; Chu, B.; Liu, C.; Rong, X.; Chen, H.; Peng, J.; Qian, Z. Injectable and thermosensitive hydrogel and pdlla electrospun nanofiber membrane composites for guided spinal fusion. *ACS Appl. Mater. Interfaces* **2018**, *10*, 4462–4470. [CrossRef] [PubMed]
8. Zhu, C.; Lei, H.; Fan, D.; Duan, Z.; Li, X.; Li, Y.; Cao, J.; Wang, S.; Yu, Y. Novel enzymatic crosslinked hydrogels that mimic extracellular matrix for skin wound healing. *J. Mater. Sci.* **2018**, *53*, 5909–5928. [CrossRef]
9. Shahriari, D.; Shibayama, M.; Lynam, D.A.; Wolf, K.J.; Kubota, G.; Koffler, J.Y.; Tuszynski, M.H.; Campana, W.M.; Sakamoto, J.S. Peripheral nerve growth within a hydrogel microchannel scaffold supported by a kink-resistant conduit. *J. Biomed. Mater. Res. Part A* **2017**, *105*, 3392–3399. [CrossRef] [PubMed]
10. Xu, H.-L.; Tian, F.-R.; Lu, C.-T.; Xu, J.; Fan, Z.-L.; Yang, J.-J.; Chen, P.-P.; Huang, Y.-D.; Xiao, J.; Zhao, Y.-Z. Thermo-sensitive hydrogels combined with decellularised matrix deliver bfgf for the functional recovery of rats after a spinal cord injury. *Sci. Rep.* **2016**, *6*, 38332. [CrossRef] [PubMed]
11. Lu, D.; Li, Y.; Li, T.; Zhang, Y.; Dou, F.; Wang, X.; Zhao, X.; Ma, H.; Guan, X.; Wei, Q.; et al. Surgical adhesive: Synthesis and properties of thermoresponsive pluronic l-31-3,4-dihydroxyphenylalanine-arginine derivatives. *J. Appl. Polym. Sci.* **2017**, *134*. [CrossRef]
12. Li, P.; Zhang, J.; Dong, C.-M. Photosensitive poly(*o*-nitrobenzyloxycarbonyl-L-lysine)-b-peo polypeptide copolymers: Synthesis, multiple self-assembly behaviors, and the photo/ph-thermo-sensitive hydrogels. *Polym. Chem.* **2017**, *8*, 7033–7043. [CrossRef]

13. Cui, H.; Shao, J.; Wang, Y.; Zhang, P.; Chen, X.; Wei, Y. PLA-PEG-PLA and its electroactive tetraaniline copolymer as multi-interactive injectable hydrogels for tissue engineering. *Biomacromolecules* **2013**, *14*, 1904–1912. [CrossRef] [PubMed]

14. Wang, D.K.; Varanasi, S.; Strounina, E.; Hill, D.J.T.; Symons, A.L.; Whittaker, A.K.; Rasoul, F. Synthesis and characterization of a POSS-PEG macromonomer and POSS-PEG-PLA hydrogels for periodontal applications. *Biomacromolecules* **2014**, *15*, 666–679. [CrossRef] [PubMed]

15. Michlovska, L.; Vojtova, L.; Humpa, O.; Kucerik, J.; Zidek, J.; Jancar, J. Hydrolytic stability of end-linked hydrogels from PLGA-PEG-PLGA macromonomers terminated by alpha,omega-itaconyl groups. *RSC Adv.* **2016**, *6*, 16808–16816. [CrossRef]

16. Wang, P.; Chu, W.; Zhuo, X.; Zhang, Y.; Gou, J.; Ren, T.; He, H.; Yin, T.; Tang, X. Modified PLGA-PEG-PLGA thermosensitive hydrogels with suitable thermosensitivity and properties for use in a drug delivery system. *J. Mater. Chem. B* **2017**, *5*, 1551–1565. [CrossRef]

17. Zhang, Y.; Zhang, J.; Chang, F.; Xu, W.; Ding, J. Repair of full-thickness articular cartilage defect using stem cell-encapsulated thermogel. *Mater. Sci. Eng. C* **2018**, *88*, 79–87. [CrossRef] [PubMed]

18. Dong, X.; Chen, H.; Qin, J.; Wei, C.; Liang, J.; Liu, T.; Kong, D.; Lv, F. Thermosensitive porphyrin-incorporated hydrogel with four-arm peg-pcl copolymer (ii): Doxorubicin loaded hydrogel as a dual fluorescent drug delivery system for simultaneous imaging tracking in vivo. *Drug. Deliv.* **2017**, *24*, 641–650. [CrossRef] [PubMed]

19. Luo, Z.; Jin, L.; Xu, L.; Zhang, Z.L.; Yu, J.; Shi, S.; Li, X.; Chen, H. Thermosensitive PEG-PCL-PEG (PECE) hydrogel as an in situ gelling system for ocular drug delivery of diclofenac sodium. *Drug. Deliv.* **2016**, *23*, 63–68. [CrossRef] [PubMed]

20. Wang, S.-J.; Zhang, Z.-Z.; Jiang, D.; Qi, Y.-S.; Wang, H.-J.; Zhang, J.-Y.; Ding, J.-X.; Yu, J.-K. Thermogel-coated poly(epsilon-caprolactone) composite scaffold for enhanced cartilage tissue engineering. *Polymers* **2016**, *8*, 200. [CrossRef]

21. Navath, R.S.; Menjoge, A.R.; Dai, H.; Romero, R.; Kannan, S.; Kannan, R.M. Injectable PAMAM dendrimer-PEG hydrogels for the treatment of genital infections: Formulation and in vitro and in vivo evaluation. *Mol. Pharm.* **2011**, *8*, 1209–1223. [CrossRef] [PubMed]

22. Moon, H.J.; Choi, B.G.; Park, M.H.; Joo, M.K.; Jeong, B. Enzymatically degradable thermogelling poly(alanine-*co*-leucine)-poloxamer-poly(alanine-*co*-leucine). *Biomacromolecules* **2011**, *12*, 1234–1242. [CrossRef] [PubMed]

23. Zheng, Y.; Cheng, Y.; Chen, J.; Ding, J.; Li, M.; Li, C.; Wang, J.-C.; Chen, X. Injectable hydrogel-microsphere construct with sequential degradation for locally synergistic chemotherapy. *ACS Appl. Mater. Interfaces* **2017**, *9*, 3487–3496. [CrossRef] [PubMed]

24. Cheng, Y.; He, C.; Ding, J.; Xiao, C.; Zhuang, X.; Chen, X. Thermosensitive hydrogels based on polypeptides for localized and sustained delivery of anticancer drugs. *Biomaterials* **2013**, *34*, 10338–10347. [CrossRef] [PubMed]

25. Conde, J.; Oliva, N.; Atilano, M.; Song, H.S.; Artzi, N. Self-assembled RNA-triple-helix hydrogel scaffold for microRNA modulation in the tumour microenvironment. *Nat. Mater.* **2016**, *15*, 353. [CrossRef] [PubMed]

26. Li, X.; Ding, J.; Zhang, Z.; Yang, M.; Yu, J.; Wang, J.; Chang, F.; Chen, X. Kartogenin-incorporated thermogel supports stem cells for significant cartilage regeneration. *ACS Appl. Mater. Interfaces* **2016**, *8*, 5148–5159. [CrossRef] [PubMed]

27. Zhang, Y.; Ding, J.; Sun, D.; Sun, H.; Zhuang, X.; Chang, F.; Wang, J.; Chen, X. Thermogel-mediated sustained drug delivery for in situ malignancy chemotherapy. *Mater. Sci. Eng. C* **2015**, *49*, 262–268. [CrossRef] [PubMed]

28. Zhang, Y.-B.; Ding, J.-X.; Xu, W.-G.; Wu, J.; Chang, F.; Zhuang, X.-L.; Chen, X.-S.; Wang, J.-C. Biodegradable thermogel as culture matrix of bone marrow mesenchymal stem cells for potential cartilage tissue engineering. *Chin. J. Polym. Sci.* **2014**, *32*, 1590–1601. [CrossRef]

29. Sitoci-Ficici, K.H.; Matyash, M.; Uckermann, O.; Galli, R.; Leipnitz, E.; Later, R.; Ikonomidou, C.; Gelinsky, M.; Schackert, G.; Kirsch, M. Non-functionalized soft alginate hydrogel promotes locomotor recovery after spinal cord injury in a rat hemimyelonectomy model. *Acta Neurochir.* **2018**, *160*, 449–457. [CrossRef] [PubMed]

30. Ansari, S.; Diniz, I.M.; Chen, C.; Sarrion, P.; Tamayol, A.; Wu, B.M.; Moshaverinia, A. Human periodontal ligament- and gingiva-derived mesenchymal stem cells promote nerve regeneration when encapsulated in alginate/hyaluronic acid 3D scaffold. *Adv. Healthc. Mater.* **2017**, *6*. [CrossRef] [PubMed]

31. Liu, H.; Ding, J.; Li, C.; Wang, C.; Wang, Y.; Wang, J.; Chang, F. Hydrogel is superior to fibrin gel as matrix of stem cells in alleviating antigen-induced arthritis. *Polymers* **2016**, *8*, 182. [CrossRef]

32. Chaudhuri, O.; Gu, L.; Klumpers, D.; Darnell, M.; Bencherif, S.A.; Weaver, J.C.; Huebsch, N.; Lee, H.-P.; Lippens, E.; Duda, G.N.; et al. Hydrogels with tunable stress relaxation regulate stem cell fate and activity. *Nat. Mater.* **2016**, *15*, 326. [CrossRef] [PubMed]

33. Knight, E.; Przyborski, S. Advances in 3D cell culture technologies enabling tissue-like structures to be created in vitro. *J. Anat.* **2015**, *227*, 746–756. [CrossRef] [PubMed]

34. Fu, S.; Ni, P.; Wang, B.; Chu, B.; Zheng, L.; Luo, F.; Luo, J.; Qian, Z. Injectable and thermo-sensitive PEG-PCL-PEG copolymer/collagen/n-ha hydrogel composite for guided bone regeneration. *Biomaterials* **2012**, *33*, 4801–4809. [CrossRef] [PubMed]

35. Yuan, B.; He, C.; Dong, X.; Wang, J.; Gao, Z.; Wang, Q.; Tian, H.; Chen, X. 5-fluorouracil loaded thermosensitive PLGA-PEG-PLGA hydrogels for the prevention of postoperative tendon adhesion. *RSC Adv.* **2015**, *5*, 25295–25303. [CrossRef]

36. Wei, L.; Chen, J.; Zhao, S.; Ding, J.; Chen, X. Thermo-sensitive polypeptide hydrogel for locally sequential delivery of two-pronged antitumor drugs. *Acta Biomater.* **2017**, *58*, 44–53. [CrossRef] [PubMed]

37. Choi, Y.Y.; Jang, J.H.; Park, M.H.; Choi, B.G.; Chi, B.; Jeong, B. Block length affects secondary structure, nanoassembly and thermosensitivity of poly(ethylene glycol)-poly(L-alanine) block copolymers. *J. Mater. Chem.* **2010**, *20*, 3416–3421. [CrossRef]

38. Cheng, Y.; He, C.; Xiao, C.; Ding, J.; Zhuang, X.; Huang, Y.; Chen, X. Decisive role of hydrophobic side groups of polypeptides in thermosensitive gelation. *Biomacromolecules* **2012**, *13*, 2053–2059. [CrossRef] [PubMed]

39. Xu, W.; Ding, J.; Xiao, C.; Li, L.; Zhuang, X.; Chen, X. Versatile preparation of intracellular-acidity-sensitive oxime-linked polysaccharide-doxorubicin conjugate for malignancy therapeutic. *Biomaterials* **2015**, *54*, 72–86. [CrossRef] [PubMed]

40. Ding, J.; Shi, F.; Xiao, C.; Lin, L.; Chen, L.; He, C.; Zhuang, X.; Chen, X. One-step preparation of reduction-responsive poly(ethylene glycol)-poly(amino acid)s nanogels as efficient intracellular drug delivery platforms. *Polym. Chem.* **2011**, *2*, 2857–2864. [CrossRef]

Sample Availability: Samples of the compounds are available from the authors.

![molecules logo]

molecules

MDPI

Review

Smart and Functional Conducting Polymers: Application to Electrorheological Fluids

Qi Lu, Wen Jiao Han and Hyoung Jin Choi *

Department of Polymer Science and Engineering, Inha University, Incheon 22212, Korea;
22172314@inha.edu (Q.L.); 22151728@inha.edu (W.J.H.)
* Correspondence: hjchoi@inha.ac.kr; Tel.: +82-32-865-5178

Received: 10 September 2018; Accepted: 21 October 2018; Published: 2 November 2018

Abstract: Electro-responsive smart electrorheological (ER) fluids consist of electrically polarizing organic or inorganic particles and insulating oils in general. In this study, we focus on various conducting polymers of polyaniline and its derivatives and copolymers, along with polypyrrole and poly(ionic liquid), which are adopted as smart and functional materials in ER fluids. Their ER characteristics, including viscoelastic behaviors of shear stress, yield stress, and dynamic moduli, and dielectric properties are expounded and appraised using polarizability measurement, flow curve testing, inductance-capacitance-resistance meter testing, and several rheological equations of state. Furthermore, their potential industrial applications are also covered.

Keywords: conducting polymer; composite; electrorheological; smart fluid; viscoelastic

1. Introduction

Smart functional materials that can detect and identify external (or internal) stimuli, including electricity, light, heat, magnetic fields, stress, strain, and chemicals [1], have gained attention due to their various engineering applications, such as vibration controls [2], detection [3], electronics [4], and drug delivery [5]. Among these smart functional materials, electrorheological (ER) suspensions consisting of electro-responsive particles and an insulating fluid, such as silicone or mineral oil [6–8], play a crucial role. By engaging an electric field, the electro-responsive functional particles begin to polarize within milliseconds and the initially dispersed particles aggregate to build a chain form, following an external electrical field direction because of an induced dipolar attractive force. Thus, the microstructure of the ER suspension changes from a liquid-like to a solid-like form, as shown in Figure 1. Therefore, their physical and rheological behaviors, such as steady shear viscosity, shear stress, dynamic modulus, and yield stresses, and even stress relaxation changes according to the applied electric fields; however, these changes can be reversed by removing the field. Most ER fluids suit the Bingham fluid model, which illustrates that a fluid has its own yield stress, and the flow motion of the fluid impedes when an input external shear stress is lower than the yield stress [9]. ER fluids with this controllable and tunable phase change have a very good market prospect and wide applications, such as dampers [10], ER haptic devices [11], shock absorbers, microfluidics [12–14], clutches [15], ER finishing devices [16], and tactile displays [17].

ER fluids have been the subject of research and exploration since their discovery [18]. Furthermore, research to find fibrillation structures between two applied electrodes with an apparent increase of the viscosity of ER fluids have also been conducted. Klass and Martinek [19] introduced a dielectric principle, and a water-associated electrical double-layer model, to explain that the critical factor of the ER effect was the molecule in the wet-base ER fluids. The particles were polarized and distorted when the extra electric field was applied and then the water molecules made an attractive bridge between dispersed particles, thereby showing a higher surface tension.

Molecules **2018**, *23*, 2854

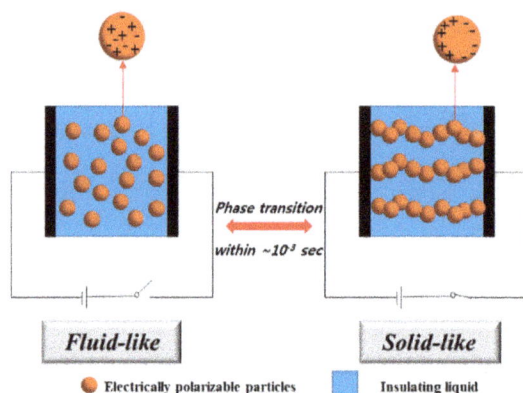

Figure 1. Schematic diagram of electrorheological (ER) behavior with an external electrical field.

Furthermore, ER materials cover a wide range of organics and inorganics particles, including polymers and their organic or inorganic composites, with their particle size ranging from nanometers to micrometers with multiple shapes, such as spherical, fibrous, sheet, and rod-like [20,21].

In this study, we focus on anhydrous ER materials with various conducting polymers and their nanocomposites. Conducting polymers possess conjugated repeating units, which can be appraised by some important variables, such as ionization potential, electron affinity, band gap, and bandwidth [22–24]. By regulating these parameters the thermal, electrical, and optical properties of polymers can be easily enhanced [25,26]. In particular, the electrical conductivity and thermal or mechanical stability of nanocomposites, which have improved mechanical property, dispersion stability, or physical characteristics, and can be easily controlled by various fabrication techniques [27].

Sundry conducting polymers have been employed as ER materials, including polyaniline (PANI) [28,29], polypyrrole (PPy) [30], poly(*p*-phenylene) (PPP) [31–33], poly(3,4-ethylenedioxythiophene) (PEDOT) [34], and so on. This study focuses on the ER effect of conducting polymeric materials and their composites, including synthesis methods and typical characteristics, using various rheological measurement methods. Furthermore, according to the viscoelastic behaviors, such as flow curves, shear viscosity, yield stresses, and dynamic modulus, the ER effect of polymeric materials dispersion is compared under an external electrical field with various strengths.

2. Polymeric ER Materials

Oxides or non-oxide inorganic particles were introduced in ER fluids at the initial stages of study. However, the dispersion stability was poor because of the large density variation between the dispersed particles and the base oil. Therefore, organic particles gained attention, especially conducting polymers, which are provided with conjugated π bonds along with a high dielectric constant, thereby exhibiting excellent polarization capabilities under applied electric field strengths. Moreover, these types of polymers can be comprehensive as ER materials because they have highly polarizable functional groups on the molecular backbone.

2.1. PPy

Among the various conductive polymers, PPy is a research hotspot because of its high electrical conductivity, simple method of preparation [35], ease of surface modification [36], excellent environmental stability [37], and ion-exchange capacity [38]. In addition, it can be easily produced in a large quantity at a room temperature using various solvents, including water. PPy is an excellent conductive polymer; thus, it can be applied in ER fluids.

Kim and Song [33] dispersed PPy in mineral oil as an ER fluid and controlled the amount of oxidant and surfactant while synthesizing PPy particles to investigate the impacts on their ER effect. They observed that as the amount of oxidant and surfactant increases the yield stress increases at the initial stage but decreases after passing culmination because of the enhanced particle polarization. Using thermo-oxidative treatment in air, Xia et al. [39] tuned the electrical conductivity of the PPy nanofibers (shown in Figure 2b) to that of a nanofiber-based ER fluid. Its yield stress and shear modulus were stronger compared to those from a typical granular-shaped PPy based ER suspension, because the fiber-like particles can ameliorate the dispersion stability of the suspension. Many studies have been conducted on preparing conducting PPy-based composite materials to surmount the poor processability of PPy particles.

Moreover, while the morphology of the conducting polymers is well-known to be related to their synthetic conditions, such as temperature, pH, oxidant species, oxidant/monomer ratio, polymerization type, and so on, shape- or size-related morphology of the dispersed particles became an important parameter in the rheological output of ER fluids. Specifically, regarding the effect of a hierarchical form on ER properties, sea urchin-like conducting polymeric particles are known to exhibit both better sedimentation stability than those with a smooth surface and better rheological behaviors. They are expected to show larger drag forces and inter-particular frictions with their increased surface area compared to that of the smooth-surfaced particles. Therefore, an improved ER efficiency with larger yield stress can be expected in the same experimental conditions. Moreover, compared to the ER fluid of spherically shaped particles, the systems with fiber-like particles were also observed to exhibit better ER characteristics and enhanced suspension stability because the elongated shape possesses a bigger induced-dipole moment and greater inter-particle interaction.

Concurrently, clays are widely used to compose polymer composites not only because of their large specific surface area and good mechanical properties, but also because of their chemical, thermal, and dimensional stability and low cost [40]. Halloysite (HNT) is a type of clay mined from natural deposits and has a predominantly hollow tubular structure. HNT has been widely used in electronics, catalysis, separation, and functional materials because of its uniquely versatile surface features. Jang et al. [41] synthesized PPy/HNT nanocomposites (Figure 2c) and observed that adding HNT is an effective method to improve PPy conductivity and properties of PPy-based ER fluids. Furthermore, montmorillonite (MMT) clay is widely used in conducting polymer nanocomposites because their particle size is smaller than 10 μm and they can be easily intercalated. Cations, such as Na^+ and Ca^{2+}, in its lamellar surface enable interactions with monomers ensuring reasonable conductivity of the associated electric signal. Following testing, Kim et al. [42] observed that PPy-dodecylbenzenesulfonic acid (DBSA)/clay nanocomposites have better insulation and thermal stability. PPy was also inserted into hexagonal channels of pore-expanded MCM-41 [43], whose insulating channel can improve the ER response of the PPy suspension (Figure 2d).

Not only clay, but also conventional polymers were introduced in conducting polymer/clay nanocomposites to overcome the limitation of poor processability, prepare conducting polymer-based composite materials, and improve the sedimentation of ER suspensions. Kim et al. [44] assembled PPy-polycaprolactone (PPy-PCL) composites (Figure 2e) to improve the mechanical and electrical properties of PPy, along with enhancing their ER response. Furthermore, Kim and Park [45] reported an ER response by coating PPy on polyethylene particles, thereby enhancing polarization by increasing the particle surface conductivity.

Moreover, compared to single-component counterparts, core-shell structured particles often exhibit better physical and chemical properties. A conducting shell can be polarized instantaneously; thus, its ER response can be improved. In addition, an insulating core or an insulating shell in the composite particles is designed to obtain a lower current density, by which the practicability and environmental protection of the materials can be improved [46,47]. Kim et al. [48] synthesized PPy-coated silica core-shell structured nanoparticles (Figure 2f), while Sedlačík et al. [49] synthesized titanium oxide (TiO_2)/PPy core-shell structured nanoparticles (Figure 2g) and analyzed the steady-state and oscillatory flow behaviors. Both sets of results showed that the novel core-shell structure possessed predominant maneuverability and ER response.

Figure 2. SEM morphologies of polypyrrole (PPy) and its composites: (**a**) granular PPy and (**b**) PPy nanofibers, reprinted with permission from [39]. Copyright Elsevier, 2011; (**c**) PPy-HNT particles, reprinted with permission from [41]. Copyright Springer Nature, 2014; (**d**) PPy/MCM-41 particles, reprinted with permission from [43]. Copyright Elsevier, 2010; (**e**) PPy-polycaprolactone (PPy-PCL) composites, reprinted with permission from [44]. Copyright Springer Nature, 2007; and, core-shell structures of (**f**) PPy-silica, reprinted with permission from [48]. Copyright RSC Publishing, 2016 and (**g**) PPy-TiO$_2$, reprinted with permission from [49]. Copyright Springer Nature, 2011.

2.2. PANI

Among various whole conductive polymers, PANI is considered an essential candidate material for ER fluids because of its ease of synthesis, low price, chemical and thermal stabilities, and sensitive response to easily controllable electrical conductivity, which can be steered by a reversible doping/dedoping process [50]. Most morphologies of a synthesized PANI are irregular. However, using some methods, such as interfacial polymerization [51] and rapid mixing method [52], and using a surfactant or template [53] to direct PANI growth, we can obtain various shapes of fibers, tubes, spheres, and even urchin-like particles (shown in Figure 3).

For ER fluid applications, the high conductivity of PANI is undesirable because of the high leakage currents passing through the suspension with the electric field [54]. Thus, the protonated PANI needs to be changed to a PANI base by treating it with an ammonium hydroxide solution. The electrical properties of the protonated PANI and its corresponding bases can be controlled at a molecular level by partial protonation and acid selection of substitution of the PANI ring, or copolymerization of aniline with substituted aniline or other monomers, as shown in Scheme 1. In addition to the powder, PANI can be deposited on almost any object in the reaction mixture during aniline oxidation, or as a colloid when the reaction mixture contains a stabilizer.

Figure 3. SEM images of synthesized polyaniline (PANI): (**a**) spheres, reprinted with permission from [55]; Copyright Spring Nature, 2013; (**b**) fibers, reprinted with permission from [52]. Copyright American Chemical Society, 2009; (**c**) urchin-like, reprinted with permission from [56]. Copyright John Wiley and Sons, 2009; and (**d**) nanotubes, reprinted with permission from [57]. Copyright American Chemical Society, 2006.

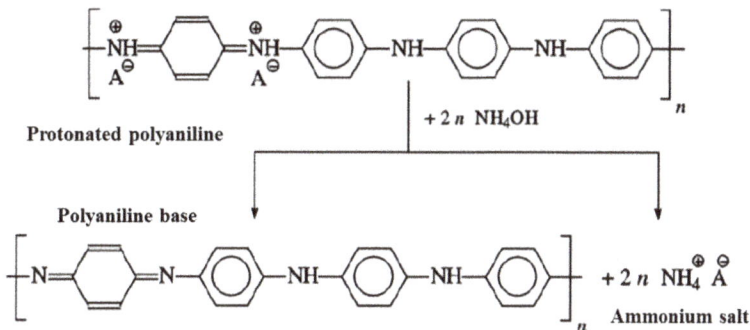

Scheme 1. Conductive protonated PANI is converted to a non-conducting PANI base by deprotonation with ammonium hydroxide.

To improve the ER response of PANI particles and obtain an appropriate semiconducting regime for an ER application, multifarious inorganics, such as clays, TiO_2, silica, and $BaTiO_3$, and even magnetic materials, such as carbonyl iron or Fe_3O_4 (shown in Figure 4) were introduced. Furthermore, some non-conducting core polymers form a core-shell structure with PANI as a shell (shown in Figure 5). For example, as one type of clay, attapulgite (ATP) or palygorskite, which is a kind of crystalline hydrated magnesium silicate with a fiber shape, has been introduced for PANI/clay composites. It not only possesses a large specific surface area but also cation exchange capacity and a reactive hydroxyl group on its surface [58]. Han et al. [59] synthesized PANI/ATP composite materials and then estimated their ER responses, observing that they were a feasible and efficacious method for obtaining good mechanical strength, and easy processing of composite ER material. Furthermore, sepiolite (SPL), another species of natural silicate clay with a fibrous morphology that has a large surface area and porosity showing its mechanical reinforcing capability [60], has also been introduced. Owing to that, Jang and Choi [61] used SPL synthetizing with PANI to obtain a neoteric ER material. As a multi-functional physical cross-linker, laponite, a synthetic clay material, which was adopted to stabilize emulsions because of its uniform and small dimensions, can lead to non-covalent cross-linking on the clay/polymer interface [62]. Jun et al. [63] combined PANI and laponite confirming that this composite displayed a typical ER behavior. Noh et al. [64] reported neoteric $Fe_3O_4/SiO_2/PANI$ nanoparticles, which can be applied in both ER and magnetorheological (MR) fluids. In addition, they observed that the ER and MR responses were significantly improved under both electric and magnetic fields.

Figure 4. SEM and TEM photos of: (**a**) PANI/ATP, reprinted with permission from [59]. Copyright Elsevier, 2017; (**b**) PANI/SPL, reprinted with permission from [61]. Copyright Elsevier, 2015; (**c**) PANI/laponite, reprinted with permission from [63]. Copyright Elsevier, 2015; and (**d**) $Fe_3O_4/SiO_2/PANI$ nanoparticles, reprinted with permission from [64]. Copyright Royal Society of Chemistry, 2017.

Moreover, PANI has also been applied comprehensively with various polymeric shells. It is not sufficiently sophisticated to form a PS/PANI nanosphere, due to π-π stacking attractive interactions between an aniline monomer and sulfonating polystyrene (PS) using concentrated sulfuric acid. Kim et al. [65] compared two types of PS/PANI-based ER fluids whose PANI shell was either smooth or urchin-like. In Lee et al.'s work [66], the glycidyl methacrylate (GMA) was used to graft an aniline monomer with poly(methyl methacrylate) PMMA to form a PANI/PMMA core-shell structure. However, in Zhang et al.'s report [67], the core-shell PGMA/PANI composites were synthesized and applied in ER fluids.

Figure 5. SEM images of: (**a**) smooth PANI@PS and (**b**) urchin-like PANI@PS, reprinted with permission from [65]. Copyright RSC Publishing, 2015; and (**c**) PGMA/PANI composite, reprinted with permission from [67]. Copyright Elsevier, 2013.

2.3. Other Conducting Polymers

In addition to either PPy or PANI, various other conducting polymers and their derivatives were applied in ER fluids. Although PANI has its own advantages in terms of controllable conductivity, environmental stability, and simple synthesis, it also exhibits limitations, such as poor solubility in various organic solvents. To overcome these defects, various derivatives were introduced [68]. Poly(diphenylamine) (PDPA) as a derivative of PANI, possesses not only a relatively low conductivity, which helps avoid cause short-circuiting and can be directly applied in an ER fluid, but also produces improved solubility in organic solvents with better thermal stability [69]. Kim et al. [70,71] at first synthesized normal PDPA nanoparticles for an ER fluid, and then further fabricated a novel core-shell-type PDPA/PS structure (shown in Figure 6), thereby observing that both these materials were suitable for ER fluids.

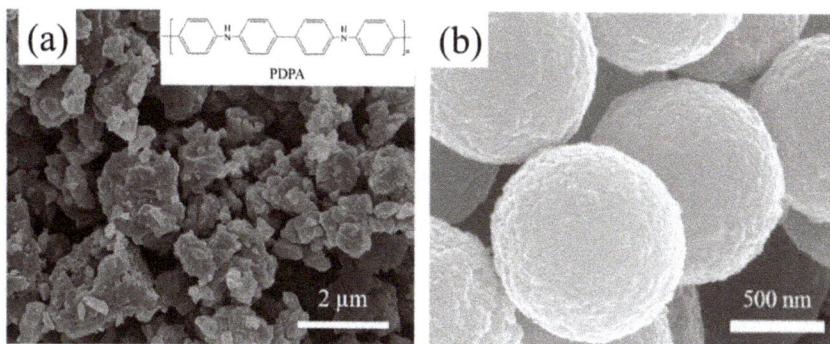

Figure 6. SEM images of (**a**) Poly(diphenylamine) (PDPA), reprinted with permission from [70]. Copyright Elsevier, 2017 and (**b**) PDPA/PS core/shell structure, reprinted with permission from [71]. Copyright Elsevier, 2017.

The chemical formula of a *p*-phenylenediamine (PDA) is similar that of an aniline. It inspires its ER application with its polymeric form [72] along with its versatile synthetic methods to prepare poly(*p*-phenylenediamine) (PPDA, whose schematic diagram is shown in Scheme 2) particles as compared to PANI. Plachy et al. [73] tried to modify PPDA by heat treatment to achieve appropriate conductivity; however, they observed that the ER efficiency decreased as the carbonization temperature increased. Cao et al. [74] fabricated poly(*p*-phenylenediamine)/graphene oxide (PPDA/GO) composite particles, and observed that both PPDA/GO particles and PPDA-based ER fluids exhibited significant ER characteristics.

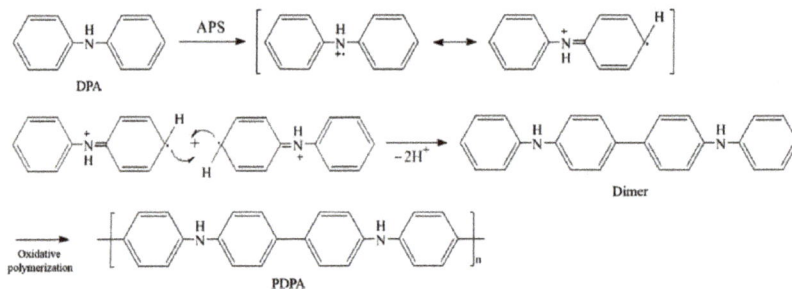

Scheme 2. Schematic diagram of PDPA synthesis via an oxidative polymerization.

It can be also noted that by introducing proper substituents on polymeric backbones, functionalized PANI could be fabricated, providing extra tuning capability of their chemical and electrical characteristics [75,76].

Recently, poly(N-methylaniline), as a derivative of PANI, has piqued the interest of many researchers as, compared to PANI, it possesses much lower electrical conductivity. Thereby, it could be directly applied to the ER fluid without further treatment. Moon et al. [77] used a simple method to fabricate a core-shell typed poly(methyl methacrylate)/poly(N-methylaniline) (PMMA-PNMA). Figure 7 shows the synthetic fabrication of the PMMA-PNMA particles. Initially, the surface of the PMMA microspheres, which were fabricated through a dispersion polymerization process, was treated with glycidyl methacrylate, ethylene glycol dimethacrylate, and oxydianiline as swelling, crosslinking, and chemical-grafting agents, respectively. Further, the PNMA shell was coated by chemical oxidation polymerization. Their ER suspension based on the PMMA-PNMA particles demonstrated conventional ER behaviors in both simple shear flow and dynamic oscillation measurements for several different electric field strengths applied.

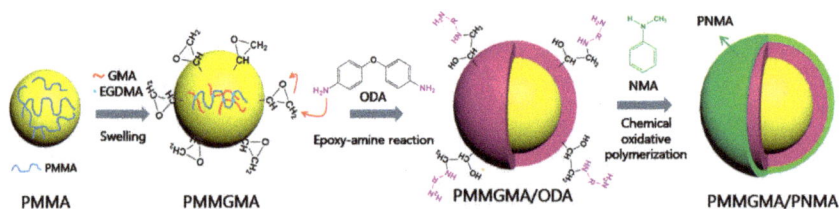

Figure 7. Schematic diagram of fabrication of poly(methyl methacrylate)/poly(N-methylaniline) (PMMA–PNMA) core-shell particles, reprinted with permission from [77]. Copyright Elsevier, 2015.

Polyindole (PIN) has attracted huge attention due to its good thermal stability and high redox activity [78]. Indole is composed of a benzene ring connected to the heterocyclic pyrolytic ring; thus, it was anticipated that it would show the properties of both PPy and PPP. Park et al. [79] synthesized emulsion-polymerized PIN nanoparticles and used these as an ER material. Sari et al. [80] synthesized PIN/polyethylene composites and observed that they were thermally more stable than PIN, and the ER suspensions was observed to be increasing with increases in the electric field.

Poly(O-anisidine) (POA), with a methoxy functional group ($-OCH_3$) at an ortho position, lowers the mobility of electric charge carriers following the polymeric backbone, thereby lowering the electrical conductivity. Recently, POA was adopted as an ER material. Cheng et al. [81] used chemical oxidation polymerization to synthesize hollow POA microspheres with 1 μm diameter in a sphere-like form. The hollow POA particles had low density, which is an advantage for a high-stability dispersed phase ER fluid. Furthermore, Lee et al. [82] fabricated a core-shell structure of POA-silica particles using *N*-[(3-trimethoxylsilyl)-propyl] aniline (PTMSPA) grafting agent and defined their ER properties (as shown in Figure 8).

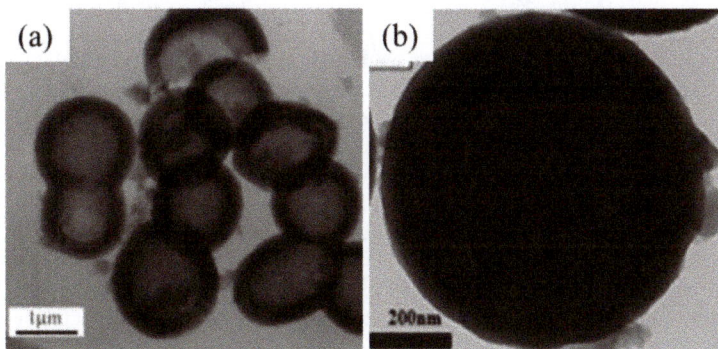

Figure 8. TEM images of (**a**) hollow poly(*O*-anisidine) (POA), reprinted with permission from [81]. Copyright IOP Publishing, 2011 and (**b**) POA/silica core/shell structure, reprinted with permission from [82]. Copyright IOP Publishing, 2017.

Polyacene quinone radicals (PAQRs) are also a type of semi-conductive polymers, which possess the p-conjugated structure. They have been widely used and studied because of their better thermal stability, environmental stability, and the characteristic that their conductivity can be controlled by polymerization conditions (such as temperature, acenes, and anhydrides). Wang et al. [83] synthesized hollow PAQR sub-microspheres by a modified solvo-thermal method and then observed that their ER fluid exhibited conventional ER properties. Poly(naphthalene quinone) (PNQR) has a similar chemical p-conjugated structure as PAQR, and so also has an electrical conductivity and can be applied as an ER fluid material. Cho et al. [84] synthesized PNQR from the Friedel–Craft acylation and its flow behaviors when suspended in the silicone oil were analyzed using rheological equations of state from Bingham, De Kee-Turcotte, and Cho-Choi-Jhon (CCJ) equations.

Meanwhile poly(3,4-ethylenedioxythiophene)(PEDOT):poly-(styrenesulfonic acid) as a derivative of polythiophene, has attracted considerable attention because of its several advantages as a conducting polymer, such as processability in aqueous solutions, and better thermal and structural stability [85]. An et al. [86] fabricated core/shell typed polystyrene (PS)/PEDOT microspheres and reported their ER behaviors under an electrical field. On the other hand, Erol et al. [87] synthesized core/shell typed nano-rods with a titania core and conducting polymeric PEDOT shell. They were fabricated via covalent bonding to obtain a thin polymeric shell, making the interfacial interactions between the two components stronger. They found that the nanorod-shaped TiO_2/PEDOT particles exhibited improved ER performance and storage moduli, as well as larger creep-recovery after stress loading, when compared to their particulate morphology.

Furthermore, to solve problems such as high conductivity and low thermal stability of PANI, aniline is often copolymerized with other conductive polymer monomers to form a more functional conductive polymer. Cho et al. [88] synthesized poly(aniline-*co*-pyrrole) (PAPP) particles (aniline:pyrrole = 1:1) via a chemical oxidation polymerization technique. It was proved that the copolymer has better thermal stability and more suitable conductivity than PANI as the dispersed phase in ER fluid. Pandey et al. [89] fabricated poly(aniline-*co*-*O*-anisidine) with a lower value of electrical conductivity by explaining the incorporated *O*-anisidine moieties in PANI backbones. Trlica et al. [90] also reported a poly(aniline-*co*-1,4-phenylenediamine) particles-based ER suspension with an applied external electric field, along with electrical characteristics of an ER suspension.

2.4. Composites

Various inorganic-conducting polymers have been introduced for their ER applications, and among these, recently graphene has attracted great interest in the fields of capacitors, electronic devices, sensors, and composites due to its unique structure and electrical characteristics, even in ER fluids [91]. Although graphene oxide (GO) which is the oxidized form of graphene has many defects during its preparation process, it contains functional components such as an hydroxyl group, epoxy group, and carboxyl group, making it more easily dispersed in water [92]. By simply compounding with other polymers, it can be modified to have multiple functions. Zhang et al. [93] synthesized PANI/GO composites by an in-situ oxidation polymerization method with a stable GO and reported its composite particle-based ER suspension showed a general ER property. On the other hand, Min et al. [94] fabricated core-shell typed PMMA/GO microspheres via a Pickering emulsion polymerization method with the GO as the solid surfactant, and studied their ER response. Cao et al. [74] synthesized poly(*p*-phenylenediamine)(PPDA)/GO composite particles via an in-situ oxidation polymerization technique of *p*-phenylenediamine in GO suspension, observing improved typical ER behavior when compared to PPDA particle-based ER fluid.

Furthermore, carbon nanotubes as another candidate for ER fluids have received a lot of consideration [95–97], although compared with single-walled carbon nanotubes (SWNT), multi-walled carbon nanotubes (MWNT) have been more widely used in the ER field due to their lower electrical conductivity. Jin et al. [98] produced MWNT-adsorbed PS and PMMA particles using different surfactants, such as anionic sodium dodecyl sulfate and sodium dodecylbenzene sulfonate, and cationic cetyl trimethylammonium bromide. Not only conductive carbon materials, but also many metal oxides, can be compounded with conductive polymers to form multifunctional materials showing remarkable ER effects. Fang et al. [99] synthesized magnetic nanoparticles (Fe_3O_4) together with a conductive PPy using a simple fabrication method, and applied it to both ER and MR fluids.

2.5. Poly(Ionic Liquid)s

Poly(ionic liquid)s (PILs), initially evaluated as ion conductive polymers, refer to a new type of polymers in which ionic liquids (ILs) are linked through polymeric backbones to form macromolecular structures. They have attracted huge scientific interest not only because they have the suitable electrical conductivity and thermal stability from the ILs, but also because, like polymers, they have the advantages of good solid morphology, simple processing and good mechanical properties [100,101]. To activate the ER effect, classical polyelectrolytes need to adsorb a tiny amount of water to boost the local ion movement which will cause chemical corrosion, dielectric breakdown, and limited operation at higher temperatures. Nevertheless, because of fluoric counter-ions including tetrafluoroborate (BF_4^-), hexafluorophosphate (PF_6^-), and fluorinated imide (($C_nF_{2n+1}SO_2)_2N^-$), PILs are generally organophilic and insoluble in water. Using a microwave-assisted dispersion polymerization technique, Dong et al. [102] synthesized mono-dispersed PIL particles possessing a styrenic backbone and different sizes of cationic/anionic parts (shown in Figure 9) and reported their enhanced anhydrous ER characteristics. Then they further synthesized several mono-dispersed poly[*p*-vinylbenzyl trimethylammonium] PIL microspheres possessing a similar diameter but different counter-anions. This showed that PILs with a smaller volume or counter-ion size exhibited higher ER behavior compared to PILs possessing organic counter-ions or inorganic counter-ions alone [103].

Figure 9. SEM photos of poly(ionic liquid)s (PIL) particles: (**a**) poly((*p*-vinylbenzyl trimethylammonium) (VBTMA)$^+$CF$_3$SO$_3$); (**b**) poly(VBTMA$^+$ (CF$_3$SO$_2$)$_2$N$^-$); and (**c**) poly((*p*-vinylbenzyltriethylammonium bis(trifluoromethane sulfonimide) (VBTEA)$^+$ (CF$_3$SO$_2$)$_2$N$^-$), reprinted with permission from [102]. Copyright Elsevier, 2016.

3. ER Characteristics

3.1. Direct Observation

The electrical stimuli response characteristics of ER fluids can be directly detected by an optical microscope (OM) with an external electric field [84]. The polarization measuring device is presented in Figure 10a. The ER suspension was placed in the center of a transparent substrate connected to a DC high voltage power supply, and micro-structural changes were captured by the OM. Sever and Unal [104] prepared an ER fluid by dispersing nanocube-TiO$_2$/poly(3-octhlthiophene) in silicone oil and observing its polarizability and potential ER performance with this OM device technique. The particles were suspended randomly similar to a liquid state when E = 0 kV/mm, while they were electrically polarized and migrated to organize a columnar structure parallel to a field direction when E $\neq$ 0 kV/mm (Figure 10b). The formation of this chain like-columnar structure caused ER fluid to exhibit a solid-like behavior, which was attributed to the polarization of the particles [8]. Furthermore, this columnar structure was found to be stronger and thicker, in accordance with the increment in applied E range from 0.5–2.5 kV/mm and could be maintained as long as an external electrical field applied, indicating that the increased E caused a tighter columnar structure. These results are consistent with OM studies carried out for various ER fluid materials [105–107].

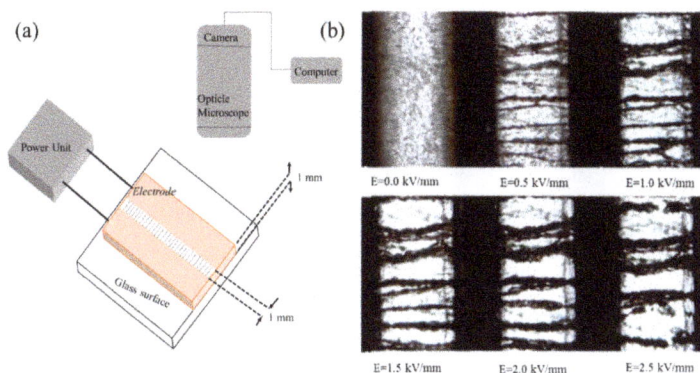

Figure 10. (**a**) Schematic illustration of polarizability measuring device; (**b**) optical microscope (OM) images showing the polarization of nanocube-TiO$_2$/poly(3-octhlthiophene)-based ER fluid, reprinted with permission from [104]. Copyright John Wiley and Sons, 2016.

3.2. Flow Curve Behaviors

Stable steady shear flow curves in ER fluids can be established via a typical rotation rheometer test, like the controlled shear rate (CSR) method, which provides a basis for observing the yield stress of ER suspensions and electrical stimulus response behavior [21,108]. The flow curve of ER fluid produced from this CSR method exhibits shear stress (τ) and shear viscosity (η) as a function of shear rate ($\dot{\gamma}$).

In general, two types of interactions occur between dispersed ER particles in a steady shear flow; the electrostatic interaction caused by an external electrical field and a hydrodynamic interaction induced by a steady shearing flow [109]. The shear flow field results in shear deformation of the chain structures formed under an electric field; however, the broken structure tends to rebuild the chain simultaneously under an applied external electrical field. These phenomena of destruction and reorganization depend on the applied shear stresses and the electric field strengths. At a low shear rate regime, the electro-static attractive interaction becomes prominent, allowing the chain structure to reduce the effect from shear deformation [110]. As a shear rate increases, the breaking rate exceeds the recombination rate, so that the chain structure will eventually be destroyed and will exhibit a liquid-like behavior. However, a stronger electric field can result in a stronger electrostatic interaction which is sufficient to resist shear deformation, allowing the solid-like behavior of the ER fluid to re-emerge.

Based on the electrical stimuli-response behaviors of ER fluids in a shear flow field, many models have been introduced to qualitatively explain the flow curve. The Bingham model (Equation (1)), the simplest equation with two parameters, has been extensively employed as a viscoelastic equation to fit steady shear behaviors exhibited in ER fluids along with various suspensions and MR fluids [111]:

$$\tau = \tau_y + \eta\dot{\gamma} \quad \tau \geq \tau_y$$
$$\dot{\gamma} = 0 \qquad \tau < \tau_y \tag{1}$$

In Equation (1), $\dot{\gamma}$ is a shear rate; η is a shear viscosity; τ and τ_y are shear and yield stresses, respectively. Piao et al. [112] explained shear stress behaviors of PS/PANI-based ER suspension using this Bingham fluid model, in which the flow curve, as shown in Figure 11a, was fitted by this model quite well. Its shear stress was observed to increase linearly with increased shear rate when E = 0 kV/mm, which is similar to that of Newtonian fluids. Under an applied electrical field, the ER suspension demonstrated increments in shear stresses and exhibited a plateau behavior because of the formation of the columnar structure.

However, in the case of Figure 11b [59], shear stresses of the PANI@ATP-based ER suspension presented a weak concave shape in a low shear rate area, passed through a minimum value, and once again increased. This behavior could be explained by the fact that the reconstituted structure is not as complete as the structure formed before the steady shearing flow applied. Thus, they adopted the Cho-Choi-Jhon (CCJ) equation [84], which is an empirical formula with six parameters to express the ER characteristics in a whole shear rate range. This CCJ equation is given as follows:

$$\tau = \frac{\tau_y}{1 + (t_1\dot{\gamma})^\alpha} + \eta_\infty \left(1 + \frac{1}{(t_2\dot{\gamma})^\beta}\right)\dot{\gamma} \tag{2}$$

The first term corresponds to the decrease in the shear stress at the low shear rate area, and the next term depicts a shear-thinning property at the high shear rate area. The t_1 and t_2 parameters are time constant, and η_∞ is a shear viscosity at the infinite shear rate. The exponent α represents the shear stress decrease, and b is located between 0 and 1. This model was observed to be a good indication of the overall behavior of various ER fluids [61,87,113], including the reduced shear stress phenomenon.

In addition, the De Kee-Turcotte equation (Equation (3)) and Herschel-Bulkley equation (Equation (4)), as a modified version of the Bingham equation, have also been applied for various ER suspensions [114,115]:

$$\tau = \tau_0 + \eta_1 \dot{\gamma} e^{-t_1 \dot{\gamma}} \tag{3}$$

$$\tau = \tau_0 + m \dot{\gamma}^n \tag{4}$$

where t_1 is a time constant, m is a consistency index, and n is a flow index. Cho et al. [84] constructed the flow curve of a poly(naphthalene quinone)-based ER suspension for different particle concentration and electric field strengths, and interpreted its steady-state flow properties using three different equations from the Bingham, De Kee-Turcotte, and CCJ models. As presented in Figure 11c, the CCJ equation fitted the decrement of shear stress better for a shear rate range from 0.01–100 s^{-1}; however, significant deviation was found at a low shear rate region compared to other two equations, especially for a lower volume fraction of poly(naphthalene quinone)-based ER fluid.

Based on the above models, Seo and colleagues [116] proposed a new rheological model (Equation (5)) that can depict the flow behavior of the ER suspensions over the full range of shear rates with four parameters.

$$\tau = \tau_{sy} \left(1 - \frac{\left(1 - \exp\left(-a\dot{\gamma}\right)\right)}{\left(1 + \left(a\dot{\gamma}\right)^{\alpha}\right)} \right) + \eta \dot{\gamma} \tag{5}$$

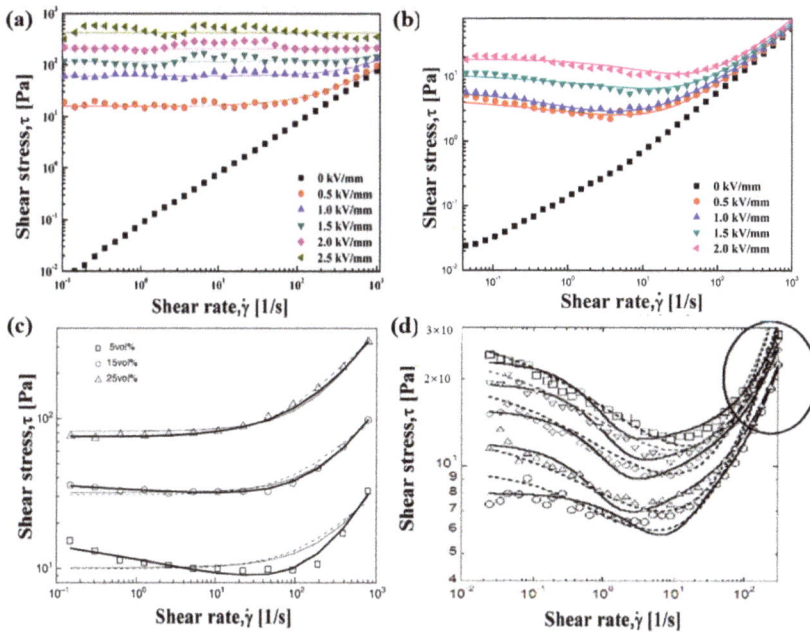

Figure 11. Shear stress vs. shear rate for: (**a**) PS/PANI-based ER suspension (solid lines for Bingham equation), reprinted with permission from [112]. Copyright Elsevier, 2017; (**b**) PANI@ATP-based ER suspension (solid lines for CCJ equation), reprinted with permission from [59]. Copyright Elsevier, 2017; (**c**) poly(naphthalene quinone)-based ER fluid (dashed lines for Bingham equation, solid lines for De Kee-Turcotte equation, and bold solid lines for CCJ equation), reprinted with permission from [84]. Copyright American Chemical Society, 2012; and (**d**) PMMA/PANI-based ER fluid (dotted and solid lines for CCJ equation and four-parameter equation), reprinted with permission from [117]. Copyright Elsevier, 2005.

This four-parameter model was introduced to predict static yield stress, while the CCJ and Bingham models were based on dynamic yield stress. In Equation (5), *a* is a time constant, τ_{sy} represents a static yield stress, and α is a power-law index relating to the shear thinning. In Figure 11d, Seo and Seo [117] analyzed the flow curves of PMMA/PANI-based ER suspension using the CCJ equation and their proposed equation. At a low shear rate regime, the shear stresses predicted by the new model converged to yield stress, while the prediction of the CCJ model continued to increase as the shear rate decreased. Moreover, at the high shear rate area, the prediction of the CCJ equation deviated from the experimental data, while the new equation prediction was considered to match the experimental data better.

3.3. Yield Stress

Yield stress, as a key parameter of rheological properties in suspensions, can provide information about the performance of the ER fluids depending on electrical field strength, volume fraction, and dielectric properties of the particles. Many test methodologies have been applied to study yield stress characteristics, corresponding to different yield stress values including dynamic, elastic, and static yield stresses [118].

As shown in Figure 12, Lim et al. [119] studied dynamic, static, and elastic yield stresses vs. electrical field strength for a PANI-nanotube-based ER suspension, estimated from controlled shear rate, controlled shear stress, and dynamic oscillatory frequency tests, respectively. The results showed that the static yield stress had the highest value because the static state that included more aggregate without shear rate control, as compared to the ER fluid under the dynamic condition. The elastic yield stress is obtained from the maximum shear stress required to achieve full recovery after the shear stress is removed, which is the lowest. Nonetheless, these three types of yield stress as a function of electric field strength presented equivalent behaviors.

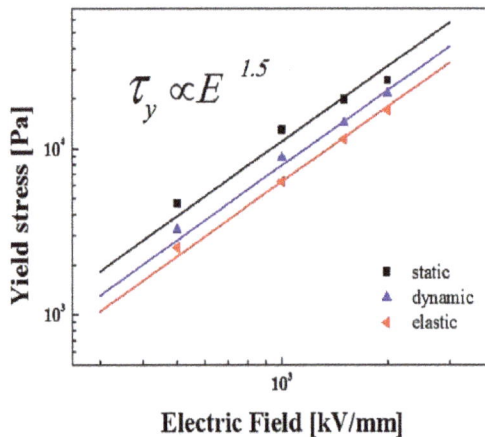

Figure 12. Static, dynamic, and elastic yield stresses as a function of electrical field strength for PANI-nanotube-based ER suspension, reprinted with permission from [119]. Copyright Elsevier, 2017.

As mentioned above, the electric field strength dependent yield stress has been the focus of many studies on ER characteristics. Klingenberg et al. [6] introduced an electrostatic polarization model and calculated the functional relation between yield stress and electrical field strength as follows:

$$\tau_0 = 18\varphi\epsilon_0\epsilon_e\beta^2 E_0^2 f_m \left[1 - \frac{(\pi/6)^{1/2}}{(L/a)\tan\theta_m \varnothing^{1/2}}\right] \tag{6}$$

where φ represents a volume fraction and L represents the electrode separation. f_m and θ_m are the maximum dimensionless restoring force and angle, respectively. This expression means that the dynamic yield stress is essentially proportional to the squared electrical field strength. However, although some experiments confirmed this result [120,121], the effects of the particular material, concentration, shape, and electric field strength will cause the index to deviate from 2.0 [122,123]. As shown in Figure 13, Sung et al. [124] observed that the yield stress of a highly substituted potato starch phosphate (HPSP)-based ER fluid was not proportional to E^2, which deviated from the polarization model because of the non-spherical shape of HPSP.

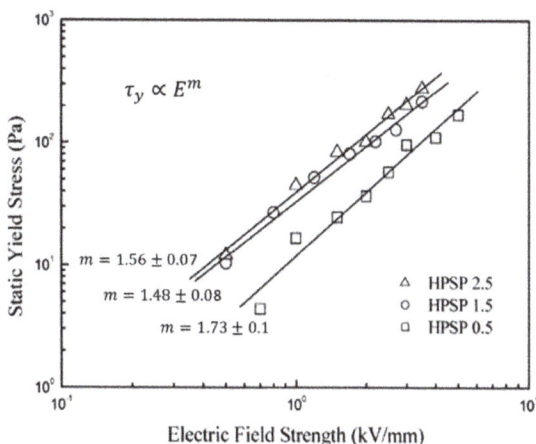

Figure 13. Yield stress vs. electrical field strength for highly substituted potato starch phosphate (HPSP)-based ER suspension with different molar ratios, reprinted with permission from [124]. Copyright American Chemical Society, 2005.

Concurrently for conducting suspensions under a high electric field strength, the ER characteristic would be affected by a conductivity mismatch and an interaction between dispersed particles and dispersing oil, rather than by their dielectric constant mismatch. Davis et al. [125] investigated the effect of field-dependent fluid conductivity and proposed that the yield stress appeared to approach $E^{3/2}$ under high electric field strength (consistent with Figure 12) [126].

Moreover, as shown in Figure 14, the yield stresses of PANI/mSiO$_2$-based ER suspensions can be separated into two different areas in the entire electric field strength ranges [127]. Choi et al. [128] proposed a critical electrical field strength (E_c) and developed the following simplified equation to represent a hybrid yield stress for a wide electrical field strength area:

$$\tau_y = \alpha E^2 \left(\frac{\tanh\sqrt{E/E_c}}{\sqrt{E/E_c}} \right) \tag{7}$$

E_c could be obtained from an intersection between two power-law slopes for a low and a high electrical field area. The parameter α is dependent on a particle volume concentration and a dielectric constant of the ER suspension. Equation (7) contains two limiting behaviors of yield stress: E^2 for $E \ll E_c$ and $E^{1.5}$ for $E \gg E_c$. As presented in Figure 14, the ER characteristics of PANI/mSiO$_2$ corresponded to the non-linear conductivity model where the indices are 2 below E_c and 1.5 above E_c. Many other studies have proved the rationality of this equation [129,130].

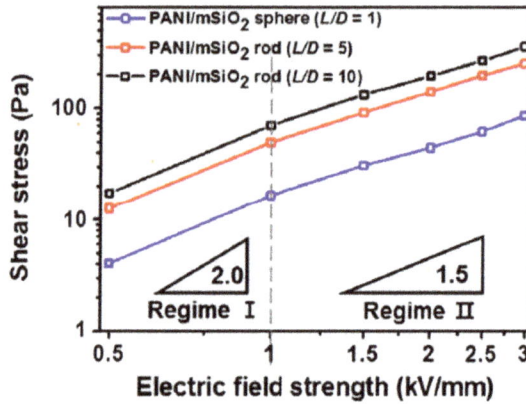

Figure 14. Yield stress of PANI/mSiO₂-based ER suspension as a function of electrical field strength, reprinted with permission from [127]. Copyright Elsevier, 2016.

4. Dielectric Analysis

A high dielectric constant shows a useful effect on the ER response because of the more effectively achievable polarizability. In general, the dielectric property is measured by an inductance capacitance resistance (LCR) tester. Figure 15 represents both the dielectric spectrum and the Cole–Cole curve of POA-silica composites [82]. The permittivity is frequency-dependent and can be expressed in its complex form involving permittivity (ε') and the loss factor (ε''): $\varepsilon^* = \varepsilon' + \varepsilon''$. The polarizability is described as the permittivity difference of $\Delta\varepsilon = \varepsilon_0 - \varepsilon_\infty$ and relaxation time ($\lambda = \frac{1}{2}\pi f_{max}$), which are obtained by fitting the dielectric spectra with the Cole–Cole formula.

$$\varepsilon^* = \varepsilon' + i\varepsilon'' = \varepsilon_\infty + \frac{\Delta\varepsilon}{(1 + i\omega\lambda)^{1-\alpha}} \tag{8}$$

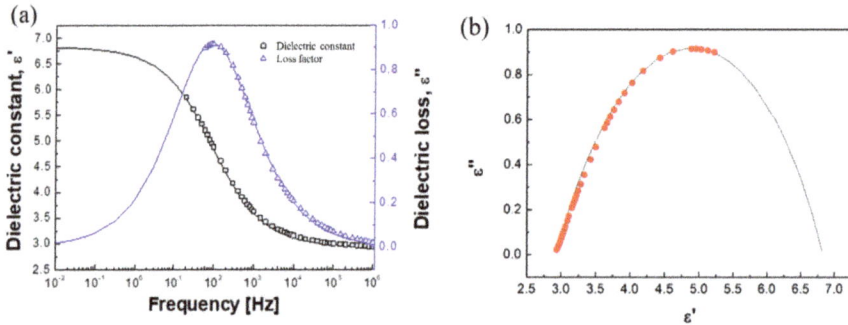

Figure 15. (a) Dielectric spectrum and (b) Cole–Cole plot of silica/POA based ER suspension, reprinted with permission from [82]. Copyright IOP Publishing, 2017.

In Equation (8), ω is a frequency and $(1 - \alpha)$ expresses the relaxation time distribution for a whole frequency area. It has been observed that a large ER efficiency is related to a large $\Delta\varepsilon$ and a dielectric loss peak is placed between 10^2 and 10^5 Hz. When α is zero, Equation (8) is reduced to the Debye's single relaxation time model. In Dong 's report [103], Equation (8) was used to compare different sizes of poly[*p*-vinylbenzyl trimethylammonium] based PILs possessing a similar diameter but different counter-anions. The results are shown in Table 1, with the counteranions of TfO, $\Delta\varepsilon$ having the largest value. Furthermore, in the subsequent ER response experiment, the PILs with TfO also exhibited the best ER effect, thereby showing that PILs with a smaller volume or size of counter-ions demonstrated a higher ER response.

Table 1. Dielectric properties of PIL suspensions [103].

Sample	ε_0	$\Delta\varepsilon$	λ
P[VBTMA][PF$_6$]	4.70	1.70	0.002
P[VBTMA][BF$_4$]	5.01	1.88	0.0061
P[VBTMA][TFSI]	5.14	1.96	0.0009
P[VBTMA][TfO]	5.11	2.01	0.00049

5. Engineering Applications

With yield stress generated under external electrical field strengths applied, ER fluids have the advantages of reversible continuous control and fast response and so have been widely adopted in vibration control, tactile display, sensor, and other applications for different modes. In a flow mode, two electrodes are set, and the vibrational control is realized by tuning the fluid motion between the two electrodes. In a shear mode, one electrode rotates freely relative to the other, thereby controlling the vibration by tuning the shear force. In a squeeze mode, the electrode gap becomes larger and the ER suspension is extruded by the normal force.

Further detailed applications using ER fluids include the torsional vibration clutch, brake, damper, optical material polishing, tactile devices, and so on. The controllable ER characteristics can be used as power transfer agents with functions similar to friction in conventional clutches; thus, much research has been devoted to the application of this special object in automotive clutches. Madeja et al. [131] investigated the practical ranges of a hydrodynamic clutch by applying an electric field,. To control braking force distribution, Choi et al. [132] introduced a vehicle ER valve-based anti-block brake system (ABS). Until now, ER fluids have relatively low yield stress under 8 kPa; thus, they are not used as hydraulic actuators. However, with the development of giant ER fluids [133], because the yield stress of the ER actuator can reach 130 kPa, it is expected to produce ER actuator.

An ER fluid can be precisely controlled by tuning the electrical field intensity; thus, it can also be used in polishing the tactile devices of optical materials. Kuriyagawa et al. [16] used an ER solution for the first time in the fabrication of optical lenses. Compared with MR polishing, ER polishing has the advantages of a uniform electric field and simple electrode structure. Su et al. [134] used an ER fluid to fabricate wheeled integrated electrodes for micro-optical components.

Moreover, Taylor et al. [135] developed a force display device using a 1-degree of freedom (1-DOF) ER brake. Recently, Han et al. [136] reported a 4-DOF haptic device using ER fluids for a minimally invasive surgery.

6. Conclusions

In this study, we discussed various modified and unmodified conductive polymers, their composites with special structural properties (fiber, core-shell spherical or non-spherical particles, and so on), and their ER responses. Considerable progress has been made in the introduction of new and synthetic methods, and in their corresponding ER characteristics. ER materials with ideal chemical and physical characteristics can be prepared by interfacial oxidation, micro-emulsion, Pickering emulsion, polymer grafting, the co-precipitation hydrothermal method, the sol-gel method, and other methods. Among these, the study of PANI and PPy is the most extensive and comprehensive, not only in terms of the morphology and properties of the polymers themselves, but also in terms of the results of their combination with other organic or inorganic compounds. In recent years, many new conductive polymers, such as PDA, PDPA, PIN, and POA, and their composites with different conductivity properties, have been widely used in ER fluids.

The power law between yield stresses and electric field strengths of various conducting polymer-based ER fluids, with its exponent between 1.5 and 2.0, is explained based on both a polarization model with an exponent of 2.0 and a conduction model with an exponent of 1.5. Nonetheless, to cover various conducting polymer-based ER suspensions in which the exponents do not follow one of the two mechanisms mentioned above, a universal yield stress equation was designed to properly correlate their dependence. In addition, both the CCJ and four-parameter models exhibited better fitting performance for the shear stress curve of the ER suspensions under applied electrical fields in the range of shear rate as compared with that of the traditional Bingham fluid model. Finally, recent applications of ER fluids, such as in damper systems, control systems, haptic devices, polishing systems, and so on, were summarized for potential commercialization of various engineering devices.

Funding: This work was supported by the National Research Foundation of Korea (2018R1A4A1025169).

Conflicts of Interest: The authors declare no conflicts of interest.

References

1. Niu, X.; Zhang, M.; Wu, J.; Wen, W.; Sheng, P. Generation and manipulation of "smart" droplets. *Soft Matter* **2009**, *5*, 576–581. [CrossRef]
2. Yarimaga, O.; Jaworski, J.; Yoon, B.; Kim, J.-M. Polydiacetylenes: Supramolecular smart materials with a structural hierarchy for sensing, imaging and display applications. *Chem. Commun.* **2012**, *48*, 2469–2485. [CrossRef] [PubMed]
3. Tsitsilianis, C. Responsive reversible hydrogels from associative "smart" macromolecules. *Soft Matter* **2010**, *6*, 2372–2388. [CrossRef]
4. Ohko, Y.; Tatsuma, T.; Fujii, T.; Naoi, K.; Niwa, C.; Kubota, Y.; Fujishima, A. Multicolour photochromism of TiO$_2$ films loaded with silver nanoparticles. *Nat. Mater.* **2003**, *2*, 29–31. [CrossRef] [PubMed]
5. Wei, K.; Bai, Q.; Meng, G.; Ye, L. Vibration characteristics of electrorheological elastomer sandwich beams. *Smart Mater. Struct.* **2011**, *20*, 055012. [CrossRef]
6. Klingenberg, D.J.; Zukoski, C.F., IV. Studies on the steady-shear behavior of electrorheological suspensions. *Langmuir* **1990**, *6*, 15–24. [CrossRef]
7. Chen, M.X.; Shang, Y.L.; Jia, Y.L.; Dong, X.Y.; Ren, J.; Li, J.R. New multifunction materials with both Electrorheological performance and luminescence property. *Korea-Australia Rheol. J.* **2017**, *29*, 29–36. [CrossRef]
8. Wen, W.; Huang, X.; Sheng, P. Electrorheological fluids: Structures and mechanisms. *Soft Matter* **2008**, *4*, 200–210. [CrossRef]
9. Lee, Y.-B. Behavior analysis of controllable electrorheology fluid plain journal bearings. *J. Dyn. Syst. Meas. Control* **2015**, *137*, 061013. [CrossRef]
10. Yamaguchi, H.; Zhang, X.-R.; Niu, X.-D.; Nishioka, K. Investigation of impulse response of an ER fluid viscous damper. *J. Intell. Mater. Syst. Struct.* **2010**, *21*, 423–435. [CrossRef]
11. Liu, Y.D.; Lee, B.M.; Park, T.-S.; Kim, J.E.; Choi, H.J.; Booh, S.W. Optically transparent electrorheological fluid with urea-modified silica nanoparticles and its haptic display application. *J. Colloid Interface Sci.* **2013**, *404*, 56–61. [CrossRef] [PubMed]
12. Niu, X.Z.; Liu, L.Y.; Wen, W.J.; Sheng, P. Microfluidic manipulation in lab-chips using electrorheological fluid. In *Electrorheological Fluids and Magnetorheological Suspensions*; World Scientific: Singapore, 2007; pp. 591–595.
13. Zhang, M.; Gong, X.; Wen, W. Manipulation of microfluidic droplets by electrorheological fluid. *Electrophoresis* **2009**, *30*, 3116–3123. [CrossRef] [PubMed]
14. Wang, L.; Zhang, M.; Li, J.; Gong, X.; Wen, W. Logic control of microfluidics with smart colloid. *Lab Chip* **2010**, *10*, 2869–2874. [CrossRef] [PubMed]
15. Choi, S.B.; Lee, D.Y. Rotational motion control of a washing machine using electrorheological clutches and brakes. *Proc. Inst. Mech. Eng. Part C J. Mech. Eng. Sci.* **2005**, *219*, 627–637. [CrossRef]
16. Kuriyagawa, T.; Saeki, M.; Syoji, K. Electrorheological fluid-assisted ultra-precision polishing for small three-dimensional parts. *Precis. Eng.* **2002**, *26*, 370–380. [CrossRef]
17. Liu, Y.; Davidson, R.; Taylor, P. Touch sensitive electrorheological fluid based tactile display. *Smart Mater. Struct.* **2005**, *14*, 1563. [CrossRef]
18. Winslow, W.M. Induced fibration of suspensions. *J. Appl. Phys.* **1949**, *20*, 1137–1140. [CrossRef]

19. Khusid, B.; Acrivos, A. Effects of conductivity in electric-field-induced aggregation in electrorheological fluids. *Phys. Rev. E* **1995**, *52*, 1669–1693. [CrossRef]

20. Yin, J.; Xia, X.; Xiang, L.; Qiao, Y.; Zhao, X. The electrorheological effect of polyaniline nanofiber, nanoparticle and microparticle suspensions. *Smart Mater. Struct.* **2009**, *18*, 095007. [CrossRef]

21. Yin, J.; Zhao, X. Electrorheology of nanofiber suspensions. *Nanoscale Res. Lett.* **2011**, *6*, 256. [CrossRef] [PubMed]

22. Ballav, N.; Sardar, P.S.; Ghosh, S.; Biswas, M. Polyaniline and polypyrrole modified conductive nanocomposites of polyfuran and polythiophene with montmorillonite clay. *J. Mater. Sci.* **2006**, *41*, 2959–2964. [CrossRef]

23. Masdarolomoor, F.; Innis, P.C.; Ashraf, S.; Kaner, R.B.; Wallace, G.G. Nanocomposites of polyaniline/poly(2-methoxyaniline-5-sulfonic acid). *Macromol. Rapid Commun.* **2006**, *27*, 1995–2000. [CrossRef]

24. Yeh, J.M.; Chin, C.P.; Chang, S. Enhanced corrosion protection coatings prepared from soluble electronically conductive polypyrrole-clay nanocomposite materials. *J. Appl. Polym. Sci.* **2003**, *88*, 3264–3272. [CrossRef]

25. Faguy, P.W.; Lucas, R.A.; Ma, W. An FT-IR-ATR spectroscopic study of the spontaneous polymerization of pyrrole in iron-exchanged montmorillonite. *Colloids Surf. A Physicochem. Eng. Asp.* **1995**, *105*, 105–112. [CrossRef]

26. Carrado, K.A.; Xu, L. In situ synthesis of polymer−clay nanocomposites from silicate gels. *Chem. Mater.* **1998**, *10*, 1440–1445. [CrossRef]

27. Fang, F.F.; Choi, H.J.; Joo, J.S. Conducting polymer/clay nanocomposites and their applications. *J. Nanosci. Nanotechnol.* **2008**, *8*, 1559–1581. [CrossRef] [PubMed]

28. Kim, S.G.; Lim, J.Y.; Sung, J.H.; Choi, H.J.; Seo, Y. Emulsion polymerized polyaniline synthesized with dodecylbenzene-sulfonic acid and its electrorheological characteristics: Temperature effect. *Polymer* **2007**, *48*, 6622–6631. [CrossRef]

29. Stěnička, M.; Pavlínek, V.; Sáha, P.; Blinova, N.V.; Stejskal, J.; Quadrat, O. Conductivity of flowing polyaniline suspensions in electric field. *Colloid Polym. Sci.* **2008**, *286*, 1403–1409. [CrossRef]

30. Shim, W.S.; Kim, J.-H.; Kim, K.; Kim, Y.-S.; Park, R.-W.; Kim, I.-S.; Kwon, I.C.; Lee, D.S. Ph-and temperature-sensitive, injectable, biodegradable block copolymer hydrogels as carriers for paclitaxel. *Int. J. Pharm.* **2007**, *331*, 11–18. [CrossRef] [PubMed]

31. Chin, B.D.; Park, O.O. Rheology and microstructures of electrorheological fluids containing both dispersed particles and liquid drops in a continuous phase. *J. Rheol.* **2000**, *44*, 397–412. [CrossRef]

32. Chin, B.D.; Park, O.O. Dispersion stability and electrorheological properties of polyaniline particle suspensions stabilized by poly(vinyl methyl ether). *J. Colloid Interface Sci.* **2001**, *234*, 344–350. [CrossRef] [PubMed]

33. Kim, Y.D.; Song, I.C. Electrorheological and dielectric properties of polypyrrole dispersions. *J. Mater. Sci.* **2002**, *37*, 5051–5055. [CrossRef]

34. Xia, Y.; Ouyang, J. Significant conductivity enhancement of conductive poly(3,4-ethylenedioxythiophene): Poly(styrenesulfonate) films through a treatment with organic carboxylic acids and inorganic acids. *ACS Appl. Mater. Interfaces* **2010**, *2*, 474–483. [CrossRef] [PubMed]

35. Street, G.B.; Clarke, T.C.; Krounbi, M.; Kanazawa, K.; Lee, V.; Pfluger, P.; Scott, J.C.; Weiser, G. Preparation and characterization of neutral and oxidized polypyrrole films. *Mol. Cryst. Liquid Cryst.* **1982**, *83*, 253–264. [CrossRef]

36. Chen, S.Y.; Sun, Z.C.; Li, L.H.; Xiao, Y.H.; Yu, Y.M. Preparation and characterization of conducting polymer-coated thermally expandable microspheres. *Chin. Chem. Lett.* **2017**, *28*, 658–662. [CrossRef]

37. Ashwell, G.J.; Jackson, P.D.; Lochun, D.; Thompson, P.A.; Crossland, W.A.; Bahra, G.S.; Brown, C.R.; Jasper, C. Second harmonic generation from alternate-layer LB films: Quadratic enhancement with film thickness. *Proc. R. Soc. Lond. A* **1994**, *445*, 385–398. [CrossRef]

38. Hailin, G.; Wallace, G.G. Electrosynthesis of chromatographic stationary phases. *Anal. Chem.* **1989**, *61*, 198–201. [CrossRef]

39. Xia, X.; Yin, J.; Qiang, P.; Zhao, X. Electrorheological properties of thermo-oxidative polypyrrole nanofibers. *Polymer* **2011**, *52*, 786–792. [CrossRef]

40. Rao, Y.; Pochan, J.M. Mechanics of polymer-clay nanocomposites. *Macromolecules* **2007**, *40*, 290–296. [CrossRef]

41. Jang, D.S.; Zhang, W.L.; Choi, H.J. Polypyrrole-wrapped halloysite nanocomposite and its rheological response under electric fields. *J. Mater. Sci.* **2014**, *49*, 7309–7316. [CrossRef]

42. Kim, B.H.; Hong, S.H.; Joo, J.; Park, I.W.; Epstein, A.J.; Kim, J.W.; Choi, H.J. Electron spin resonance signal of nanocomposite of conducting polypyrrole with inorganic clay. *J. Appl. Phys.* **2004**, *95*, 2697–2701. [CrossRef]

43. Fang, F.F.; Choi, H.J.; Ahn, W.-S. Electrorheology of a mesoporous silica having conducting polypyrrole inside expanded pores. *Microporous Mesoporous Mater.* **2010**, *130*, 338–343. [CrossRef]

44. Kim, Y.D.; Kim, J.H. Synthesis of polypyrrole–polycaprolactone composites by emulsion polymerization and the electrorheological behavior of their suspensions. *Colloid Polym. Sci.* **2008**, *286*, 631–637. [CrossRef]

45. Kim, Y.; Park, D. The electrorheological responses of suspensions of polypyrrole-coated polyethylene particles. *Colloid Polym. Sci.* **2002**, *280*, 828–834. [CrossRef]

46. Liu, Y.D.; Fang, F.F.; Choi, H.J. Silica nanoparticle decorated polyaniline nanofiber and its electrorheological response. *Soft Matter* **2011**, *7*, 2782–2789. [CrossRef]

47. Liu, Y.D.; Fang, F.F.; Choi, H.J. Silica nanoparticle decorated conducting polyaniline fibers and their electrorheology. *Mater. Lett.* **2010**, *64*, 154–156. [CrossRef]

48. Kim, M.W.; Moon, I.J.; Choi, H.J.; Seo, Y. Facile fabrication of core/shell structured SiO_2/polypyrrole nanoparticles with surface modification and their electrorheology. *RSC Adv.* **2016**, *6*, 56495–56502. [CrossRef]

49. Sedláčik, M.; Mrlík, M.; Pavlínek, V.; Sáha, P.; Quadrat, O. Electrorheological properties of suspensions of hollow globular titanium oxide/polypyrrole particles. *Colloid Polym. Sci.* **2012**, *290*, 41–48. [CrossRef]

50. Ryu, J.; Park, C.B. Synthesis of diphenylalanine/polyaniline core/shell conducting nanowires by peptide self-assembly. *Angew. Chem. Int. Ed.* **2009**, *48*, 4820–4823. [CrossRef] [PubMed]

51. Zhang, X.; Chan-Yu-King, R.; Jose, A.; Manohar, S.K. Nanofibers of polyaniline synthesized by interfacial polymerization. *Synth. Met.* **2004**, *145*, 23–29. [CrossRef]

52. Li, D.; Huang, J.; Kaner, R.B. Polyaniline nanofibers: A unique polymer nanostructure for versatile applications. *Acc. Chem. Res.* **2008**, *42*, 135–145. [CrossRef] [PubMed]

53. Yang, M.; Yao, X.; Wang, G.; Ding, H. A simple method to synthesize sea urchin-like polyaniline hollow spheres. *Colloids Surf. A Physicochem. Eng. Asp.* **2008**, *324*, 113–116. [CrossRef]

54. Stěnička, M.; Pavlínek, V.; Sáha, P.; Blinova, N.V.; Stejskal, J.; Quadrat, O. The electrorheological efficiency of polyaniline particles with various conductivities suspended in silicone oil. *Colloid Polym. Sci.* **2009**, *287*, 403–412. [CrossRef]

55. Liu, Y.D.; Choi, H.J. Electrorheological response of polyaniline and its hybrids. *Chem. Pap.* **2013**, *67*, 849–859. [CrossRef]

56. Wang, J.; Wang, J.; Wang, Z.; Zhang, F. A Template-Free Method toward Urchin-Like Polyaniline Microspheres. *Macromol. Rapid Commun.* **2009**, *30*, 604–608. [CrossRef] [PubMed]

57. Trchová, M.; Šeděnková, I.; Konyushenko, E.N.; Stejskal, J.; Holler, P.; Ćirić-Marjanović, G. Evolution of polyaniline nanotubes: The oxidation of aniline in water. *J. Phys. Chem. B* **2006**, *110*, 9461–9468. [CrossRef] [PubMed]

58. Chen, H.; Wang, A. Kinetic and isothermal studies of lead ion adsorption onto palygorskite clay. *J. Colloid Interface Sci.* **2007**, *307*, 309–316. [CrossRef] [PubMed]

59. Han, W.J.; Piao, S.H.; Choi, H.J. Synthesis and electrorheological characteristics of polyaniline@ attapulgite nanoparticles via Pickering emulsion polymerization. *Mater. Lett.* **2017**, *204*, 42–44. [CrossRef]

60. Yu, Y.; Zhong, X.; Gan, W. Conductive composites based on core–shell polyaniline nanoclay by latex blending. *Colloid Polym. Sci.* **2009**, *287*, 487–493. [CrossRef]

61. Jang, D.S.; Choi, H.J. Conducting polyaniline-wrapped sepiolite composite and its stimuli-response under applied electric fields. *Colloids Surf. A Physicochem. Eng. Asp.* **2015**, *469*, 20–28. [CrossRef]

62. Haraguchi, K.; Takehisa, T. Nanocomposite hydrogels: A unique organic–inorganic network structure with extraordinary mechanical, optical, and swelling/de-swelling properties. *Adv. Mater.* **2002**, *14*, 1120–1124. [CrossRef]

63. Jun, C.S.; Sim, B.; Choi, H.J. Fabrication of electric-stimuli responsive polyaniline/laponite composite and its viscoelastic and dielectric characteristics. *Colloids Surf. A Physicochem. Eng. Asp.* **2015**, *482*, 670–677. [CrossRef]

64. Noh, J.; Hong, S.; Yoon, C.-M.; Lee, S.; Jang, J. Dual external field-responsive polyaniline-coated magnetite/silica nanoparticles for smart fluid applications. *Chem. Commun.* **2017**, *53*, 6645–6648. [CrossRef] [PubMed]

65. Kim, D.; Tian, Y.; Choi, H.J. Seeded swelling polymerized sea urchin-like core–shell typed polystyrene/polyaniline particles and their electric stimuli-response. *RSC Adv.* **2015**, *5*, 81546–81553. [CrossRef]

66. Lee, I.S.; Cho, M.S.; Choi, H.J. Preparation of polyaniline coated poly(methyl methacrylate) microsphere by graft polymerization and its electrorheology. *Polymer* **2005**, *46*, 1317–1321. [CrossRef]

67. Zhang, W.L.; Piao, S.H.; Choi, H.J. Facile and fast synthesis of polyaniline-coated poly(glycidyl methacrylate) core–shell microspheres and their electro-responsive characteristics. *J. Colloid Interface Sci.* **2013**, *402*, 100–106. [CrossRef] [PubMed]

68. Zareh, E.N.; Moghadam, P.N. Synthesis and characterization of conductive nanoblends based on poly(aniline-*co*-3-aminobenzoic acid) in the presence of poly(styrene-*alt*-maleic acid). *J. Appl. Polym. Sci.* **2011**, *122*, 97–104. [CrossRef]

69. Hua, F.; Ruckenstein, E. Water-soluble conducting poly(ethylene oxide)-grafted polydiphenylamine synthesis through a "graft onto" process. *Macromolecules* **2003**, *36*, 9971–9978. [CrossRef]

70. Kim, M.H.; Bae, D.H.; Choi, H.J.; Seo, Y. Synthesis of semiconducting poly(diphenylamine) particles and analysis of their electrorheological properties. *Polymer* **2017**, *119*, 40–49. [CrossRef]

71. Kim, M.H.; Choi, H.J. Core–shell structured semiconducting poly(diphenylamine)-coated polystyrene microspheres and their electrorheology. *Polymer* **2017**, *131*, 120–131. [CrossRef]

72. Stejskal, J. Polymers of phenylenediamines. *Prog. Polym. Sci.* **2015**, *41*, 1–31. [CrossRef]

73. Plachy, T.; Sedlacik, M.; Pavlinek, V.; Morávková, Z.; Hajná, M.; Stejskal, J. An effect of carbonization on the electrorheology of poly(*p*-phenylenediamine). *Carbon* **2013**, *63*, 187–195. [CrossRef]

74. Cao, Y.; Choi, H.J.; Zhang, W.L.; Wang, B.; Hao, C.; Liu, J. Eco-friendly mass production of poly(*p*-phenylenediamine)/graphene oxide nanoplatelet composites and their electrorheological characteristics. *Compos. Sci. Technol.* **2016**, *122*, 36–41. [CrossRef]

75. Chen, X.P.; Jiang, J.K.; Liang, Q.H.; Yang, N.; Ye, H.Y.; Cai, M.; Shen, L.; Yang, D.G.; Ren, T.L. First-principles study of the effect of functional groups on polyaniline backbone. *Sci. Rep.* **2015**, *5*, 16907. [CrossRef] [PubMed]

76. Zhang, Y.; Duan, Y.; Liu, J.; Ma, G.; Huang, M. Wormlike Acid-Doped Polyaniline: Controllable Electrical Properties and Theoretical Investigation. *J. Phys. Chem. C* **2018**, *122*, 2032–2040. [CrossRef]

77. Moon, I.J.; Kim, H.Y.; Choi, H.J. Conducting poly(*n*-methylaniline)-coated cross-linked poly(methyl methacrylate) nanoparticle suspension and its steady shear response under electric fields. *Colloids Surf. A Physicochem. Eng. Asp.* **2015**, *481*, 506–513. [CrossRef]

78. Lin, Y.; Zhao, F.; Wu, Y.; Chen, K.; Xia, Y.; Li, G.; Prasad, S.K.K.; Zhu, J.; Huo, L.; Bin, H. Mapping polymer donors toward high-efficiency fullerene free organic solar cells. *Adv. Mater.* **2017**, *29*, 1604155. [CrossRef] [PubMed]

79. Park, I.H.; Kwon, S.H.; Choi, H.J. Emulsion-polymerized polyindole nanoparticles and their electrorheology. *J. Appl. Polym. Sci.* **2018**, *135*, 46384. [CrossRef]

80. Sari, B.; Yavas, N.; Makulogullari, M.; Erol, O.; Unal, H.I. Synthesis, electrorheology and creep behavior of polyindole/polyethylene composites. *React. Funct. Polym.* **2009**, *69*, 808–815. [CrossRef]

81. Cheng, Q.; Pavlinek, V.; He, Y.; Yan, Y.; Li, C.; Saha, P. Template-free synthesis of hollow poly(*o*-anisidine) microspheres and their electrorheological characteristics. *Smart Mater. Struct.* **2011**, *20*, 065014. [CrossRef]

82. Lee, C.J.; Choi, H.J. Fabrication of poly(*o*-anisidine) coated silica core–shell microspheres and their electrorheological response. *Mater. Res. Express* **2017**, *4*, 116310. [CrossRef]

83. Wang, B.; Tian, X.; He, K.; Ma, L.; Yu, S.; Hao, C.; Chen, K.; Lei, Q. Hollow PAQR nanostructure and its smart electrorheological activity. *Polymer* **2016**, *83*, 129–137. [CrossRef]

84. Cho, M.S.; Choi, H.J.; Jhon, M.S. Shear stress analysis of a semiconducting polymer based electrorheological fluid system. *Polymer* **2005**, *46*, 11484–11488. [CrossRef]

85. Hong, J.-Y.; Jang, J. A comparative study on electrorheological properties of various silica–conducting polymer core–shell nanospheres. *Soft Matter* **2010**, *6*, 4669–4671. [CrossRef]

86. An, J.S.; Moon, I.J.; Kwon, S.H.; Choi, H.J. Swelling-diffusion-interfacial polymerized core-shell typed polystyrene/poly(3,4-ethylenedioxythiophene) microspheres and their electro-responsive characteristics. *Polymer* **2017**, *115*, 137–145. [CrossRef]

87. Erol, O.; Unal, H.I. Core/shell-structured, covalently bonded TiO_2/poly(3,4-ethylenedioxythiophene) dispersions and their electrorheological response: The effect of anisotropy. *RSC Adv.* **2015**, *5*, 103159–103171. [CrossRef]

88. Cho, C.H.; Choi, H.J.; Kim, J.W.; Jhon, M.S. Synthesis and electrorheology of aniline/pyrrole copolymer. *J. Mater. Sci.* **2004**, *39*, 1883–1885. [CrossRef]

89. Pandey, S.S.; Annapoorni, S.; Malhotra, B.D. Synthesis and characterization of poly(aniline-*co*-*o*-anisidine). A processable conducting copolymer. *Macromolecules* **1993**, *26*, 3190–3193. [CrossRef]

90. Trlica, J.; Sáha, P.; Quadrat, O.; Stejskal, J. Electrical and electrorheological behavior of poly(aniline-*co*-1,4-phenylenediamine) suspensions. *Eur. Polym. J.* **2000**, *36*, 2313–2319. [CrossRef]

91. Abubakar, A.; Al-Wahaibi, T.; Al-Wahaibi, Y.; Al-Hashmi, A.; Al-Ajmi, A. Roles of drag reducing polymers in single-and multi-phase flows. *Chem. Eng. Res. Des.* **2014**, *92*, 2153–2181. [CrossRef]

92. Dreyer, D.R.; Park, S.; Bielawski, C.W.; Ruoff, R.S. The chemistry of graphene oxide. *Chem. Soc. Rev.* **2010**, *39*, 228–240. [CrossRef] [PubMed]

93. Zhang, W.L.; Liu, Y.D.; Choi, H.J. Fabrication of semiconducting graphene oxide/polyaniline composite particles and their electrorheological response under an applied electric field. *Carbon* **2012**, *50*, 290–296. [CrossRef]

94. Min, T.H.; Choi, H.J. Synthesis of poly(methyl methacrylate)/graphene oxide nanocomposite particles via Pickering emulsion polymerization and their viscous response under an electric field. *Macromol. Res.* **2017**, *25*, 565–571. [CrossRef]

95. Li, J.; Gong, X.; Chen, S.; Wen, W.; Sheng, P. Giant electrorheological fluid comprising nanoparticles: Carbon nanotube composite. *J. Appl. Phys.* **2010**, *107*, 093507. [CrossRef]

96. Lin, C.; Shan, J.W. Ensemble-averaged particle orientation and shear viscosity of single-wall-carbon-nanotube suspensions under shear and electric fields. *Phys. Fluids* **2010**, *22*, 022001. [CrossRef]

97. Lizcano, M.; Nava-Lara, M.R.; Alvarez, A.; Lozano, K. C_{60} structural transformation by electrorheological testing. *Carbon* **2007**, *45*, 2374–2378. [CrossRef]

98. Jin, H.-J.; Choi, H.J.; Yoon, S.H.; Myung, S.J.; Shim, S.E. Carbon nanotube-adsorbed polystyrene and poly(methyl methacrylate) microspheres. *Chem. Mater.* **2005**, *17*, 4034–4037. [CrossRef]

99. Fang, F.F.; Liu, Y.D.; Choi, H.J. Electrorheological and magnetorheological response of polypyrrole/magnetite nanocomposite particles. *Colloid Polym. Sci.* **2013**, *291*, 1781–1786. [CrossRef]

100. Nishimura, N.; Ohno, H. 15th anniversary of polymerised ionic liquids. *Polymer* **2014**, *55*, 3289–3297. [CrossRef]

101. Mecerreyes, D. Polymeric ionic liquids: Broadening the properties and applications of polyelectrolytes. *Prog. Polym. Sci.* **2011**, *36*, 1629–1648. [CrossRef]

102. Dong, Y.; Yin, J.; Yuan, J.; Zhao, X. Microwave-assisted synthesis and high-performance anhydrous electrorheological characteristic of monodisperse poly(ionic liquid) particles with different size of cation/anion parts. *Polymer* **2016**, *97*, 408–417. [CrossRef]

103. Dong, Y.; Liu, Y.; Wang, B.; Xiang, L.; Zhao, X.; Yin, J. Influence of counterion type on dielectric and electrorheological responses of poly(ionic liquid)s. *Polymer* **2017**, *132*, 273–285. [CrossRef]

104. Sever, E.; Unal, H.I. Electrorheological, viscoelastic, and creep-recovery behaviors of covalently bonded nanocube-TiO_2/poly(3-octylthiophene) colloidal dispersions. *Polym. Compos.* **2018**, *39*, 351–359. [CrossRef]

105. Cabuk, S.; Unal, H.I. Enhanced electrokinetic, dielectric and electrorheological properties of covalently bonded nanosphere-TiO_2/polypyrrole nanocomposite. *React. Funct. Polym.* **2015**, *95*, 1–11. [CrossRef]

106. Kim, S.H.; Kim, J.H.; Choi, H.J.; Park, J. Pickering emulsion polymerized poly(3,4-ethylenedioxythiophene): Poly(styrenesulfonate)/polystyrene composite particles and their electric stimuli-response. *RSC Adv.* **2015**, *5*, 72387–72393. [CrossRef]

107. Kim, D.-H.; Kim, Y.D. Electrorheological properties of polypyrrole and its composite ER fluids. *J. Ind. Eng. Chem.* **2007**, *13*, 879–894.

108. Liu, Y.D.; Choi, H.J. Electrorheological fluids: Smart soft matter and characteristics. *Soft Matter* **2012**, *8*, 11961–11978. [CrossRef]

109. Méheust, Y.; Parmar, K.P.S.; Schjelderupsen, B.; Fossum, J.O. The electrorheology of suspensions of Na-fluorohectorite clay in silicone oil. *J. Rheol.* **2011**, *55*, 809–833. [CrossRef]

110. Yalcintas, M.; Dai, H. Magnetorheological and electrorheological materials in adaptive structures and their performance comparison. *Smart Mater. Struct.* **1999**, *8*, 560. [CrossRef]

111. Engelmann, B.; Hiptmair, R.; Hoppe, R.H.W.; Mazurkevitch, G. Numerical simulation of electrorheological fluids based on an extended Bingham model. *Comput. Vis. Sci.* **2000**, *2*, 211–219. [CrossRef]
112. Piao, S.H.; Gao, C.Y.; Choi, H.J. Sulfonated polystyrene nanoparticles coated with conducting polyaniline and their electro-responsive suspension characteristics under electric fields. *Polymer* **2017**, *127*, 174–181. [CrossRef]
113. Chattopadhyay, A.; Rani, P.; Srivastava, R.; Dhar, P. Electro-elastoviscous response of polyaniline functionalized nano-porous zeolite based colloidal dispersions. *J. Colloid Interface Sci.* **2018**, *519*, 242–254. [CrossRef] [PubMed]
114. Min, T.H.; Choi, H.J.; Kim, N.-H.; Park, K.; You, C.-Y. Effects of surface treatment on magnetic carbonyl iron/polyaniline microspheres and their magnetorheological study. *Colloids Surf. A Physicochem. Eng. Asp.* **2017**, *531*, 48–55. [CrossRef]
115. Carreau, P.J. *Rheology of Polymeric Systems: Principles and Applications*; Hanser Publishers: New York, NY, USA, 1997.
116. Seo, Y.P.; Choi, H.J.; Seo, Y. A simplified model for analyzing the flow behavior of electrorheological fluids containing silica nanoparticle-decorated polyaniline nanofibers. *Soft Matter* **2012**, *8*, 4659–4663. [CrossRef]
117. Seo, Y.P.; Seo, Y. Modeling and analysis of electrorheological suspensions in shear flow. *Langmuir* **2012**, *28*, 3077–3084. [CrossRef] [PubMed]
118. De Vicente, J.; Klingenberg, D.J.; Hidalgo-Alvarez, R. Magnetorheological fluids: A review. *Soft Matter* **2011**, *7*, 3701–3710. [CrossRef]
119. Lim, G.H.; Choi, H.J. Fabrication of self-assembled polyaniline tubes and their electrorheological characteristics. *Colloids Surf. A Physicochem. Eng. Asp.* **2017**, *530*, 227–234. [CrossRef]
120. Klingenberg, D.J.; Van Swol, F.; Zukoski, C.F. The small shear rate response of electrorheological suspensions. I. Simulation in the point–dipole limit. *J. Chem. Phys.* **1991**, *94*, 6160–6169. [CrossRef]
121. Jun, C.S.; Kwon, S.H.; Choi, H.J.; Seo, Y. Polymeric nanoparticle-coated Pickering emulsion-synthesized conducting polyaniline hybrid particles and their electrorheological study. *ACS Appl. Mater. Interfaces* **2017**, *9*, 44811–44819. [CrossRef] [PubMed]
122. Song, X.; Song, K.; Ding, S.; Chen, Y.; Lin, Y. Electrorheological properties of poly [*N,N'*-(2-amino-5-carboxybutyl-1,3-phenylenedimethylene)-2,2'-diamino-4,4'-bithiazole]. *J. Ind. Eng. Chem.* **2013**, *19*, 416–420.
123. Yin, J.; Zhao, X.; Xia, X.; Xiang, L.; Qiao, Y. Electrorheological fluids based on nano-fibrous polyaniline. *Polymer* **2008**, *49*, 4413–4419. [CrossRef]
124. Sung, J.H.; Park, D.P.; Park, B.J.; Choi, H.J.; Jhon, M.S. Phosphorylation of potato starch and its electrorheological suspension. *Biomacromolecules* **2005**, *6*, 2182–2188. [CrossRef] [PubMed]
125. Davis, L.C. Time-dependent and nonlinear effects in electrorheological fluids. *J. Appl. Phys.* **1997**, *81*, 1985–1991. [CrossRef]
126. Kim, J.W.; Liu, F.; Choi, H.J. Polypyrrole/clay nanocomposite and its electrorheological characteristics. *J. Ind. Eng. Chem.* **2002**, *8*, 399–403.
127. Noh, J.; Yoon, C.-M.; Jang, J. Enhanced electrorheological activity of polyaniline coated mesoporous silica with high aspect ratio. *J. Colloid Interface Sci.* **2016**, *470*, 237–244. [CrossRef] [PubMed]
128. Choi, H.J.; Cho, M.S.; Kim, J.W.; Kim, C.A.; Jhon, M.S. A yield stress scaling function for electrorheological fluids. *Appl. Phys. Lett.* **2001**, *78*, 3806–3808. [CrossRef]
129. Cheng, Q.; Pavlinek, V.; Lengalova, A.; Li, C.; Belza, T.; Saha, P. Electrorheological properties of new mesoporous material with conducting polypyrrole in mesoporous silica. *Microporous Mesoporous Mater.* **2006**, *94*, 193–199. [CrossRef]
130. Zhang, W.L.; Jiang, D.; Wang, X.; Hao, B.N.; Liu, Y.D.; Liu, J. Growth of polyaniline nanoneedles on MoS2 nanosheets, tunable electroresponse, and electromagnetic wave attenuation analysis. *J. Phys. Chem. C* **2017**, *121*, 4989–4998. [CrossRef]
131. Madeja, J.; Kesy, Z.; Kesy, A. Application of electrorheological fluid in a hydrodynamic clutch. *Smart Mater. Struct.* **2011**, *20*, 105005. [CrossRef]
132. Choi, S.-B.; Sung, K.-G.; Cho, M.-S.; Lee, Y.-S. The braking performance of a vehicle anti-lock brake system featuring an electro-rheological valve pressure modulator. *Smart Mater. Struct.* **2007**, *16*, 1285. [CrossRef]
133. Wen, W.; Huang, X.; Yang, S.; Lu, K.; Sheng, P. The giant electrorheological effect in suspensions of nanoparticles. *Nat. Mater.* **2003**, *2*, 727–730. [CrossRef] [PubMed]

134. Su, J.; Cheng, H.; Feng, Y.; Tam, H.-Y. Study of a wheel-like electrorheological finishing tool and its applications to small parts. *Appl. Opt.* **2016**, *55*, 638–645. [CrossRef] [PubMed]
135. Taylor, P.M.; Hosseini-Sianaki, A.; Varley, C.J. Surface feedback for virtual environment systems using electrorheological fluids. *Int. J. Mod. Phys. B* **1996**, *10*, 3011–3018. [CrossRef]
136. Han, Y.-M.; Choi, S.-B.; Oh, J.-S. Tracking controls of torque and force of 4-degree-of-freedom haptic master featuring smart electrorheological fluid. *J. Intell. Mater. Syst. Struct.* **2016**, *27*, 915–924. [CrossRef]

molecules

MDPI

Review

A Perspective on Reversibility in Controlled Polymerization Systems: Recent Progress and New Opportunities

Houliang Tang [1,2], Yi Luan [1,*], Lu Yang [3] and Hao Sun [3,*]

1   School of Materials Science and Engineering, University of Science and Technology Beijing,
    30 Xueyuan Road, Haidian District, Beijing 100083, China; houliangt@smu.edu
2   Department of Chemistry, Southern Methodist University, 3215 Daniel Avenue, Dallas, TX 75275, USA
3   Department of Chemistry, University of Florida, PO Box 117200, Gainesville, FL 32611-7200, USA;
    yanglulucia@chem.ufl.edu
*   Correspondence: yiluan@ustb.edu.cn (Y.L.); Chrisun@ufl.edu (H.S.); Tel.: +01-352-281-4799 (H.S.)

Academic Editors: Jianxun Ding, Yang Li and Mingqiang Li
Received: 15 October 2018; Accepted: 2 November 2018; Published: 3 November 2018

Abstract: The field of controlled polymerization is growing and evolving at unprecedented rates, facilitating polymer scientists to engineer the structure and property of polymer materials for a variety of applications. However, the lack of degradability, particularly in vinyl polymers, is a general concern not only for environmental sustainability, but also for biomedical applications. In recent years, there has been a significant effort to develop reversible polymerization approaches in those well-established controlled polymerization systems. Reversible polymerization typically involves two steps, including (i) forward polymerization, which converts small monomers into macromolecule; and (ii) depolymerization, which is capable of regenerating original monomers. Furthermore, recycled monomers can be repolymerized into new polymers. In this perspective, we highlight recent developments of reversible polymerization in those controlled polymerization systems and offer insight into the promise and utility of reversible polymerization systems. More importantly, the current challenges and future directions to solve those problems are discussed. We hope this perspective can serve as an "initiator" to promote continuing innovations in this fairly new area.

Keywords: controlled polymerization; reversible polymerization; sustainable polymers

1. Introduction

The fundamental concept of reversibility has been widely utilized to drive the development of new polymeric materials, which can display distinct but reversible change in properties upon receiving a stimulus [1,2]. In light of this, a library of reversible materials based on polymers have recently been achieved, including self-healing materials bearing reversible-covalent linkages [3–8], recyclable materials such as vitrimers [9–11], polymer networks enabling reversible sol-gel transitions [12–14], architecture-transformable polymers [15], and covalent or metal organic frameworks harnessing reversible bonds [16–21]. Despite the tremendous success in the aforementioned polymer systems, little attention has been paid to achieving reversible polymerizations. Compared with self-immolative polymers, which can only undergo one-way depolymerization, a reversible polymerization typically features reversible transformations between polymers and original monomers [22]. Indeed, step-growth polymerizations relying on reversible-covalent chemistry, particularly Diels–Alder chemistry, have provided an approach to reversible polymers that favors forward polymerization at room temperature and tends to depolymerize at high temperatures (i.e., 120 °C) [23]. However, those polymers prepared by step-growth polymerizations typically have very broad molecular weight distributions and small molecular weights, limiting their potentials in certain applications that require precise polymer chain length, complicated architectures, and high molecular weights [24–29].

In the last two decades, the rapid advent of controlled and living polymerizations has offered polymer scientists a powerful synthetic toolbox for accessing polymers with predetermined molecular weights, well-defined architectures, and narrow distributions in molecular weights [30–35]. In addition, the excellent tolerance of functional groups in controlled polymerization systems has enabled us to achieve advanced polymer materials with desired functions and properties [36–41]. In general, controlled polymerization techniques can be categorized into three common classes. The first one involves ring-opening polymerization (ROP) of cyclized monomers (e.g., caprolactones, lactides, and *N*-carboxyanhydrides) in the presence of a nucleophilic initiator and a catalyst (metal or organic) [42–47]. The second category highlights the well-developed applications of controlled anionic/cationic polymerizations, which have led to industrial production of thermoplastic elastomers [30]. The last class focuses on reversible-deactivation radical polymerization (RDRP) methodologies, which are capable of polymerizing vinyl monomers, such as acrylates, methacrylates, and styrene in a controlled manner [48–51]. To date, three mainstream RDRP techniques, including atom-transfer radical polymerization (ATRP) [52,53], reversible addition–fragmentation transfer (RAFT) polymerization [54–57], and nitroxide-mediated polymerization have been developed [58]. Among them, ATRP and RAFT are receiving the most attention, due to their unique advantages, such as mild polymerization conditions, broad monomer scope, and ease of end-group functionalization [59]. These controlled polymerization systems are enjoying tremendous success in producing polymers that accommodate both industrial use (supported by ROP and anionic/cationic polymerizations) and academic research. However, the studies related to depolymerization are scarce, because depolymerization was typically considered as a side reaction that would lessen the performance— e.g., the mechanical properties of polymer materials [60]. While this is true in the pursuit of maximizing lifetime or long-term stability of polymer materials, serious pollution problems have arisen from the lack of degradability in commercial polymer plastics under common conditions [61]. Sustainable polymers, such as degradable polyesters deriving from biomass resources, represent a promising platform for environmental remedy. However, those polymers currently suffer from high manufacturing cost and low mechanical properties compared to vinyl polymer-based materials from petroleum resources. Moreover, the degradation of these sustainable polyesters is typically one-way, resulting in non-polymerizable fragments that are not useful for the regeneration of new materials. To fully achieve sustainability in polymer materials, recent attentions have been shifted to the development of new methods for depolymerizing polymers back into original monomers under accessible conditions [60,62–71]. These regenerated monomers can be recycled and further repolymerized to obtain new polymer materials. From this perspective, we aim to first critically assess the state-of-the-art toolbox for achieving reversible polymerization in controlled polymerization systems (Scheme 1). Recent examples on reversible polymerizations will be classified according to their controlled polymerization mechanisms, i.e., ROP and RDRP (Scheme 1 and Table 1). In addition, we believe it is important to assess the current challenges of reversible polymerizations that can trigger polymer chemists to solve this issue. Finally, potential applications deriving from this fairly new concept will be predicted and discussed.

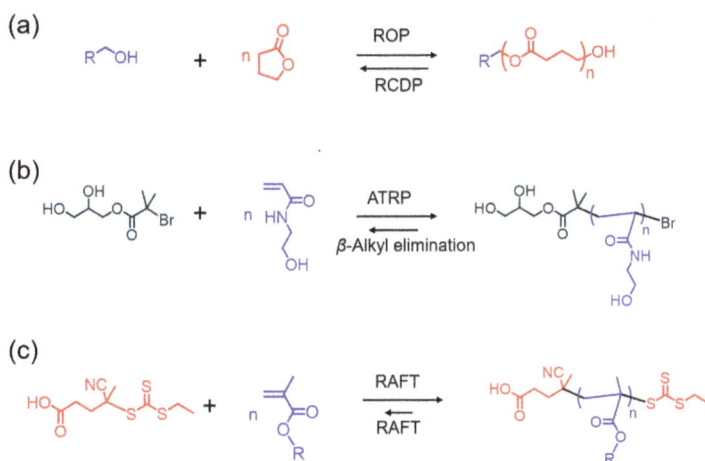

Scheme 1. Various mechanisms of reversible polymerization approaches in controlled polymerization systems; (**a**) ring-opening polymerization (ROP) and ring-closing depolymerization (RCDP); (**b**) atom transfer radical polymerization (ATRP) and β-alkyl elimination-induced depolymerization; (**c**) reversible addition–fragmentation transfer (RAFT) mediated polymerization and depolymerization.

Table 1. Summary of reversible polymerization approaches in this perspective.

Type of Polymers	Polymerzation Mechanism	Depolymerization Mechanism	Depolymerization Conversion	References
Polyester	ROP	RCDP	≥99%	Albertsson [64]
Polyester	ROP	RCDP	≥99%	Chen [68]
Polyester	ROP	RCDP	≥99%	Chen [71]
Polyester	ROP	RCDP	≥99%	Chen [69]
Polyester	ROP	RCDP	≥99%	Chen [70]
Polyester	ROP	RCDP	≥99%	Chen [67]
Polyester	ROP	RCDP	≥99%	Hoye [66]
Vinyl polymer	ATRP	β-alkyl elimination	24–34%	Zhu [62]
Vinyl polymer	ATRP	Not identified	43–71%	Haddleton [65]
Vinyl polymer	RAFT	RAFT	28%	Gramlich [60]

2. Reversible Polymerization in Ring-Opening Polymerization Systems

We begin our exploration of reversible polymerization approaches in ROP systems [63,64,66–71]. As one of the most popular controlled polymerization strategies, ROP of cyclic monomers has emerged as a useful synthetic route to prepare technologically interesting polymers with desirable architectures and specific properties. In particular, nucleophile-initiated ROP in the presence of metallic or organocatalysts allows the polymerization to proceed in a controlled manner, affording polymers with pre-determinable molecular weights and narrow molecular weight distributions. Applicable monomers include lactones, lactides, cyclic carbonates, *N*- or *O*-carboxyanhydrides, and cyclooligosiloxanes, among others [43]. To date, well-defined polymers produced from those monomers have attracted significant interest in both academic research and industry [43,72,73]

Although many of the ROP polymers are comprised of hydrolysable ester linkages in their backbones, which can cause the polymers to degrade into oligomers or possibly small molecules, it is still challenging to completely convert the polymers back to the original cyclized monomers. In 2014, Albertsson and coworkers demonstrated their pioneering work on ring-closing depolymerization to obtain a functional six-membered cyclic carbonate monomer, 2-allyloxylmethyl-2-ethyltrimethlene carbonate (AOMEC), from its oligomeric form [63]. The synthesis of AOMEC was performed in a one-pot reaction, involving the oligomerization of trimethylolpropane allyl ether, diethyl carbonate, and NaH, followed by an in-situ anionic depolymerization. It was observed that the depolymerization can occur beyond the ceiling temperature of the polyester at a certain polymer/monomer concentration, suggesting the reversible nature of the polymerization of AOMEC. However, there was still room to improve, since the oligomers were synthesized simply by condensation reaction rather than ROP, and the degree of polymerization was only from 1 to 7.

Inspired by this pioneering work, Albertsson's team continued the study on AOMEC and first demonstrated the reversible ROP of this six-membered cyclic carbonate monomer in 2016 [64]. It was concluded that the equilibrium between controlled ROP of AOMEC and controlled ring-closing depolymerization (RCDP) of poly(AOMEC) were dictated by various reaction parameters, such as monomer concentration, reaction temperature, and even solvents. In their approach, AOMEC was polymerized by ROP in the presence of an organocatalyst—that is, 1,8-diazabicyclo(5.4.0)undec-7-ene (DBU) either in bulk or in different solvents (Figure 1a). Various temperatures were used to evaluate the equilibrium conversion for calculating the ceiling temperature at certain polymerization conditions. According to the thermodynamic principle, ROP of AOMEC was favored at a temperature lower than the ceiling temperature. As the reaction temperature was higher than the ceiling temperature, RCDP of poly(AOMEC) dominated. It is worth noting that the molecular weights increased/decreased linearly as a function of monomer conversion in the cases of both ROP and RCDP (Figure 1b). Moreover, the molecular weight distribution of all the evolving polymers remained as low as 1.1, which further verified the exceptional control over the course of both polymerization and depolymerization.

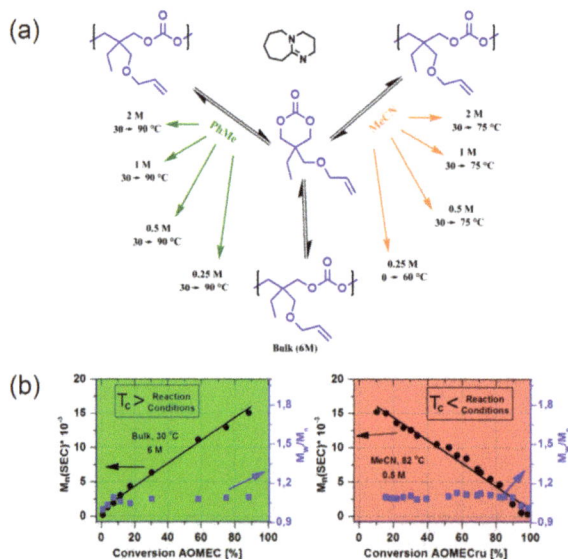

Figure 1. Reversible polymerization in ring-opening polymerization systems. (a) Reversible ROP of 2-allyloxylmethyl-2-ethyltrimethlene carbonate (AOMEC) at various reaction conditions; (b) Correlations of molecular weights, polydispersity index, and monomer conversions under two different polymerization conditions. Reproduced with permission from [64].

As one of the best suitable biomass-derived compounds to replace petroleum-derived chemicals, γ-butyrolactone (γ-BL) and its polymer PγBL have great potential as sustainable materials [68,70]. However, γ-BL has been commonly considered to be "non-polymerizable" due to its low strain energy, which has thereby rendered the ROP of γ-BL extremely challenging. Starting from 2016, Chen et al. has conducted extensive research on ROP of γ-BL, and have now developed several exciting synthetic methodologies to achieving PγBL by ROP. In agreement with Albertsson's viewpoint on the thermodynamic basis, Chen anticipated that reaction conditions should be modulated to achieve successful ROP, including a much lower reaction temperature than the ceiling temperature, a high initial monomer concentration, and most importantly, a robust catalyst. In their first work associated with γ-BL, the ROP of γ-BL was carried out at −40 °C in the presence of a lanthanide (Ln)-based coordination polymerization catalyst [68]. The polymerization was capable of affording PγBL with M_n as high as 30,000 g mol^{-1} and up to 90% monomer conversion. In addition, the polymer topology (e.g., linear or cyclic) can also be controlled by varying the initiator structure and the feeding ratio of raw materials (Figure 2). From the sustainable perspective, a quantitative depolymerization of PγBL was realized by heating the purified polymers for 1 h at higher temperatures (i.e., 220 or 300 °C) than the ceiling temperature. Interestingly, the PγBL that was dissolved in appropriate solvents depolymerized much more rapidly in the presence of an organocatalyst (e.g., 1,5,7-triazabicyclo[4.4.0]dec-5-ene) or metal catalyst (e.g., La[N(SiMe$_3$)$_2$]$_3$), even at room temperature.

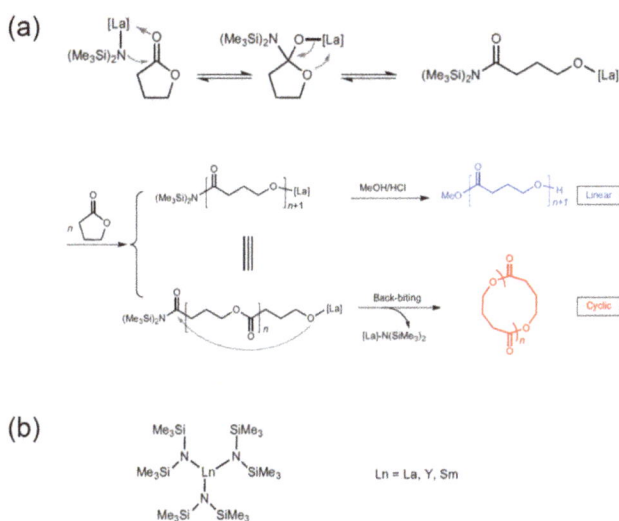

Figure 2. Reversible polymerization in ring-opening polymerization systems. (**a**) Proposed mechanism for lanthanide-catalyzed ROP of γ-BL in the preparation of both linear and cyclic PγBL; (**b**) Chemical structure of lanthanide catalyst. Reproduced with permission from [68].

In parallel, Chen and coworkers also discovered a metal-free ROP strategy to produce PγBL, using a super-basic organocatalyst with abbreviation tert-Bu-P$_4$ (Figure 3a) [71]. It should be noted that the catalyst itself was able to initiate the ROP of γ-BL at −40 °C, by abstracting protons from γ-BL to form highly reactive enolate species (Figure 3b). However, the monomer conversion was limited up to 30.4%, due to the possible interference of [tert-Bu-P4H]$^+$ and an anionic dimer. Furthermore, the ROP performance was greatly enhanced when tert-Bu-P$_4$ was added, along with a suitable alcohol serving as the initiator (e.g., BnOH). With the help of the alcoholic initiator, the monomer conversion reached as high as ca. 90%, and the corresponding PγBL possessed a molecular weight of 26,700 g mol^{-1}. Notably, the PγBL prepared by this organocatalyzed ROP was completely recyclable, and can depolymerize back to γ-BL upon heating at 260 °C for 1 h.

(a)

(b)

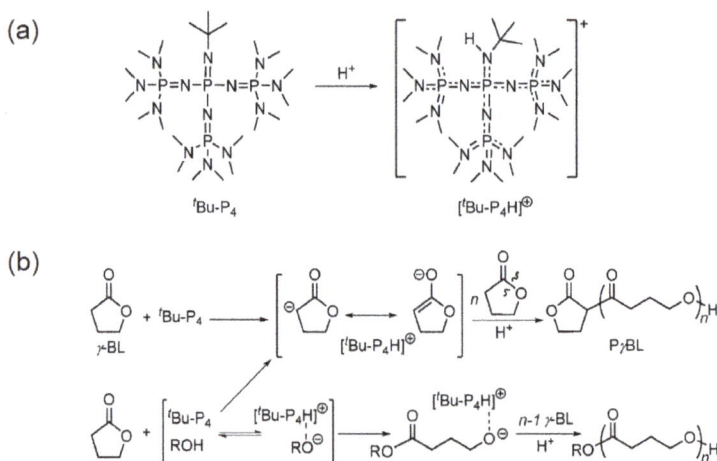

Figure 3. Reversible polymerization in ring-opening polymerization systems. (a) Structure and basicity of tert-Bu-P$_4$; (b) proposed mechanism for the tert-Bu-P$_4$-catalyzed ROP of γ-BL (i) with and (ii) without alcohol as an initiator. Reproduced with permission from [71].

Encouraged by the success of preparing fully-recyclable PγBL via ROP with either metal (La, Y) or organocatalysts, Chen's group proceeded to apply this unique polymerization process to an enriched variety of monomers. ROP of α-Methylene-γ-butyrolactone (MBL), a small molecule derived from biomass and regarded as a potential alternative to the petroleum-based MMA, was subsequently investigated (Figure 4) [69]. Since the monomer comprises a non-strained five-membered lactone and a highly reactive exocyclic C=C double bond, many knee-jerk studies were exclusively focused on traditional vinyl addition polymerization (VAP) [74,75]. In Chen's work, the lanthanide (Ln)-based coordination polymerization catalyst was utilized, leading to an unsaturated polyester P(MBL) with M$_n$ up to 21,000 g mol^{-1} through the ROP process. Remarkably, by adjusting the reaction conditions, such as the catalyst (La)/initiator (ROH) ratio and temperature, three pathways of the MBL polymerization can be realized independently, including conventional VAP, ROP, and crosslinking polymerization. As was foreseeable, only the polymers resulted from the ROP pathway were fully recyclable.

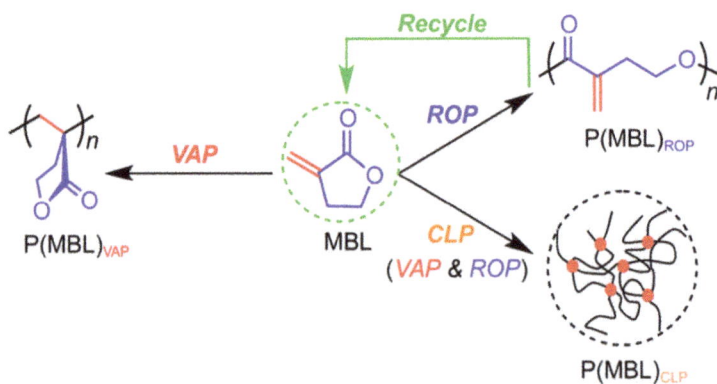

Figure 4. Reversible polymerization in ring-opening polymerization systems. Polymerization of α-Methylene-γ-butyrolactone (MBL) following three distinct mechanisms. Reproduced with permission from [69].

Despite the notable achievement in lanthanide (La, Y) or superbase (tert-Bu-P$_4$) catalyzed ROP of γ-BL and its derivatives, the undesirable low temperature (i.e., −40 °C) to implement the process still remained as one of biggest hurdles for industry use. Moreover, the as-synthesized polymers suffered from limited thermostability and crystallinity. Hence, the exploration of new materials with both superior properties and energy economy are still in demand. Recently, Chen's group proposed a γ-BL derivative (abbreviated as 3,4-T6GBL), with a cyclohexyl ring transfused to the five-membered lactone at the α and β positions [70]. This monomer can be polymerized by using the coordinative insertion ROP catalysts, producing linear or cyclic polymers with high molecular weights at room temperature (Figure 5a). Following this finding, Chen et al. extended the scope of ROP of another γ-BL derivative (4,5-T6GBL), where the cyclohexyl ring was fused at the β and γ (or 4,5) positions of the BL ring, giving rise to linear/cyclic polymers at room temperature (Figure 5b) [67]. A controlled polymerization behavior was observed as the molecular weights of evolving polymers, which increased linearly with the monomer conversions. As expected, in both studies the resulting polymers possess enhanced thermostability and could be quantitatively recycled back to their original building monomers by either thermolysis or chemolysis.

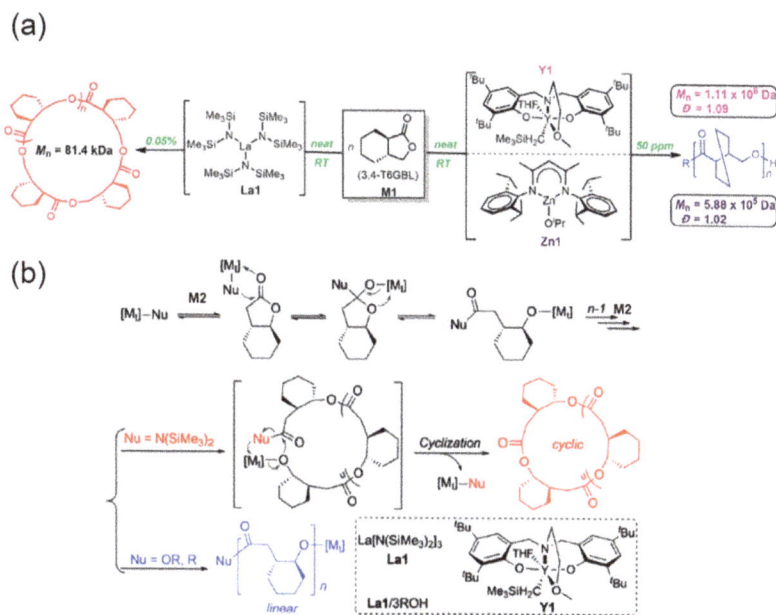

(a)

(b)

Figure 5. Reversible polymerization in ring-opening polymerization systems. (**a**) Metal catalyzed ROP of 3,4-T6GBL (M1) to linear and cyclic polymers; (**b**) Proposed pathways for metal catalyzed ROP of 4,5-T6GBL (M2) to linear and cyclic polymers. Reproduced with permission from [67,70].

Very recently, Hoye et al. described the synthesis of a novel substituted polyvalerolactone from a malic acid derived monomer, 4-carbomethoxyvalerolactone (CMVL) [66]. In their work, this six-membered ring monomer was blended with a diol (1,4-benzenedimethanol) and an organic acid (diphenyl phosphate) at an ambient temperature, finally forming a semicrystalline material with a molar mass up to 71,000 g mol^{-1}. Notably, the resulting polymer can be either depolymerized back into its original precursor monomer or degraded into acrylate-type analogues (Figure 6a). The former process was catalyzed by tin octanoate (Sn(Oct)$_2$), providing 87% monomer recovery; while the latter was promoted by DBU with a comparable yield. In particular, a substantial kinetics study showed that CMVL was polymerized smoothly and reached ca. 90% conversion after 15 h (Figure 6b). It also revealed that nearly all of the initiator was consumed within 20 min. This "fast initiation" can be regarded as one of the characteristic behaviors of controlled polymerization systems.

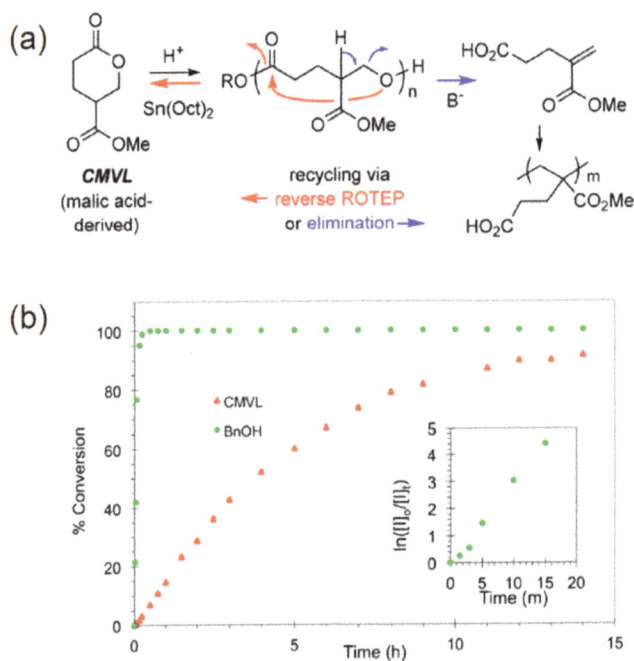

Figure 6. Reversible polymerization in ring-opening polymerization systems. (**a**) Acid-catalyzed ROP of 4-carbomethoxyvalerolactone (CMVL) and the divergent chemical recycling; (**b**) kinetics of the ROP of CMVL in the presence of BnOH as an initiator. Reproduced with permission from [66].

3. Reversible Polymerization in Reversible-Deactivation Radical Polymerization Systems

Although reversible polymerizations in ROP systems have shown great promise in next-generation sustainable polymer materials, the major market for commodity polymers is still occupied by vinyl polymers, due to their low cost in manufacturing. Vinyl polymers are derived from petroleum, a non-renewable resource. Therefore, the recycling of used vinyl polymer materials such as plastics has immense merits, not only in global waste reduction, but also for petroleum sustainability [76]. In this section, recent examples in depolymerization of vinyl polymers derived from RDRP systems will be discussed. We hope those timely developments will prompt more innovative thinking with regard to plastic recycling via a reversible polymerization approach.

In 2012, Zhu et al. reported the first example of reversible polymerizations in an RDRP system [62]. In their pioneering work, vinyl polymerizations of several acrylamide monomers, including *N*-isopropylacrylamide (NIPAM) and *N,N*-dimethylacrylamide (DMA), were successfully achieved in the presence of CuCl and tris(2-dimethylaminoethyl)amine (Me6TREN) (Figure 7a). Very intriguingly, they unexpectedly observed a phenomenon of depolymerization when radical inhibitors like 2,2,6,6-tetramethylpiperidinooxy (TEMPO) or 1,4-benzoquinone (BQ) were added to the ongoing polymerization system, with the initial purpose of terminating radical polymerizations. To further elucidate the role of the copper catalyst in the depolymerization process, a control experiment with regard to a conventional radical polymerization, using 2,2'-azobisisobutyronitrile (AIBN) as radical initiator, was carried out in the absence of a copper catalyst and ligands. TEMPO was added during the conventional radical polymerization, resulting in only the termination of polymerization, without any noticeable depolymerization. Those results unequivocally verified that a copper catalyst is essential in depolymerization. Therefore, a depolymerization mechanism based on β-alkyl elimination from the copper (II) coordination center was proposed (Figure 7b).

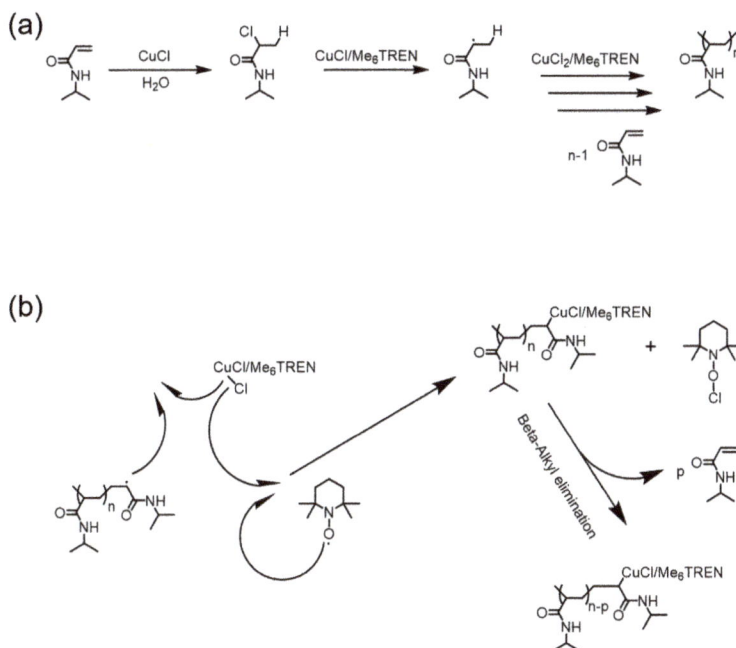

Figure 7. Reversible polymerization in copper-mediated polymerization systems. (**a**) Proposed mechanism of CuCl/ tris(2-dimethylaminoethyl)amine (Me$_6$TREN)-based radical polymerization of *N*-isopropylacrylamide (NIPAM); (**b**) 2,2,6,6-tetramethylpiperidinooxy (TEMPO)-induced depolymerization mechanism via β-alkyl elimination. Reproduced with permission from [67].

In a similar demonstration of reversible polymerization mediated by a copper catalyst, Haddleton and coworkers were able to prepare well-defined polyacrylamides and polyacrylates through aqueous copper-mediated radical polymerization in the presence of dissolved CO$_2$ (Figure 8a) [65]. In the case of a NIPAM monomer, the forward polymerization adopted rapid reaction kinetics, achieving full monomer conversion within 10 min. Thereafter, a significant in-situ depolymerization occurred to an extent of 52%, and thereby led to the regeneration of the NIPAM monomer, which was systemically confirmed by proton nuclear magnetic resonance (NMR), gel permeation chromatography (GPC), and electron ionization-mass spectroscopy. Importantly, this recycled NIPAM can be repolymerized upon deoxygenation of the resulting solutions, illustrating the reversibility of the polymerization (Figure 8b). Furthermore, the scope of reversible copper-mediated polymerization was extended to *N*-hydroxyethyl acrylamide (HEAm) and 2-hydroxyethyl acrylate (HEA), demonstrating the versatility of this system. However, it should be noted that the mechanism of depolymerization, as well as the role of CO$_2$ in the depolymerization process, was not identified in their study.

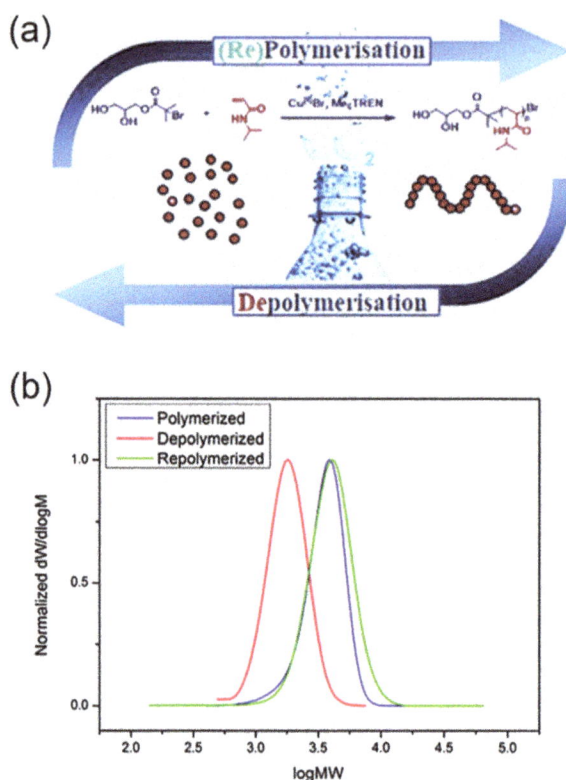

Figure 8. Reversible polymerization in copper-mediated controlled polymerization system. (a) A schematic illustration of aqueous reversible polymerizations mediated by a copper catalyst and CO_2; (b) Gel permeation chromatography GPC traces of original poly (*N*-isopropylacrylamide) (PNIPAM) (blue), depolymerized PNIPAM (red), and repolymerized PNIPAM (green). Reproduced with permission from [65].

In the aforementioned RDRP systems (Vide supra), the depolymerization phenomenon was only observed during the course of the polymerization. However, it is arguably more interesting from the materials point of view if one can depolymerize a polymer post-synthesis or after the manufacturing process. In very recent work described by Gramlich, a set of brush polymers consisting of oligo-ethylene glycol or oligo-dimethylsiloxane side chains were prepared by traditional RAFT polymerization in the presence of AIBN at 70 °C (Figure 9a) [60]. After polymerization, those polymers were thoroughly purified by repeated precipitations, to ensure the removal of residual monomers and initiators. Upon purification, thermally-induced depolymerization of the as-synthesized polymer was conducted in dilute dioxane solutions, leading to the regeneration of vinyl monomers until reaching the monomer's inherent equilibrium monomer concentration. Importantly, the residual polymers exhibited high chain-end fidelity by retaining the trithiocarbonate moiety after depolymerization, allowing for further reinitiation and repolymerization via a RAFT mechanism (Figure 9b,c).

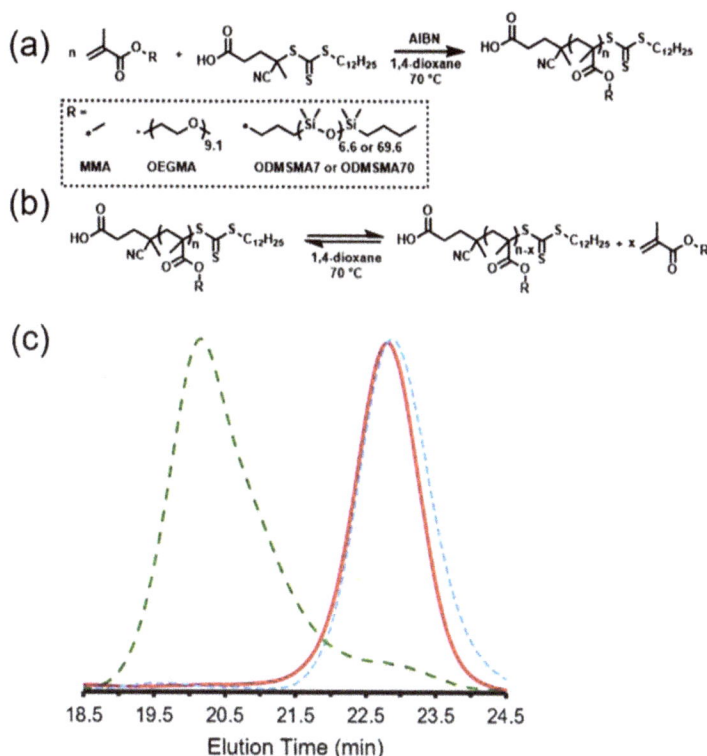

Figure 9. Reversible polymerization in a RAFT system. (**a**) Preparation of brush-type polymers via RAFT polymerizations of methacrylate monomers; (**b**) depolymerization and repolymerization of RAFT polymers; (**c**) GPC trance of original RAFT polymer (red solid line), depolymerized polymer (blue dotted line), and repolymerized polymer (green dotted line). Reproduced with permission from [60].

4. Closing Remarks

The current success in reversible polymerizations has enabled us to think about many new possibilities in future polymer science. However, many challenges still remained to be addressed, hampering the further translation of this new concept into real-world applications. One apparent hurdle is how to achieve good control over the depolymerization process. While all the examples covered in this perspective are related to controlled polymerizations, allowing for a predictable degree of forward polymerizations, little information was provided to reveal the kinetics of the depolymerization process, especially those involved in controlled radical polymerization. Indeed, previous literature has placed much focus on the start and endpoint of depolymerization (in other word, the highest degree of depolymerization). Notwithstanding, kinetic study will shed more light on the fundamental mechanisms of reversible polymerizations, and if one can predetermine and control the degree of depolymerization by changing several reaction parameters, such as time, temperature, catalyst/initiator loading, polymer/monomer concentrations, among others. Moreover, the ability to tune the depolymerization rate under normal conditions is expected to open the door to many interesting applications (in addition to sustainable materials)—for example, self-healable materials, and sustained release systems, which require slow and controllable depolymerization. It is worth noting that the concurrent depolymerization approaches are typically associated with

Molecules **2018**, *23*, 2870

harsh conditions (e.g., high temperatures, metal catalysts), significantly impeding the translation of this concept into biomedical uses. In light of this, the continuing pursuit of new depolymerization methodologies that can be implemented under mild and physiological conditions will be important towards bio-related applications.

Another challenge stems from the relatively low efficiency in depolymerizations, particularly those in RDRP systems. In comparison with ROP-based depolymerization systems, which mainly rely on breaking weak polyester backbones, the energy input necessary for reversing vinyl polymer backbones (i.e., carbon–carbon single bonds) back to vinyl monomers is considerably higher. To our best knowledge, the highest reported depolymerization conversion in RDRP systems was only 71% when N-hydroxyethyl acrylamide was involved in copper(0)-mediated reversible polymerizations [65]. In the RAFT mediated depolymerization approach, only 30% of monomers can be regenerated after heating the diluted RAFT polymer solutions at 70 °C under a vacuum for several days [60]. From the viewpoint of potential industrial applications, insufficient depolymerization could dramatically increase the cost deriving from separating regenerated monomers from residual polymers. Therefore, we envision that more attention will be paid to detailed mechanism study and the rational design of depolymerization systems, with the goal of achieving high depolymerization efficiency (such as the effort for lowering equilibrium monomer concentration). Moreover, we believe that mathematical tools, such as modeling and simulations of reversible polymerizations, should play a key role in prediction of the dynamics, final products, and optimal conditions in reversible polymerizations [77]. While the concept of reversible polymerizations is still in its infant stage, it is anticipated that the future development in this area will not only deepen our understanding of fundamental depolymerization mechanisms, but also promote many new opportunities and applications in polymer science and engineering.

Author Contributions: Conceptualization: H.S. and H.T.; literature research: H.S., H.T., and L.Y.; writing (original draft preparation): H.T. and H.S.; writing (review and editing): H.S., H.T., L.Y., and Y.L.; supervision: H.S. and H.T.; and funding acquisition: H.S.

Funding: This research received no external funding.

Acknowledgments: H.S. gratefully acknowledges the fellowship support from EASTMAN Chemical Company and the Chinese government award for outstanding self-financed students abroad.

Conflicts of Interest: The authors declare no conflict of interest.

References

1. Sun, H.; Kabb, C.P.; Dai, Y.Q.; Hill, M.R.; Ghiviriga, I.; Bapat, A.P.; Sumerlin, B.S. Macromolecular metamorphosis via stimulus-induced transformations of polymer architecture. *Nat. Chem.* **2017**, *9*, 817–823. [CrossRef] [PubMed]

2. Stuart, M.A.C.; Huck, W.T.S.; Genzer, J.; Muller, M.; Ober, C.; Stamm, M.; Sukhorukov, G.B.; Szleifer, I.; Tsukruk, V.V.; Urban, M.; et al. Emerging applications of stimuli-responsive polymer materials. *Nat. Mater.* **2010**, *9*, 101–113. [CrossRef] [PubMed]

3. Yang, Y.; Urban, M.W. Self-healing polymeric materials. *Chem. Soc. Rev.* **2013**, *42*, 7446–7467. [CrossRef] [PubMed]

4. Sun, H.; Kabb, C.P.; Sumerlin, B.S. Thermally-labile segmented hyperbranched copolymers: Using reversible-covalent chemistry to investigate the mechanism of self-condensing vinyl copolymerization. *Chem. Sci.* **2014**, *5*, 4646–4655. [CrossRef]

5. Chen, X.X.; Dam, M.A.; Ono, K.; Mal, A.; Shen, H.B.; Nutt, S.R.; Sheran, K.; Wudl, F. A thermally re-mendable cross-linked polymeric material. *Science* **2002**, *295*, 1698–1702. [CrossRef] [PubMed]

6. Yang, L.; Tang, H.L.; Sun, H. Progress in Photo-Responsive Polypeptide Derived Nano-Assemblies. *Micromachines* **2018**, *9*, 296. [CrossRef]

7. Tang, H.; Tsarevsky, N.V. Lipoates as building blocks of sulfur-containing branched macromolecules. *Polym Chem.-Uk* **2015**, *6*, 6936–6945. [CrossRef]

8. Tang, H.; Tsarevsky, N.V. Preparation and Functionalization of Linear and Reductively Degradable Highly Branched Cyanoacrylate-Based Polymers. *J. Polym. Sci. Pol. Chem.* **2016**, *54*, 3683–3693. [CrossRef]

9. Snyder, R.L.; Fortman, D.J.; De Hoe, G.X.; Hillmyer, M.A.; Dichtel, W.R. Reprocessable Acid-Degradable Polycarbonate Vitrimers. *Macromolecules* **2018**, *51*, 389–397. [CrossRef]
10. Denissen, W.; Winne, J.M.; Du Prez, F.E. Vitrimers: Permanent organic networks with glass-like fluidity. *Chem. Sci.* **2016**, *7*, 30–38. [CrossRef] [PubMed]
11. Rottger, M.; Domenech, T.; van der Weegen, R.; Nicolay, A.B.R.; Leibler, L. High-performance vitrimers from commodity thermoplastics through dioxaborolane metathesis. *Science* **2017**, *356*, 62–65. [CrossRef] [PubMed]
12. Honda, S.; Toyota, T. Photo-triggered solvent-free metamorphosis of polymeric materials. *Nat. Commun.* **2017**, *8*, 502. [CrossRef] [PubMed]
13. Zhou, H.W.; Xue, C.G.; Weis, P.; Suzuki, Y.; Huang, S.L.; Koynov, K.; Auernhammer, G.K.; Berger, R.; Butt, H.J.; Wu, S. Photoswitching of glass transition temperatures of azobenzene-containing polymers induces reversible solid-to-liquid transitions. *Nat. Chem.* **2017**, *9*, 145–151. [CrossRef] [PubMed]
14. Alfurhood, J.A.; Sun, H.; Bachler, P.R.; Sumerlin, B.S. Hyperbranched poly(N-(2-hydroxypropyl)methacrylamide) via RAFT self-condensing vinyl polymerization. *Polym Chem.* **2016**, *7*, 2099–2104. [CrossRef]
15. Sun, H.; Kabb, C.P.; Sims, M.B.; Sumerlin, B.S. Architecture-transformable polymers: Reshaping the future of stimuli-responsive polymers. *Prog. Polym. Sci.* **2018**. [CrossRef]
16. Diercks, C.S.; Yaghi, O.M. The atom, the molecule, and the covalent organic framework. *Science* **2017**, *355*, eaal1585. [CrossRef] [PubMed]
17. Evans, A.M.; Parent, L.R.; Flanders, N.C.; Bisbey, R.P.; Vitaku, E.; Kirschner, M.S.; Schaller, R.D.; Chen, L.X.; Gianneschi, N.C.; Dichtel, W.R. Seeded growth of single-crystal two-dimensional covalent organic frameworks. *Science* **2018**, *361*, 53. [CrossRef] [PubMed]
18. Du, X.; Li, X.J.; Tang, H.L.; Wang, W.Y.; Ramella, D.; Luan, Y. A facile 2H-chromene dimerization through an ortho-quinone methide intermediate catalyzed by a sulfonyl derived MIL-101 MOF. *New J. Chem.* **2018**, *42*, 12722–12728. [CrossRef]
19. Miao, Z.C.; Zhou, Z.H.; Tang, H.L.; Yu, M.D.; Ramella, D.; Du, X.; Luan, Y. Homodimerization of 2H-chromenes catalyzed by BrOnsted-acid derived UiO-66 MOFs. *Catal. Sci. Technol.* **2018**, *8*, 3406–3413. [CrossRef]
20. Zhao, J.; Wang, W.Y.; Tang, H.L.; Ramella, D.; Luan, Y. Modification of Cu2+into Zr-based metal-organic framework (MOF) with carboxylic units as an efficient heterogeneous catalyst for aerobic epoxidation of olefins. *Mol. Catal.* **2018**, *456*, 57–64. [CrossRef]
21. Deng, H.X.; Grunder, S.; Cordova, K.E.; Valente, C.; Furukawa, H.; Hmadeh, M.; Gandara, F.; Whalley, A.C.; Liu, Z.; Asahina, S.; et al. Large-Pore Apertures in a Series of Metal-Organic Frameworks. *Science* **2012**, *336*, 1018–1023. [CrossRef] [PubMed]
22. Peterson, G.I.; Larsen, M.B.; Boydston, A.J. Controlled Depolymerization: Stimuli-Responsive Self-Immolative Polymers. *Macromolecules* **2012**, *45*, 7317–7328. [CrossRef]
23. Zhou, J.W.; Guimard, N.K.; Inglis, A.J.; Namazian, M.; Lin, C.Y.; Coote, M.L.; Spyrou, E.; Hilf, S.; Schmidt, F.G.; Barner-Kowollik, C. Thermally reversible Diels-Alder-based polymerization: An experimental and theoretical assessment. *Polym Chem.-UK* **2012**, *3*, 628–639. [CrossRef]
24. Sun, H.; Dobbins, D.J.; Dai, Y.Q.; Kabb, C.P.; Wu, S.J.; Alfurhood, J.A.; Rinaldi, C.; Sumerlin, B.S. Radical Departure: Thermally-Triggered Degradation of Azo-Containing Poly(beta-thioester)s. *ACS Macro Lett.* **2016**, *5*, 688–693. [CrossRef]
25. Yokozawa, T.; Ohta, Y. Transformation of Step-Growth Polymerization into Living Chain-Growth Polymerization. *Chem. Rev.* **2016**, *116*, 1950–1968. [CrossRef] [PubMed]
26. Li, X.W.; Figg, C.A.; Wang, R.W.; Jiang, Y.; Lyu, Y.F.; Sun, H.; Liu, Y.; Wang, Y.Y.; Teng, I.T.; Hou, W.J.; et al. Cross-Linked Aptamer-Lipid Micelles for Excellent Stability and Specificity in Target-Cell Recognition. *Angew Chem. Int. Ed.* **2018**, *57*, 11589–11593. [CrossRef] [PubMed]
27. Wang, Y.Y.; Wu, C.C.; Chen, T.; Sun, H.; Cansiz, S.; Zhang, L.Q.; Cui, C.; Hou, W.J.; Wu, Y.; Wan, S.; et al. DNA micelle flares: A study of the basic properties that contribute to enhanced stability and binding affinity in complex biological systems. *Chem. Sci.* **2016**, *7*, 6041–6049. [CrossRef] [PubMed]
28. Liu, Y.; Hou, W.J.; Sun, H.; Cui, C.; Zhang, L.Q.; Jiang, Y.; Wu, Y.X.; Wang, Y.Y.; Li, J.; Sumerlin, B.S.; et al. Thiol-ene click chemistry: A biocompatible way for orthogonal bioconjugation of colloidal nanoparticles. *Chem. Sci.* **2017**, *8*, 6182–6187. [CrossRef] [PubMed]

29. Wan, S.; Zhang, L.Q.; Wang, S.; Liu, Y.; Wu, C.C.; Cui, C.; Sun, H.; Shi, M.L.; Jiang, Y.; Li, L.; et al. Molecular Recognition-Based DNA Nanoassemblies on the Surfaces of Nanosized Exosomes. *J. Am. Chem. Soc.* **2017**, *139*, 5289–5292. [CrossRef] [PubMed]

30. Grubbs, R.B.; Grubbs, R.H. 50th Anniversary Perspective: Living Polymerization-Emphasizing the Molecule in Macromolecules. *Macromolecules* **2017**, *50*, 6979–6997. [CrossRef]

31. Tan, J.B.; Sun, H.; Yu, M.G.; Sumerlin, B.S.; Zhang, L. Photo-PISA: Shedding Light on Polymerization-Induced Self-Assembly. *ACS Macro Lett.* **2015**, *4*, 1249–1253. [CrossRef]

32. Hill, M.R.; Carmean, R.N.; Sumerlin, B.S. Expanding the Scope of RAFT Polymerization: Recent Advances and New Horizons. *Macromolecules* **2015**, *48*, 5459–5469. [CrossRef]

33. Matyjaszewski, K. Atom Transfer Radical Polymerization (ATRP): Current Status and Future Perspectives. *Macromolecules* **2012**, *45*, 4015–4039. [CrossRef]

34. Zheng, Y.; Abbas, Z.M.; Sarkar, A.; Marsh, Z.; Stefik, M.; Benicewicz, B.C. Surface-initiated reversible addition-fragmentation chain transfer polymerization of chloroprene and mechanical properties of matrix-free polychloroprene nanocomposites. *Polymer* **2018**, *135*, 193–199. [CrossRef]

35. Zheng, Y.; Huang, Y.C.; Benicewicz, B.C. A Useful Method for Preparing Mixed Brush Polymer Grafted Nanoparticles by Polymerizing Block Copolymers from Surfaces with Reversed Monomer Addition Sequence. *Macromol. Rapid Comm.* **2017**, *38*, 1700300. [CrossRef] [PubMed]

36. Dong, P.; Sun, H.; Quan, D.P. Synthesis of poly(L-lactide-co-5-amino-5-methyl-1,3-dioxan-2-ones) [P(L-LA-co-TAc)] containing amino groups via organocatalysis and post-polymerization functionalization. *Polymer* **2016**, *97*, 614–622. [CrossRef]

37. Dai, Y.Q.; Sun, H.; Pal, S.; Zhang, Y.L.; Park, S.; Kabb, C.P.; Wei, W.D.; Sumerlin, B.S. Near-IR-induced dissociation of thermally-sensitive star polymers. *Chem. Sci.* **2017**, *8*, 1815–1821. [CrossRef] [PubMed]

38. Matyjaszewski, K. Architecturally Complex Polymers with Controlled Heterogeneity. *Science* **2011**, *333*, 1104–1105. [CrossRef] [PubMed]

39. Li, Z.; Tang, H.L.; Feng, A.C.; Thang, S.H. Synthesis of Zwitterionic Polymers by Living/Controlled Radical Polymerization and Its Applications. *Prog. Chem.* **2018**, *30*, 1097–1111. [CrossRef]

40. Zheng, Y.; Huang, Y.C.; Abbas, Z.M.; Benicewicz, B.C. Surface-initiated polymerization-induced self-assembly of bimodal polymer-grafted silica nanoparticles towards hybrid assemblies in one step. *Polym Chem.-UK* **2016**, *7*, 5347–5350. [CrossRef]

41. Zheng, Y.; Huang, Y.C.; Abbas, Z.M.; Benicewicz, B.C. One-pot synthesis of inorganic nanoparticle vesicles via surface-initiated polymerization-induced self-assembly. *Polym. Chem.-UK* **2017**, *8*, 370–374. [CrossRef]

42. Kamber, N.E.; Jeong, W.; Waymouth, R.M.; Pratt, R.C.; Lohmeijer, B.G.G.; Hedrick, J.L. Organocatalytic ring-opening polymerization. *Chem. Rev.* **2007**, *107*, 5813–5840. [CrossRef] [PubMed]

43. Nuyken, O.; Pask, S.D. Ring-Opening Polymerization-An Introductory Review. *Polymers* **2013**, *5*, 361–403. [CrossRef]

44. Liu, H.; Cheng, Y.L.; Chen, J.J.; Chang, F.; Wang, J.C.; Ding, J.X.; Chen, X.S. Component effect of stem cell-loaded thermosensitive polypeptide hydrogels on cartilage repair. *Acta. Biomater.* **2018**, *73*, 103–111. [CrossRef] [PubMed]

45. Wang, J.X.; Xu, W.G.; Li, S.X.; Qiu, H.P.; Li, Z.B.; Wang, C.X.; Wang, X.Q.; Ding, J.X. Polylactide-Cholesterol Stereocomplex Micelle Encapsulating Chemotherapeutic Agent for Improved Antitumor Efficacy and Safety. *J. Biomed. Nanotechnol.* **2018**, *14*, 2102–2113. [CrossRef] [PubMed]

46. Ding, J.X.; Xiao, C.S.; He, C.L.; Li, M.Q.; Li, D.; Zhuang, X.L.; Chen, X.S. Facile preparation of a cationic poly(amino acid) vesicle for potential drug and gene co-delivery. *Nanotechnology* **2011**, *22*, 494012. [CrossRef] [PubMed]

47. Ding, J.X.; Xiao, C.S.; Zhao, L.; Cheng, Y.L.; Ma, L.L.; Tang, Z.H.; Zhuang, X.L.; Chen, X.S. Poly(L-glutamic acid) Grafted with Oligo(2-(2-(2-methoxyethoxy)ethoxy) ethyl methacrylate): Thermal Phase Transition, Secondary Structure, and Self-Assembly. *J. Polym. Sci. Pol. Chem.* **2011**, *49*, 2665–2676. [CrossRef]

48. Braunecker, W.A.; Matyjaszewski, K. Controlled/living radical polymerization: Features, developments, and perspectives. *Prog. Polym. Sci.* **2007**, *32*, 93–146. [CrossRef]

49. Yeow, J.; Chapman, R.; Gormley, A.J.; Boyer, C. Up in the air: Oxygen tolerance in controlled/living radical polymerisation. *Chem. Soc. Rev.* **2018**, *47*, 4357–4387. [CrossRef] [PubMed]

50. Zetterlund, P.B.; Thickett, S.C.; Perrier, S.; Bourgeat-Lami, E.; Lansalot, M. Controlled/Living Radical Polymerization in Dispersed Systems: An Update. *Chem. Rev.* **2015**, *115*, 9745–9800. [CrossRef] [PubMed]

51. Zheng, Y.; Wang, L.; Lu, L.; Wang, Q.; Benicewicz, B.C. pH and Thermal Dual-Responsive Nanoparticles for Controlled Drug Delivery with High Loading Content. *Acs Omega.* **2017**, *2*, 3399–3405. [CrossRef] [PubMed]

52. Matyjaszewski, K. Atom Transfer Radical Polymerization: From Mechanisms to Applications. *Isr. J. Chem.* **2012**, *52*, 206–220. [CrossRef]

53. Siegwart, D.J.; Oh, J.K.; Matyjaszewski, K. ATRP in the design of functional materials for biomedical applications. *Prog. Polym. Sci.* **2012**, *37*, 18–37. [CrossRef] [PubMed]

54. Perrier, S. 50th Anniversary Perspective: RAFT Polymerization-A User Guide. *Macromolecules* **2017**, *50*, 7433–7447. [CrossRef]

55. Alfurhood, J.A.; Sun, H.; Kabb, C.P.; Tucker, B.S.; Matthews, J.H.; Luesch, H.; Sumerlin, B.S. Poly(N-(2-hydroxypropyl)-methacrylamide)-valproic acid conjugates as block copolymer nanocarriers. *Polym Chem.* **2017**, *8*, 4983–4987. [CrossRef] [PubMed]

56. Tian, X.Y.; Ding, J.J.; Zhang, B.; Qiu, F.; Zhuang, X.D.; Chen, Y. Recent Advances in RAFT Polymerization: Novel Initiation Mechanisms and Optoelectronic Applications. *Polymers* **2018**, *10*, 318. [CrossRef]

57. Almeida, C.; Costa, H.; Kadhirvel, P.; Queiroz, A.M.; Dias, R.C.S.; Costa, M.R.P.F.N. Electrochemical activity of sulfur networks synthesized through RAFT polymerization. *J. Appl. Polym. Sci.* **2016**, *133*, 43993. [CrossRef]

58. Nicolas, J.; Guillaneuf, Y.; Lefay, C.; Bertin, D.; Gigmes, D.; Charleux, B. Nitroxide-mediated polymerization. *Prog Polym Sci* **2013**, *38*, 63–235. [CrossRef]

59. Yang, L.; Sun, H.; Liu, Y.; Hou, W.; Yang, Y.; Cai, R.; Cui, C.; Zhang, P.; Pan, X.; Li, X.; et al. Self-assembled aptamer-hyperbranched polymer nanocarrier for targeted and photoresponsive drug delivery. *Angew Chem. Int. Ed.* **2018**. [CrossRef]

60. Flanders, M.J.; Gramlich, W.M. Reversible-addition fragmentation chain transfer (RAFT) mediated depolymerization of brush polymers. *Polym. Chem.-UK* **2018**, *9*, 2328–2335. [CrossRef]

61. Miller, S.A. Sustainable Polymers: Opportunities for the Next Decade. *ACS Macro. Lett.* **2013**, *2*, 550–554. [CrossRef]

62. Li, L.J.; Shu, X.; Zhu, J. Low temperature depolymerization from a copper-based aqueous vinyl polymerization system. *Polymer* **2012**, *53*, 5010–5015. [CrossRef]

63. Olsen, P.; Odelius, K.; Albertsson, A.C. Ring-Closing Depolymerization: A Powerful Tool for Synthesizing the Allyloxy-Functionalized Six-Membered Aliphatic Carbonate Monomer 2-Allyloxymethyl-2-ethyltrimethylene Carbonate. *Macromolecules* **2014**, *47*, 6189–6195. [CrossRef]

64. Olsen, P.; Undin, J.; Odelius, K.; Keul, H.; Albertsson, A.C. Switching from Controlled Ring-Opening Polymerization (cROP) to Controlled Ring-Closing Depolymerization (cRCDP) by Adjusting the Reaction Parameters That Determine the Ceiling Temperature. *Biomacromolecules* **2016**, *17*, 3995–4002. [CrossRef] [PubMed]

65. Lloyd, D.J.; Nikolaou, V.; Collins, J.; Waldron, C.; Anastasaki, A.; Bassett, S.P.; Howdle, S.M.; Blanazs, A.; Wilson, P.; Kempe, K.; et al. Controlled aqueous polymerization of acrylamides and acrylates and "in situ" depolymerization in the presence of dissolved CO2. *Chem. Commun.* **2016**, *52*, 6533–6536. [CrossRef] [PubMed]

66. Fahnhorst, G.W.; Hoye, T.R. A Carbomethoxylated Polyvalerolactone from Malic Acid: Synthesis and Divergent Chemical Recycling. *ACS Macro. Lett.* **2018**, *7*, 143–147. [CrossRef]

67. Zhu, J.B.; Chen, E.Y.X. Living Coordination Polymerization of a Six-Five Bicyclic Lactone to Produce Completely Recyclable Polyester. *Angew. Chem. Int. Edit.* **2018**, *57*, 12558–12562. [CrossRef] [PubMed]

68. Hong, M.; Chen, E.Y.X. Completely recyclable biopolymers with linear and cyclic topologies via ring-opening polymerization of gamma-butyrolactone. *Nat. Chem.* **2016**, *8*, 42–49. [CrossRef] [PubMed]

69. Tang, X.Y.; Hong, M.; Falivene, L.; Caporaso, L.; Cavallo, L.; Chen, E.Y.X. The Quest for Converting Biorenewable Bifunctional alpha-Methylene-gamma-butyrolactone into Degradable and Recyclable Polyester: Controlling Vinyl-Addition/Ring-Opening/Cross-Linking Pathways. *J. Am. Chem. Soc.* **2016**, *138*, 14326–14337. [CrossRef] [PubMed]

70. Zhu, J.B.; Watson, E.M.; Tang, J.; Chen, E.Y.X. A synthetic polymer system with repeatable chemical recyclability. *Science* **2018**, *360*, 398–403. [CrossRef] [PubMed]

71. Hong, M.; Chen, E.Y.X. Towards Truly Sustainable Polymers: A Metal-Free Recyclable Polyester from Biorenewable Non-Strained -Butyrolactone. *Angew. Chem. Int. Edit.* **2016**, *55*, 4188–4193. [CrossRef] [PubMed]

72. Albertsson, A.C.; Varma, I.K. Recent developments in ring opening polymerization of lactones for biomedical applications. *Biomacromolecules* **2003**, *4*, 1466–1486. [CrossRef] [PubMed]

73. Dechy-Cabaret, O.; Martin-Vaca, B.; Bourissou, D. Controlled ring-opening polymerization of lactide and glycolide. *Chem. Rev.* **2004**, *104*, 6147–6176. [CrossRef] [PubMed]

74. Higaki, Y.; Okazaki, R.; Takahara, A. Semirigid Biobased Polymer Brush: Poly(alpha-methylene-gamma-butyrolactone) Brushes. *ACS Macro. Lett.* **2012**, *1*, 1124–1127. [CrossRef]

75. Mosnacek, J.; Matyjaszewski, K. Atom transfer radical polymerization of tulipalin A: A naturally renewable monomer. *Macromolecules* **2008**, *41*, 5509–5511. [CrossRef]

76. Al-Salem, S.M.; Lettieri, P.; Baeyens, J. Recycling and recovery routes of plastic solid waste (PSW): A review. *Waste Manage.* **2009**, *29*, 2625–2643. [CrossRef] [PubMed]

77. Hungenburg, K.; Wulkow, B. *Modeling and Simulation in Polymer Reaction Engineering: A Modular Approach*; Wiley: Hoboken, NJ, USA, 2018; ISBN 978-3-527-33818-4.

molecules

Review

Polymeric Co-Delivery Systems in Cancer Treatment: An Overview on Component Drugs' Dosage Ratio Effect

Jiayi Pan [1], Kobra Rostamizadeh [1,2], Nina Filipczak [1,3] and Vladimir P. Torchilin [1,*]

1   Center for Pharmaceutical Biotechnology and Nanomedicine, Northeastern University, Boston, MA 02115, USA; pan.jiay@husky.neu.edu (J.P.); rostamizadeh@zums.ac.ir (K.R.); ni.filipczak@northeastern.edu (N.F.)
2   Zanjan Pharmaceutical Nanotechnology Research Center, Zanjan University of Medical Sciences, Zanjan 4513956184, Iran
3   Laboratory of Lipids and Liposomes, Department of Biotechnology, University of Wroclaw, 50-383 Wroclaw, Poland
*   Correspondence: v.torchilin@northeastern.edu; Tel.: +1-617-373-3206; Fax: +1-617-373-8886

Received: 21 February 2019; Accepted: 13 March 2019; Published: 15 March 2019

Abstract: Multiple factors are involved in the development of cancers and their effects on survival rate. Many are related to chemo-resistance of tumor cells. Thus, treatment with a single therapeutic agent is often inadequate for successful cancer therapy. Ideally, combination therapy inhibits tumor growth through multiple pathways by enhancing the performance of each individual therapy, often resulting in a synergistic effect. Polymeric nanoparticles prepared from block co-polymers have been a popular platform for co-delivery of combinations of drugs associated with the multiple functional compartments within such nanoparticles. Various polymeric nanoparticles have been applied to achieve enhanced therapeutic efficacy in cancer therapy. However, reported drug ratios used in such systems often vary widely. Thus, the same combination of drugs may result in very different therapeutic outcomes. In this review, we investigated polymeric co-delivery systems used in cancer treatment and the drug combinations used in these systems for synergistic anti-cancer effect. Development of polymeric co-delivery systems for a maximized therapeutic effect requires a deeper understanding of the optimal ratio among therapeutic agents and the natural heterogenicity of tumors.

Keywords: polymeric nanoparticles; stimuli-sensitive polymers; co-delivery systems; synergistic effect; nucleic acid delivery; chemotherapy

---

1. Introduction

Cancer, next to heart disease, ranks as the second leading illness-related cause of death worldwide with a growing incidence and mortality. It is one of the most challenging-to-treat diseases due mainly to inefficient pharmacologically active agents and the complexity of cancer. To date, chemotherapy has been widely used and has been the most efficient and successful treatment method in clinical practice. However, there are three major issues limiting the therapeutic efficacy of chemotherapy. First, most of chemotherapeutic agents have poor solubility that leads to deficiencies like low bioavailability, rapid blood/renal clearance, and nonspecific targeting, with significant undesirable side effects on healthy tissues. Second, non-uniform biodistribution limits the localization of drugs at the tumor site and leads to consequent demands for higher doses that have unacceptable toxicity. Above all, genetic variations that control survival and apoptotic pathways are involved in the development of cancers. Targeting an individual pathway with conventional chemotherapy is often unsuccessful in eradicating all cancer cells and results in multidrug resistance (MDR) over the course of treatment.

Several alternative approaches to overcome these problems associated with traditional chemotherapy have been established. Much attention has been focused directly on drug combination approaches with the aims of more effective treatment and decreased side effects [1–3]. In general, combination chemotherapy involves the simultaneous administration of two or more drugs with non-overlapping toxicities and dissimilar mechanisms of action so as to inhibit multidrug resistance. Combination therapy can overcome the toxicity of single drug therapy by targeting various signaling pathways. Lately, combination therapy regimens have been intensively studied, and the results of clinical practice have demonstrated synergistic effects that are greater than the sum of the individual drug effects and less systemic toxicity associated with the delivery of lower drug doses.

However, combination regimens are still limited by a low rate of successful outcomes and significant side effects due to low bioavailability of drugs and their nonuniform biodistribution. To take advantage of possible synergy between drugs, they must attain effective molar ratios at the tumor sites that are often hard to reach by conventional administration methods due to differences in injection schedules, pharmacokinetic properties, metabolism, and non-uniform biodistribution.

One strategy for delivery of drugs to the tumor site at the desired molar ratio involves the merging of nanotechnology with pharmacology and thereby take advantage of the nanoscale structures that carry multiple drugs, allow tuning of drug release, and modify biodistribution and pharmacokinetic characteristics of chemotherapeutic agents [4,5]. Such co-delivery systems may be used to not only regulate the dosages and the ratio of chemotherapeutic agents at the tumor site. They may also improve the efficacy of anticancer drugs through enhanced water solubility of hydrophobic molecules, lower toxicity, and higher stability, which prolongs blood circulation time to enhance accumulation in tumor tissues. Further enhancement of therapeutic efficacy can be achieved by taking the advantage of stimuli-responsive drug delivery systems equipped with target-activated moieties [6,7].

Co-delivery systems ideally possess the potential for encapsulation of both hydrophobic and hydrophilic drugs. Platforms for co-delivery systems should be designed to carry both traditional chemotherapeutics and cell regulatory molecules, such as nucleic acids [6]. Although much progress has been made with nanotechnology-based co-delivery systems, there are several problems to be considered in formulating an ideal drug delivery system including those associated with encapsulating drugs with a variety of solubilities and physicochemical properties, elevating drug concentrations within tumor tissues and regulating their sequential drug release.

To date, considerable efforts have been made to develop nano-particulate co-delivery systems for combination chemotherapy [6,8]. Various nanocarriers have been investigated, including lipid nanoparticles [9], liposomes [10–12], dendrimers [13] and polymeric nanoparticles [14]. More attention has been paid to polymeric nanoparticles mainly because of their potential to carry both hydrophobic and hydrophilic drugs, favorable controlled drug release characteristics, low toxicity, high stability and a prolonged circulation time which ultimately enhances accumulation in tumor targets.

Many research and review papers involving co-delivery of therapeutic agents by polymeric nanoparticles in cancer therapy have been published [15–17]. However, therapeutic agents are often administered separately instead of simultaneously using a polymeric nanoparticle capable of delivering both agents [18]. Additionally, to the best of our knowledge, there is no comprehensive review addressing the effects of the drug ratio in co-delivery systems, a likely significant parameter promoting a synergistic therapeutic effect. The scope of this review is distinct from generalizations of about polymeric co-delivery systems for chemotherapy. Our aim is to address the dosage, cell type, mechanism, and their efficacy relationships that need to be considered in designing co-delivery systems.

2. Polymer Types Used in Preparing Co-Delivery Systems

2.1. Block Co-Polymer Conjugates

Amphiphilic polymers self-assemble into nanoparticles that are ideal co-delivery systems for both hydrophilic and hydrophobic drugs (Table 1). These polymers are usually obtained by conjugating together polymers with diverse properties. The block co-polymers formed inherit the properties of each block, thus integrating advantages of various blocks into a single system. Depending on the chemical composition, block co-polymers are prepared by conjugating hydrophobic and hydrophilic polymers together through physical/chemical interactions or by modifying hydrophilic polymers with hydrophobic lipid moieties. These synthetic block co-polymers self-assemble into polymeric nanoparticles or micellar nanoparticles for co-delivery purposes. In order to minimize toxicity and side effects, the major materials used are biodegradable polymers: chitosan [19], poly(lactic acid) [20], gelatin [21], poly[N-(2-hydroxypropyl) methacrylamide] (HPMA) [22] and their copolymers, poly(lactide-co-glycolide-co-caprolactone) [23] and poly(lactic-co-glycolic acid) [24]. Their total degradation can occur in the body and can reach the kidney threshold for excretion [25].

Six classic methods for obtaining polymeric nanoparticles have been described: nanoprecipitation [26], emulsion-diffusion [27], emulsification-coacervation [28], double emulsification [29], surface polymerization [30], and layer-by-layer methods [31]. In addition, multifunctional polymeric structures with precisely defined morphology can be obtained by a controlled atom transfer radical polymerization (ATRP) method [32].

Polymeric systems are grouped below by sensitivity to changes in temperature, pH, light, redox potential, and other special factors in their environment. Stimuli-sensitive polymers have become one of the most prominent solutions for anti-cancer therapy. The unique properties of polymers allow them to change their accumulation and drug release profile depending on the surrounding conditions. They are used to target drugs, bioactive substances and genes. These systems selectively deliver therapeutic agents to target tissues, cells and cell compartments to release their cargos [33–35]. By doing so, the pharmacological properties, release profile and therapeutic outcomes are improved compared to delivery as free drugs.

2.2. Thermo-Sensitive Polymers

Polymeric nanoparticles formulated with thermo-sensitive polymers has been applied to activate and control the release of active ingredients after reaching the target site. For example, overheating a cancer by activating magnetic cargo-loaded polymeric nanoparticles with a local magnetic field [36]. Thermos-sensitivity has been one of the most commonly used stimulating features for biomedical applications [37]. There are two types of polymers distinguished by their phase distribution. The first type, called UCST (upper temperature of the critical solution), passes between phases during cooling. In the second type LCST (lower critical temperature of the solution), this transition occurs with increasing temperature. Systems with UCST are more prevalent for polymers soluble in an organic solvent, while systems with LCST exist mainly for polymers dissolved in aqueous solvents. The solubility of polymers in organic solvents is due to short-range van der Waals interaction. The solubility of polymers in water is related to hydrophobic and hydrophilic interactions and the formation of hydrogen bonds. For polymeric drug delivery systems, mainly aqueous solvents are used [38–40].

Table 1. Types of stimuli-sensitive polymers commonly used as co-delivery systems.

Types	Polymers	Dosage Form *	Drug 1	Drug 2	Cell Line	Ref.
pH-sensitive	PDEA-PDMA-PEG	M	siBcl-2	DOX	Hep G2	[41]
	PEG-PLL-PAsp	NP	siBcl-2	DOX	Hep G2	[42]
	Trimethyl Chitosan	NP	IL-2	DOX	SMMC7721	[43]
	PEI-PLA/PEG-PAsp	NP	siSurvivin	PCT	A549	[44]
	PEO-b-PCL	M	siMDR-1	DOX	MDA-MB-435	[45]
	PDP-PDHA	NP	shSurvivin	DOX	MCF7/ADR	[46]
Redox-sensitive	PEI-CD	NP	TRAIL pDNA	DOX	SKOV-3	[47]
	PEG2k-CLV-Dox/miRNA-34a-S-S-PE	MM	miRNA-34a	DOX	MCF7, HT1080	[48]
	mPEG-PCL-SS-DOX/mPEG-PCL-SS-DTX	MM	DOX	DTX	MCF7	[49]
	Poly(acrylic acid)	H	DOX	CDDP	MCF7/ADR	[50]
	Gambogic acid-poly(amido amine)s	M	DTX	MMP9shRNA	MCF7	[51]
	PEG-PLG-PDMAPMA	NP	siMDR-1	DOX	MCF7	[52]
Thermo-sensitive	DH700k-MF-13.5/MDocLF	HMM	DOX	DTX	CT-26	[53]
	DHmPEG-b-PELG	H	IL15	CDDP	B16F0-RFP	[54]
	PLGA-PEG-PLGA	H	DOX, MTX	CDDP	Saos-2, MG-63	[55]
	PLGA-DOX/PEO-PPO-PEO	NP	DOX	IFNγ	B16F10	[56]
	PECT	HM	DOX	131I	Hep G2	[57]
MMP-sensitive	PEG2k-CLV-Dox/miRNA-34a-S-S-PE	MM	miRNA-34a	DOX	MCF7, HT1080	[48]
	PEG-pp-PEI-DOPE	M	siSurvivin	PCT	A549 T24	[58]
	PEG-PLA, G0-C14	NP	VEGF siRNA	PCT	HT-1080, A375, PC-3	[59]
Magnetic-responsive	PLGA/TPGS/OA	NP	TPGS	DOX	MCF7, MCF7/ADR	[60]
	ASA-MNPs-CDDP/mPEG-PLL-FA	NP	CDDP	TFPI2 DNA	HNE-1, NP69	[61]
	PCL/P(NIPAAm-co-HEMA-co-MAA-co-TMSPMA)	NP	DOX	MTX	MCF7	[62]

* MM—mixed micelles; M—micelles; NP—nanoparticles; H—hydrogels; HM—micelles entrapped in hydrogel, HMM—mixed micelles entrapped in hydrogel. PECT—poly(ε-caprolactone-co-1,4,8-trioxa[4.6]spiro-9-undecanone)-poly(ethyleneglycol)-poly(ε-caprolactone-co-1,4,8-trioxa[4.6]spiro-9-undecanone).

2.3. pH-Sensitive Polymers

The sensitivity of polymers to pH has been utilized in polymeric nanoparticles designed for drug administration via the gastrointestinal tract (pH ranges from 2–4 in the stomach to about 6.8 in the gut), in solid tumors where the interstitial space is more acidic due to hypoxia, with a pH of about 6.5 compared to a blood plasma pH of about 7.4. These pH-sensitive polymers can be weak bases (more polar in an acid environment in a protonated form) or weak acids (more polar in a basic environment in deprotonated form). Polymers abundant with primary amines are sensitive to low pH (pH 5.0) and can facilitate the endosomal escape of drugs. Additionally, pH differences between normal tissue and tumor tissue can create conditions for use of pH-sensitive polymeric drug delivery systems with enhanced targeting and reduced side effects [63,64].

2.4. Redox-Sensitive Polymers

The redox potential differences in the tumor microenvironment inspired the idea of building a redox reaction-sensitive polymeric system for cancer treatment. Redox sensitivity is usually used in cases where changes of redox potential occur in inflamed tissues compared to healthy tissues. Changes in redox potential in cancer tissues are due to the production of reactive oxygen species by activated macrophages. Oxidatively degradable polymers, such as arylborone based on acid esters (which after oxidation become phenols and boric acid), or dialkyl sulphide-based polymers (which after oxidation become more hydrophilic), have been used as delivery systems for drugs to inflamed tissues [65]. The disadvantage of these polymers is that the level of reactive oxygen species is often not enough to fully oxidize the polymer so that the drug/gene cannot be released [66,67]. Drug delivery systems also use polymers that react to light exposure and the presence of certain ions or organic molecules including sodium alginate [68]. These types of polymer are applied mainly for diagnostic purposes [69].

It can be said that designing a polymeric drug system with micro-environmentally sensitive polymers is a "smart" strategy. Combining multiple therapeutic agents that inhibit tumor growth through different pathways into one system is also a "smart" strategy. Many polymeric systems have shown promising effects in cancer therapy based on these two ideas.

3. Application of Polymeric Nanoparticles as Co-Delivery Systems

3.1. Polymeric Nanoparticles for Co-Delivery of Chemotherapeutics

The aim of combining chemotherapeutics is to achieve an additive or even a synergistic effect. By targeting different pathways, combination therapy delivered by polymeric nanoparticles makes cancer cells more susceptible to the delivered therapeutic agents. Four commonly used approaches for co-delivery of therapeutic agents using polymeric nanoparticles are shown in Figure 1. Most drugs are passively loaded in polymeric nanoparticles according to their hydrophilicity. Hydrophobic molecules can be loaded in the hydrophobic moieties of micelles or polymersomes, and hydrophilic molecules are trapped in the hydrophilic compartments. Some drugs, such as nucleic acids, are co-loaded on the surface of the polymeric nanoparticles by electrostatic forces or chemical conjugations. Another approach is to directly conjugate drugs with the polymer through ester, amide or disulfide bonds [70]. Different types of polymeric nanoparticles can be further modified with targeting moieties.

Doxorubicin (DOX), a chemotherapeutic anthracycline, has been used clinically for treatment of several hematologic malignancies and solid tumors including breast cancer [71]. However, using DOX alone usually causes serious side effects in normal tissues, especially cardiotoxicity [72]. The molecular mechanism of DOX-induced cardiotoxicity is still unclear [73,74]. However, it has been postulated that it is caused by conversion of quinone into free radicals of half quinone, which in turn initiates cascading reactions leading to production of reactive oxygen and nitrogen in the body [75,76]. To increase the antitumor effect of DOX by overcoming cells resistance and, at the same time toning down the

Molecules **2019**, *24*, 1035

cardiotoxicity, DOX is often used in combination with other chemotherapeutics, nucleic acids and antibodies in cancer therapy [77,78].

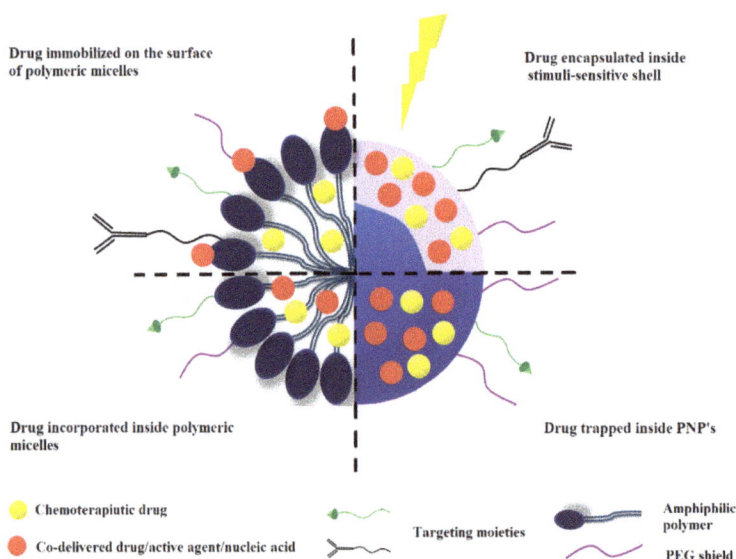

Figure 1. Drug loading in polymeric nanoparticles used as co-delivery systems in cancer treatment.

The combination of DOX with other anticancer drugs aims to achieve a synergistic effect of the combined drugs or improve their biodistribution. Doxorubicin is often combined with Paclitaxel (PCT). The hydrophobic delivery systems containing PCT are based on the entrapment of the drug in their hydrophobic core. The combination of DOX and PCT in mPEG-PLGA polymeric nanoparticles made by a dual emulsion method has allowed a better anti-cancer effect in vivo and a much faster release of the drug from the carrier in an acidic environment [79]. A similar formulation was also used to destroy tumor stem cells [80]. The double-reacting nanoparticles built of four polymers approved by the Food and Drug Administration (FDA) for medical use were composed of poly (D-, L-lactide-co-glycolide) (PLGA), Pluronic F127 (PF127), chitosan, and hyaluronic acid (HA). Combining PLGA and PF127 forms a more stable and homogeneous nanoparticles than with PLGA or PF127 alone. HA was used as a targeting moiety towards cancer stem cells to reduce drug resistance associated with the dormant metabolic state. As a result of the combination of both drugs, the anti-cancer effect was amplified ~500 times compared with a simple mix of the two drugs [80]. Similar polymeric co-delivery systems were also used that enclosed the hydrophobic irinotecan (CPT) and hydrophilic DOX HCl, which inhibited the activity of topoisomerase I and II and exhibited an enhanced therapeutic effect on breast and brain tumors [81].

Another strategy, used to deliver camptothecin (CPT) and doxorubicin, was to form a graft copolymer with side drug segments that form nanostructures using a protein folding pathway [82]. The graft copolymer was constructed by direct polymerization of g-camptothecin-glutamate N-carboxyhydride (Glu (CPT)-NCA) at multiple sites on the PEG-based main chain via open ring polymerization (ORP). Only the conjugated CPT is hydrophobic and served as the main driving force during the assembly process. When exposed to water, the copolymer, together with DOX, curls to form monodisperse nanolayers for the delivery of the two drugs. PEG coated nanocarriers exhibit good stability and are internalized by various cancer cell lines through the lipid raft and clathrin-mediated endocytosis pathways without premature leakage. These nanolayers generated high synergistic activity of the CPT and DOX in various tumor cell lines. The in vivo study confirmed that the nanolayers can show strong accumulation at tumor sites and result in significant anti-tumor activity in a lung cancer xenograft model compared to free drugs [82].

To overcome the drug resistance of tumors, redox-sensitive polymeric nanogels (<100 nm) based on poly(acrylic acid) were designed. Doxorubicin and cisplatin were enclosed in the nanogels by chelation, electrostatic interaction and π-π stacking interaction. Compare to free drugs, nanogels delivered more drugs to MCF-7/ADR cells. Considerable accumulation in cancerous tissues was observed in biodistribution experiments. In vitro anticancer studies showed better cell killing activity of the nano drug delivery system. All the data indicated that the combination therapy was more effective with reduced side effects [50]. Combinational delivery of DOX and verapamil in pH-sensitive polymeric nanoparticles based on co-polymer methoxy-poly(ethylene glycol)2k-poly(e-caprolactone)4k-poly(glutamic acid)1k (mPEG2k-PCL4k-PGA1k-FA) demonstrated a high drug release efficiency in tumor environment. The system was reported to overcome the multidrug resistance and improve the anti-cancer effect in MCF7/ADR cells (Figure 2) [83]. Other strategies to overcome the drug resistance of a tumor include regulation of the level of multidrug resistant protein [84]. Combining chemotherapeutics with gene therapy in polymeric nanoparticles for cancer treatment has received extensive attention [85,86].

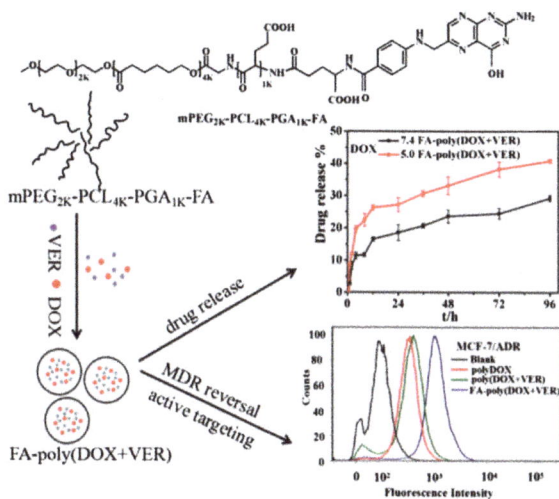

Figure 2. Folate-modified pH-sensitive co-delivery system of FA-poly(DOX+VER) polymer assembly exhibits obvious pH-sensitivity, high active targeting ability, strong multidrug resistance reversal and the enhanced therapeutic effect. Reproduced with permission from Li et al., Journal of Colloid and Interface; Elsevier, 2016 [83].

3.2. Polymeric Nanoparticles for Co-Delivery of Nucleic Acid Therapeutics and Chemotherapeutics

Multiple genetic targets have been established for cancer treatment over the past several decades. Based on the genetic links associated with tumor progression and development, nucleic acid therapeutics, such as siRNA, plasmid DNA (pDNA), miRNA, and antisense oligonucleotides, have provided highly attractive approaches to downregulate tumor-associated proteins or to recover the function of tumor-suppressing pathways. However, because of their high molecular weight, large amounts of negative charge, and enzyme-induced degradation, nucleic acids are very unstable in the systemic circulation and can barely penetrate the cellular membrane. Thus, intensive efforts have been made to develop delivery systems for nucleic acid therapy [87]. Here, combinations of chemotherapeutics together with siRNA, pDNA, and miRNA delivered by polymeric nanoparticles for cancer treatment are discussed.

3.2.1. Co-Delivery of siRNA and Chemotherapeutics with Polymeric Nanoparticles

One of the most widely used nucleic acid molecules used in preclinical and clinical studies has been small interfering RNA (siRNA). Cationic charged polymers, including chitosan [88,89], poly(ethylenimine) [90], poly(amidoamine) (PAMAM) dendrimer [13,91], and poly [2-(*N*,*N*-dimethyl aminoethyl) methacrylate] (PDMAEMA) [92,93], are capable of complexing with siRNA through electrostatic interaction, preventing degradation of siRNA and enhancing delivery of siRNA across the cell membrane. These polymers can be further modified by conjugating with other polymeric compartments to form an amphiphilic polymer with the ability to deliver siRNA and chemotherapeutics simultaneously.

Chitosan is a non-toxic and efficient vector for siRNA delivery. Surface modification with cationic chitosan by either absorption or covalent binding is a good strategy to enable the traditional drug delivery system to deliver siRNA as well [94]. Wang et al. coated chitosan on PLGA (50:50) nanoparticles loaded with DOX and grafted with a co-delivery system for both DOX and siRNA for an epidermal growth factor receptor [95]. Targeted with angiopep-2, the co-delivery system induced a 13% higher cell apoptosis rate than a PLGA nanoparticle loaded with DOX alone.

Cao et al. conjugated low molecular weight PEI with poly(ε-caprolactone) (PCL) through disulfide or ester covalent linkages [96]. This amphiphilic PEI-PCL self-assembled into a micellar structure. Doxorubicin was loaded into the PEI-PCL micelles using a chloroform/water mixture with sonication. Bcl-2 siRNA was complexed with PEI-PCL micelles followed by further modification with PEG chains to decrease the cytotoxicity of the nanoparticles. This DOX and siBcl-2 co-delivery system induced a 60% decrease in cell viability 96 h after treatment in a Bel7402 cell line. However, the cell viability decreased by only 40% when siScramble was used instead of siBcl-2, indicating a synergistic effect with the co-delivery of DOX and siBcl-2. With further modification of the DOX plus siBcl-2 loaded nanocarrier using folic acid, the cell viability decreased to 5%.

Another co-delivery system for siRNA and chemotherapy based on PEI was reported by Navarro et al. [97]. To prepare the amphiphilic molecule, PEI was conjugated with a dioleoylphosphatidylethanolamine (DOPE) moiety. Micellar nanoparticles formed by PEI-DOPE reversed the multidrug resistance in MCF7/ADR cells with delivery of P-glycoprotein siRNA (siP-gp) with DOX. Similarly, stearic acid was attached to PEI (1.8 kDa and 10 kDa) as a hydrophobic compartment by Huang et al. [98]. The combination of DOX and VEGF siRNA (siVEGF) co-delivered by PEI-SA micelles produced a promising in vivo Huh-7 tumor growth inhibition effect over 30 days. Tang et al. constructed an amphiphilic polymer, polyethyleneimine-block-poly((1,4-butanediol)-diacrylate-b-5-hydroxyamylamine) (PEI-PDHA), based on PEI 1.8 kDa [46]. Together with PEG-PDHA, this polymer self-assembled into nanoparticles co-loaded with siSnail, siTwist, and PCT. Significantly, an enhanced cytotoxic effect was observed at a PCT concentration of 50 μg/mL. The IC50 of the nanoparticles loaded with the three agents was 54.7-fold lower than that of free PCT.

PAMAM dendrimer is another candidate for siRNA delivery. The siRNA molecules are complexed via the primary amines on the surface of PAMAM dendrimers. Hydrophobic compartments used to load hydrophobic drugs can also be attached on the molecules through chemical conjugation. Biswas et al. prepared a tri-block co-polymer PAMAM-PEG2k-DOPE by conjugation between G4 PAMAM and DOPE-modified PEG [99]. PAMAM-PEG2k-DOPE self-assembled into micellar nanoparticles that complexed siRNA on the surface via the PAMAM moiety and encapsulated DOX base into the hydrophobic core. This co-delivery system efficiently delivered siRNA and DOX into cells and downregulated green fluorescent proteins (GFP) used to monitor the transfection efficacy in C166-GFP cells.

Another promising polymer for siRNA and chemotherapeutics co-delivery is PDMAEMA. Modification in the backbone of PDMAEMA not only decreased the cytotoxicity of the polymer, but also gave it the ability to load both siRNA and hydrophobic drugs [93,100]. Zhu et al. synthesized the block co-polymer PDMAEMA-PCL-PDMAEMA by free radical reversible addition-fragmentation chain transfer polymerization and assembled it into cationic micelles. PCT and siBcl-2 were delivered simultaneously using this system to PC3 cells, resulting in a 20% increased cytotoxicity than with free PCT after 24 h' treatment [100]. Wang et al. also delivered siBcl-2 together with DOX using a PEG-PDEA-PDMA-DDAT triblock co-polymer. Nanoparticles loaded with both DOX and siBcl-2 increased the cytotoxicity by 27.5% and 19.8% compared to nanoparticles loaded only with Dox or only siBcl-2, respectively [41].

3.2.2. Co-Delivery of pDNA and Chemotherapeutics with Polymeric Nanoparticles

Plasmid DNA (pDNA) was also delivered into cells after forming complexes with cationic polymers such as PEI, chitosan and PDMAEMA [101]. After modifying polymers with hydrophobic moieties, these polymers were widely applied for co-delivery of pDNA and hydrophobic chemotherapeutic agents. In 2006, Wang et al. developed PCT and interleukin-12 encoding pDNA co-delivery systems based on p(MDS-co-CES), which is poly{(*N*-methyldietheneamine sebacate)-co-[(cholesteryl oxocarbonylamido ethyl) methyl bis(ethylene) amoonium bromide] sebacate}. PCT was loaded inside the hydrophobic core during self-assembly, and pDNA was complexed with the cationic moieties. P(MDS-co-CES) micelles co-loaded with PCT and pDNA resulted in a greater tumor reduction than treatment with the gene or drug alone in a 4T1 mouse breast cancer model [102].

An abundance of primary amines makes branched PEI an ideal candidate for interaction with the large number of phosphate groups found on pDNA. Similar to the PEI-PCL used for co-delivery of siRNA and DOX, PEI1.8k-PCL was also reported as a co-delivery system for pDNA and hydrophobic drug by Qiu et al. [103]. After loading with both DOX and pDNA encoding luciferase reporter gene, PEI-PCL self-assembled into nanoparticles. Higher gene transfection efficacy than PEI25k and enhanced cytotoxicity compared to DOX alone were achieved in HepG2 [103]. Based on this structure, Shi et al. crafted a triblock co-polymer which consisted of mPEG5k-PCL2k-PEI2k for the delivery of DOX and Msurvivin T34A encoding pDNA [104]. By reducing the proliferation of tumor cells through Msurvivin T34A-induced caspase-mediated apoptosis, the author showed a higher anti-tumor effect compared to micelles loaded with DOX or pDNA alone. The author also suggested that treatment with dual-loaded micelles may allow lower doses of chemotherapeutics while maintaining a similar therapeutic outcome and help mitigate the toxic effect associated with high dose chemotherapeutics [104].

3.2.3. Co-Delivery of miRNA and Chemotherapeutics with Polymeric Nanoparticles

MiRNAs regulate the multiple molecular pathways associated with cancer development with a high degree of biochemical specificity and potency [85]. Combination of miRNA with small molecule therapeutics has provided an unprecedented opportunity to improve therapeutic efficacy in a broad range of human cancers [105–108]. Using vectors identical for siRNA delivery, modified cationic polymers have been widely applied for co-delivery of miRNA and small molecule drugs. Mitall et al. conjugated gemcitabine (GEM) and complexed miR-205 mimics on a mPEG-b-PCC-g-GEM-g-DC-g-CAT co-polymer, which self-assembled into polymeric micelles. GEM and miR-205 mimics were co-delivered at concentrations of 500 nM and 100 nM in vitro. Co-delivery of both drugs in polymeric micelles reduced tumor cell migration and restored chemosensitivity to gemcitabine in resistant MIA PaCa-2R pancreatic cancer cells. Intratumor injection of miR-205 (1 mg/kg)/gemcitabine (40 mg/kg) micelles to mice bearing MIA PaCa-2R xenografts potently arrested tumor growth, whereas free gemcitabine or gemcitabine micelles had only a modest effect [109].

Hyaluronic acid-chitosan nanoparticles used to co-encapsulate DOX and miR-34a were reported by Deng et al. for treatment of triple negative breast cancer. Combinations of 100 nM of miR-34a and 0.1/0.5/2.5 μM of DOX were investigated for their cytotoxic effect on MDA-MD-231 cells. Nanoparticles loaded with both drugs significantly increased cytotoxicity at all three DOX concentrations. A superior in vivo efficacy of the combination therapy with 5 mg/kg of DOX and

2 mg/kg of miR-34a was further demonstrated in mice bearing MDA-MB-231 subcutaneous xenografts following intravenous administration of the co-delivery nanoparticles every two days [110]. Similarly, Wang et al. prepared hyaluronic acid coated PEI-PLGA nanoparticles as a polymeric co-delivery system for miR-542-3p and DOX. A range of DOX concentrations from 0.25 µg/mL to 2 µg/mL was investigated for their therapeutic efficacy with 100 nM of miR-542-3p. The highest cytotoxicity increase, compared to nanoparticles loaded with DOX only, was at a combination of 2 µg/mL of DOX and 100 nM of miR-542-3p in triple negative breast cancer cell lines [111].

In addition to nanoparticles composed of cationic polymers, neutral nanoparticles have also been used to co-encapsulate miRNA and chemotherapeutics. As reported by Salzano et al., miRNA-34a was conjugated with phospholipids through a disulfide bond, and DOX was conjugated with PEG through a metalloproteinase 2 (MMP-2)-sensitive peptide (Figure 3). Conjugates were formulated into dual-stimuli sensitive micellar nanoparticles that can simultaneously deliver DOX and miRNA-34a. The combination of both drugs reduced HT1080 cell viability to 40% and 50% in monolayer and 3D spheroids tumor models, respectively [48]. Liu et al. incorporated miR-200c and docetaxel (DTX) into the PEG-gelatinase cleavable peptide-poly(ε-caprolactone) (PEG-pep-PCL) nanoparticles. The concurrent delivery of 10 mg/kg of DTX and 10 mg/kg of miR-200c markedly potentiated the anti-tumor efficacy of DTX in vivo [112].

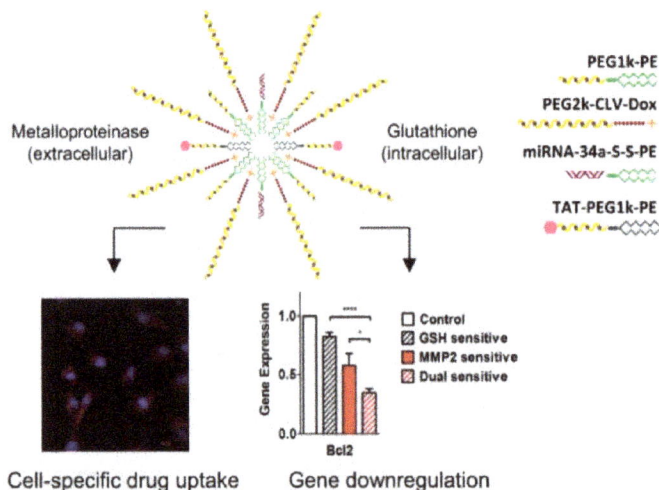

Figure 3. MMP-2 and glutathione sensitive polymeric nanoparticles used for co-delivery of DOX and miRNA-34a. Reproduced with permission from Salzano et al., Small; John Wiley and sons, 2016 [48]. * $p \leq 0.05$, **** $p \leq 0.0001$, $n = 3$, error bars represent mean $\pm$ SD.

4. Dose/Efficacy Relationship within Co-Delivery Systems

Different combination delivery systems of anti-cancer drugs within a single polymeric vesicle have been discussed in previous sections. The aim of these polymeric system is to leverage the different mechanisms of the individual therapeutic agent for additive or synergistic therapeutic effects. However, eventual success is determined by several other important factors that deserve consideration in designing polymeric co-delivery systems. First, binding of pDNA onto polymeric nanoparticles affects the DOX loading and vice versa. Xu et al. reported that co-encapsulation of p53 pDNA and DOX within PLA coated PLGA microsphere resulted in significantly reduced encapsulation efficiency of DOX [113]. However, similar hurdles were less obvious in carriers loaded with RNA and DOX. Second, another factor that influences the therapeutic outcome can be ascribed to the drug release profiles of different payloads. Depending on the interaction between drugs and polymers, drug delivered by hydrophobic interaction may be released much faster than those conjugated with polymers under physiological conditions [114]. Thus, the drug release profile of polymeric nanoparticles is one of the most crucial factors to consider when designing polymeric co-delivery systems [79,115]. Third, due to different mechanisms of therapeutic agents, the therapeutic efficacy of one drug may disguise the effect of others. Zhang et al. screened the IC50 of DOX and curcumin (CUR) at different ratios using SMMC 7721 cells after 48 h treatments. The IC50 value for DOX and CUR in cells treated with a combination of DOX/CUR at ratio of 1:10 decreased to 0.46 μM and 4.65 μM, respectively, compared to IC50's of free DOX at 1.30 μM and free CUR at 25.7 μM [115]. Much research requires a focus on optimizing the drug ratio for the optimal synergistic effect prior to co-loading multiple drugs into nanoparticles.

We would like to further elaborate on how drug combinations co-delivered are related to their therapeutic outcomes by concentrating on the dosing ratios between different therapeutic agents loaded in polymeric nanoparticles. Among the research published within the past decade, certain drug combinations have been widely used as models for developing co-delivery systems, including DOX, PCT, CUR, siBcl-2, siMDR-1. Although identical in drug combination, the therapeutic outcomes of polymeric co-delivery systems vary between cells and types of polymers. Here, we reviewed the factors that influence the therapeutic outcomes, with the aim of providing guidance in the design of polymeric co-delivery systems.

4.1. Dose Combinations of Chemotherapeutics in Polymeric Co-Delivery Systems

4.1.1. Doxorubicin and Paclitaxel Combinations

Combining two chemotherapeutics in a single carrier has been a popular approach in designing polymeric co-delivery systems. To validate the efficacy of these systems, a lot of work has to been done to characterize the relationship between the therapeutic outcome with two agents and the ratio of the two agents. In Table 2, drug combinations, at which synergistic anti-cancer effect was achieved, were listed. DOX and PCT are two classic chemotherapeutic agents that have been applied extensively in various polymeric systems. Combination delivery of both DOX and PCT are attractive for their achievement of a higher therapeutic outcome. Wang et al. loaded hydrophilic DOX and hydrophobic PCT in methoxyl PEG-PLGA co-polymer nanoparticles. The mPEG-PLGA co-loaded with both DOX and PCT produced greater tumor growth inhibition in vitro than mPEG-PLGA loaded with either DOX or PCT at the same concentrations. Moreover, the highest anti-tumor efficacy was achieved when DOX and PCT were loaded at a concentration ratio of 2:1 using three different types of tumor cell line [79].

Xu et al. prepared an amphiphilic poly(ethylene glycol)-poly(L-lactic acid) (PEG-PLA) diblock co-polymer and incorporated DOX and PCT into the ultrafine PEG-PLA fibers by an "emulsion-electrospinning" method [116]. The authors showed a lower cell viability and higher percentage of cell cycle arrest when DOX and PCT were delivered at a concentration ratio of 1:1 to rat glioma C6 cells 72 h after treatment. Chen et al. conjugated Pluronic P105 with DOX and encapsulated PCT into the hydrophobic core formed by P105-DOX and Pluronic F127 as a co-delivery system for hydrophobic DOX and PCT (Figure 4). The ratio between DOX and PCT was 2:3, when a higher in vitro cytotoxicity was observed in both MCF7/ADR and KBv cell lines. An increased apoptotic event, S phase cell cycle arrest and enhanced spheroid growth inhibition were observed in MCF7/ADR cells. The in vivo study also effected an efficient tumor growth inhibition over 14 days at this drug ratio [117]. Similarly, Ma et al. investigated the performance of pH-sensitive Pluronic F127-grafted chitosan for delivery of DOX together with PCT in vivo. In their study, 25 mg/kg of DOX and 20 mg/kg of PCT were used. All these studies used a concentration ratio between DOX and PCT of about 1:1 [118].

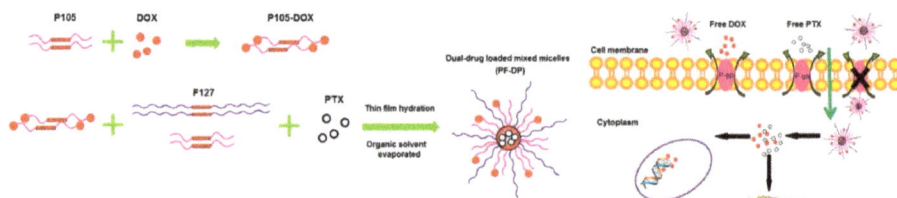

Figure 4. Co-delivery of DOX and PCT in polymeric nanoparticles consist of P105 and Pluronic F127 into MCF7/ADR cells. Reproduced with permission from Chen et al., International Journal of Pharmaceutics; Elsevier, 2015 [117].

However, other groups have held a different opinion on the optimal drug ratio between DOX and PCT loaded in polymeric vehicles. Duong et al. also prepared a PEG-PLGA based co-polymer system for delivery of hydrophobic DOX base and PCT. In addition, folic acid and TAT peptide were included to enhance the cellular interaction between PEG-PLGA micelles and a human oral cavity carcinoma KB cell line. After screening several different combinations of DOX and PCT and calculated their effectiveness using Calcusyn software, they demonstrated a better synergistic effect of DOX and PCT at a concentration ratio of 1:0.2 than of a concentration ratio of 1:1 [119]. This ratio is supported by Lv et al.'s study. In their case, an amphiphilic deoxycholate decorated methoxy poly(ethylene glycol)-b-poly(l-glutamic acid)-b-poly(l-lysine) triblock co-polymer (mPEG-b-PLG-b-PLL) was synthesized and developed as a nano-vehicle for co-delivery of DOX and PCT. The DOX and PCT co-loaded nanoparticles at a concentration ratio of 4:1 showed an obvious synergistic effect. The CI50 value was approximately 0.57, indicating co-delivery of DOX and PCT had evident superiority in tumor growth inhibition as compared with free drug combinations [120]. In an animal study, 4 mg/kg of DOX and 1 mg/kg of PCT loaded in nanoparticles were given to animals every four days. An efficient tumor growth inhibition was observed over 18 days in a A549 xenograft tumor model. The tumor volume of the co-loaded nanoparticles-treated group was only 9.0% of the control group at the 18th day, which was 3.2-fold, 6.3-fold, and 2.4-fold smaller than when treated with free DOX, free PCT and free DOX + PCT, respectively [120].

Although various drug combinations were used among these different cases to achieve a synergistic effect, one conclusion shared by these results is that control of the amount of DOX used is a prominent factor. DOX has a faster release rate than PCT, and the release of DOX facilitates the release of PCT. No significant synergistic effect was observed when PCT was used in excess [119]. Additionally, PCT inhibited the tumor growth by stabilizing the microtubule during cell mitosis. Cells were arrested rather than entering an apoptotic pathway when treated with DOX. Thus, synergistic effects can hardly be observed 24–48 h after treatment because the effect of DOX overrode the effect of PCT [79].

Table 2. Combinations of chemotherapeutics delivered in polymeric co-delivery systems.

Polymers	Drug 1	Concentration 1	Drug 2	Concentration 2	Cell line	Ref.
			In vitro			
HA	CPT	0.05–5.0 µM	DOX	0.22–0.5 µM	BT-474,	[81]
HA	CPT	0.04–0.45 µM	DOX	0.02–0.4 µM	bEnd.3	[81]
PLL-PTX [1], HA30k-GEM [2]	GEM	1×10^{-4}–1.0 mM	PCT	1×10^{-4}–1.0 mM	SCK HutCCT1	[121]
PEG-soyPC-PLGA	DOX	1–50.0 ng/mL	triptolide (TPL)	0.05–1 folds over DOX	KB	[122]
PEG-PLGA	CUR	5–15 µM	Chrysin	15–45 µM	Caco-2	[123]
DOX-PEG-GEM	DOX	0.001–100 µM	GEM	0.001–100 µM	SKOV-3, MCF-7, MDA-MB-231	[124]
TPGS-PAE [3]	DOX	0.031–1.0 µM	CUR	0.312–10 µM	SMMC7721	[115]
PEG-P(Glu)-P(Phe) [4]	PCT	0.0041–3.0 µg/mL	CDDP	0.94–60 µg/mL	HeLa, A549	[125]
PSn(P2VP-b-(PAA-g-PNIPAM)) [5]	PCT	1–15 µg/mL	camptothecin	1–15 µg/mL	A549	[126]
PLGA	PCT	10 µM	tetrandrine	10 µM	A2780	[127]
Polyphosphazene	DOX	12.5 µg/mL	CQ	1:1 and 2:1 over DOX	MCF7/ADR and HL60/ADR	[128]
mPEG-PLGA	DOX	Various ratios	PCT	Various ratios	A549, HepG2, B16	[79]
P(MeOx-b-BuOx-b-MeOx) [6]	PCT	0–1 µg/mL	alkylated CDDP	0–2.5 µg/mL	LCC-6-MDR, A2780 A2780/CisR	[129]
PEG-P(Glu)-P(Phe)	PCT	CDDP/PCT = 10:1	CDDP	0–10 µg/ml	A2780	[130]
			In vivo			
PHBV-PLGA	OXa	5 mg/kg	5-FU	25 mg/kg	CT26	[131]
HA	CPT	2 mg/kg	DOX	1.05 mg/kg	4T1	[81]
PLL-PTX, HA30k-GEM	GEM	108.8 ug/animal	PCT	54 ug/animal	HutCCT1	[121]
Pluronic F127-chitosan	DOX	25 mg/kg	PCT	20 mg/kg	Healthy rat	[118]
TPGS-PAE	DOX	1 mg/kg	CUR	10 mg/kg	SMMC 7721	[115]
PEG-P(Glu)-P(Phe)	PCT	3 mg/kg	CDDP	10 mg/kg	A549	[125]
Polyphosphazene	DOX	15 ng/animal	CQ	15 ng/animal	MCF7/ADR	[128]
PPBV [7]	PCT	4 mg/kg	CUR	10 mg/kg	MCF7	[132]
P(MeOx-b-BuOx-b-MeOx)	PCT	20 mg/kg	alkylated CDDP	20 mg/kg	A2780/CisR	[129]
PEG-P(Glu)-P(Phe)	PCT	4 mg/kg	CDDP	4 mg/kg	A2780/Luc	[130]

[1] Poly (l-lysine)–carboxylate paclitaxel. [2] Hyaluronic acid–gemcitabine. [3] D-α-tocopheryl poly(ethylene glycol) 1000-block-poly (β-amino ester). [4] Poly(ethylene glycol)-b-poly (l-glutamic acid)-b-poly(l-phenylalanine). [5] Star-graft quarterpolymers, composed of hydrophobic polystyrene and pH-sensitive poly(2-vinylpyridine)-b-poly(acrylic acid). [6] Poly(2-methyl-2-oxazoline-block-2-butyl-2-oxazoline-block-2-methyl-2-oxazoline). [7] Poly(ethylene glycol)-benzoic imine-poly(γ-benzyl-l-aspartate)-b-poly(1-vinylimidazole) block copolymer.

4.1.2. Doxorubicin and Curcumin Combinations

Another classis combination is DOX and CUR. CUR was believed to inhibit the tumor growth by causing cell cycle arrest [133,134], inducing an apoptotic signal [135,136], reversing multidrug resistance [137] and inhibiting the activation of NF-κB [138,139]. The exact molecular mechanisms of curcumin-induced apoptosis in cancer cells varied and depended on the cell type and dose used [140]. Application of CUR as an adjuvant in co-delivery systems has aroused great interest [141].

Zhang et al. combined pro-apoptotic, anti-angiogenic activities in pH-sensitive nanoparticles prepared with D-α-tocopherol poly(ethylene glycol)1k-block-poly (β-amino ester) (TPGS-PAE) polymers. The authors optimized the concentration of DOX and CUR at a 1:10 ratio. When exposed to 0.25 μM of DOX together with 2.5 μM of CUR, a 45% increased cytotoxicity occurred over that treated with DOX alone after 48 h in human SMMS 7721 liver cancer cells. The same ratio was also used in an in vivo study, where 1 mg/kg of DOX and 10 mg/kg of CUR were delivered within TPGS-PAE nanoparticles given intravenously once every other day. The CUR + DOX loaded nanoparticles induced a tumor weight suppression of 73.4%, compared to the 32.6% in a free CUR + DOX group [115].

Yan et al. also investigated the performance of a DOX CUR co-delivery system in a human liver cancer cell line, Hep G2. They came to a contradictory conclusion from Zhang's. regarding the drug ratio between DOX and CUR needed to reach a synergistic effect [142]. They prepared a redox-responsive micelle composed of a glycyrrhetinic acid-modified chitosan-cystamine-poly(ε-caprolactone) co-polymer (PCL-SS-CTS-GA). DOX and CUR were loaded in the PCL-SS-CTS-GA nanoparticles at mole ratios of DOX:CUR = 1:1, 2:1 and 3:1. A synergistic effect was observed only at ratio of 2:1 and 3:1 in a Hep G2 cell line [142]. This result was similar to that reported by Zhang et al., who conjugated DOX with methoxy-poly(ethylene glycol)-aldehyde (mPEG-CHO) and encapsulated CUR into the hydrophobic core formed within the mPEG-DOX micelles. They used a 2.5-fold higher concentration of DOX compared to CUR corresponding to a mole ratio of DOX:CUR = 1.6:1. In both Hep G2 and HeLa cell lines, an enhanced cytotoxic effect was observed in the mPEG-DOX-CUR nanoparticle-treated group compared to cells treated with DOX or CUR alone [143].

In other studies where CUR was utilized in an attempt to overcome multidrug resistance, Wang et al. studied the effect of co-delivery of DOX and CUR in mice bearing MCF7/ADR or 4T1 tumors. The polymeric micelles for DOX and CUR co-delivery were based on two diblock co-polymers, D-α-tocopherol polyethylene glycol succinate (TPGS) and PEG2k-DSPE. Cells were incubated with polymeric micelles containing both drugs at a 1:1 ratio (mole ratio: DOX:CUR = 1:1.6) for 48 h. Over a concentration range of DOX of 0.5 μg/mL to 40 μg/mL, the maximum cytotoxicity increase was observed at 10 μg/mL. An 18.3-fold increase of apoptotic events was also observed at 10 μg/mL of DOX in the group treated with polymeric micelles loaded with DOX and CUR compared with the one treated with DOX only. A formulation co-loaded with 5 mg/kg of DOX and 5 mg/kg of CUR injected intravenously every two days resulted in a significantly lower tumor volume after a 10-day treatment than those treated with either DOX or CUR alone [144]. Duan et al. co-encapsulated 0.12 μg/mL of DOX and 0.2 μg/mL of CUR (mole ratio: DOX: CUR = 1:2.6) in chitosan/poly(butyl cyanoacrylate) nanoparticles to reverse the multidrug resistance in MCF7 ADR cell lines. They reported that polymeric nanoparticles loaded with both drugs induced a higher cell growth inhibition and a significantly greater downregulation of MDR-1 protein at 48 h after treatment [145].

Overall, the mole ratio of CUR and DOX reported for these polymeric delivery systems had a broad range, suggesting that in the presence of multiple tumor inhibiting pathways, CUR acts differently depending on the cell line. Thus, it's hard to predict an effective universal ratio for combination treatment with CUR plus DOX. Additionally, the release profile of CUR differs markedly between polymeric delivery systems [146,147]. In some cases, DOX is directly conjugated with the polymer, while in other cases, DOX and CUR are both passively encapsulated. The release profile and types of tumor are key points to be taken into consideration during the design of polymeric co-delivery systems using CUR.

4.1.3. Paclitaxel and Cisplatin Combination

The use of a combination of PCT and cisplatin (CDDP) is another classic and popular co-delivery strategy. Cisplatin, a member of the platinum-containing anticancer drugs, has been used intensively for the treatment of various solid tumors, particularly in advanced stages. There are different mechanisms that account for the effect of CDDP. It is well-known as a non-specific DNA-modifying agent that induces cell apoptosis by interaction with nuclear DNA to inhibit the transcription and replication of DNA. The clinical application of CDDP is limited because of its high dose-dependent toxicity, drug resistance, and low bioavailability. PCT, on the other hand, a number of the taxane family, works as a microtubule-stabilizing chemotherapeutic agent. It is known that PCT can also inhibit platinum−DNA adduct repair and enhance apoptosis of cisplatin-resistant tumor cells [148]. However, it enhances the nephrotoxicity of CDDP. Nowadays, the combination therapy including CDDP and PCT is still a popular clinical regimen used as a the first-line chemotherapeutic agent for advanced cancer treatment.

To date, few works have dwelt on the study of co-delivery of the CDDP/PCT, due mainly to the hydrophilic nature of CDDP and hydrophobicity of PCT, which has made the loading step very challenging [149,150]. One strategy to overcome this problem is to use different cisplatin prodrugs to facilitate drug loading [151,152]. Polymeric nanoparticles, particularly micelles with the ability to carry both hydrophilic and hydrophobic drugs represent a promising candidate for the co-delivery of CDDP and PCT. Mi et al. used a D-α-tocopherol-co-poly(ethylene glycol)1k succinate (TPGS)-cisplatin prodrug (TPGS-CDDP) along with docetaxel to improve CDDP loading into Herceptin-decorated PLA-TPGS nanoparticles. The highest drug loadings achievable for CDDP, PCT, and Herceptin using this approach were 3.5 ± 0.1, 9.0 ± 0.5, and 73.1 ± 5.8 µg/mg, respectively [151]. The presence of polymers with the ability to form a chelate with CDDP, such as PEG or poly-glutamic acid in the copolymer structure, can also enhance drug loading capacity [125,129,130]. In He et al.'s work, the drug loading capacity of PCT and CDDP attainable with FA-PLGA-PEG micelles has been reported to be 5.83 ± 0.04%, 4.68 ± 0.07%, respectively, at a polymer/drug ratio of 40:2:4 (polymer:CDDP:PCT) [153].

It is important to note that, the ratio of drug loading in the carrier plays a determinant role in achieving the highest therapeutic efficacy for combination therapy. According to Wan et al., the actual drug ratios in the tumor did not differ significantly from the drug ratios in the initial co-loaded drug formulations [129]. Thus, assuring the drugs loading for polymeric nanoparticles in an appropriate ratio would be of great importance in formulating a treatment. He et al. showed that in A549 and M109 cells, the effectiveness of co-delivery of NPs, with a CDDP/PCT concentration ratio of 1:2, was approximately twice that of CDDP [153]. Moreover, co-delivery of NPs, with a CDDP/PCT concentration ratio of 1:2, exhibited the greatest anti-tumor activity among the two varieties of lung cancer cell. Although in general, dual drug-loaded nanoparticles with a higher ratio of PCT exhibited the greatest response, there are some reports indicating the best response corresponds to an equal drug ratio [152].

Time-dependency is another factor affecting the synergism of CDDP/PCT co-delivery systems [125]. This phenomenon was related to the release behavior of CDDP and PCT from the carrier. In single drug-loaded nanoparticles, drug is quickly released, while in a dual drug-loaded carrier system, chelation of PCT with CDDP prevented an initial burst release and resulted in a lower growth inhibition effect. However, over time, more PCT was released, and the combination effect became manifest. The in vivo studies on co-delivery systems for CDDP and PCT also clearly revealed improved pharmacokinetics and biodistribution in the blood and tumor of either one or both drugs compared to single drug micelles [129].

4.2. Dose Combinations of Nucleic Acid Therapeutics and Chemotherapeutics in Polymeric Co-Delivery Systems

The rationale for combination of nucleic acid therapy with chemotherapeutics in a single platform to ensure that cells will be simultaneously exposed to two types of damage has been discussed previously. The dose combination between nucleic acids and chemotherapeutics delivered has large effects on the synergistic anti-cancer efficacy of such co-delivery systems (Table 3). Nucleic acid therapy induces tumor inhibitory effects via different molecular pathways than traditional chemotherapeutics. Plasmid DNA requires delivery into the nucleus followed by translation into therapeutic proteins and the oligonucleotide requires interference with the mRNA or proteins in various pathways to generate their effect. Thus, nucleic acid therapy takes a longer time than chemotherapeutics before the tumor growth inhibition effect can be observed. The effect of chemotherapeutic agents can also disguise the one from nucleic acid therapy. Therefore, controlling the ratio between nucleic acid molecules and chemotherapeutics involving different mechanisms and types of pathways in polymeric systems is critical for an ideal synergistic effect.

4.2.1. Inducing Apoptosis through Delivery of TRAIL

Interfering with the apoptotic pathway by either enhancing the pro-apoptotic effect or inhibiting the anti-apoptotic effect with nucleic acid therapies is a promising cancer treatment. One of the most attractive pro-apoptotic pathways for the treatment of cancer is through the tumor necrosis factor (TNF)-related apoptosis-inducing ligand (TRAIL). TRAIL is a potent stimulator of apoptosis, and tumor cells are significantly more sensitive to TRAIL-induced apoptosis than normal cells [154]. However, numerous cancer cells exhibit a certain amount of resistance to TRAIL-induced apoptosis. Thus, combining TRAIL protein with traditional chemotherapeutics could increase the therapeutic efficacy in TRAIL-resistant cancer cells [155,156]. Lee et al. co-loaded both TRAIL protein and DOX in polymeric nanoparticles self-assembled from a biodegradable cationic co-polymer P(MDS-co-CES) to achieve a synergistic cytotoxic effect in cancer cells. A synergistic effect of a 40% enhanced cytotoxicity was observed with nanoparticles loaded with 10 nM of TRAIL and 0.8 µM of DOX compared to nanoparticles loaded with each drug alone in TRAIL-resistant SW480 colorectal carcinoma cells [157]. The same group also used P(MDS-co-CES) for co-delivery of TRAIL and PCT. In the study, various human breast cancer cell lines were exposed to 10 nM of TRAIL and 1.67 µM of PCT co-loaded in P(MDS-co-CES) polymeric nanoparticles. A 25% enhanced cytotoxic effect was achieved compared to single drug-loaded treatments [158].

In addition to delivering TRAIL in polymeric nanoparticles, Fan et al. delivered plasmid DNA encoded for TRAIL together with DOX in β-cyclodextrin-polyethyleneimine (PEI-CD) supramolecular nanoparticles (SNP). The PEI-CD SNP loaded with 0.5 µg/mL of DOX and 2.5 µg/mL of TRAIL pDNA (pTRAIL) induced many more apoptotic events than individual drug treatment in SKOV-3 cells after 48 h treatment. At the same drug ratio, SKOV-3 ovarian tumor-bearing mice received a combination treatment of 6 µg of DOX and 30 µg of pTRAIL twice per week. After 18 days, a significantly higher tumor growth inhibition was observed in those mice [47]. In another study, DOX was complexed with pORF-hTRAIL, which was then complexed with polyamidoamine-PEG-T7 (PAMAM-PEG-T7) through electrostatic interaction. A significant synergistic effect occurred both in vitro and in vivo when 12.5-fold more DNA than DOX by weight was loaded into this platform. Each Bel-7402 tumor-bearing mouse was treated with 50 µg of DNA in combination with 4 µg of DOX [159].

Table 3. Combinations of nucleic acids therapeutics and chemotherapeutics delivered in polymeric co-delivery systems.

Polymers	Drug 1	Concentration 1	Drug 2	Concentration 2	Cell Line	Ref.
In vitro						
P(MDS)-co-CES	TRAIL	10 nM	DOX	0.8 µM	SW480-TR	[157]
P(MDS)-co-CES	TRAIL	10 nM	PCT	1.67 µM	MCF7, T47D, MDA-MB-231	[158]
PEI-CD	TRAIL pDNA	2.5 µg/mL	DOX	0.5 µg/mL	SKOV-3	[47]
PLGA-PLA	p53 gene	2 µg/mL	DOX	0.9 µg/mL	Hep G2	[160]
PDMAEMA-PMPD	p53 gene	4 µg/mL	DOX	3 µg/mL	MCF7	[161]
PEI-PCL/FA-PEG-PGA	siBcl-2	20 nM	DOX	50 nM	Bel-7402	[96]
PEI-PCL/FA-PEG-PGA	siBcl-2	25 nM	DOX	0.5 µg/mL	C6	[162]
PDEA-PDMA-PEG	siBcl-2	20 nM	DOX	1.69 µM	Hep G2	[41]
PEG-PLL-PAsp	siBcl-2	100 nM	DOX	0.6 µg/mL	Hep G2	[42]
P(MDS)-co-CES	siBcl-2	20 nM	PCT	100/400 nM	MDA-MB-231	[102]
PDMAEMA-PCL-PDMAEMA	siBcl-2	188 nM	PCT	0.58 µM	PC3	[100]
PEI-CyD	shSurvivin	2 µg/mL	PCT	0.6 µg/mL	SKOV-3	[163]
PEG-pp-PEI-DOPE	siSurvivin	150 nM	PCT	12–24 nM	A549 T24	[58]
PEI-PLA/PEG-PAsp	siSurvivin	20 nM	PCT	0.096 µg/mL	A549	[44]
PEO-b-PCL	siMDR-1	100 nM	DOX	5 µg/mL	MDA-MB-435	[45]
PEI-DOPE	siMDR-1	100 nM	DOX	1 µg/mL	MCF7/ADR	[97]
PAMAM-PEG-DOPE	siMDR-1	125 nM	DOX	1.7 µg/mL	A2780/ADR	[91]
PAMAM-PEG-DOPE	siMDR-1	125 nM	DOX	0.43 µg/mL	MCF7/ADR	[91]
NSC-PLL-PA	siMDR-1	100 nM	DOX	5 µg/mL	Hep G2/ADM	[164]
PEG-PLG-PDMAPMA	siMDR-1	100 nM	DOX	3 µg/mL	MCF7	[52]
DMAB-PLGA	siMDR-1	100 nM	DOX	11.6 µg/mL	MCF7/ADR	[165]
HA/PEI-PLGA	miR-542-3p	100 nM	DOX	2 µg/mL	MDA-MB-231	[111]
PEG-PLGA-PLL	miR-7	100 nM	PCT	0.01 µg/mL	HO8910pm	[166]
PEG-PCC-GEM-DC-CAT	miR-205	100 nM	GEM	500 nM	MIA PaCa-2R, CAPAN-1R	[109]
PCL-PEG-PHIS	siVEGF	100 nM	PCT	2 µg/mL	MCF7	[167]
In vivo						
PEI-CD	TRAIL pDNA	30 µg/animal	DOX	6 µg/animal	SKOV-3	[47]
PAMAM-PEG-T7	pORF-hTRAIL	50 µg/animal	DOX	4 µg/animal	Bel-7402	[159]
PDMAEMA-PMPD	p53	1.5 mg/kg	DOX	2 mg/kg	MCF7	[161]
PEI-PCL/FA-PEG-PGA	siBcl-2	1.6 mg/kg	DOX	22.5 µg/kg	C6	[162]
PEG-PLL-PAsp	siBcl-2	200 µg/kg	DOX	1 mg/kg	Hep G2	[42]
mPEG-PCL-g-PEI	Msurvivin	5 mg/kg	DOX	4 mg/kg	B16-F10	[104]
PEI-CyD	shSurvivin	6 µg/animal	PCT	20 µg/animal	SKOV-3	[163]
P85-PEI/TGPS	shSurvivin	2 mg/kg	PCT	10 mg/kg	A549	[168]
PDP-PDHA	shSurvivin	2 mg/kg	DOX	6 mg/kg	MCF7/ADR	[169]
PEI-PLA/PEG-PAsp	siSurvivin	20 nM	PCT	0.096 µg/kg	A549	[44]
NSC-PLL-PA	siMDR-1	0.2 mg/kg	DOX	0.5 mg/kg	Hep G2/ADM	[164]
DOPE-PEI	siMDR-1	1.2 mg/kg	DOX	2 mg/kg	MCF7/ADR	[170]
POEG-st-Pmor	IL-36γ	50 µg/animal	DOX	5 mg/kg	4T1.2	[171]
Trimethyl Chitosan	IL-2	1.2 µg/animal	DOX	2 mg/kg	SMMC7721	[43]
P(MDS)-co-CES	IL-2	5 µg/animal	PCT	10 µg/animal	4T1	[102]
PLGA/Pluronic F127	IL-2	2.5 µg/kg	PCT	10 mg/kg	B16-F10	[172]
PEG-PLGA-PLL	miR-7	2 mg/kg	PCT	6 mg/kg	HO8910pm	[166]
PEG-PCC-GEM-DC-CAT	miR-205	1 mg/kg	GEM	40 mg/kg	MIA PaCa-2R	[109]
PCL-PEG-PHIS	siVEGF	5 mg/kg	PCT	1.2 mg/kg	MCF7	[167]
PEI-SA	siVEGF	0.3 mg/kg	DOX	0.45 mg/kg	Huh7	[98]

4.2.2. Increasing Apoptosis by Restoring p53

A similar combination was also used to restore the function of the tumor suppressor, p53. Xu et al. treated Hep G2 cells with a combination of 2 µg/mL of p53 gene and 0.9 µg/mL of DOX. Chitosan-p53 nanoparticles and DOX molecules were co-loaded in double-walled microspheres that consisted of a PLGA core surrounded by a poly(L-lactic acid) (PLA) shell. Overall, the combined DOX and p53 treatment enhanced cytotoxicity and increased activation of caspase-3 compared to either DOX or p53 treatment alone [160].

Li et al. used a combination of DOX and p53 plasmid at 3 µg/mL and 4 µg/mL, respectively, to induce a synergistic anti-cancer effect in breast cancer. Both drugs were loaded in micelles formed from a star-shaped polymer consisting of a cationic poly[2-(dimethylamino) ethyl methacrylate] (PDMAEMA) shell and a zwitterionic poly{N-[3-(methacryloylamino) propyl]-N,N-dimethyl-N -(3-sulfopropyl) ammonium hydroxide} (PMPD) corona that was grafted from a polyhedral oligomeric silsesquioxane (POSS)-based initiator. A high tumor cell apoptosis in MCF7 cells occurred in vitro and extensive tumor growth inhibition was observed over 28 days with 1.5 mg/kg of DOX and 2 mg/kg of p53 plasmid administrated every five days [161].

Usually, 2–4 µg/mL of plasmid DNA was used in co-delivery with chemotherapeutics to achieve a synergistic effect. But the ratio between chemotherapeutics and pDNA ranged from 1–10. Chen et al. investigated the response of cell viability in MCF7 cells to DOX-loaded polymeric nanoparticles complexed with different ratios of p53 pDNA (Figure 5) [173]. An enhanced cell growth inhibition was observed at weight ratios between DOX-NP and p53 pDNA from 5–10. This enhanced effect was not observed at weight ratios higher than 10 or lower than 5. These results suggested that the ratio between chemotherapy and pDNA for a synergistic effect could be affected by a diversity of cell lines and types of cationic polymers. Especially because that, as previously discussed, the complexation of pDNA with polymeric nanoparticles could affect the loading of chemotherapeutics [113]. Investigating the pDNA complexation effect on drug loading, drug release profile is essential in evaluating the anti-cancer performance of pDNA and chemotherapeutics-loaded polymeric co-delivery systems.

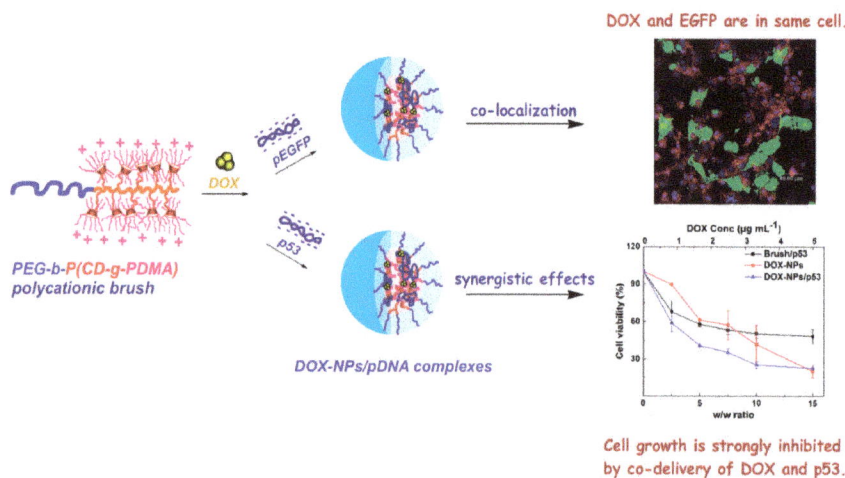

Figure 5. Co-delivery of DOX and pDNA in cationic polymeric nanoparticles with co-localization of cargos and enhanced tumor cell growth inhibition. Reproduced with permission from Chen et al., Polymers; MDPI, 2019 under the license CC BY 4.0 [173].

216

4.2.3. Decreasing Anti-Apoptotic Effect through Downregulation of Bcl-2

Another popular candidate used within co-delivery systems with pro-apoptotic effects has been siBcl-2. Overexpression of Bcl-2 family proteins suppresses the cell death induced by chemotherapeutics [174]. Thus, it's particularly interesting to pursue possible synergistic effects derived from the downregulation of Bcl-2 and administration of chemotherapeutics. Cao et al. proposed loading DOX and siBcl-2 into PEI-PCL micelles using a chloroform/water mixture under sonication. The micelles were then coated with folic acid-conjugated poly(ethylene glycol)-block-poly(glutamic acid) (FA-PEG-PGA) after complexation of siBcl-2. Together with 20 nM of siBcl-2, 0.05 µM of DOX was delivered to Bel-7402 liver cancer cells. The PEI-PCL/FA-PEG-PGA micelles co-loaded with DOX and siBcl-2 induced a 60% increased cytotoxicity 96 h after treatment [96]. From the same group, Cheng et al. investigated the therapeutic efficacy of PEI-PCL/FA-PEG-PGA micelles loaded with siBcl-2 and DOX using C6 glioma cells. They demonstrated that 25 nM of siBcl-2 achieved a higher knockdown effect than that at 12.5 nM. Most importantly, the knockdown effect of Bcl-2 became saturated at concentrations higher than 25 nM. The combined therapeutic outcome of DOX (22.5 µg/kg) and siBcl-2 (1.6 µg/kg) treatment in vivo highlighted the importance of combined therapy of DOX and siRNA for tumor growth inhibition [162].

Wang et al. also investigated the combination of DOX and siBcl-2 in another human liver cancer cell line, Hep G2, using a PDEA-PDMA-PEG co-polymer. They also used 20 nM of siBcl-2 but increased the DOX concentration to 1.69 µM. Under this circumstance, co-delivery of both drugs increased the cytotoxicity by 27.5% and 19.8% compared to nanoparticles loaded with DOX and siBcl-2, respectively [41]. Instead of DOX, PCT has also been delivered together with siBcl-2. Wang et al. studied the synergistic effect between siBcl-2 and PCT in triple negative breast adenocarcinoma MDA-MB-231 cells. In the presence of 20 nM siBcl-2, cell viability decreased from 78% to 59% and from 58% to 39% at PCT concentrations of 100 nM and 400 nM, respectively [102].

However, other groups have used a higher concentration of siBcl-2. For example, Zhu et al. treated PC3 human prostate cancer cells with a combination of 188 nM of siBcl-2 and 0.58 µM of PCT. The drugs were delivered using micellar nanoparticles composed of PDMAEMA-PCL-PDMAEMA triblock co-polymer. They reported a synergistic effect of about 20% increased cytotoxicity compared to free PCT at 24 h after treatment [100]. Such differences in the concentration of siBcl-2 used may have resulted from the responses to siBcl-2 seen in different cell lines. Instead of screening for the response of cells to siBcl-2, most groups have used a standard siRNA concentration of 100 nM in their studies [42,175].

4.2.4. Decreasing the Anti-Apoptotic Effect through Downregulation of Survivin

Survivin is one of the most frequently occurring antiapoptotic proteins seen in cancerous tissues (i.e., breast, colon, pancreas, and lung). Its main mechanism of action depends on inhibition of caspase activation [176]. Through its action, survivin leads to increased proliferation of tumor cells [177]. Wang et al. developed a DOX, PCT, and survivin co-delivery system using a nano-emulsion composed of a methoxy-poly (ethylene glycol) block copolymer (mPEG-PLGA) and e-polylysine (EPL). The core of the nano-emulsion was DOX, and the PCT was enclosed in the hydrophobic layer. EPL on the surface of the nano-emulsion complexed siRNA by electrostatic interaction. Experiments in mice bearing a B16-F10 melanoma tumor showed a synergistic tumor growth inhibition effect from DOX (8.6 mmol/kg), PCT (17.2 mmol/kg), and survivin-siRNA (1.5 mg/kg) [178].

In Shi et al.'s report where the block copolymer mPEG-PCL-g-PEI was used for co-delivery of doxorubicin and Msurvivin T34A plasmid, a synergistic effect of DOX (4 mg/kg DOX) and Msurvivin T34A plasmid (5 mg/kg) was demonstrated in mice bearing a B16-F10 melanoma, both in subcutaneous and lung metastases models. Although they obtained only a slightly higher antitumor activity when compared to free DOX, they effectively reduced systemic toxicity of the treatment [104]. Survivin shRNA-encoding plasmid was also delivered to SKOV-3 cells by self-assembled supramolecular micelles composed of b-cyclodextrin-polyethylenimine (PEI600-CyD) and 2-amineadamantine-conjugated PCT (Ada-PCT) by Hu et al. They proved that simultaneously administrated PCT and shRNA at concentrations of 0.6 μg/mL, 2 μg/mL in vitro and 6 μg/animal, 20 μg/animal in vivo, respectively, induces significantly higher cell apoptosis and inhibits tumor growth [163].

Another co-delivery pluronic system P85-PEI/TPGS/PCT/shSur containing survivin hairpin RNA was developed to treat A549 human lung cancer. The purpose of this study was to overcome paclitaxel resistance. Simultaneously administrated PCT (10 mg/kg) and shSur (2 mg/kg) showed enhanced efficacy of anticancer activity including higher PTX-induced apoptosis and cells arrested in G2/M phase [168]. A combination of DOX and survivin shRNA was also investigated by Tang's group for its effect in overcoming multidrug resistance. In their work, a pH-sensitive polymer based on poly(b-amino ester), poly[(1,4-butanediol)-diacrylate-b-5-polyethylenimine]-block-poly[(1,4-butanediol)-diacrylate-b-5-hydroxy amylamine] (PDP-PDHA) was synthesized. Nanoparticles containing 6 mg/kg DOX and 2 mg/kg shRNA were administered to MCF7/ADR tumor-bearing mice. The authors successfully raised the accumulation of DOX and shSur in the tumor tissue, resulting in a tumor growth inhibition of 95.9% after 21 days [169].

4.2.5. Increasing Intracellular Drug Accumulation by Inhibiting Drug Efflux

P-glycoprotein (P-gp), encoded by the MDR-1 gene, overexpressed in many types of human cancers, contributes to the multidrug resistant effect. Downregulation of P-gp has been associated particularly with enhancing the therapeutic outcome of chemotherapeutics. Thus, it is a popular inclusion in co-delivery systems for its synergistic effect. Xiong et al. reported the co-delivery of siMDR-1 and DOX using polymeric micelles formed by poly(ethylene oxide)-block-poly(ε-caprolactone) (PEO-b-PCL) amphiphilic block co-polymers to improve the anticancer effect in the multidrug drug resistant human breast cancer cell line (MDA-MB-435/LCC6MDR1). Micelles containing 5 μg/mL of DOX and 100 nM of siMDR-1 led to a maximum of ~70% cell growth inhibition at 72 h after treatment [45]. Navarro et al. also demonstrated that the downregulation of P-gp led to the inhibition of DOX efflux activity resulting in an enhanced cytotoxicity of DOX in the MCF7/ADR cell line. The combination used in their study was 1 μg/mL of DOX and 100 nM of siMDR-1. Drugs were delivered in polymeric nanoparticles consisting of PEI modified DOPE [97].

Zhang et al. prepared polymeric micelles based on N-succinyl chitosan-poly-L-lysine-palmitic acid (NSC-PLL-PA) for co-delivery of DOX and siRNA targeting P-gp. The study revealed that the therapeutic efficacy was close to the maximum when the siRNA concentration reached about 100 nM. This finding indicated that 100 nM was sufficient to downregulate P-gp expression, increase intracellular DOX concentration, and maximize the therapeutic effects. The cytotoxicity results at 48 h after treatment also indicated that a synergistic effect was achieved at 5 μg/mL of DOX and 100 nM of siMDR-1. Increasing the concentration of DOX disguised the effect of siMDR-1, leading to a more than 80% cytotoxic effect among all groups. Additionally, micelles loaded with 0.5 mg/kg of DOX and 0.2 mg/kg of siMDR-1 given to tumor-bearing mice every three days showed a significant tumor growth inhibition over 24 days [164]. Other groups also reported synergistic effects derived from co-delivery of siMDR-1 with DOX using polymeric nanoparticles at similar concentration combinations. For example, Xu et al. co-delivered 3 μg/mL of DOX and 100 nM of siP-gp in polymeric vehicles prepared from triblock copolymers, folate/methoxy-poly(ethylene glycol)-block-poly(L-glutamate-hydrazide)-block-poly(N,N-dimethylaminopropyl methacrylamide) (FA/m-PEG-b-P(LG-Hyd)-b-PDMAPMA) to MCF7 breast cancer cells [52]. And Misra et al. overcame multidrug resistance in cells by co-delivering 11.6 μg/mL of DOX and 100 nM of siMDR-1 using dimethyldidodecylammonium bromide (DMAB)-coated PLGA nanoparticles in the MCF7 ADR cell line [165]. All these studies demonstrated the enhanced therapeutic efficacy against cancer using polymeric co-delivery systems.

In our previous work, micellar nanoparticles consisting of PAMAM-PEG2k-DOPE and PEG5k-DOPE were investigated as a co-delivery system for both hydrophobic drugs and siRNA

(Figure 6). Combinations of DOX and siMDR-1 at different concentrations were applied to multidrug resistant cell lines A2780/ADR and MCF7/ADR [91,99]. The synergistic effect of the co-delivery system was observed in MCF7/ADR and A2780/ADR when treated with 125 nM of siMDR-1, 0.43 μM of DOX, or 125 nM of siMDR-1, 1.7 μM of DOX, respectively. The cytotoxicity results also suggested that co-delivery of siMDR-1 and DOX achieved an increased anti-cancer effect when delivery of siMDR-1 was followed by DOX treatment separately. Since downregulation of P-gp alone does not cause significant tumor growth inhibition, delivery of siMDR-1 together with intracellular delivery of chemotherapeutics is required for an ideal therapeutic effect. However, inadequate downregulation amount of P-gp or excess of chemotherapeutics could impair the performance of co-delivery systems targeting inhibition of drug-efflux. An optimal concentration ratio should be established for desired therapeutic outcomes.

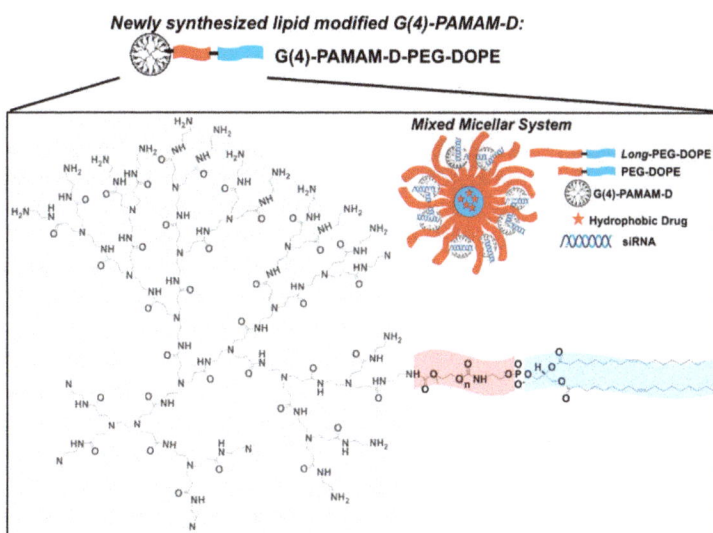

Figure 6. Schematic structure of mixed dendrimer micelles composed of PAMAM-PEG2k-DOPE and PEG5k-DOPE in co-delivery of DOX and siMDR-1. Reproduced with permission from Pan et al., European Journal of Pharmaceutics and Biopharmaceutics; Elsevier, 2019 [91].

4.2.6. Inhibiting Tumor Growth by Altering Immune Responses

Among many strategies for co-delivery systems, being able to modulate the natural immune response against cancer cells is one of the most promising challenges. Chen et al. designed a co-delivery system of DOX + IL-36γ/POEG-st-Pmor with an improved anti-metastatic effect in a mouse breast cancer lung metastasis model with 4T1.2 cells. DOX 5 mg/kg and IL-36γ plasmid 50 μg per animal synergistically enhanced the immune response (type I) by increasing the IFN-γ positive CD4+ and CD8+ T cells [171]. Wu et al. also delivered the IL-2 immunoactivator together with DOX, but at a much lower dose. N,N,N-Trimethyl chitosan-based polymeric nanoparticles loaded with 2 mg/kg of DOX and 1.2 μg/animal of IL-2 were injected intravenously into SMMC 7721-bearing mice every two days. When further modified with 8.45% *w/w* folic acid, the polymeric co-delivery system resulted in a five-fold smaller tumor volume than those treated with DOX alone after 14 days [43].

Combination of IL-2 and PCT also has produced a synergistic effect in several reported studies. Wang et al. delivered IL-2 at 5 μg in combination with 10 μg of PCT per animal using P(MDS)-co-CES polymeric nanoparticles. Co-delivery of IL-2 and PCT resulted in a 2.5-fold lower tumor volume after 17 days of treatment than with individual drug loaded nanoparticles [102]. In another study, Zhao et al. delivered an IL-2 and PCT combination of 2.5 μg/kg and 10 mg/kg, respectively, in PLGA/Pluronic

F127-based nanoparticles. The in vivo study using a murine melanoma B16-F10 cell line showed significant tumor growth and metastasis inhibition. Additionally, a prolonged overall survival rate was demonstrated in tumor-bearing mice treated by co-delivery of IL-2 and PCT [172].

The release profile of ILs is critical in designing co-delivery systems for IL and chemotherapeutic agents. Systemic administration of IL-2 may initiate an auto-immune response, resulting in side effects from the therapy. Previously, combination of local administration of IL-2 and systemic administration of PCT was used as a combination strategy [179]. IL-2 loaded in polymeric nanoparticles generated a controlled release profile in the systemic circulation. Co-encapsulation of IL-2 with chemotherapeutic agents not only induced a synergistic effect, but also minimized the auto-immune response. Although the co-delivery approach using a single delivery system offers many advantages, pharmacokinetics and cytotoxic side effect still need to be considered. The release profile of IL-2 is critical for the design of co-delivery systems for IL-2 and chemotherapeutic agents.

5. Conclusions and Perspectives

In this review, we have discussed the different types of block co-polymers that have been formulated into polymeric co-delivery systems for cancer treatment. We also focused on the different drug combinations used in co-delivery systems and the influence of the combinations on their therapeutic outcomes. As discussed earlier, targeting two multidrug resistance mechanisms simultaneously with a single delivery platform is a promising strategy that can provide synchronized pharmacokinetics and doses to the same cell population. Many co-delivery systems have been developed that exhibit promising anti-tumor efficacy, especially in multidrug resistant tumors. Nevertheless, it has been well-established that cancer cells have the inherent ability to avoid cell death by activation of various anti-apoptotic pathways. These pathways seem to be cell type specific and assault type specific. Therefore, it appears that the effectiveness of decreasing cell viability by double sensitization is cell-specific. That means a certain combination of drugs may work perfectly in one cell line, but may not work at all in another cell line without modification of the combination of drugs. A "one type fit all" formulation seems unlikely for eradication of all types of tumors clinically. Individualized therapy designed specifically for each type of cancer cell with different drug combination may be more likely to be necessary for an effective therapeutic outcome [180].

Polymeric nanoparticles are ideal platforms for co-delivery because of the multicompartments they contain. Amphiphilic co-polymers integrated with different properties have been synthesized to encapsulate hydrophilic agents as well as hydrophobic agents. Cationic moieties in co-polymers also provide the possibility for complexed nucleic acid molecules. Although there is much research focusing on combination delivery using polymeric nanoparticles, little emphasis has been put on the optimization of drug ratios to promote synergistic effects. The synergistic effect of polymeric co-delivery systems depends on multiple factors. First of all, it is important to ensure that the chemistries of delivery, carrier and therapeutic agents do not interact detrimentally with each other [86]. Another challenge is to ensure that the presence of one therapeutic agent does not interfere with the action of others. Additionally, the ratio between two therapeutic agents can directly determine the outcome of a co-delivery system. The release profile of small molecule drug and nucleic acid molecules is another major challenge that requires considerable research.

Overall, a deeper understanding of the optimal ratio between therapeutic agents and the natural heterogenicity of the tumor is necessary for development of polymeric co-delivery systems that maximize therapeutic effects. Additionally, multifunctional polymeric nanoparticles, especially stimuli-sensitive polymeric nanoparticles, allow high drug loading, optimized release profiles, and enhanced in vivo stability for the co-delivery of distinctly different classes of therapeutic molecules. There is a need to continue efforts at understanding the relationship between the mechanism of action of encapsulated therapeutic agents and their pharmacological activities. Resolving these challenges should result in multifunctional polymeric nanoparticles with significantly enhanced therapeutic efficacy.

Funding: This research was funded by National Institutes of Health (NIH), United States, grant number [1R01CA200844].

Acknowledgments: The authors would like to acknowledge William Hartner for helpful comments on the manuscript. The authors would also like to thank for funding a post-doctoral scholarship for one of the authors under the Mobilnosc Plus program.

Conflicts of Interest: The authors declare no conflict of interest.

References

1. Editorial. Rationalizing combination therapies. *Nat. Med.* **2017**, *23*, 1113. [CrossRef] [PubMed]
2. Tolcher, A.W.; Mayer, L.D. Improving combination cancer therapy: The CombiPlex® development platform. *Future Oncol.* **2018**, *14*, 1317–1332. [CrossRef] [PubMed]
3. Hodge, J.W.; Ardiani, A.; Farsaci, B.; Kwilas, A.R.; Gameiro, S.R. The Tipping Point for Combination Therapy: Cancer Vaccines with Radiation, Chemotherapy, or Targeted Small Molecule Inhibitors. *Semin. Oncol.* **2012**, *39*, 323–339. [CrossRef] [PubMed]
4. Sanjay, K.; Anchal, S.; Uma, N.; Sweta, M.; Pratibha, K. Recent progresses in Organic-Inorganic Nano technological platforms for cancer therapeutics. *Curr. Med. Chem.* **2019**, *26*. [CrossRef]
5. Qi, S.-S.; Sun, J.-H.; Yu, H.-H.; Yu, S.-Q. Co-delivery nanoparticles of anti-cancer drugs for improving chemotherapy efficacy. *Drug Deliv.* **2017**, *24*, 1909–1926. [CrossRef] [PubMed]
6. Mahira, S.; Kommineni, N.; Husain, G.M.; Khan, W. Cabazitaxel and silibinin co-encapsulated cationic liposomes for CD44 targeted delivery: A new insight into nanomedicine based combinational chemotherapy for prostate cancer. *Biomed. Pharmacother.* **2019**, *110*, 803–817. [CrossRef]
7. Gozde, U.; Ufuk, G. Smart Drug Delivery Systems in Cancer Therapy. *Curr. Drug Targets* **2018**, *19*, 202–212.
8. Kang, L.; Gao, Z.; Huang, W.; Jin, M.; Wang, Q. Nanocarrier-mediated co-delivery of chemotherapeutic drugs and gene agents for cancer treatment. *Acta Pharm. Sin. B* **2015**, *5*, 169–175. [CrossRef]
9. Wu, Y.; Gu, W.; Xu, Z.P. Enhanced combination cancer therapy using lipid-calcium carbonate/phosphate nanoparticles as a targeted delivery platform. *Nanomedicine* **2018**, *14*, 77–92. [CrossRef]
10. Caliskan, Y.; Dalgic, A.D.; Gerekci, S.; Gulec, E.A.; Tezcaner, A.; Ozen, C.; Keskin, D. A new therapeutic combination for osteosarcoma: Gemcitabine and Clofazimine co-loaded liposomal formulation. *Int. J. Pharm.* **2019**, *557*, 97–104. [CrossRef]
11. Chuanmin, Z.; Shubiao, Z.; Defu, Z.; Jingnan, C. Cancer Treatment with Liposomes Based Drugs and Genes Co-delivery Systems. *Curr. Med. Chem.* **2018**, *25*, 3319–3332.
12. Sriraman, S.K.; Pan, J.; Sarisozen, C.; Luther, E.; Torchilin, V. Enhanced Cytotoxicity of Folic Acid-Targeted Liposomes Co-Loaded with C6 Ceramide and Doxorubicin: In Vitro Evaluation on HeLa, A2780-ADR, and H69-AR Cells. *Mol. Pharm.* **2016**, *13*, 428–437. [CrossRef]
13. Palmerston Mendes, L.; Pan, J.; Torchilin, V.P. Dendrimers as Nanocarriers for Nucleic Acid and Drug Delivery in Cancer Therapy. *Molecules* **2017**, *22*, 1401. [CrossRef] [PubMed]
14. Sarisozen, C.; Pan, J.; Dutta, I.; Torchilin, V.P. Polymers in the co-delivery of siRNA and anticancer drugs to treat multidrug-resistant tumors. *J. Pharm. Investig.* **2017**, *47*, 37–49. [CrossRef]
15. Afsharzadeh, M.; Hashemi, M.; Mokhtarzadeh, A.; Abnous, K.; Ramezani, M. Recent advances in co-delivery systems based on polymeric nanoparticle for cancer treatment. *Artif. Cells Nanomed. Biotechnol.* **2018**, *46*, 1095–1110. [CrossRef] [PubMed]
16. Shen, S.; Liu, M.; Li, T.; Lin, S.; Mo, R. Recent progress in nanomedicine-based combination cancer therapy using a site-specific co-delivery strategy. *Biomater. Sci.* **2017**, *5*, 1367–1381. [CrossRef]
17. Eldar-Boock, A.; Polyak, D.; Scomparin, A.; Satchi-Fainaro, R. Nano-sized polymers and liposomes designed to deliver combination therapy for cancer. *Curr. Opin. Biotechnol.* **2013**, *24*, 682–689. [CrossRef] [PubMed]
18. Zhan, C.; Wei, X.; Qian, J.; Feng, L.; Zhu, J.; Lu, W. Co-delivery of TRAIL gene enhances the anti-glioblastoma effect of paclitaxel in vitro and in vivo. *J. Control. Release* **2012**, *160*, 630–636. [CrossRef] [PubMed]
19. Mazzarino, L.; Travelet, C.; Ortega-Murillo, S.; Otsuka, I.; Pignot-Paintrand, I.; Lemos-Senna, E.; Borsali, R. Elaboration of chitosan-coated nanoparticles loaded with curcumin for mucoadhesive applications. *J. Colloid Interface Sci.* **2012**, *370*, 58–66. [CrossRef]
20. Xu, J.; Zhao, J.H.; Liu, Y.; Feng, N.P.; Zhang, Y.T. RGD-modified poly(D,L-lactic acid) nanoparticles enhance tumor targeting of oridonin. *Int. J. Nanomed.* **2012**, *7*, 211–219.

21. Saraogi, G.K.; Gupta, P.; Gupta, U.D.; Jain, N.K.; Agrawal, G.P. Gelatin nanocarriers as potential vectors for effective management of tuberculosis. *Int. J. Pharm.* **2010**, *385*, 143–149. [CrossRef] [PubMed]

22. Krakovicova, H.; Etrych, T.; Ulbrich, K. HPMA-based polymer conjugates with drug combination. *Eur. J. Pharm. Sci.* **2009**, *37*, 405–412. [CrossRef] [PubMed]

23. Qu, X.; Wan, Y.; Zhang, H.; Cui, W.; Bei, J.; Wang, S. Porcine-derived xenogeneic bone/poly(glycolide-co-lactide-co-caprolactone) composite and its affinity with rat OCT-1 osteoblast-like cells. *Biomaterials* **2006**, *27*, 216–225. [CrossRef] [PubMed]

24. Park, J.; Fong, P.M.; Lu, J.; Russell, K.S.; Booth, C.J.; Saltzman, W.M.; Fahmy, T.M. PEGylated PLGA nanoparticles for the improved delivery of doxorubicin. *Nanomedicine* **2009**, *5*, 410–418. [CrossRef] [PubMed]

25. Des Rieux, A.; Fievez, V.; Garinot, M.; Schneider, Y.J.; Preat, V. Nanoparticles as potential oral delivery systems of proteins and vaccines: A mechanistic approach. *J. Control. Release* **2006**, *116*, 1–27. [CrossRef] [PubMed]

26. Kashi, T.S.; Eskandarion, S.; Esfandyari-Manesh, M.; Marashi, S.M.; Samadi, N.; Fatemi, S.M.; Atyabi, F.; Eshraghi, S.; Dinarvand, R. Improved drug loading and antibacterial activity of minocycline-loaded PLGA nanoparticles prepared by solid/oil/water ion pairing method. *Int. J. Nanomed.* **2012**, *7*, 221–234.

27. Quintanar-Guerrero, D.; Tamayo-Esquivel, D.; Ganem-Quintanar, A.; Allemann, E.; Doelker, E. Adaptation and optimization of the emulsification-diffusion technique to prepare lipidic nanospheres. *Eur. J. Pharm. Sci.* **2005**, *26*, 211–218. [CrossRef]

28. Natrajan, D.; Srinivasan, S.; Sundar, K.; Ravindran, A. Formulation of essential oil-loaded chitosan-alginate nanocapsules. *J. Food Drug Anal.* **2015**, *23*, 560–568. [CrossRef] [PubMed]

29. Cohen-Sela, E.; Teitlboim, S.; Chorny, M.; Koroukhov, N.; Danenberg, H.D.; Gao, J.; Golomb, G. Single and double emulsion manufacturing techniques of an amphiphilic drug in PLGA nanoparticles: Formulations of mithramycin and bioactivity. *J. Pharm. Sci.* **2009**, *98*, 1452–1462. [CrossRef] [PubMed]

30. Ding, F.; Lu, Z.; Zou, R.; Zhang, Y.; Guo, Q.; Li, S.; Yang, J. Evaluation of a novel paclitaxel-eluting stent with a bioabsorbable polymeric surface coating in the porcine artery injury model. *Acta Cardiol.* **2011**, *66*, 765–772. [CrossRef] [PubMed]

31. Azzaroni, O.; Lau, K.H. Layer-by-Layer Assemblies in Nanoporous Templates: Nano-Organized Design and Applications of Soft Nanotechnology. *Soft Matter* **2011**, *7*, 8709–8724. [CrossRef]

32. Siegwart, D.J.; Oh, J.K.; Matyjaszewski, K. ATRP in the design of functional materials for biomedical applications. *Prog. Polym. Sci.* **2012**, *37*, 18–37. [CrossRef] [PubMed]

33. Gogoi, M.; Sarma, H.D.; Bahadur, D.; Banerjee, R. Biphasic magnetic nanoparticles-nanovesicle hybrids for chemotherapy and self-controlled hyperthermia. *Nanomedicine* **2014**, *9*, 955–970. [CrossRef] [PubMed]

34. Kelley, E.G.; Albert, J.N.; Sullivan, M.O.; Epps, T.H., 3rd. Stimuli-responsive copolymer solution and surface assemblies for biomedical applications. *Chem. Soc. Rev.* **2013**, *42*, 7057–7071. [CrossRef] [PubMed]

35. Zhu, L.; Torchilin, V.P. Stimulus-responsive nanopreparations for tumor targeting. *Integr. Biol.* **2013**, *5*, 96–107. [CrossRef]

36. Kobayashi, T. Cancer hyperthermia using magnetic nanoparticles. *Biotechnol. J.* **2011**, *6*, 1342–1347. [CrossRef] [PubMed]

37. Mura, S.; Nicolas, J.; Couvreur, P. Stimuli-responsive nanocarriers for drug delivery. *Nat. Mater.* **2013**, *12*, 991–1003. [CrossRef]

38. Burkhart, A.; Ritter, H. Influence of cyclodextrin on the UCST- and LCST-behavior of poly(2-methacrylamido-caprolactam)-co-(N,N-dimethylacrylamide). *Beilstein J. Org. Chem.* **2014**, *10*, 1951–1958. [CrossRef] [PubMed]

39. Hocine, S.; Li, M.-H. Thermoresponsive self-assembled polymer colloids in water. *Soft Matter* **2013**, *9*, 5839–5861. [CrossRef]

40. Meiswinkel, G.; Ritter, H. Polymers from 1-Vinyl-2-(hydroxymethyl)imidazole in Water: Altering from UCST to LCST Behavior via O-Ethylation. *Macromol. Chem. Phys.* **2014**, *215*, 682–687. [CrossRef]

41. Wang, Y.; Fang, J.; Cheng, D.; Wang, Y.; Shuai, X. A pH-sensitive micelle for codelivery of siRNA and doxorubicin to hepatoma cells. *Polymer* **2014**, *55*, 3217–3226. [CrossRef]

42. Sun, W.; Chen, X.; Xie, C.; Wang, Y.; Lin, L.; Zhu, K.; Shuai, X. Co-Delivery of Doxorubicin and Anti-BCL-2 siRNA by pH-Responsive Polymeric Vector to Overcome Drug Resistance in In Vitro and In Vivo HepG2 Hepatoma Model. *Biomacromolecules* **2018**, *19*, 2248–2256. [CrossRef] [PubMed]

43. Wu, J.; Tang, C.; Yin, C. Co-delivery of doxorubicin and interleukin-2 via chitosan based nanoparticles for enhanced antitumor efficacy. *Acta Biomater.* **2017**, *47*, 81–90. [CrossRef]

44. Jin, M.; Jin, G.; Kang, L.; Chen, L.; Gao, Z.; Huang, W. Smart polymeric nanoparticles with pH-responsive and PEG-detachable properties for co-delivering paclitaxel and survivin siRNA to enhance antitumor outcomes. *Int. J. Nanomed.* **2018**, *13*, 2405–2426. [CrossRef] [PubMed]

45. Xiong, X.B.; Lavasanifar, A. Traceable multifunctional micellar nanocarriers for cancer-targeted co-delivery of MDR-1 siRNA and doxorubicin. *ACS Nano* **2011**, *5*, 5202–5213. [CrossRef] [PubMed]

46. Tang, S.; Yin, Q.; Su, J.; Sun, H.; Meng, Q.; Chen, Y.; Chen, L.; Huang, Y.; Gu, W.; Xu, M.; et al. Inhibition of metastasis and growth of breast cancer by pH-sensitive poly (beta-amino ester) nanoparticles co-delivering two siRNA and paclitaxel. *Biomaterials* **2015**, *48*, 1–15. [CrossRef]

47. Fan, H.; Hu, Q.D.; Xu, F.J.; Liang, W.Q.; Tang, G.P.; Yang, W.T. In vivo treatment of tumors using host-guest conjugated nanoparticles functionalized with doxorubicin and therapeutic gene pTRAIL. *Biomaterials* **2012**, *33*, 1428–1436. [CrossRef]

48. Salzano, G.; Costa, D.F.; Sarisozen, C.; Luther, E.; Mattheolabakis, G.; Dhargalkar, P.P.; Torchilin, V.P. Mixed Nanosized Polymeric Micelles as Promoter of Doxorubicin and miRNA-34a Co-Delivery Triggered by Dual Stimuli in Tumor Tissue. *Small* **2016**, *12*, 4837–4848. [CrossRef]

49. Wu, J.; Zhang, H.; Hu, X.; Liu, R.; Jiang, W.; Li, Z.; Luan, Y. Reduction-sensitive mixed micelles assembled from amphiphilic prodrugs for self-codelivery of DOX and DTX with synergistic cancer therapy. *Colloids Surf. B Biointerfaces* **2018**, *161*, 449–456. [CrossRef]

50. Wu, H.; Jin, H.; Wang, C.; Zhang, Z.; Ruan, H.; Sun, L.; Yang, C.; Li, Y.; Qin, W.; Wang, C. Synergistic Cisplatin/Doxorubicin Combination Chemotherapy for Multidrug-Resistant Cancer via Polymeric Nanogels Targeting Delivery. *ACS Appl. Mater. Interfaces* **2017**, *9*, 9426–9436. [CrossRef] [PubMed]

51. Kang, Y.; Lu, L.; Lan, J.; Ding, Y.; Yang, J.; Zhang, Y.; Zhao, Y.; Zhang, T.; Ho, R.J.Y. Redox-responsive polymeric micelles formed by conjugating gambogic acid with bioreducible poly(amido amine)s for the co-delivery of docetaxel and MMP-9 shRNA. *Acta Biomater.* **2018**, *68*, 137–153. [CrossRef] [PubMed]

52. Xu, M.; Qian, J.; Suo, A.; Cui, N.; Yao, Y.; Xu, W.; Liu, T.; Wang, H. Co-delivery of doxorubicin and P-glycoprotein siRNA by multifunctional triblock copolymers for enhanced anticancer efficacy in breast cancer cells. *J. Mater. Chem. B* **2015**, *3*, 2215–2228. [CrossRef]

53. Sheu, M.T.; Jhan, H.J.; Su, C.Y.; Chen, L.C.; Chang, C.E.; Liu, D.Z.; Ho, H.O. Codelivery of doxorubicin-containing thermosensitive hydrogels incorporated with docetaxel-loaded mixed micelles enhances local cancer therapy. *Colloids Surf. B Biointerfaces* **2016**, *143*, 260–270. [CrossRef] [PubMed]

54. Wu, X.; Wu, Y.; Ye, H.; Yu, S.; He, C.; Chen, X. Interleukin-15 and cisplatin co-encapsulated thermosensitive polypeptide hydrogels for combined immuno-chemotherapy. *J. Control. Release* **2017**, *255*, 81–93. [CrossRef] [PubMed]

55. Ma, H.; He, C.; Cheng, Y.; Yang, Z.; Zang, J.; Liu, J.; Chen, X. Localized Co-delivery of Doxorubicin, Cisplatin, and Methotrexate by Thermosensitive Hydrogels for Enhanced Osteosarcoma Treatment. *ACS Appl. Mater. Interfaces* **2015**, *7*, 27040–27048. [CrossRef]

56. Yin, Y.; Hu, Q.; Xu, C.; Qiao, Q.; Qin, X.; Song, Q.; Peng, Y.; Zhao, Y.; Zhang, Z. Co-delivery of Doxorubicin and Interferon-gamma by Thermosensitive Nanoparticles for Cancer Immunochemotherapy. *Mol. Pharm.* **2018**, *15*, 4161–4172. [CrossRef]

57. Huang, P.; Zhang, Y.; Wang, W.; Zhou, J.; Sun, Y.; Liu, J.; Kong, D.; Liu, J.; Dong, A. Co-delivery of doxorubicin and (131)I by thermosensitiv.e micellar-hydrogel for enhanced in situ synergetic chemoradiotherapy. *J. Control. Release* **2015**, *220*, 456–464. [CrossRef]

58. Zhu, L.; Perche, F.; Wang, T.; Torchilin, V.P. Matrix metalloproteinase 2-sensitive multifunctional polymeric micelles for tumor-specific co-delivery of siRNA and hydrophobic drugs. *Biomaterials* **2014**, *35*, 4213–4222. [CrossRef]

59. Li, X.; Hong, E.Y.; Chan, A.K.; Poon, C.T.; Li, B.; Wu, L.; Yam, V.W. Amphiphilic Carbazole-Containing Compounds with Lower Critical Solution Temperature Behavior for Supramolecular Self-Assembly and Solution-Processable Resistive Memories. *Chem. Asian J.* **2018**, *13*, 2626–2631. [CrossRef]

60. Metin, E.; Mutlu, P.; Gunduz, U. Co-delivery of Doxorubicin and D-alpha-Tocopherol Polyethylene Glycol 1000 Succinate by Magnetic Nanoparticles. *Anticancer Agents Med. Chem.* **2018**, *18*, 1138–1147. [CrossRef]

61. Li, H.; Fu, C.; Miao, X.; Li, Q.; Zhang, J.; Yang, H.; Liu, T.; Chen, X.; Xie, M. Multifunctional magnetic co-delivery system coated with polymer mPEG-PLL-FA for nasopharyngeal cancer targeted therapy and MR imaging. *J. Biomater. Appl.* **2017**, *31*, 1169–1181. [CrossRef] [PubMed]

62. Hosseini Sadr, S.; Davaran, S.; Alizadeh, E.; Salehi, R.; Ramazani, A. Enhanced anticancer potency by thermo/pH-responsive PCL-based magnetic nanoparticles. *J. Biomater. Sci. Polym. Ed.* **2018**, *29*, 277–308. [CrossRef]
63. Dai, S.; Ravi, P.; Tam, K.C. pH-Responsive polymers: Synthesis, properties and applications. *Soft Matter* **2008**, *4*, 435–449. [CrossRef]
64. Lalles, J.P.; Bosi, P.; Janczyk, P.; Koopmans, S.J.; Torrallardona, D. Impact of bioactive substances on the gastrointestinal tract and performance of weaned piglets: A review. *Animal* **2009**, *3*, 1625–1643. [CrossRef] [PubMed]
65. Huo, M.; Yuan, J.; Tao, L.; Wei, Y. Redox-responsive polymers for drug delivery: From molecular design to applications. *Polym. Chem.* **2014**, *5*, 1519–1528. [CrossRef]
66. Broaders, K.E.; Grandhe, S.; Frechet, J.M. A biocompatible oxidation-triggered carrier polymer with potential in therapeutics. *J. Am. Chem. Soc.* **2011**, *133*, 756–758. [CrossRef]
67. Song, C.-C.; Ji, R.; Du, F.-S.; Li, Z.-C. Oxidation-Responsive Poly(amino ester)s Containing Arylboronic Ester and Self-Immolative Motif: Synthesis and Degradation Study. *Macromolecules* **2013**, *46*, 8416–8425. [CrossRef]
68. Thankam, F.G.; Muthu, J. Infiltration and sustenance of viability of cells by amphiphilic biosynthetic biodegradable hydrogels. *J. Mater. Sci. Mater. Med.* **2014**, *25*, 1953–1965. [CrossRef]
69. Avci, P.; Erdem, S.S.; Hamblin, M.R. Photodynamic therapy: One step ahead with self-assembled nanoparticles. *J. Biomed. Nanotechnol.* **2014**, *10*, 1937–1952. [CrossRef]
70. Khandare, J.; Minko, T. Polymer–drug conjugates: Progress in polymeric prodrugs. *Prog. Polym. Sci.* **2006**, *31*, 359–397. [CrossRef]
71. Weiss, R.B. The anthracyclines: Will we ever find a better doxorubicin? *Semin. Oncol.* **1992**, *19*, 670–686. [PubMed]
72. Smith, L.; Watson, M.B.; O'Kane, S.L.; Drew, P.J.; Lind, M.J.; Cawkwell, L. The analysis of doxorubicin resistance in human breast cancer cells using antibody microarrays. *Mol. Cancer Ther.* **2006**, *5*, 2115–2120. [CrossRef] [PubMed]
73. Mordente, A.; Meucci, E.; Silvestrini, A.; Martorana, G.E.; Giardina, B. New developments in anthracycline-induced cardiotoxicity. *Curr. Med. Chem.* **2009**, *16*, 1656–1672. [CrossRef] [PubMed]
74. Singal, P.K.; Li, T.; Kumar, D.; Danelisen, I.; Iliskovic, N. Adriamycin-induced heart failure: Mechanism and modulation. *Mol. Cell. Biochem.* **2000**, *207*, 77–86. [CrossRef] [PubMed]
75. Gutierrez, P.L. The role of NAD(P)H oxidoreductase (DT-Diaphorase) in the bioactivation of quinone-containing antitumor agents: A review. *Free Radic. Biol. Med.* **2000**, *29*, 263–275. [CrossRef]
76. Shadle, S.E.; Bammel, B.P.; Cusack, B.J.; Knighton, R.A.; Olson, S.J.; Mushlin, P.S.; Olson, R.D. Daunorubicin cardiotoxicity: Evidence for the importance of the quinone moiety in a free-radical-independent mechanism. *Biochem. Pharmacol.* **2000**, *60*, 1435–1444. [CrossRef]
77. Espelin, C.W.; Leonard, S.C.; Geretti, E.; Wickham, T.J.; Hendriks, B.S. Dual HER2 Targeting with Trastuzumab and Liposomal-Encapsulated Doxorubicin (MM-302) Demonstrates Synergistic Antitumor Activity in Breast and Gastric Cancer. *Cancer Res.* **2016**, *76*, 1517–1527. [CrossRef] [PubMed]
78. Zhang, R.X.; Wong, H.L.; Xue, H.Y.; Eoh, J.Y.; Wu, X.Y. Nanomedicine of synergistic drug combinations for cancer therapy—Strategies and perspectives. *J. Control. Release* **2016**, *240*, 489–503. [CrossRef]
79. Wang, H.; Zhao, Y.; Wu, Y.; Hu, Y.L.; Nan, K.; Nie, G.; Chen, H. Enhanced anti-tumor efficacy by co-delivery of doxorubicin and paclitaxel with amphiphilic methoxy PEG-PLGA copolymer nanoparticles. *Biomaterials* **2011**, *32*, 8281–8290. [CrossRef]
80. Wang, H.; Agarwal, P.; Zhao, S.; Xu, R.X.; Yu, J.; Lu, X.; He, X. Hyaluronic acid-decorated dual responsive nanoparticles of Pluronic F127, PLGA, and chitosan for targeted co-delivery of doxorubicin and irinotecan to eliminate cancer stem-like cells. *Biomaterials* **2015**, *72*, 74–89. [CrossRef]
81. Camacho, K.M.; Kumar, S.; Menegatti, S.; Vogus, D.R.; Anselmo, A.C.; Mitragotri, S. Synergistic antitumor activity of camptothecin-doxorubicin combinations and their conjugates with hyaluronic acid. *J. Control. Release* **2015**, *210*, 198–207. [CrossRef]
82. Tai, W.; Mo, R.; Lu, Y.; Jiang, T.; Gu, Z. Folding graft copolymer with pendant drug segments for co-delivery of anticancer drugs. *Biomaterials* **2014**, *35*, 7194–7203. [CrossRef]
83. Li, N.; Huang, C.; Luan, Y.; Song, A.; Song, Y.; Garg, S. Active targeting co-delivery system based on pH-sensitive methoxy-poly(ethylene glycol)2K-poly(epsilon-caprolactone)4K-poly(glutamic acid)1K for enhanced cancer therapy. *J. Colloid Interface Sci.* **2016**, *472*, 90–98. [CrossRef]

84. Skatrud, P.L. The impact of multiple drug resistance (MDR) proteins on chemotherapy and drug discovery. *Prog. Drug Res.* **2002**, *58*, 99–131.

85. Dai, X.; Tan, C. Combination of microRNA therapeutics with small-molecule anticancer drugs: Mechanism of action and co-delivery nanocarriers. *Adv. Drug Deliv. Rev.* **2015**, *81*, 184–197. [CrossRef]

86. Teo, P.Y.; Cheng, W.; Hedrick, J.L.; Yang, Y.Y. Co-delivery of drugs and plasmid DNA for cancer therapy. *Adv. Drug Deliv. Rev.* **2016**, *98*, 41–63. [CrossRef]

87. Navarro, G.; Pan, J.; Torchilin, V.P. Micelle-like nanoparticles as carriers for DNA and siRNA. *Mol. Pharm.* **2015**, *12*, 301–313. [CrossRef]

88. Alinejad, V.; Hossein Somi, M.; Baradaran, B.; Akbarzadeh, P.; Atyabi, F.; Kazerooni, H.; Samadi Kafil, H.; Aghebati Maleki, L.; Siah Mansouri, H.; Yousefi, M. Co-delivery of IL17RB siRNA and doxorubicin by chitosan-based nanoparticles for enhanced anticancer efficacy in breast cancer cells. *Biomed. Pharmacother.* **2016**, *83*, 229–240. [CrossRef]

89. Wei, W.; Lv, P.P.; Chen, X.M.; Yue, Z.G.; Fu, Q.; Liu, S.Y.; Yue, H.; Ma, G.H. Codelivery of mTERT siRNA and paclitaxel by chitosan-based nanoparticles promoted synergistic tumor suppression. *Biomaterials* **2013**, *34*, 3912–3923. [CrossRef]

90. Zakeri, A.; Kouhbanani, M.A.J.; Beheshtkhoo, N.; Beigi, V.; Mousavi, S.M.; Hashemi, S.A.R.; Karimi Zade, A.; Amani, A.M.; Savardashtaki, A.; Mirzaei, E.; et al. Polyethylenimine-based nanocarriers in co-delivery of drug and gene: A developing horizon. *Nano Rev. Exp.* **2018**, *9*, 1488497. [CrossRef]

91. Pan, J.; Palmerston Mendes, L.; Yao, M.; Filipczak, N.; Garai, S.; Thakur, G.A.; Sarisozen, C.; Torchilin, V.P. Polyamidoamine dendrimers-based nanomedicine for combination therapy with siRNA and chemotherapeutics to overcome multidrug resistance. *Eur. J. Pharm. Biopharm.* **2019**. [CrossRef] [PubMed]

92. Cheng, Q.; Du, L.; Meng, L.; Han, S.; Wei, T.; Wang, X.; Wu, Y.; Song, X.; Zhou, J.; Zheng, S.; et al. The Promising Nanocarrier for Doxorubicin and siRNA Co-delivery by PDMAEMA-based Amphiphilic Nanomicelles. *ACS Appl. Mater. Interfaces* **2016**, *8*, 4347–4356. [CrossRef]

93. Wang, X.; Liow, S.S.; Wu, Q.; Li, C.; Owh, C.; Li, Z.; Loh, X.J.; Wu, Y.L. Codelivery for Paclitaxel and Bcl-2 Conversion Gene by PHB-PDMAEMA Amphiphilic Cationic Copolymer for Effective Drug Resistant Cancer Therapy. *Macromol. Biosci.* **2017**, *17*, 1700186. [CrossRef] [PubMed]

94. Chen, H.; Yang, W.; Chen, H.; Liu, L.; Gao, F.; Yang, X.; Jiang, Q.; Zhang, Q.; Wang, Y. Surface modification of mitoxantrone-loaded PLGA nanospheres with chitosan. *Colloids Surf. B Biointerfaces* **2009**, *73*, 212–218. [CrossRef]

95. Wang, L.; Hao, Y.; Li, H.; Zhao, Y.; Meng, D.; Li, D.; Shi, J.; Zhang, H.; Zhang, Z.; Zhang, Y. Co-delivery of doxorubicin and siRNA for glioma therapy by a brain targeting system: Angiopep-2-modified poly(lactic-co-glycolic acid) nanoparticles. *J. Drug Target* **2015**, *23*, 832–846. [CrossRef] [PubMed]

96. Cao, N.; Cheng, D.; Zou, S.; Ai, H.; Gao, J.; Shuai, X. The synergistic effect of hierarchical assemblies of siRNA and chemotherapeutic drugs co-delivered into hepatic cancer cells. *Biomaterials* **2011**, *32*, 2222–2232. [CrossRef] [PubMed]

97. Navarro, G.; Sawant, R.R.; Biswas, S.; Essex, S.; Tros de Ilarduya, C.; Torchilin, V.P. P-glycoprotein silencing with siRNA delivered by DOPE-modified PEI overcomes doxorubicin resistance in breast cancer cells. *Nanomedicine* **2012**, *7*, 65–78. [CrossRef] [PubMed]

98. Huang, H.Y.; Kuo, W.T.; Chou, M.J.; Huang, Y.Y. Co-delivery of anti-vascular endothelial growth factor siRNA and doxorubicin by multifunctional polymeric micelle for tumor growth suppression. *J. Biomed. Mater. Res. A* **2011**, *97*, 330–338. [CrossRef] [PubMed]

99. Biswas, S.; Deshpande, P.P.; Navarro, G.; Dodwadkar, N.S.; Torchilin, V.P. Lipid modified triblock PAMAM-based nanocarriers for siRNA drug co-delivery. *Biomaterials* **2013**, *34*, 1289–1301. [CrossRef] [PubMed]

100. Zhu, C.; Jung, S.; Luo, S.; Meng, F.; Zhu, X.; Park, T.G.; Zhong, Z. Co-delivery of siRNA and paclitaxel into cancer cells by biodegradable cationic micelles based on PDMAEMA-PCL-PDMAEMA triblock copolymers. *Biomaterials* **2010**, *31*, 2408–2416. [CrossRef] [PubMed]

101. Yue, X.; Qiao, Y.; Qiao, N.; Guo, S.; Xing, J.; Deng, L.; Xu, J.; Dong, A. Amphiphilic methoxy poly(ethylene glycol)-b-poly(epsilon-caprolactone)-b-poly(2-dimethylaminoethyl methacrylate) cationic copolymer nanoparticles as a vector for gene and drug delivery. *Biomacromolecules* **2010**, *11*, 2306–2312. [CrossRef] [PubMed]

102. Wang, Y.; Gao, S.; Ye, W.H.; Yoon, H.S.; Yang, Y.Y. Co-delivery of drugs and DNA from cationic core-shell nanoparticles self-assembled from a biodegradable copolymer. *Nat. Mater.* **2006**, *5*, 791–796. [CrossRef] [PubMed]

103. Qiu, L.Y.; Bae, Y.H. Self-assembled polyethylenimine-graft-poly(epsilon-caprolactone) micelles as potential dual carriers of genes and anticancer drugs. *Biomaterials* **2007**, *28*, 4132–4142. [CrossRef]

104. Shi, S.; Shi, K.; Tan, L.; Qu, Y.; Shen, G.; Chu, B.; Zhang, S.; Su, X.; Li, X.; Wei, Y.; et al. The use of cationic MPEG-PCL-g-PEI micelles for co-delivery of Msurvivin T34A gene and doxorubicin. *Biomaterials* **2014**, *35*, 4536–4547. [CrossRef] [PubMed]

105. Li, H.; Hui, L.; Xu, W. miR-181a sensitizes a multidrug-resistant leukemia cell line K562/A02 to daunorubicin by targeting BCL-2. *Acta Biochim. Biophys. Sin.* **2012**, *44*, 269–277. [CrossRef]

106. Nishida, N.; Mimori, K.; Fabbri, M.; Yokobori, T.; Sudo, T.; Tanaka, F.; Shibata, K.; Ishii, H.; Doki, Y.; Mori, M. MicroRNA-125a-5p is an independent prognostic factor in gastric cancer and inhibits the proliferation of human gastric cancer cells in combination with trastuzumab. *Clin. Cancer Res.* **2011**, *17*, 2725–2733. [CrossRef] [PubMed]

107. Pezzolesi, M.G.; Platzer, P.; Waite, K.A.; Eng, C. Differential expression of PTEN-targeting microRNAs miR-19a and miR-21 in Cowden syndrome. *Am. J. Hum. Genet.* **2008**, *82*, 1141–1149. [CrossRef] [PubMed]

108. Weeraratne, S.D.; Amani, V.; Neiss, A.; Teider, N.; Scott, D.K.; Pomeroy, S.L.; Cho, Y.J. miR-34a confers chemosensitivity through modulation of MAGE-A and p53 in medulloblastoma. *Neuro Oncol.* **2011**, *13*, 165–175. [CrossRef]

109. Mittal, A.; Chitkara, D.; Behrman, S.W.; Mahato, R.I. Efficacy of gemcitabine conjugated and miRNA-205 complexed micelles for treatment of advanced pancreatic cancer. *Biomaterials* **2014**, *35*, 7077–7087. [CrossRef]

110. Deng, X.; Cao, M.; Zhang, J.; Hu, K.; Yin, Z.; Zhou, Z.; Xiao, X.; Yang, Y.; Sheng, W.; Wu, Y.; et al. Hyaluronic acid-chitosan nanoparticles for co-delivery of MiR-34a and doxorubicin in therapy against triple negative breast cancer. *Biomaterials* **2014**, *35*, 4333–4344. [CrossRef]

111. Wang, S.; Zhang, J.; Wang, Y.; Chen, M. Hyaluronic acid-coated PEI-PLGA nanoparticles mediated co-delivery of doxorubicin and miR-542-3p for triple negative breast cancer therapy. *Nanomedicine* **2016**, *12*, 411–420. [CrossRef] [PubMed]

112. Liu, Q.; Li, R.T.; Qian, H.Q.; Wei, J.; Xie, L.; Shen, J.; Yang, M.; Qian, X.P.; Yu, L.X.; Jiang, X.Q.; et al. Targeted delivery of miR-200c/DOC to inhibit cancer stem cells and cancer cells by the gelatinases-stimuli nanoparticles. *Biomaterials* **2013**, *34*, 7191–7203. [CrossRef]

113. Xu, Q.; Xia, Y.; Wang, C.H.; Pack, D.W. Monodisperse double-walled microspheres loaded with chitosan-p53 nanoparticles and doxorubicin for combined gene therapy and chemotherapy. *J. Control. Release* **2012**, *163*, 130–135. [CrossRef] [PubMed]

114. Patri, A.K.; Kukowska-Latallo, J.F.; Baker, J.R., Jr. Targeted drug delivery with dendrimers: Comparison of the release kinetics of covalently conjugated drug and non-covalent drug inclusion complex. *Adv. Drug Deliv. Rev.* **2005**, *57*, 2203–2214. [CrossRef] [PubMed]

115. Zhang, J.; Li, J.; Shi, Z.; Yang, Y.; Xie, X.; Lee, S.M.; Wang, Y.; Leong, K.W.; Chen, M. pH-sensitive polymeric nanoparticles for co-delivery of doxorubicin and curcumin to treat cancer via enhanced pro-apoptotic and anti-angiogenic activities. *Acta Biomater.* **2017**, *58*, 349–364. [CrossRef]

116. Xu, X.; Chen, X.; Wang, Z.; Jing, X. Ultrafine PEG-PLA fibers loaded with both paclitaxel and doxorubicin hydrochloride and their in vitro cytotoxicity. *Eur. J. Pharm. Biopharm.* **2009**, *72*, 18–25. [CrossRef]

117. Chen, Y.; Zhang, W.; Huang, Y.; Gao, F.; Sha, X.; Fang, X. Pluronic-based functional polymeric mixed micelles for co-delivery of doxorubicin and paclitaxel to multidrug resistant tumor. *Int. J. Pharm.* **2015**, *488*, 44–58. [CrossRef] [PubMed]

118. Ma, Y.; Fan, X.; Li, L. pH-sensitive polymeric micelles formed by doxorubicin conjugated prodrugs for co-delivery of doxorubicin and paclitaxel. *Carbohydr. Polym.* **2016**, *137*, 19–29. [CrossRef]

119. Duong, H.H.; Yung, L.Y. Synergistic co-delivery of doxorubicin and paclitaxel using multi-functional micelles for cancer treatment. *Int. J. Pharm.* **2013**, *454*, 486–495. [CrossRef] [PubMed]

120. Lv, S.; Tang, Z.; Li, M.; Lin, J.; Song, W.; Liu, H.; Huang, Y.; Zhang, Y.; Chen, X. Co-delivery of doxorubicin and paclitaxel by PEG-polypeptide nanovehicle for the treatment of non-small cell lung cancer. *Biomaterials* **2014**, *35*, 6118–6129. [CrossRef]

121. Noh, I.; Kim, H.-O.; Choi, J.; Choi, Y.; Lee, D.K.; Huh, Y.-M.; Haam, S. Co-delivery of paclitaxel and gemcitabine via CD44-targeting nanocarriers as a prodrug with synergistic antitumor activity against human biliary cancer. *Biomaterials* **2015**, *53*, 763–774. [CrossRef]

122. Wu, B.; Lu, S.-T.; Zhang, L.-J.; Zhuo, R.-X.; Xu, H.-B.; Huang, S.-W. Codelivery of doxorubicin and triptolide with reduction-sensitive lipid-polymer hybrid nanoparticles for in vitro and in vivo synergistic cancer treatment. *Int. J. Nanomed.* **2017**, *12*, 1853–1862. [CrossRef] [PubMed]

123. Pilehvar-Soltanahmadi, Y.; Dadashpour, M.; Alipour, S.; Farajzadeh, R.; Javidfar, S.; Zarghami, N. Co-Delivery of Curcumin and Chrysin by Polymeric Nanoparticles Inhibit Synergistically Growth and hTERT Gene Expression in Human Colorectal Cancer Cells AU—Lotfi-Attari, Javid. *Nutr. Cancer* **2017**, *69*, 1290–1299.

124. Liu, D.; Chen, Y.; Feng, X.; Deng, M.; Xie, G.; Wang, J.; Zhang, L.; Liu, Q.; Yuan, P. Micellar nanoparticles loaded with gemcitabine and doxorubicin showed synergistic effect. *Colloids Surf. B Biointerfaces* **2014**, *113*, 158–168. [CrossRef]

125. Song, W.; Tang, Z.; Li, M.; Lv, S.; Sun, H.; Deng, M.; Liu, H.; Chen, X. Polypeptide-based combination of paclitaxel and cisplatin for enhanced chemotherapy efficacy and reduced side-effects. *Acta Biomater.* **2014**, *10*, 1392–1402. [CrossRef] [PubMed]

126. Iatridi, Z.; Angelopoulou, A.; Voulgari, E.; Avgoustakis, K.; Tsitsilianis, C. Star-Graft Quarterpolymer-Based Polymersomes as Nanocarriers for Co-Delivery of Hydrophilic/Hydrophobic Chemotherapeutic Agents. *ACS Omega* **2018**, *3*, 11896–11908. [CrossRef]

127. Zhang, J.; Wang, L.; Fai Chan, H.; Xie, W.; Chen, S.; He, C.; Wang, Y.; Chen, M. Co-delivery of paclitaxel and tetrandrine via iRGD peptide conjugated lipid-polymer hybrid nanoparticles overcome multidrug resistance in cancer cells. *Sci. Rep.* **2017**, *7*, 46057. [CrossRef]

128. Xu, J.; Zhu, X.; Qiu, L. Polyphosphazene vesicles for co-delivery of doxorubicin and chloroquine with enhanced anticancer efficacy by drug resistance reversal. *Int. J. Pharm.* **2016**, *498*, 70–81. [CrossRef]

129. Wan, X.; Beaudoin, J.J.; Vinod, N.; Min, Y.; Makita, N.; Bludau, H.; Jordan, R.; Wang, A.; Sokolsky, M.; Kabanov, A.V. Co-delivery of paclitaxel and cisplatin in poly(2-oxazoline) polymeric micelles: Implications for drug loading, release, pharmacokinetics and outcome of ovarian and breast cancer treatments. *Biomaterials* **2019**, *192*, 1–14. [CrossRef]

130. Desale, S.S.; Soni, K.S.; Romanova, S.; Cohen, S.M.; Bronich, T.K. Targeted delivery of platinum-taxane combination therapy in ovarian cancer. *J. Control. Release* **2015**, *220*, 651–659. [CrossRef]

131. Handali, S.; Moghimipour, E.; Rezaei, M.; Saremy, S.; Dorkoosh, F.A. Co-delivery of 5-fluorouracil and oxaliplatin in novel poly(3-hydroxybutyrate-co-3-hydroxyvalerate acid)/poly(lactic-co-glycolic acid) nanoparticles for colon cancer therapy. *Int. J. Biol. Macromol.* **2019**, *124*, 1299–1311. [CrossRef]

132. Yang, Z.; Sun, N.; Cheng, R.; Zhao, C.; Liu, Z.; Li, X.; Liu, J.; Tian, Z. pH multistage responsive micellar system with charge-switch and PEG layer detachment for co-delivery of paclitaxel and curcumin to synergistically eliminate breast cancer stem cells. *Biomaterials* **2017**, *147*, 53–67. [CrossRef]

133. Liu, H.S.; Ke, C.S.; Cheng, H.C.; Huang, C.Y.; Su, C.L. Curcumin-induced mitotic spindle defect and cell cycle arrest in human bladder cancer cells occurs partly through inhibition of aurora A. *Mol. Pharmacol.* **2011**, *80*, 638–646. [CrossRef]

134. Tima, S.; Ichikawa, H.; Ampasavate, C.; Okonogi, S.; Anuchapreeda, S. Inhibitory effect of turmeric curcuminoids on FLT3 expression and cell cycle arrest in the FLT3-overexpressing EoL-1 leukemic cell line. *J. Nat. Prod.* **2014**, *77*, 948–954. [CrossRef]

135. Balasubramanian, S.; Eckert, R.L. Curcumin suppresses AP1 transcription factor-dependent differentiation and activates apoptosis in human epidermal keratinocytes. *J. Biol. Chem.* **2007**, *282*, 6707–6715. [CrossRef]

136. Moragoda, L.; Jaszewski, R.; Majumdar, A.P. Curcumin induced modulation of cell cycle and apoptosis in gastric and colon cancer cells. *Anticancer Res.* **2001**, *21*, 873–878.

137. Limtrakul, P.; Anuchapreeda, S.; Buddhasukh, D. Modulation of human multidrug-resistance MDR-1 gene by natural curcuminoids. *BMC Cancer* **2004**, *4*, 13. [CrossRef]

138. Das, L.; Vinayak, M. Long-term effect of curcumin down-regulates expression of tumor necrosis factor-alpha and interleukin-6 via modulation of E26 transformation-specific protein and nuclear factor-κB transcription factors in livers of lymphoma bearing mice. *Leuk. Lymphoma* **2014**, *55*, 2627–2636. [CrossRef]

139. Kuttan, G.; Kumar, K.B.; Guruvayoorappan, C.; Kuttan, R. Antitumor, anti-invasion, and antimetastatic effects of curcumin. *Adv. Exp. Med. Biol.* **2007**, *595*, 173–184.

140. Tuorkey, M.J. Curcumin a potent cancer preventive agent: Mechanisms of cancer cell killing. *Interv. Med. Appl. Sci.* **2014**, *6*, 139–146. [CrossRef]
141. Hussain, Z.; Thu, H.E.; Amjad, M.W.; Hussain, F.; Ahmed, T.A.; Khan, S. Exploring recent developments to improve antioxidant, anti-inflammatory and antimicrobial efficacy of curcumin: A review of new trends and future perspectives. *Mater. Sci. Eng. C Mater. Biol. Appl.* **2017**, *77*, 1316–1326. [CrossRef]
142. Yan, T.; Li, D.; Li, J.; Cheng, F.; Cheng, J.; Huang, Y.; He, J. Effective co-delivery of doxorubicin and curcumin using a glycyrrhetinic acid-modified chitosan-cystamine-poly(epsilon-caprolactone) copolymer micelle for combination cancer chemotherapy. *Colloids Surf. B Biointerfaces* **2016**, *145*, 526–538. [CrossRef] [PubMed]
143. Zhang, Y.; Yang, C.; Wang, W.; Liu, J.; Liu, Q.; Huang, F.; Chu, L.; Gao, H.; Li, C.; Kong, D.; et al. Co-delivery of doxorubicin and curcumin by pH-sensitive prodrug nanoparticle for combination therapy of cancer. *Sci. Rep.* **2016**, *6*, 21225. [CrossRef] [PubMed]
144. Wang, J.; Ma, W.; Tu, P. Synergistically Improved Anti-tumor Efficacy by Co-delivery Doxorubicin and Curcumin Polymeric Micelles. *Macromol. Biosci.* **2015**, *15*, 1252–1261. [CrossRef] [PubMed]
145. Duan, J.; Mansour, H.M.; Zhang, Y.; Deng, X.; Chen, Y.; Wang, J.; Pan, Y.; Zhao, J. Reversion of multidrug resistance by co-encapsulation of doxorubicin and curcumin in chitosan/poly(butyl cyanoacrylate) nanoparticles. *Int. J. Pharm.* **2012**, *426*, 193–201. [CrossRef] [PubMed]
146. Bisht, S.; Feldmann, G.; Soni, S.; Ravi, R.; Karikar, C.; Maitra, A.; Maitra, A. Polymeric nanoparticle-encapsulated curcumin ("nanocurcumin"): A novel strategy for human cancer therapy. *J. Nanobiotechnol.* **2007**, *5*, 3. [CrossRef]
147. Khalil, N.M.; do Nascimento, T.C.; Casa, D.M.; Dalmolin, L.F.; de Mattos, A.C.; Hoss, I.; Romano, M.A.; Mainardes, R.M. Pharmacokinetics of curcumin-loaded PLGA and PLGA-PEG blend nanoparticles after oral administration in rats. *Colloids Surf. B Biointerfaces* **2013**, *101*, 353–360. [CrossRef]
148. Jones, N.A.; Turner, J.; McIlwrath, A.J.; Brown, R.; Dive, C. Cisplatin- and paclitaxel-induced apoptosis of ovarian carcinoma cells and the relationship between bax and bak up-regulation and the functional status of p53. *Mol. Pharmacol.* **1998**, *53*, 819–826. [PubMed]
149. Yang, J.; Ju, Z.; Dong, S. Cisplatin and paclitaxel co-delivered by folate-decorated lipid carriers for the treatment of head and neck cancer. *Drug Deliv.* **2017**, *24*, 792–799. [CrossRef]
150. Liu, B.; Han, L.; Liu, J.; Han, S.; Chen, Z.; Jiang, L. Co-delivery of paclitaxel and TOS-cisplatin via TAT-targeted solid lipid nanoparticles with synergistic antitumor activity against cervical cancer. *Int. J. Nanomed.* **2017**, *12*, 955–968. [CrossRef]
151. Mi, Y.; Zhao, J.; Feng, S.-S. Targeted co-delivery of docetaxel, cisplatin and herceptin by vitamin E TPGS-cisplatin prodrug nanoparticles for multimodality treatment of cancer. *J. Control. Release* **2013**, *169*, 185–192. [CrossRef]
152. Tian, J.; Min, Y.; Rodgers, Z.; Au, K.M.; Hagan, C.T.; Zhang, M.; Roche, K.; Yang, F.; Wagner, K.; Wang, A.Z. Co-delivery of paclitaxel and cisplatin with biocompatible PLGA–PEG nanoparticles enhances chemoradiotherapy in non-small cell lung cancer models. *J. Mater. Chem. B* **2017**, *5*, 6049–6057. [CrossRef]
153. He, Z.; Huang, J.; Xu, Y.; Zhang, X.; Teng, Y.; Huang, C.; Wu, Y.; Zhang, X.; Zhang, H.; Sun, W. Co-delivery of cisplatin and paclitaxel by folic acid conjugated amphiphilic PEG-PLGA copolymer nanoparticles for the treatment of non-small lung cancer. *Oncotarget* **2015**, *6*, 42150–42168. [CrossRef]
154. Merino, D.; Lalaoui, N.; Morizot, A.; Solary, E.; Micheau, O. TRAIL in cancer therapy: Present and future challenges. *Expert Opin. Ther. Targets* **2007**, *11*, 1299–1314. [CrossRef]
155. Grotzer, M.A.; Eggert, A.; Zuzak, T.J.; Janss, A.J.; Marwaha, S.; Wiewrodt, B.R.; Ikegaki, N.; Brodeur, G.M.; Phillips, P.C. Resistance to TRAIL-induced apoptosis in primitive neuroectodermal brain tumor cells correlates with a loss of caspase-8 expression. *Oncogene* **2000**, *19*, 4604–4610. [CrossRef]
156. Kagawa, S.; He, C.; Gu, J.; Koch, P.; Rha, S.J.; Roth, J.A.; Curley, S.A.; Stephens, L.C.; Fang, B. Antitumor activity and bystander effects of the tumor necrosis factor-related apoptosis-inducing ligand (TRAIL) gene. *Cancer Res.* **2001**, *61*, 3330–3338.
157. Lee, A.L.; Dhillon, S.H.; Wang, Y.; Pervaiz, S.; Fan, W.; Yang, Y.Y. Synergistic anti-cancer effects via co-delivery of TNF-related apoptosis-inducing ligand (TRAIL/Apo2L) and doxorubicin using micellar nanoparticles. *Mol. Biosyst.* **2011**, *7*, 1512–1522. [CrossRef]
158. Lee, A.L.; Wang, Y.; Pervaiz, S.; Fan, W.; Yang, Y.Y. Synergistic anticancer effects achieved by co-delivery of TRAIL and paclitaxel using cationic polymeric micelles. *Macromol. Biosc.i* **2011**, *11*, 296–307. [CrossRef]

159. Han, L.; Huang, R.; Li, J.; Liu, S.; Huang, S.; Jiang, C. Plasmid pORF-hTRAIL and doxorubicin co-delivery targeting to tumor using peptide-conjugated polyamidoamine dendrimer. *Biomaterials* **2011**, *32*, 1242–1252. [CrossRef]

160. Xu, Q.; Leong, J.; Chua, Q.Y.; Chi, Y.T.; Chow, P.K.; Pack, D.W.; Wang, C.H. Combined modality doxorubicin-based chemotherapy and chitosan-mediated p53 gene therapy using double-walled microspheres for treatment of human hepatocellular carcinoma. *Biomaterials* **2013**, *34*, 5149–5162. [CrossRef]

161. Li, Y.; Xu, B.; Bai, T.; Liu, W. Co-delivery of doxorubicin and tumor-suppressing p53 gene using a POSS-based star-shaped polymer for cancer therapy. *Biomaterials* **2015**, *55*, 12–23. [CrossRef]

162. Cheng, D.; Cao, N.; Chen, J.; Yu, X.; Shuai, X. Multifunctional nanocarrier mediated co-delivery of doxorubicin and siRNA for synergistic enhancement of glioma apoptosis in rat. *Biomaterials* **2012**, *33*, 1170–1179. [CrossRef]

163. Hu, Q.; Li, W.; Hu, X.; Hu, Q.; Shen, J.; Jin, X.; Zhou, J.; Tang, G.; Chu, P.K. Synergistic treatment of ovarian cancer by co-delivery of survivin shRNA and paclitaxel via supramolecular micellar assembly. *Biomaterials* **2012**, *33*, 6580–6591. [CrossRef]

164. Zhang, C.G.; Zhu, W.J.; Liu, Y.; Yuan, Z.Q.; Yang, S.D.; Chen, W.L.; Li, J.Z.; Zhou, X.F.; Liu, C.; Zhang, X.N. Novel polymer micelle mediated co-delivery of doxorubicin and P-glycoprotein siRNA for reversal of multidrug resistance and synergistic tumor therapy. *Sci. Rep.* **2016**, *6*, 23859. [CrossRef]

165. Misra, R.; Das, M.; Sahoo, B.S.; Sahoo, S.K. Reversal of multidrug resistance in vitro by co-delivery of MDR1 targeting siRNA and doxorubicin using a novel cationic poly(lactide-co-glycolide) nanoformulation. *Int. J. Pharm.* **2014**, *475*, 372–384. [CrossRef]

166. Cui, X.; Sun, Y.; Shen, M.; Song, K.; Yin, X.; Di, W.; Duan, Y. Enhanced Chemotherapeutic Efficacy of Paclitaxel Nanoparticles Co-delivered with MicroRNA-7 by Inhibiting Paclitaxel-Induced EGFR/ERK pathway Activation for Ovarian Cancer Therapy. *ACS Appl. Mater. Interfaces* **2018**, *10*, 7821–7831. [CrossRef]

167. Yang, Y.; Meng, Y.; Ye, J.; Xia, X.; Wang, H.; Li, L.; Dong, W.; Jin, D.; Liu, Y. Sequential delivery of VEGF siRNA and paclitaxel for PVN destruction, anti-angiogenesis, and tumor cell apoptosis procedurally via a multi-functional polymer micelle. *J. Control. Release* **2018**, *287*, 103–120. [CrossRef]

168. Shen, J.; Yin, Q.; Chen, L.; Zhang, Z.; Li, Y. Co-delivery of paclitaxel and survivin shRNA by pluronic P85-PEI/TPGS complex nanoparticles to overcome drug resistance in lung cancer. *Biomaterials* **2012**, *33*, 8613–8624. [CrossRef]

169. Tang, S.; Yin, Q.; Zhang, Z.; Gu, W.; Chen, L.; Yu, H.; Huang, Y.; Chen, X.; Xu, M.; Li, Y. Co-delivery of doxorubicin and RNA using pH-sensitive poly (β-amino ester) nanoparticles for reversal of multidrug resistance of breast cancer. *Biomaterials* **2014**, *35*, 6047–6059. [CrossRef]

170. Essex, S.; Navarro, G.; Sabhachandani, P.; Chordia, A.; Trivedi, M.; Movassaghian, S.; Torchilin, V.P. Phospholipid-modified PEI-based nanocarriers for in vivo siRNA therapeutics against multidrug-resistant tumors. *Gene Ther.* **2015**, *22*, 257–266. [CrossRef]

171. Chen, Y.; Sun, J.; Huang, Y.; Liu, Y.; Liang, L.; Yang, D.; Lu, B.; Li, S. Targeted codelivery of doxorubicin and IL-36gamma expression plasmid for an optimal chemo-gene combination therapy against cancer lung metastasis. *Nanomedicine* **2019**, *15*, 129–141. [CrossRef] [PubMed]

172. Zhao, Y.; Song, Q.; Yin, Y.; Wu, T.; Hu, X.; Gao, X.; Li, G.; Tan, S.; Zhang, Z. Immunochemotherapy mediated by thermosponge nanoparticles for synergistic anti-tumor effects. *J. Control. Release* **2018**, *269*, 322–336. [CrossRef] [PubMed]

173. Chen, W.; Zhang, M.; Shen, W.; Du, B.; Yang, J.; Zhang, Q. A Polycationic Brush Mediated Co-Delivery of Doxorubicin and Gene for Combination Therapy. *Polymers* **2019**, *11*, 60. [CrossRef]

174. Yip, K.W.; Reed, J.C. Bcl-2 family proteins and cancer. *Oncogene* **2008**, *27*, 6398–6406. [CrossRef] [PubMed]

175. Qian, J.; Xu, M.; Suo, A.; Xu, W.; Liu, T.; Liu, X.; Yao, Y.; Wang, H. Folate-decorated hydrophilic three-arm star-block terpolymer as a novel nanovehicle for targeted co-delivery of doxorubicin and Bcl-2 siRNA in breast cancer therapy. *Acta Biomater.* **2015**, *15*, 102–116. [CrossRef] [PubMed]

176. Shin, S.; Sung, B.J.; Cho, Y.S.; Kim, H.J.; Ha, N.C.; Hwang, J.I.; Chung, C.W.; Jung, Y.K.; Oh, B.H. An anti-apoptotic protein human survivin is a direct inhibitor of caspase-3 and -7. *Biochemistry* **2001**, *40*, 1117–1123. [CrossRef]

177. Lu, B.; Mu, Y.; Cao, C.; Zeng, F.; Schneider, S.; Tan, J.; Price, J.; Chen, J.; Freeman, M.; Hallahan, D.E. Survivin as a therapeutic target for radiation sensitization in lung cancer. *Cancer Res.* **2004**, *64*, 2840–2845. [CrossRef]

178. Wang, H.; Wu, Y.; Zhao, R.; Nie, G. Engineering the assemblies of biomaterial nanocarriers for delivery of multiple theranostic agents with enhanced antitumor efficacy. *Adv. Mater.* **2013**, *25*, 1616–1622. [CrossRef]
179. Janat-Amsbury, M.M.; Yockman, J.W.; Lee, M.; Kern, S.; Furgeson, D.Y.; Bikram, M.; Kim, S.W. Combination of local, nonviral IL12 gene therapy and systemic paclitaxel treatment in a metastatic breast cancer model. *Mol. Ther.* **2004**, *9*, 829–836. [CrossRef]
180. Creixell, M.; Peppas, N.A. Co-delivery of siRNA and therapeutic agents using nanocarriers to overcome cancer resistance. *Nano Today* **2012**, *7*, 367–379. [CrossRef]

![molecules logo] *molecules*

Review

Recent Development of pH-Responsive Polymers for Cancer Nanomedicine

Houliang Tang [1], Weilong Zhao [2], Jinming Yu [3], Yang Li [4],* and Chao Zhao [3],*

1   Department of Chemistry, Southern Methodist University, 3215 Daniel Avenue, Dallas, TX 75275, USA;
     houliangt@smu.edu
2   Global Research IT, Merck & Co., Inc., Boston, MA 02210, USA; weilong.zhao@merck.com
3   Department of Chemical and Biological Engineering, the University of Alabama, Tuscaloosa, AL 35487, USA;
     jyu56@crimson.ua.edu
4   Boston Children's Hospital, Harvard Medical School, 300 Longwood Avenue, Boston, MA 02115, USA
*   Correspondence: Yang.Li2@childrens.harvard.edu (Y.L.); czhao15@eng.ua.edu (C.Z.); Tel.: +1-(205)348-9869 (C.Z.)

Academic Editor: Derek J. McPhee
Received: 21 November 2018; Accepted: 17 December 2018; Published: 20 December 2018

Abstract: Cancer remains a leading cause of death worldwide with more than 10 million new cases every year. Tumor-targeted nanomedicines have shown substantial improvements of the therapeutic index of anticancer agents, addressing the deficiencies of conventional chemotherapy, and have had a tremendous growth over past several decades. Due to the pathophysiological characteristics that almost all tumor tissues have lower pH in comparison to normal healthy tissues, among various tumor-targeted nanomaterials, pH-responsive polymeric materials have been one of the most prevalent approaches for cancer diagnosis and treatment. In this review, we summarized the types of pH-responsive polymers, describing their chemical structures and pH-response mechanisms; we illustrated the structure-property relationships of pH-responsive polymers and introduced the approaches to regulating their pH-responsive behaviors; we also highlighted the most representative applications of pH-responsive polymers in cancer imaging and therapy. This review article aims to provide general guidelines for the rational design of more effective pH-responsive nanomaterials for cancer diagnosis and treatment.

Keywords: pH responsive polymers; nanomedicine; tumor imaging; drug delivery

---

1. Introduction

Many diseases originate from abnormal biological processes at the molecular level, such as gene mutation, protein misfolding, and cell malfunctions as a result of infections [1]. Advances in genomics, proteomics, and regenerative medicine have inspired development of numerous therapies and technologies for the treatment of various diseases. With over 10 million new cases every year globally, cancer remains a difficult disease to treat and one of the leading causes for death [2]. Cancer is a prevalent disease that involves a series of genome mutations over time and results in both genetic and phenotypic variations in different tumor cells [3,4]. Targeting cancer-specific biomarkers offers great opportunities for precise tumor detection and efficacious drug delivery, and more importantly, causes minimal side effects on normal cells [5]. However, it is impractical to apply a certain cell surface receptor to a broader range of cancers due to their genetic or phenotypic heterogeneity. In contrast, aerobic glycolysis, also known as the Warburg effect, represents a dysregulated energy metabolism in many types of cancer, where glucose is preferentially taken up and converted into lactic acid [6]. As a result, acidosis has occurred concurrently and evolved as a ubiquitous characteristic of cancer [7].

With comparable scales to biologic molecules, nanomaterials have been extensive investigated for biomedical applications over past several decades [8,9]. The nanomaterials are present in small sizes ranging from 1 to 1000 nanometers with correspondingly large surface-to-volume ratio. A multitude of targeting moieties can be incorporated to the surface of nanostructures via diverse engineering procedures [10,11]. The physiochemical properties such as composition, size, and shape can be sophisticatedly tailored for disease diagnosis, treatment, and prevention [12]. Different physiochemical properties like optical, electrical, and magnetic responsiveness have been introduced to nanostructures to facilitate biological diagnostics [13]. One extraordinary property of nanomaterials is that they are generally multi-component systems. Different components or subunits of the nanostructures can be delicately engineered to address the same specific challenge in medicine, resulting in strong cooperative effect that is absent in monomolecular therapeutics [14]. A recent report revealed that over 250 nanomaterial-based technologies or therapies have been approved by the Food and Drug Administration (FDA) or are currently in different stage of clinical trials [15].

There has been explosive development of numerous nanotechnologies to diagnose and treat cancer [16]. The tumor heterogeneity leads to the growing emphasis on customized nanomedicine [17]. Molecular imaging platforms are used to locate tumor and facilitate image-guided surgery [18]. Due to enhanced permeability and retention (EPR) effect, nanoparticles are more likely to accumulate in tumors and spare the surrounding benign tissues [19]. Since abnormal pH has been recognized as a universally diagnostic hallmark of cancer, the design of pH responsive polymers that are capable of altering their chemical or physical properties are of particular interest for cancer theranostics [20–23]. Moreover, nanoscale polymers with pH responsive segments can strengthen the targeting specificity and hence promote the uptake of nanomedicine into tumor cells [24].

In this perspective, we will review the recent advances in the development of pH responsive nanomaterials for cancer diagnosis and treatment. This article aims to provide general guidelines for the rational design of pH-responsive nanomaterials for the diagnosis and treatment of cancer. Main focus is placed on the polymer designing, mechanistic insights, and specific applications.

2. pH-Responsive Polymers

Fabricating pH responsive polymeric nanomaterials has been rapidly developed after the emerging of advanced techniques in polymer synthesis, especially reversible-deactivation radical polymerization such as ATRP and RAFT [25,26]. The technical state of the art allows for precise control over molecular weights and molecular weight distributions, affording polymers with well-defined topology structure [27–29]. The pH responsive polymers can be mainly classified into two categories: (1) polymers with ionizable moieties and (2) polymers that contain acid-labile linkages [30]. The former strategy employs a noncovalent transition to achieve pH responsivity. Typical ionizable moieties include amines and carboxylic acids, which can be protonated or deprotonated at different pH values. The alteration of solubility in aqueous medium serves as an amplified signal to represent the change of environmental pH. For the latter, the backbone of polymers usually comprises acid-labile covalent linkages. The decrease of pH is able to trigger the cleavage of these bonds, causing a degradation of polymer chains or a dissociation of polymer aggregates. Compared to polymers with ionizable moieties, polymers containing acid-labile linkages often present a slower internal structural transition due to the nature of covalent bonding, which facilitate their applications in the drug delivery systems.

2.1. pH Responsive Polymers with Ionizable Groups

The pH response arises from the reversible protonation and deprotonation of ionizable groups at the molecular level. The pKa of a polyelectrolyte, which is defined as the pH with equal concentration of the protonated and deprotonated forms, can serve as a critical benchmark to reflect the polymer ionization behaviors at various pH levels [31]. In general, there are two types of ionizable polymers: basic polymers that accept protons at a relatively low pH and acidic polymers that release protons at a relatively high pH. Consequently, these polybases or polyacids can form positively or negatively charged polymer chains at different pH levels. Common basic moieties include amines, pyridines, morpholines, piperazines; and common acidic groups include carboxylic acids, sulfonic acids, phosphoric acids, boronic acids, etc. [30]. Among them, amines, especially tertiary amines, have drawn particular attention due to their ease to prepare and the feasibility to finely tune their pKa [32,33]. It was reported that the amines can present a marginally lower pKa when substituted with longer hydrophobic chains [34].

Recently, Gao et al. designed a series of ultra-pH sensitive (UPS) polymers for real-time tumor imaging [35–38]. These nanoprobes can sharply respond to and amplify in vivo pH signals in a very narrow pH span. The UPS nanoparticles are comprised of an amphiphilic block copolymer: PEG-*b*-PMA, where PEG stands for hydrophilic poly(ethylene glycol) and PMA is a hydrophobic segment based on polymethacrylates with tertiary amine substituent. At physiological pH (7.4), the block copolymers stay as core-shell micelles driven by self-assembly. If the environmental pH is lower than the pKa of the pendent tertiary amines, the amines will get protonated and the micelles will dissociate rapidly into unimers. It is worth noting that the process that comprises assembly and dissociation is completely reversible, and fully determined by ambient pH. Acid-triggered dissociation of micelles enable the increase of fluorescence intensity for molecular imaging and release of cargo for drug delivery.

Yan et al. further expand the UPS polymer design based on biodegradable polypeptides (Figure 1) [39]. They modified the peptides with ionizable tertiary amines for pH responsive behavior. Hydrophilic and hydrophobic blocks were synthesized independently and covalently connected by the copper(I)-catalyzed alkyne-azide cycloaddition (CuAAC) "click" chemistry [40]. The copolymer's pKa can be readily tuned by altering the ratio of amino substituents. Similar to aforementioned UPS designing, a fluorescent cyanine dye (Cy5.5) was attached at the chain end of the amphiphilic copolymer to convert the subtle pH variation into significant fluorescence intensity change. This study suggested that the ultra-pH sensitivity resulted from the reversible protonation of ionizable amines rather than the peptide backbone. It also inspired the design of degradable ultra-pH sensitive polymers.

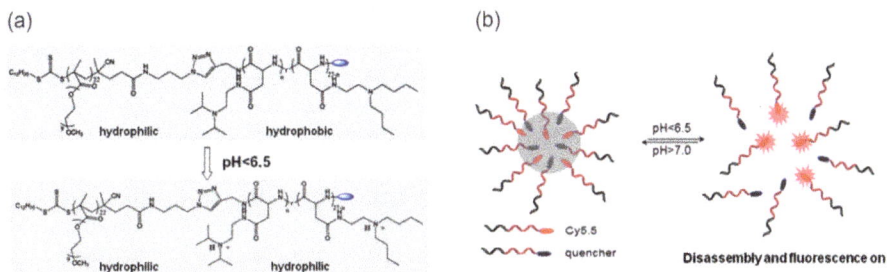

Figure 1. Design of ultra-pH-sensitive polypeptide micelles. (**a**) Structure of amphiphilic copolymers with ionizable tertiary amines and their transition at lower pH. (**b**) pH-triggered disassembly and fluorescence. Reproduced with permission from Liyi Fu, Pan Yuan, Zheng Ruan, Le Liu, Tuanwei Li, and Lifeng Yan, Polymer Chemistry; published by Royal Society of Chemistry, 2017.

Natural macromolecules with ionizable amino acid residues have also been investigated for pH-triggered delivery of imaging and therapeutic agents [41–45]. Engelman and colleagues have been working on the development of novel pH-responsive transmembrane peptides, pH low insertion peptide (pHLIPs), for basic research and translational applications in membrane biophysics and medicine (Figure 2) [46]. These pH-responsive peptides were derived from the C-helix of the protein bacteriorhodopsin [47]. pHLIPs spontaneously self-assemble into a helix structure and insert across the membrane upon exposure to acidic environment [48]. In physiological condition, where the pH is around 7.4, the ionizable acid residues of the pHLIP (red circles) stay negatively charged and the peptide will be weakly bound to the surface of membrane. Once encountering acidic condition like tumor microenvironment, carboxyl group will be protonated and neutralized. Increased lipophilicity (green circles) as a result of ionization drastically enhances the affinity of pHLIP to the hydrophobic inner core of the cellular membrane and triggers the formation of a helix and ensuing insertion across the membrane. When the protonatable carboxyl groups are exposed to the normal intracellular pH, pHLIP gets reversibly deprotonated and anchors in the membrane. The pH-responsive behavior of pHLIP can be easily modified by replacing ionizable aspartic acid residues with glutamic acid residues [43,49] or positively-charged lysine residues [41], or other protonatable amino acids [44].

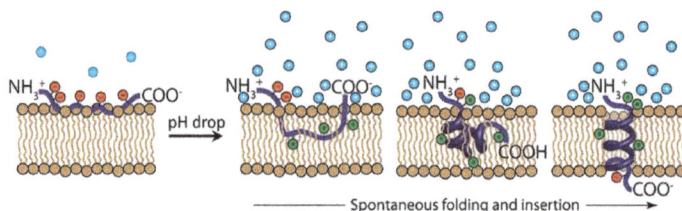

Figure 2. Mechanism of pH (Low) Insertion Peptide (pHLIP) Insertion into the Cellular Membrane. In healthy tissue where the pH is around 7.4, the ionizable residues of the pHLIP (red circles) remain deprotonated and negatively charged, and the peptide resides at or near the hydrophilic surface of the cellular membrane. Upon exposure to acidic tumor microenvironment, the ionizable residues and negatively charged C-terminal carboxyl group of the pHLIP become neutrally charged (green circles). The protonation results in an increase in the hydrophobicity of the pHLIP, triggering the pHLIP to spontaneously fold into a helix and insert across the hydrophobic lipid bilayer of cell membrane, resulting in the formation of a transmembrane helix. Following internalization and exposure to cytosol with pH above 7, C-terminal ionizable residues are deprotonated again and anchor the pHLIP into the membrane. Reproduced with permission from Linden C. Wyatt, Jason S. Lewis, Oleg A. Andreev, Yana K. Reshetnyak, and Donald M. Engelman, Trends in Biotechnology; published by Elsevier, 2017.

Polyacids have also been extensively investigated for the design of pH-responsive nanoplatforms. Hyaluronic acid (HA) is a key component of the extracellular matrix and is known to bind to CD44 proteins as a surface receptor on cancer cells [50]. Kono and coworkers reported HA-based pH-sensitive polymer-modified liposomes for tumor-targeted delivery of chemotherapeutics [51]. Instead of simply using HA as targeting moiety, these authors introduced 3-methyl glutarylated (MGlu) units and 2-carboxycyclohexane-1-carboxylated (CHex) units for the design of a new class of pH-responsive polymers with transition pH around 5.4–6.7. Carboxyl group-introduced HA derivatives were prepared by reaction with various dicarboxylic anhydrides. Terminal alkyl chains served as anchor units inserting into the hydrophobic lipid bilayers of liposomes (Figure 3). Upon exposure to acidic endosome pH, both MGlu and CHex modified HA were protonated, enabling lipid membrane disruption and intracellular release of chemotherapeutics. Mechanistic investigation demonstrated that the Mglu/CHex significantly affected the pH-responsive behavior of HA derivatives. Increased hydrophobicity led to a higher transition pH.

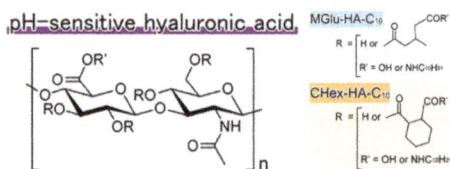

Figure 3. Structures of pH-sensitive modified hyaluronic acid derivatives. Reproduced with permission from Maiko Miyazaki, Eiji Yuba, Hiroshi Hayashi, Atsushi Harada, and Kenji Kono, Bioconjugate Chemistry; published by American Chemical Society, 2018.

2.2. pH Responsive Dissociation Based on Acid-Labile Linkages

Polymers containing acid-labile or base-labile linkages can respond to the change of pH by degradation. Since the tumor microenvironment is slightly acidic compared to normal physiological environment, polymeric nanoparticles containing base-labile linkages have seldom been used for cancer therapy and will not be covered in the perspective. In contrast, polymers with acid-labile linkages have been extensively utilized for the designing of anticancer drug delivery systems [52]. The most commonly used acid-labile linkages are listed in Table 1. The different degradation mechanisms of the linkages and the products after cleavage are described [53,54].

Table 1. Summary of acid-labile linkages.

Type of Linkages	Structure	Product after Acid Cleavage
Hydrozone		
Acetal		
Ketal		
Boronate ester		

Hydrazone is one of the most explored acid-labile linkages used for acid-responsive dissociation release due to its easy synthesis and moderate sensitivity [55]. Incorporation of hydrazine into the backbone of polymers represents an ideal strategy for the design of tumor-targeted drug delivery systems. Nie group designed a pH-sensitive drug-gold nanoparticle system for tumor chemotherapy and surface enhanced Raman scattering (SERS) imaging (Figure 4) [56]. This multifunctional system comprised of poly(ethylene glycol), doxorubicin (Dox), hydrazone linker, and gold nanoparticles (Au–Dox–PEG). 3-[2-pyridyldithio]propionyl hydrazide (PDPH) was conjugated with Dox in methanol. PDPH acted as a pH-sensitive linker and introduced thiol groups to ensure the anchoring of drug conjugates onto the surface of gold nanoparticles. Gao and coworkers developed a pH-responsive polypeptide–drug nanoparticles for targeted cancer therapy based on well-defined elastin-like polypeptides and acid-labile hydrazone linker [57]. Wang reported a hydrazone-based multifunctional sericin nanoparticles for pH-sensitive subcellular delivery of anti-tumor chemotherapeutics [58].

Acetal and ketal are also commonly used acid-labile linkages, which are very stable under basic conditions, but can be readily hydrolyzed to corresponding carbonyl compound (aldehyde and ketone) and alcohols through a caboxonium ion intermediate upon acidic cleavage [59]. Lu and coworkers reported a novel envelop-like mesoporous silica nanoparticle platform [60]. This system immobilized acetals on the surface of silica (yellow, left) before coupling to gate-keeper nanoparticle (purple, right) (Figure 5). At acidic pH, the acetal was effectively hydrolyzed to remove the gate keepers, allowing the escape of entrapped drug molecules. Liu et al. reported the facile fabrication of acid-sensitive polymersomes for intracellular release of drug over several days [61]. The polymersomes compromising cyclic benzylidene acetals in the hydrophobic bilayers were stable under neutral pH, whereas were hydrolyzed into hydrophilic diol moieties upon exposure to acidic pH milieu. The pH-triggered hydrolysis can be easily monitored by many experimental methods including UV/Vis spectroscopy and TEM. A novel class of acid degradable poly(acetal urethane) was also reported for the construction of acid-degradable micelles for delivery of hydrophobic anti-tumor therapeutics [62]. More recently, Wang group designed a new hyperbranched amphiphilic acetal polymers for pH-sensitive drug delivery [63]. Under neutral conditions, the block copolymers self-assembled into well-defined core-shell micelles. This pH-induced acetal cleavage resulted in the drastic decrease of hydrophobicity and dissociation of micelles. De Geest and coworkers synthesized ketal-containing block copolymers as pH-responsive nanocarriers for the hydrophobic anticancer drug paclitaxel (PTX) [64]. The hydrolysis of the ketal groups in the block copolymer side chains at pH < 5 lead to decomposition of the block copolymer nanoparticles and the PTX release.

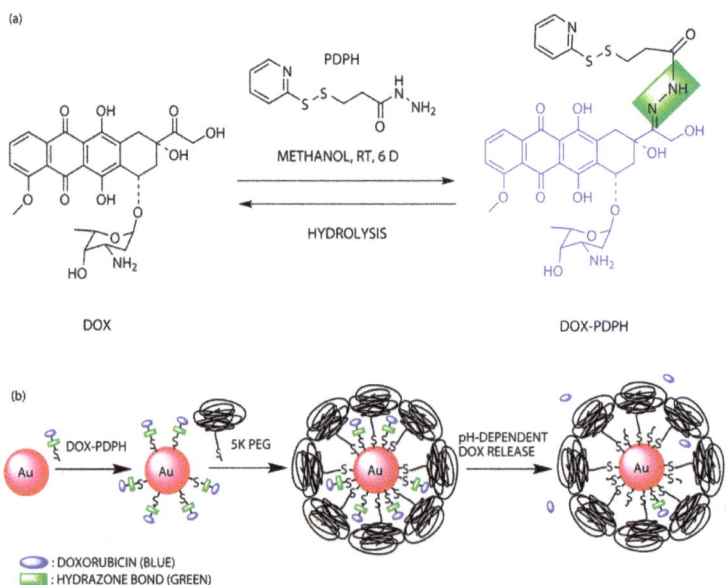

Figure 4. A pH-sensitive drug-gold nanoparticle system. (**a**) Chemical synthesis of the doxorubicin-hydrazone linker conjugate (dox–PDPH). (**b**) Schematic illustration for the synthesis of the multifunctional drug delivery system and its pH-dependent doxorubicin release. Reproduced with permission from Kate Y. J. Lee, Yiqing Wang, and Shuming Nie, RSC Advances; published by Royal Society of Chemistry, 2015.

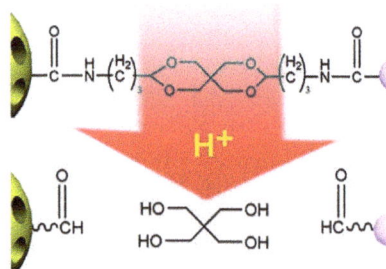

Figure 5. Schematic illustration of the envelope-type mesoporous silica nanoparticle for pH-responsive drug delivery. Reproduced with permission from Yan Chen, Kelong Ai, Jianhua Liu, Guoying Sun, Qi Yin, and Lehui Lu, Biomaterials; published by Elsevier, 2015.

Boronate esters, formed between a boronic acid and an alcohol, are stable at neutral or alkaline pH and readily dissociate to boronic acid and alcohol groups in a low-pH environment. Boronate esters have been widely employed to exploit pH-sensitive polymeric carriers for anticancer drug delivery. Messersmith et al. conjugated the boronic acid containing anticancer drug bortezomib (BTZ) to catechol-containing polymers via the boronate ester (Figure 6) [65]. Under neutral or basic condition, BTZ and catechol formed stable conjugates via boronate ester linkers, deactivating the cytotoxicity of BTZ. Upon exposure to lower pH, the conjugates readily release free drugs. Levkin and coworkers reported a dextran-based pH-sensitive nanoparticle system by modifying vicinal diol of dextran with boronate esters [66]. pH-triggered hydrolysis of ester linkers resumed the hydrophilic hydroxyl groups of dextran and destabilized the dug-encapsulated nanoparticles. Kim et al. reported a pH-sensitive nanocomplex by grafting phenylboronic acid (PBA) onto the backbone of poly(maleic anhydride) [67]. Dox and PBA readily formed boronate esters following simple mixing.

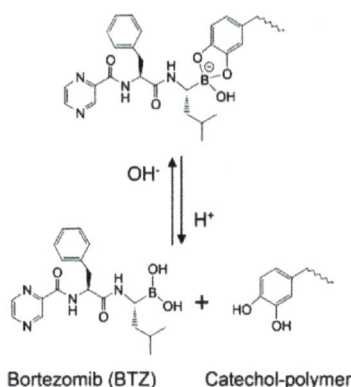

Figure 6. Catechol-containing polymers conjugated the boronic acid containing anticancer drug bortezomib via acid-labile boronate esters. Boronate esters are stable at neutral or alkaline pH and readily dissociate to boronic acid and alcohol groups in acidic environments to release the free active drug. Reproduced with permission from Jing Su, Feng Chen, Vincent L. Cryns, and Phillip B. Messersmith, Journal of the American Chemical Society; published by American Chemical Society, 2011.

3. Tunable pH-Responsive Behavior

A fundamental challenge in nanomedicine is the specific delivery the therapeutic or diagnostic agents to the targeted tissues or cells [68,69]. Targeted delivery has shown promise in reducing off-target effect and lowering toxicity [70]. One major consideration in the design of pH-responsive nanomaterials is choosing polymers with pKa values matching the desired pH range. The acidity of gastrointestinal track is drastically different from that of blood [71]. Intracellular compartments such as mitochondria, endosomes, and lysosomes also have slightly different pH values [72]. Even the level of acidosis shows small variation among different types of tumors [6]. Therefore, a key consideration in the design of tumor-targeted pH-responsive polymers is to ensure the polymers are able to differentiate acidic tumor microenvironment from surrounding normal tissues. The variation in the potency of diverse anti-tumor therapeutics also requires tunable release kinetics of payloads.

Reversible protonation of pH responsive polymers leads to the change of the hydrodynamic volume, chain conformation, water solubility and maybe supramolecular self-assembly. Ionization also allows us to modulate the pH-responsive behavior of these polymers. Mechanistic investigation suggests that the acid-base equilibrium in the ionization process is controlled by the balance between hydrophobic interaction and electrostatic repulsion [73]. The transition pH can be tuned by altering the structural factors that affect the hydrophobic or electrostatic interactions. Environmental factors that shift the transition pH such as ionic strength and species are beyond the scope of this review.

3.1. Hydrophobic Modification

Hydrophobic modification is an important approach to modulate the transition pH of ionizable polymers. Incorporating different hydrophobic groups or changing the length of hydrophobic chains can lead to the shift of pKa. Li et al. systemically investigated key structural parameters that affected the transition pH of a series of polymers containing ionizable tertiary amines (Figure 7a) [34]. The hydrophobic interactions can be strengthened by increasing the hydrophobicity of the amine substituents. To accomplish this goal, they synthesized a series of pH sensitive block copolymers with an identical poly(methacrylate) backbone and similar chain length but different linear terminal alkyl groups on the side chain. The polymer with the most hydrophobic pentyl group yielded the lowest pKa at 4.4. Meanwhile, the one with the least hydrophobic isopropyl group as an amine substituent showed the highest pKa, close to 6.6. These results demonstrated that ionizable pH-sensitive copolymers with more hydrophobic amine substituents have a lower pKa. They calculated the octanol–water partition coefficients (LogP) of the repeating unit of hydrophobic segment and used them as a quantitative measure of the strength of hydrophobic interactions. The plot of pKa values as a function of LogP (Figure 7b) showed a linear correlation. To confirm the observation, they synthesized another series of block copolymers with the same backbone and similar chain length, but different cyclic terminal alkyl groups. A plot of transition pH as a function of LogP also showed a similar linear correlation.

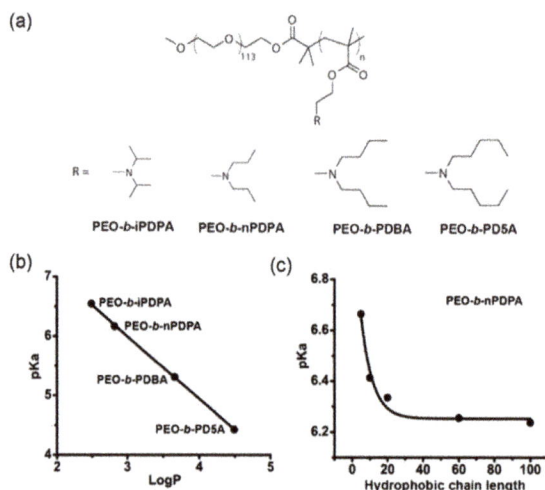

Figure 7. Tunable pKa based on hydrophobic modification. (**a**) Structure of methacrylate-based block copolymers with different alkyl substituents. (**b**) Increase of the hydrophobicity of the alkyl substituents resulted in the linear decrease of pKa. (**c**) Increased hydrophobic chain length lead to decreased pKa in a representative poly(ethylene glycol)-*b*-poly(2-(diisopropylamino)ethyl methacrylate) (PEG-*b*-PDPA) polymer. Reproduced with permission from Yang Li, Zhaohui Wang, Qi Wei, Min Luo, Gang Huang, Baran D. Sumer, and Jinming Gao. Biomaterials; published by Elsevier, 2016.

To prove the concept that hydrophobic chain length has a significant effect on the pKa value, Li et al. synthesized a series of poly(methacrylate)-poly(ethylene oxide) block copolymers with the fixed hydrophilic poly(ethylene oxide) chains but variable chain lengths of the hydrophobic poly(methacrylate) block (x = 5, 10, 20, 60, and 100), and evaluated the pKa value shift caused by the change of the hydrophobic chain length. Results indicated that the pKa values were inversely proportional to the length of the hydrophobic chain. The transition pH of block copolymers, with the longest hydrophobic segment, yielded the lowest pKa at 6.2 (Figure 7c). In contrast, polymers with the shortest hydrophobic chain length, displayed the highest transition pH around 6.7. The plot of pKa values as a function of the hydrophobic chain length showed a dramatic hydrophobic chain length-dependent transition pH shift.

Hydrophobic modification has also been demonstrated to be a practical strategy in tuning the degradation rate of polymeric systems containing acid-labile linkers. Ramakrishnan et al. investigated how the hydrophobic end-groups affected the degradation of hyperbranched polyacetals (Figure 8) [74]. They changed the structure of hyperbranched polymers by altering the pendant alkyl groups of monomers. Results indicated that the degradation rates of the acid-responsive polymers were significantly affected by the hydrophobic nature the terminal alkyl substituents, where more hydrophobic alkyl groups generally contributed to slower degradation and prolonged release of cargos.

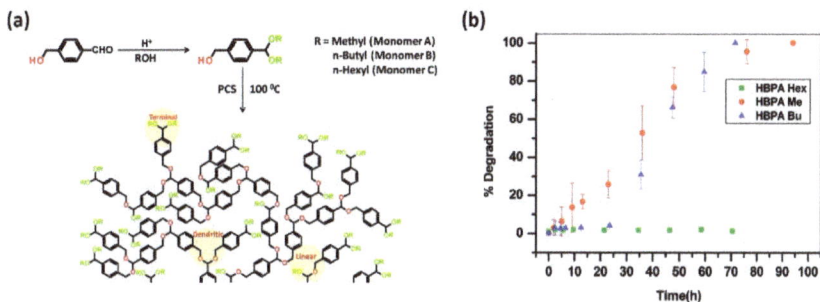

Figure 8. Hydrophobic modification in tuning the degradation rate of hyperbranched polyacetals. (a) Structure and synthetic route of hyperbranched polyacetals. Monomers with different hydrophobic substituents (methyl, butyl, and hexyl) were used in the preparation of three model branched polyacetals. (b) Branched polyacetal with least hydrophobic methyl groups showed fast degradation within 10 h, whereas polyacetal with most hydrophobic hexyl group showed almost no significant degradation after 4 days. Reproduced with permission from Saptarshi Chatterjee and S. Ramakrishnan. Macromolecules; published by American Chemical Society, 2011.

3.2. Copolymerization with Non-Ionizable Polymers

Stayton group reported the development of a series of pH-responsive block copolymers for the delivery of siRNA [75]. These polymers were composed of a positively-charged block of dimethylaminoethyl methacrylate (DMAEMA) to mediate siRNA condensation, and a second endosomal releasing block composed of DMAEMA and propylacrylic acid (PAA) in roughly equimolar ratios, together with butyl methacrylate (BMA). The polymers self-organized into micelles at physiological pH while rendered pH-induced disassembly upon exposure to acidic endosomal pH. The transition pH where reversible micellization occurred could be precisely tuned by systemically changing the fraction of non-ionizable hydrophobic segment (Figure 9), with increased BMA ratio exhibited lower pKa values [76]. Moreover, experimental results also showed that higher fraction of BMA also lead to enhanced homolytic activity, indicating that the transition pH was critical for therapeutic effect associated with pH-triggered drug release.

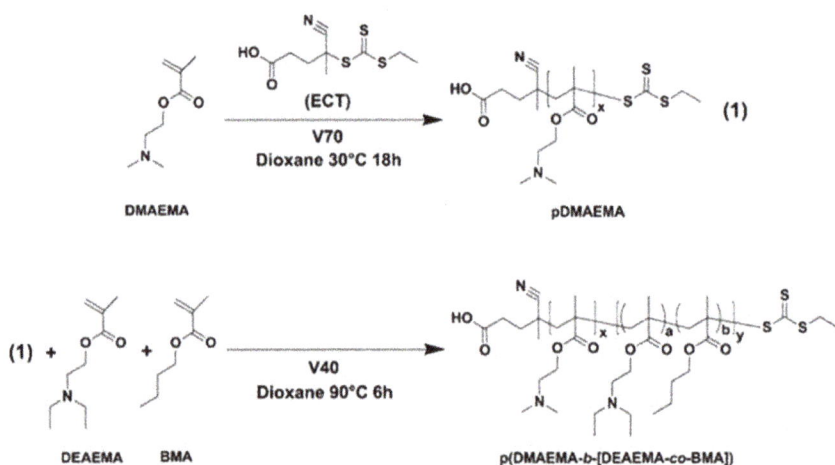

Figure 9. Preparation of a representative copolymer consisting of a cationic poly(DMAEMA) (dimethylaminoethyl methacrylate) block and an endosomolytic hydrophobic block incorporating DEAEMA and butyl methacrylate (BMA) at varying molar feed ratios. Reproduced with permission from Matthew J. Manganiello, Connie Cheng, Anthony J. Convertine, James D. Bryers, and Patrick S. Stayton. Biomaterials; published by Elsevier, 2012.

3.3. Copolymerization with Ionizable Polymers

Incorporation of additional polyelectrolytes that changes both hydrophobic interactions and electrostatic repulsion has also been reported for the modification of the transition pH of polymers. Gao et al. reported a strategy to design ultra-pH sensitive nanoparticles with broad tunability via a copolymerization method [35]. A series of PEO-*b*-P(R1-*r*-R2) block copolymers were prepared. The hydrophobic block P(R1-*r*-R2) contained two randomly distributed ionizable monomers R1 and R2 with different hydrophobicity. The molar fraction of the two monomers (R1 and R2) can be precisely controlled prior to polymerization. A series of ionizable monomers with different alkyl groups (e.g., ethyl, propyl, butyl, and pentyl) were applied in their study. The pKa values of PEO-*b*-PDPA (propyl) and PEO-*b*-PDBA (butyl) were 6.2 and 5.3, respectively. They were able to get a series of PEO-*b*-P(DPA-*r*-DBA) copolymers with pKa values between 6.2 and 5.3 successfully. A plot of pKa values of the obtained block copolymers as a function of the molar fraction of monomers with butyl group yielded a linear correlation. It is worth noting that matching of the hydrophobicity of the two monomers is critical to maintain the ultra-pH sensitivity. Based on this method, they established a library of nanoprobes with pH transitions cover the entire physiologic range of pH (4.0−7.4) (Figure 10). Compared to simple molecular mixture method, this copolymerization strategy achieves robust and broad tunability in transition pH.

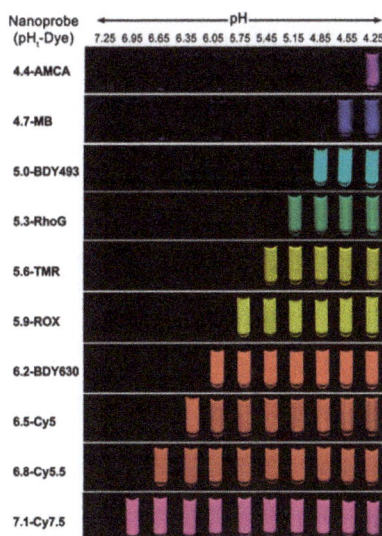

Figure 10. Ultra-pH sensitive (UPS) library based on a copolymerization strategy spanning a wide pH range. Polymers prepared from different monomers ionizable tertiary amines were encoded with different fluorophores. UPS polymers with more hydrophobic repeating unit showed lower transition pH as quantified by fluorescence intensity. Reproduced with permission from Xinpeng Ma, Yiguang Wang, Tian Zhao, Yang Li, Lee-Chun Su, Zhaohui Wang, Gang Huang, Baran D. Sumer, and Jinming Gao. Journal of the American Chemical Society; published by American Chemical Society, 2014.

3.4. Mechanistic Insights into the Tunable pH-Responsive Behavior

Compared to small molecules, polymerization and supramolecular self-assembly represent powerful strategies to produce high-performance materials at nanoscale [77–80]. Despite the promise in a multitude of biomedical applications, polymeric materials also introduce significantly increased complexity inherent to the multiple interacting components within the systems. The lack of molecular understanding of polymeric systems frequently hampers our ability to rationally design nanomaterials for medicine and healthcare.

In a recent study, Li and coworkers elucidated the molecular pathway of pH-triggered supramolecular self-assembly, which may pave the way for future design of pH-sensitive polymers with easily tunable pKa and pH transition sharpness. They found that hydrophobic phase separation was critical in tuning the pKa values of amphiphilic pH-responsive polymers (Figure 11) [73]. For pH responsive polymers that are not hydrophobic enough at neutral state to self-assemble into nanoparticles, the pH responsive behavior was very similar to that of small molecular pH sensors. The hydrophobic phase separation changes the molecular pathway of protonation in ionizable polymers that can form self-associated nanoparticles. For polyamines, the hydrophobic phase separation significantly shifted the pKa values to lower pH ranges. It is reasonable to hypothesize that the increased hydrophobicity in polyacid systems shifts the transition pH to more basic pH, which is actually validated by hyaluronic acid modified liposomes as described previously [51]. A molecular cooperativity was observed in pH-induced protonation and phase separation, as observed in thermo-responsive polymers and many natural macromolecular systems. This suggests that supramolecular cooperativity may serve as a general strategy in tuning the responsive behavior of stimuli-sensitive nanomaterials [73].

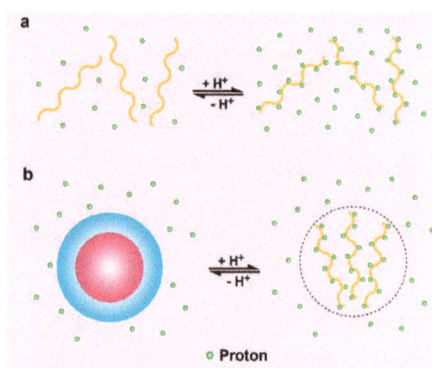

Figure 11. Schematic illustration of phase separation induced cooperativity in pH-triggered supramolecular self-assembly. (**a**) Hydrophilic polymers with ionizable amines were protonated homogeneously, showing no cooperativity. (**b**) Hydrophobic phase separation (e.g., micellization) drove cooperative protonation of amphiphilic pH-responsive polymers. Phase separation also drove the pKa values of polymers to lower pH because of hydrophobic barrier, which required higher proton concentration (lower pH) to initiate the protonation process. Reproduced with permission from Yang Li, Tian Zhao, Chensu Wang, Zhiqiang Lin, Gang Huang, Baran D. Sumer, and Jinming Gao. Nature Communications; published by Nature Publishing Group, 2016.

Thayumanavan and coworkers systematically investigated the substituent effects on the pH-sensitivity of acetals and ketals [59]. It is well established that the degradation of acetals and ketals operated through a resonance-stabilized caboxonium ion intermediate [81]. Their study validated that electron-withdrawing substituents generally decreased the hydrolysis rate due to the cationic nature of the intermediate. Benzylidene acetals that were unsubstituted or substituted with electron-withdrawing groups, exhibited drastically slowed hydrolysis under the pH 5 conditions (Figure 12). For examples, half of phenyl-substituted acetal groups were hydrolyzed after 4 min. Incorporating an electron-withdrawing trifluoromethyl moiety increased the half-life to around 12.3 h. The presence of a resonance-based electron-withdrawing functionality further increased the half-life to 51.8 h. The structural fine-tuning of the linkers offers powerful and practical method in future design of acid-degradable polymeric drug delivery systems.

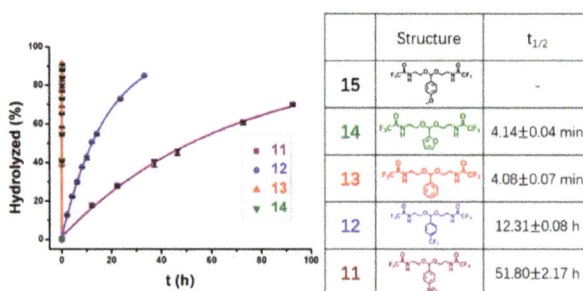

	Structure	$t_{1/2}$
15		-
14		4.14±0.04 min
13		4.08±0.07 min
12		12.31±0.08 h
11		51.80±2.17 h

Figure 12. Substituent effects upon the hydrolysis of benzylidene acetals. Electron-withdrawing substituents significantly contributed to decreased hydrolysis rate of acetals. As compared to phenyl substituent with half-life of 4 min, trifluoromethyl increased the half-life of acetal to around 12.3 h. The presence of a resonance-based electron-withdrawing functionality further increased the half-life to 51.8 h. Reproduced with permission from Bin Liu and S. Thayumanavan, Journal of the American Chemical Society; published by American Chemical Society, 2017.

4. Applications of pH-Responsive Nanomaterials in Cancer Diagnosis and Treatment

Cancer remains one of the leading causes of death around the world. Nanotechnology has marked a new dimension in fighting against cancer. pH-responsive nanomaterials that target tumor acidosis offer a new paradigm in addressing the deficiencies of conventional chemotherapy in the diagnosis and treatment of various types of cancer.

4.1. pH-Sensitive Nanoprobe-Based Fluorescent Imaging and Image-Guided Surgery

Surgery is one of the main options for cancer treatment. When resecting a tumor, a surgeon needs to precisely identify the spread of the cancer. If tumor is not completely removed, there is a great chance of recurrence of cancer. The patients may suffer from crucial organ dysfunction or damage if normal tissues are cut. Thus, intraoperative technologies that can help surgeons visualize the tumor margins during the operations may significantly improve the long-term survival of cancer patients. There is an urgent clinical need for the cheap and practical technology that can universally differentiate tumors from adjacent normal tissues, which could potentially help the surgeons save numerous lives [82–84].

Wang et al. reported a pH-responsive nanoparticle-based strategy for the imaging of a broad range of tumors by nonlinear amplification of tumor acidosis signals [85]. As shown in Figure 13, the ionizable amines were neutralized and the amphiphilic block copolymers stayed at micelles state with conjugated fluorescent dyes quenched due a self-quenching Forster resonance energy transfer (HomoFRET) effect. Upon access to acidic tumor microenvironment or internalized into endocytic organelles in the tumor endothelial cells. This system demonstrated a broad tumor specificity with an extraordinary tumor-to-blood ratio (>300-fold) in a variety of tumor models. Following the exceptional imaging outcome, they continued the study in image-guided surgery with several clinical-compatible fluorescent cameras [86]. The real-time tumor-acidosis-guided detection and resection of tumors significantly improved the long-term survival of tumor-bearing mice.

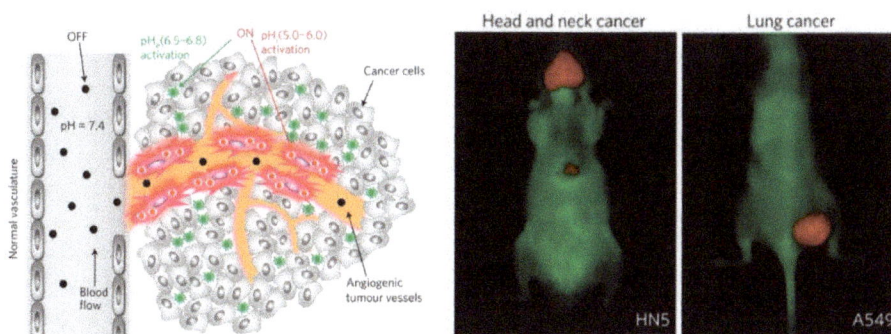

Figure 13. pH-activatable nanoparticles for tumor-specific fluorescent imaging. (Left panel) The nanoparticles stayed in micelle state in normal physiological pH and the fluorescence was quenched as a result of self-quenching Forster resonance energy transfer (HomoFRET) effect. Upon exposure to acidic tumor microenvironment, protonation of tertiary amines lead to the dissociation of micelles and resume of fluorescence. (Right panel) pH-responsive nanoparticles selectively light up tumors instead of surrounding normal tissues in various tumor models. Reproduced with permission from Yiguang Wang, Kejin Zhou, Gang Huang, Christopher Hensley, Xiaonan Huang, Xinpeng Ma, Tian Zhao, Baran D. Sumer, Ralph J. DeBerardinis, and Jinming Gao. Nature Materials; published by Nature Publishing Group, 2014.

4.2. pH-Sensitive Nanoprobe-Based Magnetic Resonance Imaging

Magnetic resonance imaging (MRI) with a contrast agent has been widely used as a powerful tool for cancer diagnosis [87]. However, most MRI contrast agents are non-targeted for cancer and passively distributed throughout the body, which results in a low efficiency and a need for high doses.

One approach to addressing these problems is to use pH-responsive amphiphilic block copolymers that would work as platforms for the tumor-targeting delivery of MRI contrast agents and MRI signal enhancement in the tumor region.

Kun Na and coworkers developed a cancer-recognizable MRI contrast agents (CR-CAs) using pH-responsive polymeric micelles [88]. The micelles were self-assembled from copolymers of methoxy poly(ethylene glycol)-*b*-poly(L-histidine) (PEG–p(L-His)) and methoxy poly(ethylene glycol)-*b*-poly(L-lactic acid)–diethylenetriaminopentaacetic acid dianhydride–gadolinium chelate (PEG–p(L-LA)–DTPA-Gd) (Figure 14a–c). In the micelles, p(L-His) blocks were the pH-responsive component. The imidazole groups of p(L-His) blocks were protonated, causing the broken of the micellar structure in acidic tumoral environment. Paramagnetic gadolinium (Gd^{3+}) chelates were the MRI contrast agents that enhance the signal intensity upon the exposure to water molecules as the micelles broken. In addition, the CR-CAs' core was positively charged by protonation of the imidazole groups of p(L-His) blocks after extravasation, which rapidly facilitated accumulation of CR-CAs compared with pH-insensitive micelle-based CAs (Ins-CAs) due to the strengthened interaction between the positively charge CR-CAs and the negatively charged cellular membrane. In vivo, the CR-CAs exhibit highly effective MR contrast enhancement in the CT26 murine tumor region of Balb/c mice, while the MR contrast of the tumor treated with Ins-CAs did not show a significant change over time. CR-CAs enabled the detection of small tumors (3 mm^3) in vivo within a few minutes (Figure 14d).

Lee et al. encapsulated Fe_3O_4 nanoparticles, which are frequently used as a contrast agent for MRI, into the pH-responsive polymeric micelle [89]. This technique has been tested on mice and shown an increased signal intensity over a 24-h period compared to a pH-insensitive contrast agent which did not change in signal intensity.

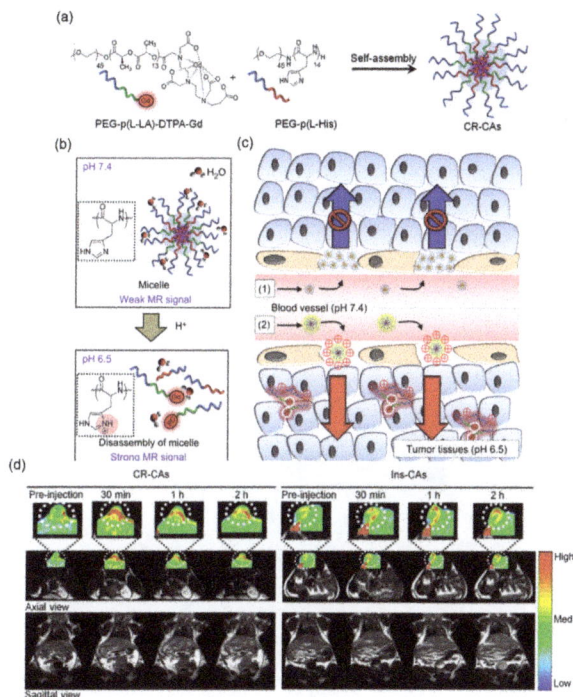

Figure 14. A cancer-recognizable MRI contrast agents (CR-CAs) based on pH-responsive polymeric micelles. (**a**) Schematic representation of the preparation of cancer-recognizable CR-CAs. (**b**) Schematic representation of the pH-dependent structural transformation and related MR signal change in CR-CAs. Inset: Chemical structural representation of the protonation of imidazole groups in PEG-p(L-His) at acidic pH. (**c**) Schematic representation of the tumor-accumulation behavior of (1) conventional micelle-based CAs and (2) CR-CAs. (**d**) Temporal color-coded in vivo longitudinal relaxation time (T1)-weighted MR images of CT26 murine tumor bearing Balb/c mice after the intravenous injection of CR-CAs and pH-insensitive micelle-based CAs (Ins-CAs). Reproduced with permission from Kyoung Sub Kim, Wooram Park, Jun Hu, You Han Bae, and Kun Na. Biomaterials; published by Elsevier, 2014.

4.3. pH-Responsive Polymeric siRNA Carriers for Cancer Treatment

The use of RNA interference (RNAi) as a tumor-specific gene therapy has attracted increasing attention, and is considered one of the most promising platforms for cancer therapy [90]. However, naked siRNA is unstable, and the effective intracellular delivery of siRNA into the cytoplasm remains a significant challenge. After cellular uptake by passive or receptor-mediated endocytosis, siRNA predominantly locates in endosomes and is degraded by specific enzymes in the lysosome. Thus, to achieve an effective treatment efficiency, escape of siRNA from endosomes to reach cytoplasm is desired. pH-responsive polymeric carriers that facilitate the endosomal escape have been demonstrated one effective approach to mediate intracellular siRNA delivery.

Shi et al. developed several pH-responsive nanoparticle (NP) platforms, captaining Poly(2-(diisopropylamino)ethyl methacrylate) (PDPA) components, for cancer-specific in vivo siRNA delivery [91,92]. PDPA is a type of polycations containing low pKa amines. Low pKa amine group have been shown to exhibit "proton sponge effect" to induce the endosomal escape [93]. Polycations induce the endosomal escape by binding to the oppositely charged cellular membrane and perturbing membrane integrity. In addition, to synergize the endosomal escape, cationic membrane-penetrating oligoarginine grafts or cationic lipid-like grafts were also incorporated into the nanoparticles. After cellular uptake, the rapid protonation of the PDPA segment causes the fast disassembly of the nanoparticles, endosomal swelling, and the exposed membrane-penetrating oligoarginine grafts or cationic lipid-like grafts lead to efficient endosomal escape (Figure 15a). In vivo, the siRNA NPs showed efficient gene silencing and significant inhibition of tumor growth (Figure 15b–d).

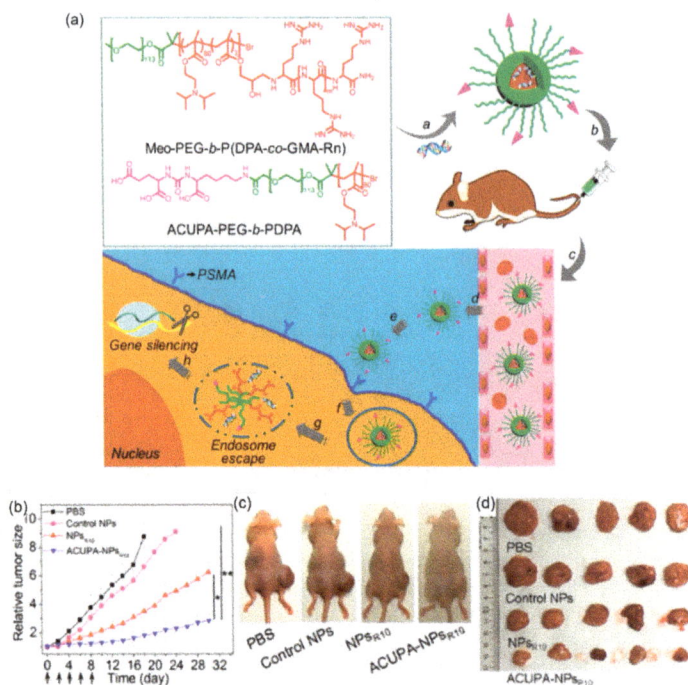

Figure 15. A pH-responsive nanoparticle (NP) platform for cancer-specific in vivo siRNA delivery. (**a**) Molecular structures of the oligoarginine functionalized sharp pH-responsive polymer methoxyl-polyethylene glycol-b-poly(2-(diisopropylamino) ethyl methacrylate-co-glycidyl methacrylate) (Meo-PEG-b-P(DPA-co-GMA-Rn, n = 6, 8, 10, 20, and 30)) and S,S-2-[3–[5–amino-1-carboxypentyl]-ureido]pentanedioic acid functionalized poly(ethylene glycol)-b-poly(2-(diisopropylamino)ethyl methacrylate) (ACUPA-PEG-b-PDPA) and a schematic illustration of the multifunctional envelope-type NP platform for in vivo siRNA delivery and therapy. (**b**) NPs prepared from Meo-PEG-b-P(DPA-co-GMA-R10) was denoted as NPs_{R10}. Relative tumor size of the Luc-HeLa and PCa cell lines (LNCaP) xenograft tumor-bearing nude mice ($n = 5$) after treatment by Luc siRNA-loaded NPs_{R10} (control NPs) and PHB1 siRNA-loaded NPs_{R10} (NPs_{R10}) and siRNA-loaded ACUPA-NPs_{R10} (ACUPA-NPs_{R10}). * $p < 0.05$; ** $p < 0.01$. (**c**) Representative photograph of the LNCaP xenograft tumor-bearing nude mice in each group at day 18. (**d**) Photograph of the harvested LNCaP tumors after a 30-day evaluation. Reproduced with permission from Xiaoding Xu, Jun Wu, Yanlan Liu, Phei Er Saw, Wei Tao, Mikyung Yu, Harshal Zope, Michelle Si, Amanda Victorious, Jonathan Rasmussen, Dana Ayyash, Omid C. Farokhzad, and Jinjun Shi. ACS Nano; published by American Chemical Society, 2017.

Stayton et al. synthesized a diblock polymer composed of PDMAEMA block to condense siRNA and a second block composed of DMAEMA, BMA, and PAA for endosomal-releasing [75].

PAA induced the endosomal escape as a pH responsive membrane-destabilizing polymer, which respond to changes in pH by transitioning from an ionized, hydrophilic structure at physiologic pH (~7.4) to a hydrophobic, membrane-destabilizing conformation at endosomal pH values (<6.6). The diblock copolymers condensed siRNA into 80–250 nm particles. In HeLa cells, the siRNA-mediated knockdown of glyceraldehyde 3-phosphate dehydrogenase (GAPDH) increased as the percentage of BMA in the second block increased.

4.4. pH-Responsive Polymeric Anti-Cancer Drug Carriers for Cancer Treatment

Among the various therapeutic strategies for cancer, such as surgery, chemotherapy, radiotherapy, and gene therapy discussed above, chemotherapy is the most often used method in clinical practice [94,95]. However, the side-effects, low therapeutic efficacy and cytotoxicity of the traditional chemical drugs such as Dox, camptothecin and PTX, have hindered the treatment efficiency of cancer chemotherapy [96]. Polymeric nanoparticles have been developed and used as approaches to remove the problems because they can improve pharmacokinetics and biodistribution profiles of anti-cancer drugs via the EPR effect [97]. However, the EPR impact can only improve the accumulation of NPs in tumor tissues, the efficiency of cancer chemotherapy has always still been hindered by the insufficient drug release that induced the concentration of anticancer drugs to the level below the therapeutic window [98]. pH-responsive polymeric drug carriers, which enhance the triggered release of anti-cancer drugs by responding to the tumour acidic microenvironment, have been demonstrated a pathway to address this problem.

For tumor environment triggered drug release, anti-cancer drugs could be ether physically encapsulated into pH-responsive nanoparticles or conjugated to polymer through acid-liable bonds. Kim group physically encapsulated Dox into a variety of pHis-based polymeric micelles for the CT26 tumor treatment. In vitro, the destabilized pH-responsive pHis core enhanced the triggered release of Dox into the cancer cell. In vivo, the Dox-loaded micelles showed a higher CT26 tumour suppression than free Dox did [99]. Etrych et al. conjugated Dox onto an amphiphilic *N*-(2-hydroxypropyl) methacrylamide (HPMA)-based polymer by hydrazone bonds [100]. HPMA copolymer−Dox conjugates were stable in a buffer at pH 7.4, Dox was released in a mild acidic conditions of the tumor microenvironment. In vivo, HPMA copolymer−Dox conjugates significantly reduced the toxic side effects of Dox and enhanced the anti-tumor efficacy.

Liu group reported a drug delivery system with the combination of physical encapsulation and covalent conjugation of anti-cancer drugs for cancer treatment [101]. Dox was conjugated to PEG by Schiff's base reaction. PEG-Dox prodrug formed stable nanoparticles (PEG-Dox NPs) in water at physiological pH, and encapsulated curcumin (Cur) into the core through hydrophobic interaction (Figure 16a). When the formed nanoparticles, denoted as PEG-Dox-Cur NPs, are internalized by tumor cells, the Schiff's base linker between PEG and Dox would break in the acidic environment that is often observed in tumors, causing disassembling of the PEG-Dox-Cur NPs and releasing both Dox and Cur into the nuclei and cytoplasma of the tumor cells, respectively. The PEG-Dox-Cur NPs demonstrated enhanced anti-tumor activity in animals compared with free Dox-Cur combination (Figure 16b).

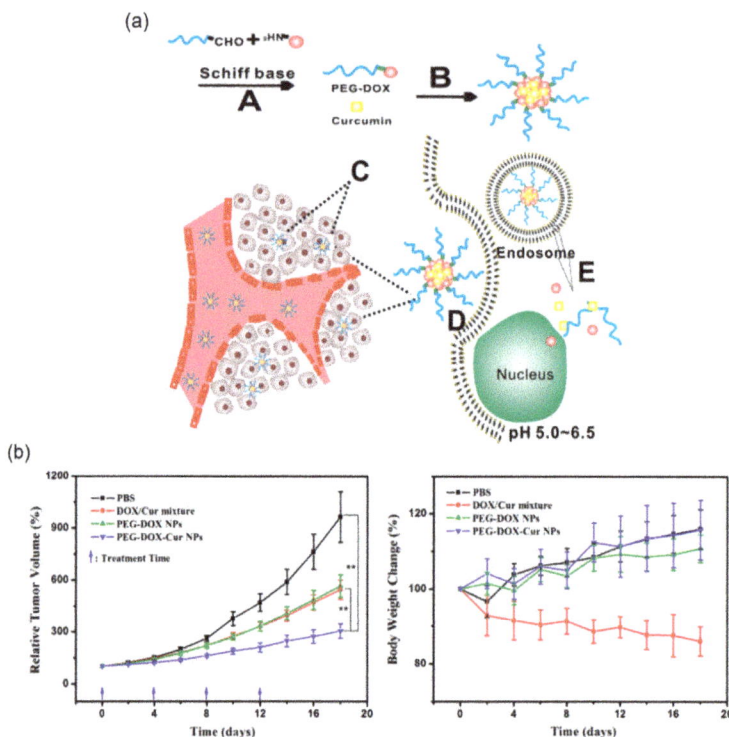

Figure 16. pH-responsive nanoparticles with the combination of physical encapsulation and covalent conjugation of anti-cancer drugs for cancer treatment. (**a**) Schematic illustration of the synthesis and working principle of PEG-Dox-Cur NPs. **A** Synthesis of PEG-Dox NPs via Schiff's base. **B** Preparation of PEG-Dox-Cur NPs by nanopreciptated technique. **C** Passive tumor targeting was achieved by EPR effect. **D** The PEG-Dox-Cur NPs could be internalized by cancer cells through endocytosis. **E** Dox and Cur were released with the cleavage of the Schiff's base in tumor cells and diffused into nucleus. (**b**) Relative tumor volume and bodyweight change in Balb/C nude mice after treatment by PBS, free DOX/Cur mixture, PEG-Dox NPs and PEG-Dox-Cur NPs. * $p < 0.05$; ** $p < 0.01$. Reproduced with permission from Yumin Zhang, Cuihong Yang, Weiwei Wang Jinjian Liu, Qiang Liu, Fan Huang, Liping Chu, Honglin Gao, Chen Li, Deling Kong, Qian Liu, and Jianfeng Liu. Scientific Reports; published by Nature Publishing Group, 2016.

4.5. Challenges and Opportunities of Translating pH-Responsive Nanomaterials

Despite the wide range of success evidences from pre-clinical studies, there remains significant challenges for pH-responsive polymeric nanomaterials to enter clinical path. In addition to the specific cases aforementioned, several general aspects should be taken into account. First of all, the formulation of the polymeric nanomaterial applications is much more complex than conventional technology, such as tablets and injections; this brings difficulties in achieving Good Manufacturing Practice (GMP) standards in large-scale production and quality control. Secondly, developing and standardizing biological assays for human toxicity evaluation and monitoring in clinical trials has to adopt the potential novel biology coming from these materials. Besides toxicity, assays are also needed for clinically relevant biomarker identification and validation. In addition, the pharmacokinetics and pharmacodynamics (PK/PD) study is obligated to grasp a deep understanding of the physiochemical properties of polymeric theranostics, adding layers of complexity to high-throughput data acquisition and modeling. From a regulatory perspective, the clear guidelines for the nanomaterials to be developed facing forward translation still requires collaborative efforts.

Promising efforts have been invested towards these challenges. The development of reproducible and stable polymeric nanomaterial batch synthesis protocol can potentially benefit from advanced nanoparticle preparation technology [102,103]. Human organs-on-a-chip platform, mimicking key functions in a human body, offers a great opportunity to address the limited predictive value of conventional pre-clinical cell and animal models, which may be readily applied to polymeric nanomaterials for safety and efficacy evaluations [104].

5. Summary and Future Perspective

Various physiological abnormalities, including pH, reactive oxygen species, overexpressed proteins or enzymes, associated with cancer offer great opportunities for the design of stimuli-responsive polymeric materials for tumor-targeted delivery of diagnostic and therapeutic agents. There has been rapid growing interest in the design of pH-responsive polymeric materials to target tumor acidosis, a ubiquitous characteristic shared by almost all types of cancers.

Despite obvious promise in pH-responsive nanomaterials, several challenges may warrant further exploration. First, design of pH-sensitive polymeric materials that are safe for medical and pharmaceutical applications remains a major challenge. It is still highly desirable to explore biodegradable pH-responsive polymers with improved biocompatibility and reduced toxicity. Second, additional efforts should also be directed toward improved tumor targeting efficacy. pH-sensitive polymeric materials that can specifically deliver payload to tumor's microenvironment with minimal accumulation in normal tissues or endosomal organelles of tumor cells can drastically improve the therapeutic efficacy and decrease the side effects. However, the pH variation between tumor and surrounding normal tissues is not very significant, and different types of tumors may have slightly different level of acidosis. To meet the requirements of various applications that target tumor microenvironment and intracellular organelles pH-responsive polymers with m tunable pKa values are required. Incorporation of multiple modalities into the same polymer or nanoparticle has also shown promise in enhancing tumor targeting efficiency. Intrinsic heterogeneity of tumors may require the design of polymeric nanomaterial system with multiple transition pH values. Moreover, lack of comprehensive mechanistic understanding of pH-triggered responsive behaviors, especially interactions between nanomaterials and in vivo host environment, still hampers our capability in rational design of more effective pH-responsive nanomaterials for cancer treatment. While continuous development of pH-responsive polymers with improved therapeutic efficacy is of great importance, the modification and optimization of existing polymeric materials with validated biocompatibility may represent a more straightforward and efficient strategy.

It's worth noting that nanomedicine is an interdisciplinary field that requires close collaboration between physicists, chemists, biologists, and physicians. Many reviews' proof-of-concept studies need a substantial amount of work before potential successful clinical translation. Nevertheless, recent advances in the polymer design, structure-property correlation, mechanistic understanding, and applications of a multitude of pH-sensitive nanomaterials reviewed here provide general guidelines for future rational design of more effective pH-responsive nanomaterials for cancer diagnosis and treatment.

Author Contributions: Conceptualization: Y.L., H.T., W.Z., and C.Z.; literature research: Y.L., H.T., W.Z., J.Y., and C.Z.; writing (original draft preparation): Y.L., H.T., and C.Z.; writing (review and editing): Y.L., H.T., and C.Z.; supervision: Y.L., H.T., and C.Z.

Funding: This research received no external funding.

Acknowledgments: C.Z. acknowledges the start-up support from the University of Alabama.

Conflicts of Interest: W.Z. is employed by Merck & Co., Inc. The authors declare no conflict of interest.

References

1. Kim, B.Y.S.; Rutka, J.T.; Chan, W.C.W. Current Concepts: Nanomedicine. *N. Engl. J. Med.* **2010**, *363*, 2434–2443. [CrossRef] [PubMed]
2. Shi, J.J.; Kantoff, P.W.; Wooster, R.; Farokhzad, O.C. Cancer nanomedicine: Progress, challenges and opportunities. *Nat. Rev. Cancer* **2017**, *17*, 20–37. [CrossRef] [PubMed]
3. Bozic, I.; Reiter, J.G.; Allen, B.; Antal, T.; Chatterjee, K.; Shah, P.; Moon, Y.S.; Yaqubie, A.; Kelly, N.; Le, D.T. Evolutionary dynamics of cancer in response to targeted combination therapy. *Elife* **2013**, *2*, e00747. [CrossRef] [PubMed]
4. Stratton, M.R.; Campbell, P.J.; Futreal, P.A. The cancer genome. *Nature* **2009**, *458*, 719–724. [CrossRef] [PubMed]
5. Carmeliet, P.; Jain, R.K. Angiogenesis in cancer and other diseases. *Nature* **2000**, *407*, 249–257. [CrossRef] [PubMed]
6. Neri, D.; Supuran, C.T. Interfering with pH regulation in tumours as a therapeutic strategy. *Nat. Rev. Drug Discov.* **2011**, *10*, 767–777. [CrossRef] [PubMed]
7. Corbet, C.; Feron, O. Tumour acidosis: From the passenger to the driver's seat. *Nat. Rev. Cancer* **2017**, *17*, 577–593. [CrossRef]
8. Langer, R.; Tirrell, D.A. Designing materials for biology and medicine. *Nature* **2004**, *428*, 487–492. [CrossRef]
9. Tang, H.; Tsarevsky, N.V. Preparation and functionalization of linear and reductively degradable highly branched cyanoacrylate-based polymers. *J. Polym. Sci. Pol. Chem.* **2016**, *54*, 3683–3693. [CrossRef]
10. Zhao, J.; Wang, W.; Tang, H.; Ramella, D.; Luan, Y. Modification of Cu^{2+} into Zr-based metal–organic framework (MOF) with carboxylic units as an efficient heterogeneous catalyst for aerobic epoxidation of olefins. *Mol. Catal.* **2018**, *456*, 57–64. [CrossRef]
11. Du, X.; Li, X.; Tang, H.; Wang, W.; Ramella, D.; Luan, Y. A facile 2H-chromene dimerization through an ortho-quinone methide intermediate catalyzed by a sulfonyl derived MIL-101 MOF. *New J. Chem.* **2018**, *42*, 12722–12728. [CrossRef]
12. Freitas, R.A.J. Nanotechnology, nanomedicine and nanosurgery. *Int. J. Surg.* **2005**, *3*, 243–246. [CrossRef] [PubMed]
13. Yang, L.; Sun, H.; Liu, Y.; Hou, W.; Yang, Y.; Cai, R.; Cui, C.; Zhang, P.; Pan, X.; Li, X.; et al. Self-assembled aptamer-hyperbranched polymer nanocarrier for targeted and photoresponsive drug delivery. *Angew. Chem. Int. Edit.* **2018**. [CrossRef]
14. Zhuang, J.; Gordon, M.R.; Ventura, J.; Li, L.; Thayumanavan, S. Multi-stimuli responsive macromolecules and their assemblies. *Chem. Soc. Rev.* **2013**, *42*, 7421–7435. [CrossRef] [PubMed]
15. Etheridge, M.L.; Campbell, S.A.; Erdman, A.G.; Haynes, C.L.; Wolf, S.M.; McCullough, J. The big picture on nanomedicine: The state of investigational and approved nanomedicine products. *Nanomed. Nanotechnol. Biol. Med.* **2013**, *9*, 1–14. [CrossRef] [PubMed]
16. Peer, D.; Karp, J.M.; Hong, S.; Farokhzad, O.C.; Margalit, R.; Langer, R. Nanocarriers as an emerging platform for cancer therapy. *Nat. Nanotechnol.* **2007**, *2*, 751–760. [CrossRef]
17. Conde, J.; Oliva, N.; Artzi, N. Revisiting the 'One Material Fits All' Rule for Cancer Nanotherapy. *Trends Biotechnol.* **2016**, *34*, 618–626. [CrossRef]
18. Keereweer, S.; Kerrebijn, J.D.F.; van Driel, P.; Xie, B.W.; Kaijzel, E.L.; Snoeks, T.J.A.; Que, I.; Hutteman, M.; van der Vorst, J.R.; Mieog, J.S.D. Optical Image-guided Surgery-Where Do We Stand? *Mol. Imaging. Biol.* **2011**, *13*, 199–207. [CrossRef]
19. Sawyers, C.J.N. Targeted cancer therapy. *Nature* **2004**, *432*, 294–297. [CrossRef]
20. Schmaljohann, D. Thermo-and pH-responsive polymers in drug delivery. *Adv. Drug Deliv. Rev.* **2006**, *58*, 1655–1670. [CrossRef]
21. Mura, S.; Nicolas, J.; Couvreur, P. Stimuli-responsive nanocarriers for drug delivery. *Nat. Mater.* **2013**, *12*, 991–1003. [CrossRef] [PubMed]
22. Gu, L.; Mooney, D.J. Biomaterials and emerging anticancer therapeutics: Engineering the microenvironment. *Nat. Rev. Cancer* **2016**, *16*, 56–66. [CrossRef] [PubMed]
23. Alfurhood, J.A.; Sun, H.; Kabb, C.P.; Tucker, B.S.; Matthews, J.H.; Luesch, H.; Sumerlin, B.S. Poly(*N*-(2-Hydroxypropyl) Methacrylamide)-Valproic Acid Conjugates as Block Copolymer Nanocarriers. *Polym. Chem.* **2017**, *8*, 4983–4987. [CrossRef] [PubMed]

24. Bae, Y.; Fukushima, S.; Harada, A.; Kataoka, K.J.A.C. Design of environment-sensitive supramolecular assemblies for intracellular drug delivery: Polymeric micelles that are responsive to intracellular pH change. *Angew. Chem.-Int. Edit.* **2003**, *115*, 4788–4791. [CrossRef]

25. Dai, S.; Ravi, P.; Tam, K.C. pH-Responsive polymers: Synthesis, properties and applications. *Soft Matter* **2008**, *4*, 435–449. [CrossRef]

26. Tang, H.; Luan, Y.; Yang, L.; Sun, H. A Perspective on Reversibility in Controlled Polymerization Systems: Recent Progress and New Opportunities. *Molecules* **2018**, *23*, 2870. [CrossRef]

27. Sun, H.; Kabb, C.P.; Dai, Y.; Hill, M.R.; Ghiviriga, I.; Bapat, A.P.; Sumerlin, B.S. Macromolecular metamorphosis via stimulus-induced transformations of polymer architecture. *Nat. Chem.* **2017**, *9*, 817–823. [CrossRef]

28. Sun, H.; Kabb, C.P.; Sumerlin, B.S. Thermally-labile segmented hyperbranched copolymers: Using reversible-covalent chemistry to investigate the mechanism of self-condensing vinyl copolymerization. *Chem. Sci.* **2014**, *5*, 4646–4655. [CrossRef]

29. Sun, H.; Kabb, C.P.; Sims, M.B.; Sumerlin, B.S. Architecture-transformable polymers: Reshaping the future of stimuli-responsive polymers. *Prog. Polym. Sci.* **2018**, in press.

30. Kocak, G.; Tuncer, C.; Bütün, V. pH-Responsive polymers. *Polym. Chem.* **2017**, *8*, 144–176. [CrossRef]

31. Bazban-Shotorbani, S.; Hasani-Sadrabadi, M.M.; Karkhaneh, A.; Serpooshan, V.; Jacob, K.I.; Moshaverinia, A.; Mahmoudi, M. Revisiting structure-property relationship of pH-responsive polymers for drug delivery applications. *J. Control. Release.* **2017**, *253*, 46–63. [CrossRef]

32. Ranneh, A.H.; Takemoto, H.; Sakuma, S.; Awaad, A.; Nomoto, T.; Mochida, Y.; Matsui, M.; Tomoda, K.; Naito, M.; Nishiyama, N. An Ethylenediamine-based Switch to Render the Polyzwitterion Cationic at Tumorous pH for Effective Tumor Accumulation of Coated Nanomaterials. *Angew. Chem. Int. Ed.* **2018**, *57*, 5057–5061. [CrossRef] [PubMed]

33. Mizuhara, T.; Saha, K.; Moyano, D.F.; Kim, C.S.; Yan, B.; Kim, Y.K.; Rotello, V.M. Acylsulfonamide-Functionalized Zwitterionic Gold Nanoparticles for Enhanced Cellular Uptake at Tumor pH. *Angew. Chem. Int. Ed.* **2015**, *54*, 6567–6570. [CrossRef] [PubMed]

34. Li, Y.; Wang, Z.; Wei, Q.; Luo, M.; Huang, G.; Sumer, B.D.; Gao, J. Non-covalent interactions in controlling pH-responsive behaviors of self-assembled nanosystems. *Polym. Chem.* **2016**, *7*, 5949–5956. [CrossRef] [PubMed]

35. Ma, X.; Wang, Y.; Zhao, T.; Li, Y.; Su, L.-C.; Wang, Z.; Huang, G.; Sumer, B.D.; Gao, J. Ultra-pH-Sensitive Nanoprobe Library with Broad pH Tunability and Fluorescence Emissions. *J. Am. Chem. Soc.* **2014**, *136*, 11085–11092. [CrossRef] [PubMed]

36. Li, Y.; Wang, Y.; Huang, G.; Gao, J. Cooperativity Principles in Self-Assembled Nanomedicine. *Chem. Rev.* **2018**, *118*, 5359–5391. [CrossRef]

37. Li, Y.; Wang, Y.; Huang, G.; Ma, X.; Zhou, K.; Gao, J. Chaotropic-Anion-Induced Supramolecular Self-Assembly of Ionic Polymeric Micelles. *Angew. Chem. Int. Edit.* **2014**, *53*, 8074–8078. [CrossRef]

38. Wang, C.; Zhao, T.; Li, Y.; Huang, G.; White, M.A.; Gao, J. Investigation of endosome and lysosome biology by ultra pH-sensitive nanoprobes. *Adv. Drug Deliv. Rev.* **2017**, *113*, 87–96. [CrossRef]

39. Fu, L.; Yuan, P.; Ruan, Z.; Liu, L.; Li, T.; Yan, L. Ultra-pH-sensitive polypeptide micelles with large fluorescence off/on ratio in near infrared range. *Polym. Chem.* **2017**, *8*, 1028–1038. [CrossRef]

40. Kolb, H.C.; Finn, M.G.; Sharpless, K.B. Click Chemistry: Diverse Chemical Function from a Few Good Reactions. *Angew. Chem. Int. Edit.* **2001**, *40*, 2004–2021. [CrossRef]

41. Weerakkody, D.; Moshnikova, A.; Thakur, M.S.; Moshnikova, V.; Daniels, J.; Engelman, D.M.; Andreev, O.A.; Reshetnyak, Y.K. Family of pH (low) insertion peptides for tumor targeting. *P. Natl. Acad. Sci. USA.* **2013**, *110*, 5834–5839. [CrossRef]

42. Segala, J.; Engelman, D.M.; Reshetnyak, Y.K.; Andreev, O.A. Accurate analysis of tumor margins using a fluorescent pH low insertion peptide (pHLIP). *Int. J. Mol. Sci.* **2009**, *10*, 3478–3487. [CrossRef] [PubMed]

43. Barrera, F.N.; Fendos, J.; Engelman, D.M. Membrane physical properties influence transmembrane helix formation. *P. Natl. Acad. Sci. USA.* **2012**, *109*, 14422–14427. [CrossRef] [PubMed]

44. Nguyen, V.P.; Alves, D.S.; Scott, H.L.; Davis, F.L.; Barrera, F.N. A novel soluble peptide with pH-responsive membrane insertion. *Biochemistry.* **2015**, *54*, 6567–6575. [CrossRef] [PubMed]

45. Emmetiere, F.; Irwin, C.; Viola-Villegas, N.T.; Longo, V.; Cheal, S.M.; Zanzonico, P.; Pillarsetty, N.; Weber, W.A.; Lewis, J.S.; Reiner, T. 18F-labeled-bioorthogonal liposomes for in vivo targeting. *Bioconjugate Chem.* **2013**, *24*, 1784–1789. [CrossRef] [PubMed]

46. Wyatt, L.C.; Lewis, J.S.; Andreev, O.A.; Reshetnyak, Y.K.; Engelman, D.M. Applications of pHLIP technology for cancer imaging and therapy. *Trends Biotechnol.* **2017**, *35*, 653–664. [CrossRef] [PubMed]

47. Hunt, J.F.; Rath, P.; Rothschild, K.J.; Engelman, D.M. Spontaneous, pH-dependent membrane insertion of a transbilayer α-helix. *Biochemistry* **1997**, *36*, 15177–15192. [CrossRef] [PubMed]

48. Reshetnyak, Y.K.; Andreev, O.A.; Lehnert, U.; Engelman, D.M. Translocation of molecules into cells by pH-dependent insertion of a transmembrane helix. *P. Natl. Acad. Sci. USA.* **2006**, *103*, 6460–6465. [CrossRef]

49. Musial-Siwek, M.; Karabadzhak, A.; Andreev, O.A.; Reshetnyak, Y.K.; Engelman, D.M. Tuning the insertion properties of pHLIP. *BBA-Biomembranes.* **2010**, *1798*, 1041–1046. [CrossRef]

50. Platt, V.M.; Szoka Jr, F. C Anticancer therapeutics: Targeting macromolecules and nanocarriers to hyaluronan or CD44, a hyaluronan receptor. *Mol. Pharm.* **2008**, *5*, 474–486. [CrossRef]

51. Miyazaki, M.; Yuba, E.; Hayashi, H.; Harada, A.; Kono, K. Hyaluronic acid-based pH-sensitive polymer-modified liposomes for cell-specific intracellular drug delivery systems. *Bioconjugate Chem.* **2017**, *29*, 44–55. [CrossRef]

52. Wei, H.; Zhuo, R.; Zhang, X. Design and development of polymeric micelles with cleavable links for intracellular drug delivery. *Prog. Polym. Sci.* **2013**, *38*, 503–535. [CrossRef]

53. Zhang, X.; Malhotra, S.; Molina, M.; Haag, R. Micro- and nanogels with labile crosslinks – from synthesis to biomedical applications. *Chem. Soc. Rev.* **2015**, *44*, 1948–1973. [CrossRef] [PubMed]

54. Tang, H.; Tsarevsky, N.V. Lipoates as building blocks of sulfur-containing branched macromolecules. *Polym. Chem.* **2015**, *6*, 6936–6945. [CrossRef]

55. Suvarapu, L.N.; Seo, Y.K.; Baek, S.-O.; Ammireddy, V.R. Review on analytical and biological applications of hydrazones and their metal complexes. *E-J Chem.* **2012**, *9*, 1288–1304. [CrossRef]

56. Lee, K.Y.; Wang, Y.; Nie, S. In vitro study of a pH-sensitive multifunctional doxorubicin–gold nanoparticle system: Therapeutic effect and surface enhanced Raman scattering. *RSC Adv.* **2015**, *5*, 65651–65659. [CrossRef]

57. Hu, J.; Xie, L.; Zhao, W.; Sun, M.; Liu, X.; Gao, W. Design of tumor-homing and pH-responsive polypeptide–doxorubicin nanoparticles with enhanced anticancer efficacy and reduced side effects. *Chem Commun.* **2015**, *51*, 11405–11408. [CrossRef]

58. Huang, L.; Tao, K.; Liu, J.; Qi, C.; Xu, L.; Chang, P.; Gao, J.; Shuai, X.; Wang, G.; Wang, Z. interfaces, Design and fabrication of multifunctional sericin nanoparticles for tumor targeting and pH-responsive subcellular delivery of cancer chemotherapy drugs. *ACS Appl. Mater. Inter.* **2016**, *8*, 6577–6585. [CrossRef] [PubMed]

59. Liu, B.; Thayumanavan, S. Substituent effects on the pH sensitivity of acetals and ketals and their correlation with encapsulation stability in polymeric nanogels. *J. Am. Chem. Soc.* **2017**, *139*, 2306–2317. [CrossRef] [PubMed]

60. Chen, Y.; Ai, K.; Liu, J.; Sun, G.; Yin, Q.; Lu, L. Multifunctional envelope-type mesoporous silica nanoparticles for pH-responsive drug delivery and magnetic resonance imaging. *Biomaterials* **2015**, *60*, 111–120. [CrossRef] [PubMed]

61. Wang, L.; Liu, G.; Wang, X.; Hu, J.; Zhang, G.; Liu, S. Acid-disintegratable polymersomes of pH-responsive amphiphilic diblock copolymers for intracellular drug delivery. *Macromolecules* **2015**, *48*, 7262–7272. [CrossRef]

62. Huang, F.; Cheng, R.; Meng, F.; Deng, C.; Zhong, Z. Micelles based on acid degradable poly (acetal urethane): Preparation, pH-sensitivity, and triggered intracellular drug release. *Biomacromolecules* **2015**, *16*, 2228–2236. [CrossRef] [PubMed]

63. Cao, H.; Chen, C.; Xie, D.; Chen, X.; Wang, P.; Wang, Y.; Song, H.; Wang, W. A hyperbranched amphiphilic acetal polymer for pH-sensitive drug delivery. *Polym. Chem.* **2018**, *9*, 169–177. [CrossRef]

64. Louage, B.; Zhang, Q.; Vanparijs, N.; Voorhaar, L.; Vande Casteele, S.; Shi, Y.; Hennink, W.E.; Van Bocxlaer, J.; Hoogenboom, R.; De Geest, B.G. Degradable Ketal-Based Block Copolymer Nanoparticles for Anticancer Drug Delivery: A Systematic Evaluation. *Biomacromolecules.* **2015**, *16*, 336–350. [CrossRef] [PubMed]

65. Su, J.; Chen, F.; Cryns, V.L.; Messersmith, P.B. Catechol polymers for pH-responsive, targeted drug delivery to cancer cells. *J. Am. Chem. Soc.* **2011**, *133*, 11850–11853. [CrossRef] [PubMed]

66. Li, L.; Bai, Z.; Levkin, P.A. Boronate–dextran: An acid-responsive biodegradable polymer for drug delivery. *Biomaterials* **2013**, *34*, 8504–8510. [CrossRef] [PubMed]
67. Lee, J.; Kim, J.; Lee, Y.M.; Park, D.; Im, S.; Song, E.H.; Park, H.; Kim, W.J. Self-assembled nanocomplex between polymerized phenylboronic acid and doxorubicin for efficient tumor-targeted chemotherapy. *ACTA Pharmacol. Sin.* **2017**, *38*, 848–858. [CrossRef] [PubMed]
68. Wang, Y.; Kohane, D.S. External triggering and triggered targeting strategies for drug delivery. *Nat. Rev. Mater.* **2017**, *2*, 17020. [CrossRef]
69. Bae, Y.H.; Park, K. Targeted drug delivery to tumors: Myths, reality and possibility. *J. Control. Release.* **2011**, *153*, 198–205. [CrossRef]
70. Brannon-Peppas, L.; Blanchette, J.O. Nanoparticle and targeted systems for cancer therapy. *Adv. Drug Deliv. Rev.* **2012**, *64*, 206–212. [CrossRef]
71. Zhang, S.; Bellinger, A.M.; Glettig, D.L.; Barman, R.; Lee, Y.-A.L.; Zhu, J.; Cleveland, C.; Montgomery, V.A.; Gu, L.; Nash, L.D. A pH-responsive supramolecular polymer gel as an enteric elastomer for use in gastric devices. *Nat. Mater.* **2015**, *14*, 1065–1071. [CrossRef]
72. Casey, J.R.; Grinstein, S.; Orlowski, J. Sensors and regulators of intracellular pH. *Nat. Rev. Mol. Cell Biol.* **2010**, *11*, 50–61. [CrossRef] [PubMed]
73. Li, Y.; Zhao, T.; Wang, C.; Lin, Z.; Huang, G.; Sumer, B.D.; Gao, J. Molecular basis of cooperativity in pH-triggered supramolecular self-assembly. *Nat. Commun.* **2016**, *7*, 13214. [CrossRef] [PubMed]
74. Chatterjee, S.; Ramakrishnan, S. Hyperbranched polyacetals with tunable degradation rates. *Macromolecules.* **2011**, *44*, 4658–4664. [CrossRef]
75. Convertine, A.J.; Benoit, D.S.; Duvall, C.L.; Hoffman, A.S.; Stayton, P.S. Development of a novel endosomolytic diblock copolymer for siRNA delivery. *J. Control. Release.* **2009**, *133*, 221–229. [CrossRef] [PubMed]
76. Manganiello, M.J.; Cheng, C.; Convertine, A.J.; Bryers, J.D.; Stayton, P.S. Diblock copolymers with tunable pH transitions for gene delivery. *Biomaterials* **2012**, *33*, 2301–2309. [CrossRef] [PubMed]
77. Nie, S. Understanding and overcoming major barriers in cancer nanomedicine. *Nanomedicine* **2010**, *5*, 523–528. [CrossRef] [PubMed]
78. Hubbell, J.A.; Chilkoti, A. Nanomaterials for drug delivery. *Science* **2012**, *337*, 303–305. [CrossRef]
79. Allen, T.M.; Cullis, P.R. Drug delivery systems: Entering the mainstream. *Science* **2004**, *303*, 1818–1822. [CrossRef]
80. Qiu, L.Y.; Bae, Y.H. Polymer architecture and drug delivery. *Pharm. Res.* **2006**, *23*, 1–30. [CrossRef]
81. Cordes, E.; Bull, H. Mechanism and catalysis for hydrolysis of acetals, ketals, and ortho esters. *Chem Rev.* **1974**, *74*, 581–603. [CrossRef]
82. Wang, C.; Wang, Z.; Zhao, T.; Li, Y.; Huang, G.; Sumer, B.D.; Gao, J. Optical molecular imaging for tumor detection and image-guided surgery. *Biomaterials* **2018**, *157*, 62–75. [CrossRef] [PubMed]
83. Wang, Z.; Luo, M.; Mao, C.; Wei, Q.; Zhao, T.; Li, Y.; Huang, G.; Gao, J. A Redox-Activatable Fluorescent Sensor for the High-Throughput Quantification of Cytosolic Delivery of Macromolecules. *Angew. Chem. Int. Edit.* **2017**, *56*, 1319–1323. [CrossRef] [PubMed]
84. Wang, Y.; Wang, C.; Li, Y.; Huang, G.; Zhao, T.; Ma, X.; Wang, Z.; Sumer, B.D.; White, M.A.; Gao, J. Digitization of Endocytic pH by Hybrid Ultra-pH-Sensitive Nanoprobes at Single-Organelle Resolution. *Adv. Mater.* **2017**, *29*, 1603794. [CrossRef] [PubMed]
85. Wang, Y.; Zhou, K.; Huang, G.; Hensley, C.; Huang, X.; Ma, X.; Zhao, T.; Sumer, B.D.; De Berardinis, R.J.; Gao, J. A nanoparticle-based strategy for the imaging of a broad range of tumours by nonlinear amplification of microenvironment signals. *Nat. Mater.* **2014**, *13*, 204–212. [CrossRef] [PubMed]
86. Zhao, T.; Huang, G.; Li, Y.; Yang, S.; Ramezani, S.; Lin, Z.; Wang, Y.; Ma, X.; Zeng, Z.; Luo, M. A transistor-like pH nanoprobe for tumour detection and image-guided surgery. *Nat. Biomed. Eng.* **2017**, *1*, 0006. [CrossRef] [PubMed]
87. Estelrich, J.; Sánchez-Martín, M.J.; Busquets, M.A. Nanoparticles in magnetic resonance imaging: From simple to dual contrast agents. *Int. J. Nanomed.* **2015**, *10*, 1727–1741.
88. Kim, K.S.; Park, W.; Hu, J.; Bae, Y.H.; Na, K. A cancer-recognizable MRI contrast agents using pH-responsive polymeric micelle. *Biomaterials* **2014**, *35*, 337–343. [CrossRef] [PubMed]

89. Gao, G.H.; Im, G.H.; Kim, M.S.; Lee, J.W.; Yang, J.; Jeon, H.; Lee, J.H.; Lee, D.S. Magnetite-Nanoparticle-Encapsulated pH-Responsive Polymeric Micelle as an MRI Probe for Detecting Acidic Pathologic Areas. *Small* **2010**, *6*, 1201–1204. [CrossRef] [PubMed]

90. Uchino, K.; Ochiya, T.; Takeshita, F. RNAi Therapeutics and Applications of MicroRNAs in Cancer Treatment. *Jpn. J. Clin. Oncol.* **2013**, *43*, 596–607. [CrossRef] [PubMed]

91. Xu, X.; Wu, J.; Liu, Y.; Saw, P.E.; Tao, W.; Yu, M.; Zope, H.; Si, M.; Victorious, A.; Rasmussen, J.; et al. Multifunctional Envelope-Type siRNA Delivery Nanoparticle Platform for Prostate Cancer Therapy. *ACS Nano* **2017**, *11*, 2618–2627. [CrossRef] [PubMed]

92. Xu, X.; Wu, J.; Liu, Y.; Yu, M.; Zhao, L.; Zhu, X.; Bhasin, S.; Li, Q.; Ha, E.; Shi, J.; et al. Ultra-pH-Responsive and Tumor-Penetrating Nanoplatform for Targeted siRNA Delivery with Robust Anti-Cancer Efficacy. *Angew. Chem.-Int. Edit.* **2016**, *55*, 7091–7094. [CrossRef]

93. Chen, J.; Lin, L.; Guo, Z.; Xu, C.; Li, Y.; Tian, H.; Tang, Z.; He, C.; Chen, X. N-Isopropylacrylamide Modified Polyethylenimines as Effective siRNA Carriers for Cancer Therapy. *J. Nanosci. Nanotechnol.* **2016**, *16*, 5464–5469. [CrossRef]

94. Banerjee, S.; Norman, D.D.; Lee, S.C.; Parrill, A.L.; Pham, T.C.T.; Baker, D.L.; Tigyi, G.J.; Miller, D.D. Highly Potent Non-Carboxylic Acid Autotaxin Inhibitors Reduce Melanoma Metastasis and Chemotherapeutic Resistance of Breast Cancer Stem Cells. *J. Med. Chem.* **2017**, *60*, 1309–1324. [CrossRef]

95. Zhou, X.X.; Jin, L.; Qi, R.Q.; Ma, T. pH-responsive polymeric micelles self-assembled from amphiphilic copolymer modified with lipid used as doxorubicin delivery carriers. *R. Soc. Open Sci.* **2018**, *5*, 171654. [CrossRef]

96. Zhang, C.Y.; Yang, Y.Q.; Huang, T.X.; Zhao, B.; Guo, X.D.; Wang, J.F.; Zhang, L.J. Self-assembled pH-responsive MPEG-b-(PLA-co-PAE) block copolymer micelles for anticancer drug delivery. *Biomaterials* **2012**, *33*, 6273–6283. [CrossRef]

97. Cheng, R.; Meng, F.; Deng, C.; Klok, H.-A.; Zhong, Z. Dual and multi-stimuli responsive polymeric nanoparticles for programmed site-specific drug delivery. *Biomaterials* **2013**, *34*, 3647–3657. [CrossRef]

98. Yang, X.; Grailer, J.J.; Rowland, I.J.; Javadi, A.; Hurley, S.A.; Matson, V.Z.; Steeber, D.A.; Gong, S. Multifunctional Stable and pH-Responsive Polymer Vesicles Formed by Heterofunctional Triblock Copolymer for Targeted Anticancer Drug Delivery and Ultrasensitive MR Imaging. *ACS Nano* **2010**, *4*, 6805–6817. [CrossRef]

99. John, J.V.; Uthaman, S.; Augustine, R.; Chen, H.; Park, I.-K.; Kim, I. pH/redox dual stimuli-responsive sheddable nanodaisies for efficient intracellular tumour-triggered drug delivery. *J. Mat. Chem. B* **2017**, *5*, 5027–5036. [CrossRef]

100. Chytil, P.; Šírová, M.; Kudláčová, J.; Říhová, B.; Ulbrich, K.; Etrych, T. Bloodstream Stability Predetermines the Antitumor Efficacy of Micellar Polymer–Doxorubicin Drug Conjugates with pH-Triggered Drug Release. *Mol. Pharm.* **2018**, *15*, 3654–3663. [CrossRef]

101. Zhang, Y.; Yang, C.; Wang, W.; Liu, J.; Liu, Q.; Huang, F.; Chu, L.; Gao, H.; Li, C.; Kong, D.; et al. Co-delivery of doxorubicin and curcumin by pH-sensitive prodrug nanoparticle for combination therapy of cancer. *Sci. Rep.* **2016**, *6*, 21225. [CrossRef]

102. Xu, J.; Wong, D.H.C.; Byrne, J.D.; Chen, K.; Bowerman, C.; DeSimone, J.M. Future of the Particle Replication in Nonwetting Templates (PRINT) Technology. *Angew. Chem. Int. Ed.* **2013**, *52*, 6580–6589. [CrossRef]

103. Lim, J.M.; Swami, A.; Gilson, L.M.; Chopra, S.; Choi, S.; Wu, J.; Langer, R.; Karnik, R.; Farokhzad, O.C. Ultra-High Throughput Synthesis of Nanoparticles with Homogeneous Size Distribution Using a Coaxial Turbulent Jet Mixer. *ACS Nano* **2014**, *8*, 6056–6065. [CrossRef]

104. Zhang, B.Y.; Korolj, A.; Lai, B.F.L.; Radisic, M. Advances in organ-on-a-chip engineering. *Nat. Rev. Mater.* **2018**, *3*, 257–278. [CrossRef]

molecules

MDPI

Review

Recent Advances in Phenylboronic Acid-Based Gels with Potential for Self-Regulated Drug Delivery

Chenyu Wang [1], Bozhong Lin [2], Haopeng Zhu [2], Fei Bi [2], Shanshan Xiao [2], Liyan Wang [2,*], Guangqing Gai [2] and Li Zhao [2,*]

[1] Department of Orthopedics, Hallym University, 1 Hallymdaehak-gil, Chuncheon, Gangwon-do 200-702, Korea; cathywang0111@hotmail.com

[2] Laboratory of Building Energy-Saving Technology Engineering, College of Material Science and Engineering, Jilin Jianzhu University, Changchun 130118, China; bzlin2018@163.com (B.L.); hpzhu2018@163.com (H.Z.); bifei1224@163.com (F.B.); xiaoshanshan@jlju.edu.cn (S.X.); gaigq@163.com (G.G.)

* Correspondence: wlynzy@163.com (L.W.); zhaolizdl@163.com (L.Z.)

Academic Editors: Jianxun Ding, Yang Li and Mingqiang Li
Received: 2 February 2019; Accepted: 15 March 2019; Published: 19 March 2019

Abstract: Glucose-sensitive drug platforms are highly attractive in the field of self-regulated drug delivery. Drug carriers based on boronic acid (BA), especially phenylboronic acid (PBA), have been designed for glucose-sensitive self-regulated insulin delivery. The PBA-functionalized gels have attracted more interest in recent years. The cross-linked three-dimensional (3D) structure endows the glucose-sensitive gels with great physicochemical properties. The PBA-based platforms with cross-linked structures have found promising applications in self-regulated drug delivery systems. This article summarizes some recent attempts at the developments of PBA-mediated glucose-sensitive gels for self-regulated drug delivery. The PBA-based glucose-sensitive gels, including hydrogels, microgels, and nanogels, are expected to significantly promote the development of smart self-regulated drug delivery systems for diabetes therapy.

Keywords: phenylboronic acid; gel; glucose sensitivity; drug delivery; diabetes therapy

---

1. Introduction

Diabetes mellitus threatens human health seriously with a series of complications, such as cardiovascular disease [1]. The worldwide prevalence of diabetics is predicted to be about 366 million in 2030 with increasing attention for the treatment of diabetes [2]. The frequent injection of exogenous insulin is the major treatment of diabetes. In spite of quickly decreasing the blood glucose level, frequent insulin injection causes inevitable injection suffering and decline of the quality of life of patients. Even though implantable insulin pumps are the current optimal therapy for type 1 diabetic patients, their high cost limits their wide clinical application [3]. Smart insulin modified with an aliphatic domain and a phenylboronic acid (PBA) has been developed to tune the pharmacokinetics of insulin activity for personalized therapy, however, it must interface with insulin pumps, infusion devices, or controlled release materials to further improve performance [4]. An alternative treatment of diabetes with continuous and automatic regulation of drug release triggered by glucose directly is required. Therefore, glucose-sensitive materials, a kind of "intelligent" polymer, have greatly been used in self-regulated systems. Self-regulated systems, which are called an artificial pancreas, control insulin release triggered by elevated level of blood glucose continuously and automatically. By integrating glucose-triggered drug delivery with minimal patient intervention and improved diabetic life quality, glucose-sensitive drug delivery systems may prove valuable in diabetes therapy and replace frequent insulin injection [5].

Three kinds of glucose-sensitive materials have attracted growing scientific attention and are considered novel candidates to serve as self-regulated drug delivery systems. One strategy is based on glucose oxidase (GOD), which converts glucose to gluconic acid upon a pH change of the microenvironment, resulting in physicochemical changes of GOD-incorporated carriers [6]. The physicochemical changes of GOD-incorporated platforms induce glucose-triggered payload release [7–10]. Concanavalin A (Con A) is also be used to design glucose-sensitive drug delivery systems due to the specific binding capacity for glucose, glycopolymers, and polysaccharides [11–14]. However, GOD and Con A are protein-based components, and the instability and biotoxicity of GOD and Con A during fabrication and storage restrict their application in self-regulated drug delivery [15,16].

An alternative mechanism based on totally synthetic components, boronic acid (BA) and especially PBA, and their derivatives, have been largely investigated for self-regulated drug delivery systems [17–20]. Since Okano and coworkers firstly reported PBA-based glucose-triggered drug release, PBA and its derivatives have been greatly employed to design glucose-triggered drug delivery systems [21,22]. PBA-containing platforms are much more stable and suitable for long-term storage than GOD- and Con A-based drug carriers [23]. The principle of the PBA-mediated glucose-sensitive drug delivery system is based on the reversible reaction between PBA and *cis*-diol compounds [24]. There are two states of PBA in aqueous solution. One is the neutral trigonal-planar form and the other is the negatively charged tetrahedral boronate form. Between the two states of PBA moieties there is an equilibrium (Scheme 1). When in aqueous solution with pH above the pK_a of PBA (i.e., 8.2–8.6), most PBA moieties are negative and relatively hydrophilic. However, the PBA moieties are neutral and hydrophobic when the pH of the solution is below the pK_a of PBA [25,26]. In the presence of glucose or other 1,2- or 1,3-diols, both kinds of PBA moieties form a 5- or 6-cyclic boronic ester. However, the cyclic boronic ester between neutral trigonal-planar PBA and a diol cannot be formed with easy hydrolysis. The negatively charged PBA state forms a more stable cyclic boronic ester with *cis*-diol compounds, resulting in the increase of the negatively charged PBA content and improved hydrophilicity of PBA-containing materials [27,28]. The increased hydrophilicity of PBA-modified materials induces the swelling, disassembly, and/or destruction of PBA-mediated platforms with subsequent release of the payload. Glucose can also form bis-bidendate complexes with PBA resulting from the reaction between one glucose molecule and two cis-diols. Therefore, the bis-bidendate complexes induce the shrinking of PBA-functionalized platforms [29]. All the swelling and shrinking of platforms alter the diffusion behavior of the payload, triggered by glucose-sensitive volume phase transition. PBA-functionalized materials present great potential applications in self-regulated drug delivery [29,30].

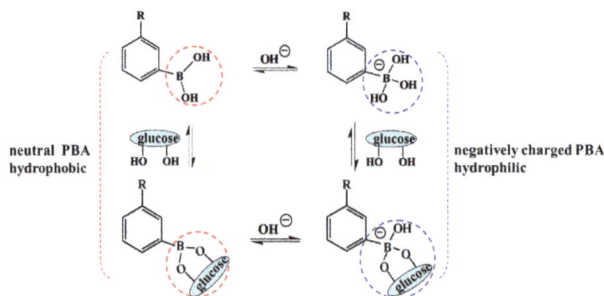

Scheme 1. Complexation equilibrium between phenylboronic acid (PBA) and glucose.

There are many forms of PBA-functionalized platforms which have been exploited to investigate glucose sensitivity. Polymer micelles, vesicles, and capsules are obtained by the self-assembly of PBA-modified amphiphilic polymers, while layer-by-layer (LbL) films are obtained by the alternating adsorption of polymers with opposite charges. Even though these devices have good glucose sensitivity, the concentration-dependent disintegration, unstably long-term storability, and weakened mechanical strength restrict their applications in glucose-sensitive drug delivery. PBA-functionalized mesoporous silica nanoparticles (MSN) have attracted improving interests due to their characteristics of high biocompatibility, improved dug loading, and functionalized surface. The glucose-sensitive nanocarriers and LbL films and microcapsules for self-regulated drug delivery have been reviewed [26,31]. Besides, PBA-based hydrogels, microgels, and nanogels with chemically or physically cross-linked structures have great potential applications in self-regulated drug delivery systems. The 3-D structures endow the hydrogels, microgels, and nanogels with reversible physicochemical changes that result in payload release. Usually, hydrogels are macroscopic hydrogels. Microgels are microsized hydrogel particles, while nanogels are hydrogel nanoparticles or nanohydrogels with nano-scale dimensions (typically 20~250 nm) [32,33]. The Zhang group and Catargi group have discussed the fabrication and application of PBA-containing macroscopic hydrogels and microgels with discussion of their further development [29,34]. However, studies of PBA-based hydrogels, microgels, and nanogels for self-regulated drug delivery are rare. In summary, this article reviews the recent development of self-regulated drug delivery based on PBA-mediated glucose-sensitive gels.

2. PBA-Based Hydrogels

Hydrogels have received considerable attention for drug delivery applications. There have been extensive studies on PBA-functionalized hydrogels for self-regulated drug delivery systems.

Low-molecular-weight hydrogels (LMWG) have been largely investigated for controlled drug delivery due to their super-sensitivity to external stimuli. LMWG were formed by the self-assembly of gelators with low-molecular-weight, where the gel-sol transition was easily achieved by changing stimuli. An injectable glucose-sensitive LMWG was prepared by the self-aggregation of the gelator which was synthesized with a pyrene moiety coupled with PBA by L-phenylalanine and a 2,2'-(ethylenedioxy) bis(ethylamine) linker [35]. The pyrene unit and PBA moiety endowed the physical cross-linked LMWG with glucose sensitivity, and the resulting compound was utilized to detect glucose and control insulin release at physiological pH. Differently from chemically cross-linked hydrogels, physically cross-linked networks have transient junctions arising from hydrogen bonding, π–π stacking, and van der Waals interactions. The self-aggregation of the gelator provides a new strategy for the design of PBA-based glucose-sensitive hydrogels.

Gao and coworkers investigated the dual-responsiveness of a PBA-based low-molecular-weight organogel to both glucose and pH [36]. In addition, the group studied the gelation properties and glucose sensitivity of a PBA-containing low-molecular-weight organogel based on an alkyl chain (C2–C11), where the glucose sensitivity of the organogel was associated with the molecular structure of the gelator [37]. Integrating an oligopeptide with PBA, the same group demonstrated that the LMWG self-assembled from a gelator exhibited dual-responsive and long-lasting drug delivery [38]. The structure of the gelator is shown in Figure 1. The hydrophilicity of gelator was regulated by the oligopeptide which was made of the natural amino acids L-phenylalanine, glycylglycine, and L-glutamine. The alkyl chain of lauroyl chloride was introduced to adjust the hydrophobicity of the gelator in favor of the sol-gel translation. PBA unit endowed the gelator with glucose sensitivity. When the sol was cooled to body temperature, LMWG was obtained very quickly within two minutes. The sol-gel translation resulted from the enhanced π–π stacking owing to the benzene units as well as hydrogen bonding and van der Waals force within the gelators. The LMWG was very stable during the drug release even for three months or with sever shaking, indicating its long-lasting drug delivery. The pH-sensitivity of the hydrogels, a result of the ionization of the PBA and amide moieties in acid or alkali solutions, was used for doxorubicin delivery for cancer therapy. Importantly, PBA endowed the hydrogels with glucose sensitivity. The release of phenformin, an antidiabetic drug, was regulated in response to changes in glucose concentration. The fact that increasing glucose concentration induced much more drug release indicated its promising application in a self-regulated drug delivery system. Integrating the pH-sensitivity with glucose-triggered drug release, the long-lasting glucose/pH-responsive hydrogels will have notable importance in the development of self-tuning controlled-release systems for diabetes and cancer therapy.

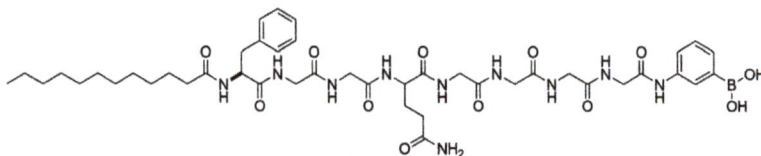

Figure 1. The structure of the gelator used for self-regulated drug delivery. (Reprinted from ref. [38]).

Even though LMWG used as a drug carrier possesses good glucose-sensitive drug delivery, the stability of LMWG is not sustainable enough for long-term drug delivery compared to the supramolecular hydrogels with chemical cross-linked structure. A multiresponsive hydrogel was synthesized by copolymerization of (2-dimethylamino) ethyl methacrylate (DMAEMA) and 3-acrylamidephenylboronic acid (AAPBA) using *N,N′*-methylenebisacrylamide (NNMBA) as a cross-linker [39]. The P(DMAEMA-*co*-AAPBA) interpenetrating (IPN) hydrogels possessed glucose-triggered drug release under physiological pH and temperature. The increased Lewis acidity of the boron center lowered the pK_a of the PBA residues. The Lewis acid–base interactions between the electron-poor boron atom and the electron-rich nitrogen atom on poly(DMAEMA) endowed the hydrogel with glucose sensitivity under physiological pH. These multiresponsive hydrogels have important application in self-regulated drug delivery.

Also based on poly(DMAEMA), a novel triple-responsive semi-interpenetrating (semi-IPN) hydrogel was exploited [40]. Poly(3-acrylamidephenylboronic acid-*co*-(2-dimethylamino) ethyl methacrylate) (P(AAPBA-*co*-DMAEMA)) has a cross-linked structure incorporated with interpenetrating β-cyclodextrin-epichlorohydrin (β-CD-EPI)—P(AAPBA-*co*-DMAEMA)/(β-CD-EPI) semi-IPN hydrogel. The higher content of β-CD decreased the equilibrium swelling ratios (ESRs) of semi-IPN hydrogels due to the complexation between PBA groups and the dihydroxyl of β-CD. Besides, pH, temperature, ionic strength, and glucose concentration of the media significantly affected the ESRs of the hydrogels. Using ibuprofen and aminophylline as hydrophobic and hydrophilic model drugs, the drug loading and release profiles were investigated. The structure of the hydrogels affected the drug loading ratio. Increased β-CD content led to of ibuprofen and a lower drug loading ratio of aminophylline. The lower drug loading ratio of hydrophilic aminophylline in hydrogels with high β-CD content resulted from the lower ESRs, which was induced by the increased complexation between PBA and β-CD. However, the loading ratio of hydrophobic ibuprofen increased with the increase of β-CD content in hydrogels. The reason was that hydrophobic ibuprofen formed inclusion complexes with β-CD via host–guest interactions. The release profiles of drug from the hydrogels could be adjusted by pH, temperature, glucose concentration, and type release medium. The semi IPN hydrogels may be useful as a guideline for the optimal design of glucose-sensitive drug delivery systems.

To enhance the response of PBA-based hydrogel to blood glucose concentration, a comb-type grafted poly(*N*-isopropylacrylamide-*co*-3-acrylamidophenylboronic acid) (poly(NIPAM-*co*-AAPBA)) hydrogel was exploited [41]. The hydrogel was introduced by grafting poly(NIPAM-*co*-AAPBA) side chains onto cross-linked poly(NIPAM-*co*-AAPBA) networks. The comb-type hydrogel presented a rapid response to the blood glucose concentration owing to the effects of the freely mobile ends of the grafted poly(NIPAM-*co*-AAPBA). Grafted poly(NIPAM-*co*-AAPBA) side chains formed complexes with glucose quickly without any restrictions than that between cross-linked poly(NIPAM-*co*-AAPBA) hydrogel and glucose. The rapid response rate endowed the grafted hydrogel with attractive applications in self-regulated drug delivery systems.

Besides the structure of hydrogels, the composition of hydrogels also affects the glucose sensitivity. Magda and coworkers studied the effect of chemical composition on the response of zwitterionic glucose-sensitive hydrogels using DOE (design of experiments) methods [42]. Zwitterionic glucose-sensitive hydrogels were prepared by the copolymeration of N-(3-(dimethylamino)propyl) acrylamide (DMAPAA) and AAPBA. The molar ratio of AAPBA/DMAPAA and the wt % of the monomer in the pregel solution were the primary factor for determining the value of the inverse of the 1st order rate constant, where a decreasing amount of cross-linker obtained faster glucose responses.

In addition to stable chemical cross-linked hydrogels using a small molecular as the cross-linker, reversible dynamic cross-linked hydrogels have also been designed as glucose-sensitive drug carriers. Hydrogels were obtained by the reversible covalent complexation of PBA with *cis*-1,2 or *cis*-1,3-diol compounds as a dynamic link [43,44]. Glucose-sensitive solid-like hydrogels based on dynamic covalent chemistry and inclusion complexation were also designed [45]. By simply mixing the solutions of poly (ethylene oxide)-*b*-poly vinyl alcohol diblock polymer (PEO-*b*-PVA), α-cyclodextrin (α-CD), and a double PBA-terminated PEO cross-linker, hydrogels were easily obtained. As shown in Figure 2, the inclusion complexation between PEO and α-CD, and the dynamic covalent bonds between PBA and PVA, strengthened the hydrogel network. In addition, the formation of hydrogels was the cooperative interaction of the inclusion complexation and dynamic covalent chemistry. The increase of α-CD content shortened the gelation time and enhanced the structural recovery ability with the significant contribution of increased dynamic covalent chemistry to the cross-linking density of hydrogel network. The release of FITC-labeled BSA from the hydrogels had no burst release property at physiological pH, indicating that the FITC-labeled BSA was loaded inside the hydrogels instead of being adsorbed on the surfaces of the hydrogels. These hydrogels have great potential as self-regulated drug delivery vehicles with tunable glucose sensitivity.

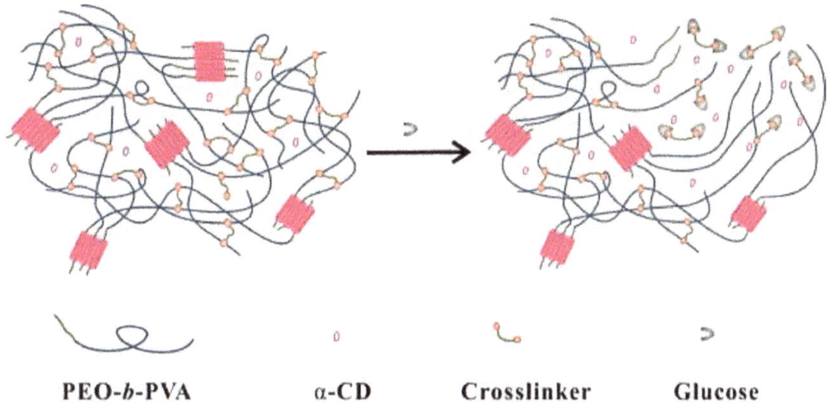

PEO-*b*-PVA   **α-CD**   **Crosslinker**   **Glucose**

Figure 2. Schematic and the glucose-sensitive mechanism of the hydrogel composed of PEO-*b*-PVA, α-CD, and double PBA-terminated poly (ethylene oxide) (PEO) cross-linker. (Reprinted from ref. [45].)

Via dynamic boronic ester bonds, glycopolymer hydrogels were exploited to control insulin release triggered by glucose [46]. The copolymerization of AAPBA and 2-lactobionamidoethyl methacrylate (LAMA) was conducted by the reversible addition–fragmentation chain transfer (RAFT) method and the obtained block glycopolymer formed hydrogels by phenylboronate-diol crosslinked binding (Figure 3). The LAMA content played an important role in the swelling of the hydrogels. For the hydrogels with the highest LAMA content, the equilibrium swelling ratio was up to 1856%. In addition, with high LAMA content, the insulin loading capacity of the hydrogels increased. Higher content of LAMA enhanced the charged phenylborates, which resulted in enhanced hydrophilicity of the hydrogels and subsequently promoted drug permeation and adsorption. The introduction of carbohydrate moieties improved the cytocompatibility of the glycopolymer hydrogels. The insulin release from the glycopolymer hydrogels exhibited glucose sensitivity due to glucose-induced dissociation of boronic ester linkages of hydrogels. Importantly, the released insulin possessed an original conformation compared with standard insulin. PBA-based glycopolymer hydrogels are an alternative design of glucose-sensitive drug delivery systems.

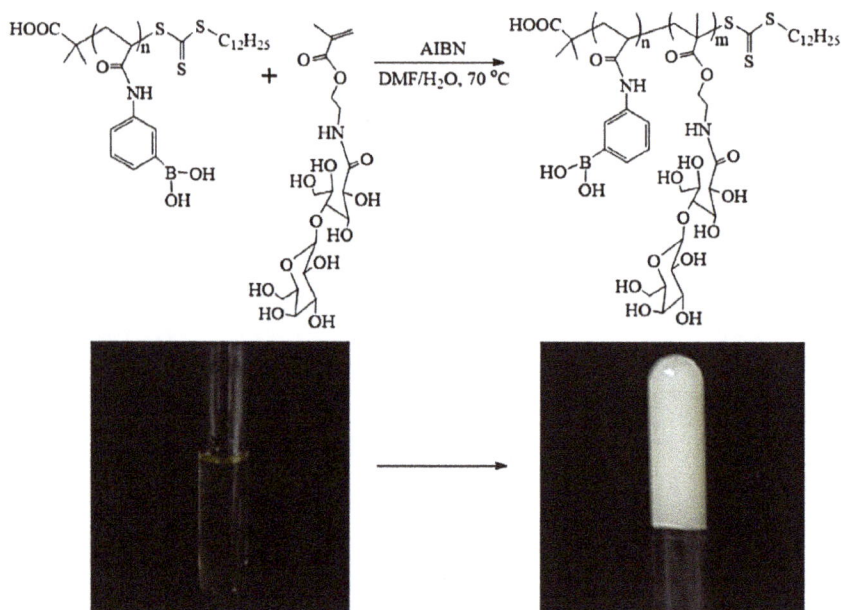

Figure 3. The formation of PBA-containing glycopolymer hydrogel before and after. (Reprinted from ref. [46]).

Another series of glucose-sensitive block glycopolymer hydrogels based on dynamic boronic ester bonds were exploited for glucose-sensitive drug delivery [47]. As shown in Figure 4a, block glycopolymer, (3-propionamidophenyl)boronic acid (*N*-(3-((2,3,4,5,6-pentahydroxyhexyl) amino)propyl)propionamide) (noted polymer BG), was cross-linked through PBA–glucose complexation within the glycopolymer and injectable self-healing hydrogels were obtained. The glycopolymer with 10–60% content of PBA formed self-supporting hydrogels at 5% weight in water, which could be easily loaded and extruded through a needle with the reformation of hydrogels (Figure 4b). The hydrogels possessed rapid recovery with shear-thinning and self-healing behaviors which were consistent with the injectable hydrogel. In addition, the glycopolymer hydrogels exhibited glucose-triggered rhodamine B release. Integrating injectable and self-healing properties with glucose-induced drug release, glycopolymer hydrogels have potential application in the treatment of diabetes.

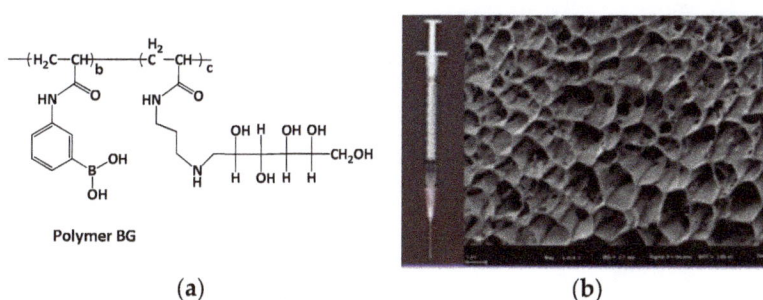

(a)   (b)

Figure 4. (a) Structure of PBA-containing glycopolymer BG; (b) the soft and injectable hydrogel formed by polymer BG with 50% PBA, and the SEM image of the obtained hydrogel. Scale bar: 1 μm. (Reprinted from ref. [47]).

Hydrogels based on PBA-containing polymers for self-regulated drug delivery system have been studied widely, while the bioconjugates of PBA-functionalized polymer and protein or peptide have been rarely used for glucose-sensitive drug delivery. Multiresponsive hydrogels were prepared from the bioconjugates of end-functionalized PBA containing copolymers and rod-like M13 viruses [48]. The end-functionalized PBA containing copolymer poly(NIPAM-*co*-PBA)-NHS was synthesized by two steps (Figure 5a). Firstly, poly(NIPAM-*co*-PBA)-COOH was synthesized by chain transfer free radical copolymerization of NIPAM and 4-(1,6-dioxo-2,5-diaza-7-oxamyl) phenylboronic acid (DDOPBA) using 3-mercaptopropionic acid (MPA) as the chain transfer agent. Then, the −COOH of the obtained poly(NIPAM-*co*-PBA)-COOH was transferred into *N*-hydroxysuccinidic ester. The hybrid virus–polymer bioconjugate was prepared by the conjugation of poly(NIPAM-*co*-PBA)-NHS to rod-like M13 virus, which was a natural protein assembly with a large amount of functional groups. The gelation behavior of virus–polymer bioconjugates was multiresponsive and reversible. The gelation of virus–polymer bioconjugates was owing to the collapsed hydrophobic state of poly(NIPAM-*co*-PBA), which conferred the attractive interactions between the viruses and drove the viruses into the hydrogels. AFM revealed that the rod-like virus was interconnected with itself inside the hydrogel with large pores (Figure 5b). The virus–polymer bioconjugates were in sol state when the temperature was below the critical gelation temperate (T_g), and in the gel state when the temperature was above T_g. The T_g of virus–polymer bioconjugates was sensitive to pH. At pH 7.10, the T_g was 15 °C, and increased when the pH increased. Furthermore, the T_g of the hybrid virus–polymer bioconjugates was 18 °C in the absence of glucose, while it was 23 °C in the presence of glucose at pH 7.65. In addition, the negligible hysteresis for the sol-to-gel and opposite gel-to-sol transition cycles indicated the reversible gelation behavior of the virus–polymer bioconjugates. Besides, the temperature-sensitive gelation behavior was in favor of the encapsulation of bioactive species inside the hydrogels with high loading capacity. Insulin-loaded hydrogels were obtained by the injection of a mixture of insulin and polymer-grafted virus at 4 °C into PBS buffer at 37 °C (inset of Figure 5c). Glucose-sensitive insulin release was also observed (Figure 5c). Glucose-triggered insulin release, combined with reversible temperature-, pH-, and glucose-regulated gelation behavior, endowed the PBA-containing polymer-virus bioconjugates with great potential in self-regulated drug delivery.

Cross-linking glucose-sensitive hydrogels based on PBA provides suitable semiwetting with reversible swelling/shrinking changes which are suitable for self-regulated drug delivery. However, the glucose sensitivity of bulk hydrogels hysteretically results from the slow permeation of glucose into the hydrogel network. In addition, the practical application of bulk hydrogels for the regulation of blood glucose levels is restricted by the implantation or other surgical approach. Even though injectable hydrogels are convenient to administrate, a lot of effort is needed to promote the practical application of glucose-sensitive hydrogels for diabetes therapy.

(a)

(b)

(c)

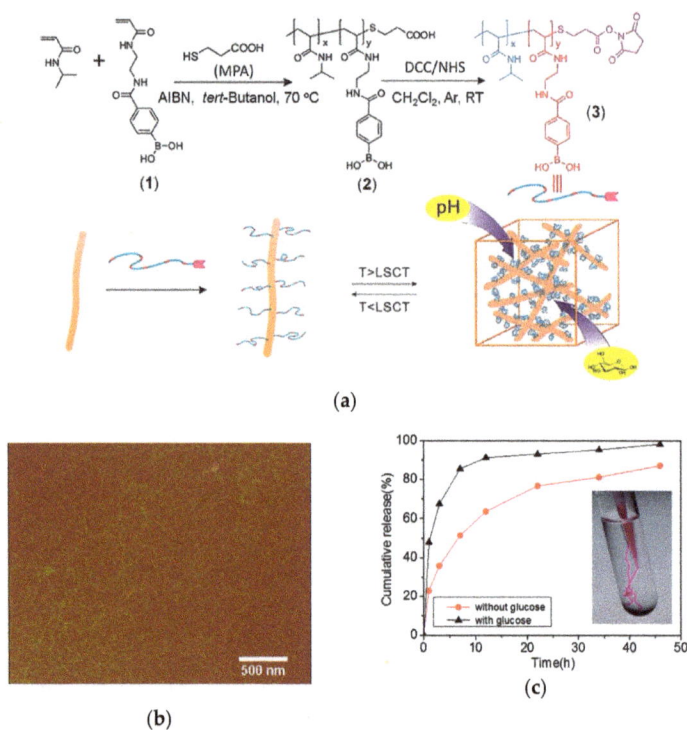

Figure 5. (**a**) Schematic representation of the preparation of the hybrid virus–polymer bioconjugate of end-functionalized PBA- and *N*-isopropylacrylamide (NIPAM)-containing copolymer to the rod-like M13 virus, and the multiresponsiveness of the hybrid virus–polymer bioconjugate; (**b**) Internal structure of the gel as revealed by AFM; (**c**) insulin release behavior of the virus based hydrogel in the presence or absence of glucose. Inset in (**c**) hydrogel forms instantly when the polymer grafted virus in the sol state was injected into the aqueous solution at 37 °C. (Reprinted from ref. [48]).

3. PBA-Functionalized Microgels

Even though PBA-based hydrogels are glucose-sensitive, the implantation administration of bulk hydrogels and the inconvenient injection of injectable hydrogels restrict the practical application of glucose-sensitive hydrogels in self-regulated drug delivery. Compared with hydrogels, PBA-based microgels have promising application in glucose-sensitive drug delivery [49]. Microgels are gel particle dispersions with average diameters ranging between 50 nm and 5 mm [50]. The small size endows the microgels with rapid swelling or shrinking in response to environmental changes. The glucose-response time of PBA-containing microgels are much faster than that of hydrogels. Kataoka and coworkers reported that gel beads containing AAPBA and NIPAM need a very long time (as long as 400 min) for glucose sensitivity [51]. However, microgels with the same components need only 10^2 s for glucose-induced swelling [52]. Besides faster glucose-response time, the administration of microgels using a minimally invasive method is another advantage of microgels for self-regulated drug delivery. In addition, microgels display high stability, which is promising for glucose-sensitive drug delivery vehicles.

A range of PNIPAM-based and PBA-functionalized microgels have been synthesized via the copolymerization of NIPAM and PBA-containing comonomers or the postpolymerization of functional PBA derivatives to PNIPAM microgels [53–57]. Poly(*N*-isopropylacrylamide-*co*-acrylic acid)

P(NIPAM-*co*-AA) was modified by 3-aminophenylboronic acid (3-APBA), resulting in PBA-mediated glucose-sensitive microgels [58,59].

Glucose-induced swelling of the poly(*N*-isopropylacrylamide-*co*-3-acrylamidophenylboronic acid) (P(NIPAM-3-AAPBA)) microgel was explained by the formation of a 1:1 glucose–phenylboronate complex, which increased the degree of ionization and created a Donnan potential between the gel phase and the liquid phase [52,60].

For P(NIPAM-PBA) microgels, the swelling response of microgels to glucose relates to the microgel composition and glucose concentration. Ravaine studied the effects of PBA content and glucose concentration on the swelling of monodispersed PNIPAM submicrometric microgels modified with AAPBA at different temperatures [61]. Compared to native PNIPAM particles, the incorporation of AAPBA decreased the volume phase transition temperature (VPTT) of the P(NIPAM-PBA) microgels. With the increase of PBA content, the swelling response of the microgels increased. However, the swollen state of the microgels that was achieved with same PBA content was strongly dependent on the initial temperature of the suspension with a constant glucose concentration. As shown in Figure 6, when the initial temperature was above VPTT, the microgels were smaller and the swelling was slight. In this case, glucose was unable to diffuse inside the microgel particle when it was initially collapsed. Only a small part of glucose linked to the PBA which was incorporated on the surface of microgels. However, when the initial temperature was below VPTT, a slight swelling of the microgels was aided the permeation and diffusion of glucose to the inside of the microgel particle. As a result, all the PBA was linked by glucose to form more hydrophilic boronic ester, inducing a much higher swelling degree of the microgels.

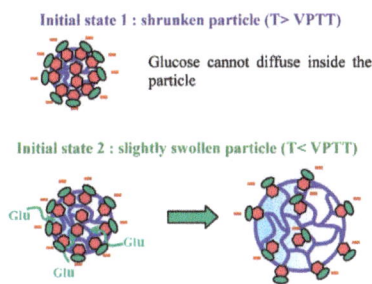

Figure 6. Schematic representation of microgel swelling upon glucose addition, showing the importance of the initial state. (Reprinted from ref. [61]).

Zhang and coworkers also studied the volume phase transitions of glucose-sensitive PBA-functionalized P(NIPAM-3-AAPBA) microgels which were synthesized by the coupling of 3-APBA to P(NIPAM-*co*-AA) microgels [62]. The microgels presented a two-stage thermosensitive volume phase transition in the presence of glucose. When temperature increased, the microgels underwent a small degree of collapse and a large volume change followed. In addition, the glucose sensitivity of the microgels was affected by pH. At pH = 7.5, which is below the pK_a of PBA, the glucose-induced size changes of the microgels were negligible. In this case, most PBA groups were hydrophobic with few boronate esters produced. However, when the pH increased, the glucose sensitivity of the microgels was enhanced as a result of the improved hydrophilicity of the microgels with resultant boronate ester. The glucose-induced size expansion of the microgels was depressed by high ionic strength due to the weakening of Donnan potential. At room temperature, remarkable glucose-induced swelling was observed in the microgels with high content of PBA.

Most PBA-based microgels exhibit glucose-induced expansion due to the production of glucose–mono(boronate) complexes [57,63]. The glucose–mono(boronate) complexes make the ionization equilibrium of PBA shift from uncharged and hydrophobic state to the charged and hydrophilic state with the formation of tetrahedral boronate esters. The tetrahedral boronate esters further increase the hydrophilicity of the microgels and induce the swelling of microgels due to the glucose-induced increase in the Donnan potential. Besides glucose-induced swelling of microgels, glucose-induced shrinking of microgels also has been reported. As shown in Figure 7, glucose exhibits the unique property to form glucose–bis(boronate) complexes through its furanose form [64–66]. However, most physiological relevant saccharides, such as galactose, mannose, and fructose, can only bind to boronic acid as monodentate complexes [67]. Zhang designed a poly(N-isopropylacrylamide-*co*-2-acrylamidophenylboronic acid) (P(NIPAM-2-AAPBA)) microgel, which was synthesized by modification of the poly(*N*-isopropylacrylamide-*co*-acrylic acid) (P(NIPAM-AAc)) microgel with 2-aminophenylboronic acid (2-AAPBA) [17]. In the presence of glucose, the size of the P(NIPAM-2-AAPBA) microgels decreased, showing contraction-type glucose sensitivity. The glucose-induced shrinking of the microgels was related with the glucose–bis(boronate) complexes in which one glucose molecule complexed with two PBA groups in a 1:2 binding model. The resultant glucose–bis(boronate) complexes increased the cross-linking density and reduced the degree of swelling of the microgels. The contraction-type glucose-sensitive microgels offer a new strategy for the design of glucose-sensitive microgels.

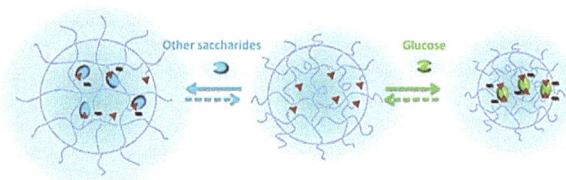

Figure 7. Schematic representation of glucose-induced microgels shrinking with glucose–bis(boronate) complexation (Reprinted from ref. [66]).

Except for contraction-type glucose-sensitive microgels, glucose-induced shrinking/swelling microgels were also designed. Ravaine and coworkers firstly designed the PBA-containing microgels cross-linked with bis-boronate complexes which either swelled or shrank selectively depending on glucose concentration under physiological conditions [66]. The microgels were obtained by copolymerization of DDOPBA with an alkylacrylamide (NIPAM, N-isopropylmethacrylamide (NIPMAM), or N-ethylmethacrylamide (NEMAM)) and a cross-linking agent (N,N'-methylenebis(acrylamide) (BIS), or ethylene glycol dimethacrylate (EGDMA)). The effect of monomer composition on the swelling/shrinking behavior of the microgels was studied. For microgels based on NIPAM and PBA (P(NIPAM-PBA) microgels), the VPTT was associated with the PBA content. More PBA content lowed the VPTT of P(NIPAM-PBA) microgels, which was conformed in the author's earlier work [61]. However, high glucose concentration increased the VPTT of P(NIPAM-PBA) microgels. In addition, at constant PBA content, the VPTT of NIPAM microgels was lower than that of NIPMAM microgels, which was lower than that of NEMAM microgels. All kinds of microgels displayed glucose-sensitive behaviors. When at a pH above the pK_a of PBA, EGDMA cross-linked NEMAM and NIPMAM microgels exhibited a two-step behavior wherein low glucose concentration induced microgel shrinkage and high glucose concentration induced microgel swelling. The shrinking behavior was associated with the additional cross-linking junctions due to glucose-bis(boronate) complexes at low glucose concentration. When the glucose concentration increased, the glucose–bis(boronate) complexes were converted to glucose–mono(boronate) complexes owing to the glucose-induced cross-link disruption. As a result, the microgels swelled under high glucose concentration due to the increased polymer charges and hydrophilicity. The glucose-response behavior of the microgels was related to the nature of the cross-linker and the modification of the PBA. The shrinking/swelling behavior was selective for glucose because the saccharide–bis(boronate) complexes were highly selective for glucose and could not occur with other sugars, such as fructose, which was also demonstrated by Asher and coworkers [67].

As mentioned above, polymers based on NIPAM, NEMAM, and EGDMA are thermosensitive and the VPTT of P(NIPAM-PBA) microgels are adjusted by glucose concentration and PBA content. Some other kinds of PBA-based microgels also possess volume phase transition behavior. The glucose-responsive volume phase transition behavior of poly(phenylboronic acid) (pPBA) microgels was switchable [68]. The preparation of the pPBA microgels was comprised 3-VAPBA covalently bonded onto the microgels of oligo(ethylene glycol)-based polymers (Figure 8). The stability of the pPBA-2 microgels was improved by further addition of a poly(acrylamide) (poly(AAm)) gel layer onto pPBA-1 microgels. In the presence of glucose, the glucose-response behaviors of pPBA microgels were different under different temperatures. At lower temperatures below 29.0 °C, the microgels shrunk when glucose was added, while at higher temperatures above 33.0 °C, the microgels swelled when glucose was added. Around 31.0 °C, negligible volume change of the microgels was recorded upon adding glucose. The pPBA microgels with switchable glucose-responsive volume phase transition behavior provide guidelines for the design of glucose-sensitive microgels.

Figure 8. Illustration of the synthesis of the proposed pPBA microgels. (Reprinted from ref. [68]).

Another multifunctional microgel with high stability and degradability was designed [69]. As shown in Figure 9a, the microgels were prepared by free radical polymerization of NIPAM, DMAEMA, and AAPBA through a precipitation emulsion method using reductive degradable *N,N'*-bis(arcyloyl)cystamine (BAC) as the cross-linker. The as-synthesized microgels exhibited pH-, temperature-, and glucose sensitivity at physiological conditions and gradual degradation. The porous network structure of the microgels was in favor of the trapping of insulin. Higher glucose concentration triggered the faster release of a larger amount of insulin. The microgels were degraded in the presence of dithiothreitol (DTT), which has similar function to glutathione tripeptide (GSH). As a result, a higher amount of insulin was released (90%) from the multiresponsive microgels in the medium (Figure 9b). The degradable multifunctional microgels may provide a new strategy for the design of PBA-functionalized self-regulated drug delivery systems.

Additionally, core–shell microgels based on NIPAM and PBA were exploited to investigate their glucose sensitivity. PNIPAM (core)/P(NIPAM-AAPBA) (shell) microgels were prepared by the modification of PNIPAM (core)/P(NIPAM-AA) (shell) microgels with 3-APBA [70]. The core–shell microgels exhibited three structure-related phase transitions when heating (Figure 10). The first phase transition was assigned to the P(NIPAM-AAPBA) shell, while the second and the third phase transitions were related to the PNIPAM core. The structure of the PNIPAM core was heterogeneous. The cross-linking density decreased gradually from the core towards the periphery due to the higher polymerization rate of cross-linking monomer BIS than that of the PNIPAM monomer. As a result, the

PNIPAM core had a quasi-core–shell-like structure with a "quasi-core" with a higher cross-linking density and a "shell-like" with a lower cross-linking density. The phase transition temperature of the "shell-like" structure was lower than that of the "quasi-core", providing the two phase transitions of the PNIPAM core. Besides, the core–shell microgels exhibited glucose-triggered swelling due to the PBA moieties on the P(NIPAM-AAPBA) shell.

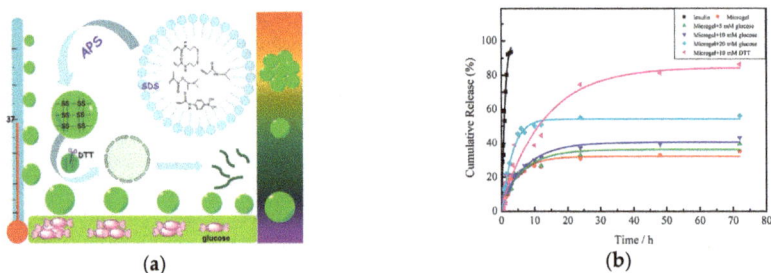

(a)   (b)

Figure 9. (a) Schematic illustration of smart multifunctional microgels with pH-, temperature-, and glucose sensitivity at physiological conditions and gradual degradation; (b) Release profiles of microgels in the presence of glucose in PBS of 7.4 at 37 °C. (Reprinted from ref. [69]).

Figure 10. Schematic illustration of the structure of PNIPAM (core)/P(NIPAM-AAPBA) (shell) microgel particles, where the PNIPAM core can be further divided as highly cross-linked "core" with BIS-rich and lightly cross-linked "shell" that is BIS-poor. (Reprinted from ref. [70]).

The glucose-induced shell permeability of core–shell microgels with the same components of NIPAM and PBA was also demonstrated [71]. Differently, the core of P(NIPAM-AAPBA) microgels was cross-linked by degradable cross-linker N,N'-(1,2-dihydroxyethylene)bisacrylamide (DHEA), while the shell was cross-linked by BIS. The cross-linked core of the microgels could be degraded by addition of stoichiometric amount of $NaIO_4$. The permeability of the P(NIPAM-PBA) shell was controlled by temperature and pH change. More importantly, glucose concentration also could tune the shell permeability. The shell permeability increased with increased glucose concentration, which resulted from the increased borate ester. The fundamental understanding of the glucose-induced permeability control is very important for the design and potential application of PBA-based glucose-sensitive microgels.

PBA-functionalized cross-linking microgels exhibit glucose-induced swelling or shrinking depending on the structure of borate ester between PBA and glucose. At lower glucose concentration, the glucose–bis(boronate) complexes induce the shrinking of microgels with a swollen state. In this condition, there is contraction-type glucose sensitivity. At higher glucose concentration, glucose–mono(boronate) complexes lead to glucose-induced expansion of microgels. Even though the investigations were performed with in vitro studies, it is still worth emphasizing that the reversible glucose-triggered swelling/shrinking changes of PBA-based microgels have potential application in self-regulated drug delivery.

4. PBA-Functionalized Nanogels

Even though microgels are most commonly used as the glucose-sensitive drug delivery carrier, the current studies are achieved with the VPTT and in vitro drug delivery. In contrast, nanogels with cross-linking structure are more stable and have received much more attention in drug delivery systems [72]. Nanogels, nanosized particles with cross-linking polymer networks, combine the properties of both hydrogels and nanomaterials [73]. Nanogels have enhanced stability with a long blood circulation time because the particle size and surface properties can be manipulated to avoid rapid clearance by phagocytic cells [74,75]. In addition, a larger surface area of the nanogel is convenient for functionalization of the nanogel [76].

There are several methods for the preparation of glucose-sensitive nanogels based on PBA. Li prepared glucose-sensitive nanogels by one-pot copolymerization of polyethylene glycol methylacrylate (PEGMEM) and AAPBA with BIS as a cross-linking agent [77]. Hollow nanogels composed of PNIPAM and poly(N-phenylboronic acid acrylamide) were obtained by two-step colloidal template polymerization [78]. The nanogels had interpenetrating polymer network structure which provided the stability of the nanogels. The morphological structure of inner cavity increased the drug loading content of the nanogels. More importantly, the nanogels were both temperature- and glucose-responsive.

By one-pot copolymerization of pentaerythritoltetra (3-mercaptopropionate) (QT), poly(ethylene glycol) diacrylate (PEGDA), poly(ethylene glycol) acrylate (mPEGA), and AAPBA, a nanogel with a disulfide cross-linked core and PEG shell was obtained [79]. The copolymerization of the monomers was performed via thiol-ene click reaction between thiol and double bond. The obtained nanogels had a core–shell structure. The core was performed by tetrathiol functional QT and bifunctional PEGDA with the free mercapto groups in the nanogel, and the free mercapto groups were terminated by AAPBA and mPEGA (Figure 11). The introduction of PBA group endowed the nanogels with remarkable glucose sensitivity, which was confirmed by fluorescence spectrometry using Alizarin red S (ARS) as a fluorescent probe. Insulin, a model drug, was released from the nanogels with highly glucose concentration dependence. Additionally, methyl thiazolyl tetrazolium, lactate dehydrogenase, and hemolysis assays confirmed the nontoxicity and biocompatibility of the nanogels. The glucose-sensitive nanogel with good biocompatibility has promising potential for application in self-regulated drug delivery.

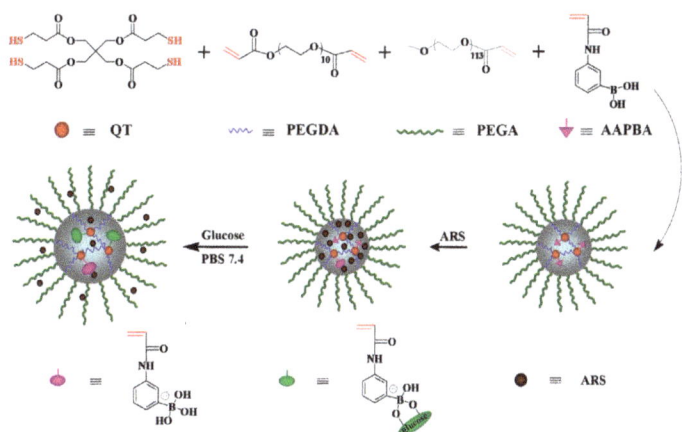

Figure 11. Structure of the nanogel and glucose-sensitive behavior of ARS-loaded nanogel in PBS at pH 7.4. (Reprinted from ref. [79]).

Nondegradable materials are unfavorable for blood clearance after drug delivery, which limits their clinical application in self-regulated drug delivery. The glucose-sensitive nanogels with improved biocompatibility and degradability have attracted more interest. A novel multifunctional chitosan and PBA-based nanohydrogel with enhanced glucose sensitivity was designed, which was prepared by the modification of chitosan-poly (acrylamide-co-methacrylic acid) nanohydrogel with 3-APBA [80]. The glucose-triggered volume phase transition and release profile of the model drug ARS (comparative to insulin as a drug as well as a dye for bioseparation) were studied at various glucose concentrations, pH, and ionic strengths. The nanohydrogel may find applications in bioseparation and glucose-induced drug delivery with enhanced sensitivity toward glucose.

Besides chitosan, polypeptides are greatly used in drug delivery systems. Chen and coworkers exploited a novel kind of glucose-sensitive polypeptide nanogels via a two-step procedure. The glycopolypeptide, methoxy poly(ethylene glycol)-*block*-poly(γ-benzyl-L-glutamate-*co*-(γ-propargyl-L-glutamate-graft-glucose) (mPEG-*b*-P(BLG-*co*-(PLG-g-Glu))) was fabricated by clicking 2′-azidoethyl-O-α-D-glucopyranoside to the PLG unit in mPEG-*b*-P(BLG-*co*-PLG). Then, the mPEG-*b*-P(BLG-*co*-(PLG-*g*-Glu) was cross-linked through boronate esters between the glucose moieties on glycopolypeptide with adipoylamidophenylboronic acid resulting in the glucose-sensitive polypeptide nanogels [81]. Insulin, a model drug, was loaded in the nanogels and the insulin release from the nanogels possessed excellent glucose sensitivity (Figure 12). There was a competitive binding mechanism for glucose-triggered insulin release. When free glucose was added, the complexes between PBA and glucose moieties on glycopolypeptide were destroyed due to the formed complexes between PBA and free glucose. Therefore, more free glucose entered into the nanogel core and the cross-linking density of the nanogels decreased, endowing the nanogels with more hydrophilicity and swelling. As a result, the preloaded insulin was released, induced by glucose, and high glucose concentration triggered more insulin release with a higher release rate. In addition, the polypeptide nanogels exhibited good cytocompatibility and hemocompatibility. The biocompatible nanogels with intelligent glucose-induced insulin release ability may have potential applications in diabetes therapy.

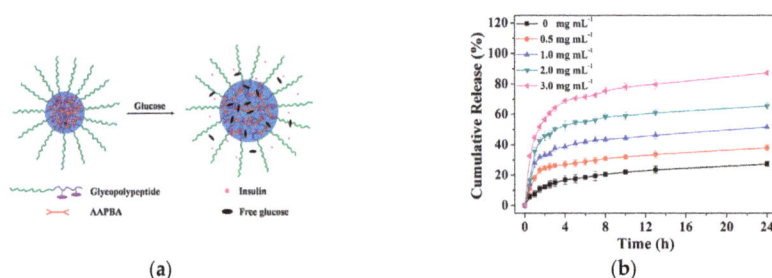

(a)　　　　　　　　　　　　　　　　　　(b)

Figure 12. (a) Schematic illustration of glucose-triggered insulin release from PBA-functionalized polypeptide nanogel; (b) Cumulative insulin release from the insulin-loaded nanogel in PBS with various glucose concentrations at pH 7.4, 37 °C. (Reprinted from ref. [81]).

Also using polypeptide, glycol chitosan (GC)/sodium alginate(SA)-poly(L-glutmate-*co*-N-3-L-glutamylphenylboronic acid) (PGGA) graft polymer (GC/SA-PGGA) double-layered nanogels were prepared by an isotropic gelation method and electrostatic interactions between GC and SA-PGGA (Figure 13a) [82]. In glucose solution, the binding between glucose molecules and PBA moieties on PGGA converted the PGGA chains to hydrophilic structures. Therefore, the hydrophilicity of nanogels was enhanced, resulting in the swelling of nanogels and consequent insulin release triggered by glucose. Furthermore, a mouse study was conducted to demonstrate the controlled insulin release capability of GC/SA-PGGA double-layered nanogel in vivo. Glucose was administrated to mice by retro-orbital injection to raise the blood glucose levels to diabetic glucose ranges at 0 min. Excluding the blank GC/SA-PGGA group, other groups were administrated glucose again at 70 min (Figure 13b). The blood glucose levels of the groups (excluding blank GC/SA-PGGA group) possessed a similar behavior with progressively lowered glucose levels before the second glucose injection. However, after the second glucose injection, the blood glucose levels of the free insulin and insulin-loaded GC/SA groups no longer had significant decrease. In contrast, the blood glucose level of insulin-loaded GC/SA-PGGA nanogels group was kept relatively low for a longer time, indicating that the nanogels exhibited controlled insulin release with high pharmacological activity to decrease the blood glucose levels. The glucose-triggered insulin release makes the GC/SA-PGGA double-layered nanogels a promising approach for diabetes treatment.

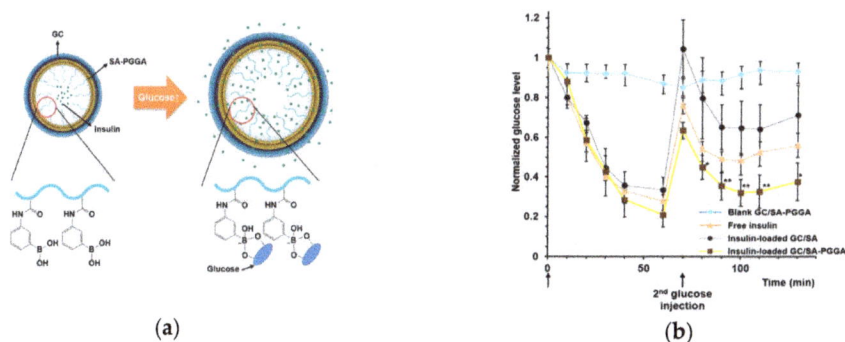

(a)　　　　　　　　　　　　　　　　　　(b)

Figure 13. (a) Schematic illustration of glucose-sensitive GC/SA-PGGA double-layered nanogel controlled insulin release by complexation between PBA derivatives and glucose; (b) Normalized blood glucose levels in mice by retro-orbital administration of blank GC/SA-PGGA nanogels, free insulin (0.5 IU kg^{-1}), insulin-loaded GC/SA (0.5 IU kg^{-1}), and insulin-loaded GC/SA-PGGA nanogels (0.5 IU kg^{-1}). (Reprinted from ref. [82]).

An injectable nanogel with interpenetrating polymer networks of PNIPAM, dextran, and poly(3-acrylamidophenylboronic acid) (P(NIPAM-Dex-PBA)) was exploited using maleic acid-dextran as a cross-linker [83]. The nanogels presented reversible glucose sensitivity under physiological conditions, which was related to dextran content. The preloaded insulin was released from the nanogels with high dextran content triggered by glucose. More importantly, the considerable hypoglycemic effect of the insulin-loaded nanogels was also confirmed. As shown in Figure 14, in vivo experiments demonstrated that the blood glucose level of diabetic rats treated with insulin-loaded nanogels was maintained in a low state for almost two hours. The reduction of blood glucose level for insulin-loaded nanogels treated diabetic rats was 51% of the baseline level, which was associated with long-term controlled insulin release triggered by glucose. In addition, insulin-loaded nanogels maintained stable blood glucose levels without remarkable fluctuations of blood sugar. The insulin-loaded nanogels with prolonged and stable blood glucose reduction effect may have potential applications for diabetes treatment.

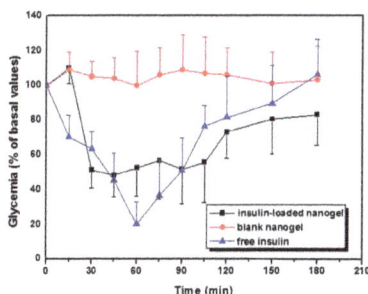

Figure 14. Profiles of glycemia after a subcutaneous administration of free insulin (2.0 IU kg^{-1}), insulin-loaded nanogels (4 IU kg^{-1}), and blank nanogels in fed diabetic rats. Before the injections, glycemia was 426 ± 13 mg mL^{-1}. Results are expressed as mean $\pm$ SD ($n = 8$). (Reprinted from ref. [83]).

Even though most PBA-functionalized nanogels are exploited to achieve the glucose sensitivity and glucose-triggered hypoglycemic effect, the particular and comprehensive studies on the blood compatibility of the nanogels have been reported rarely. Zhou and coworkers explored the glucose-sensitive insulin controlled release and blood compatibility of PBA-based nanogels [84]. The nanogels were prepared through a thermally initiated precipitation copolymerization of methacrylic acid (MAA) and AAPBA monomers using ethylene glycol dimethacrylate (EGDMA) as a cross-linker. By changing the molar ratio of MAA/AAPBA, nanogels with different PBA contents were obtained and noted as Glu(a/b), in which a/b represent the molar ratio of MAA/AAPBA. The release of FITC-insulin from Glu(2/3) was dependent on the glucose concentrations without initial burst release. In particular, the blood glucose levels of diabetic rats treated with insulin-loaded Glu(2/3) nanogels were maintained at a low concentration (below the 70% of initial blood glucose level) with a long-term stable hypoglycemic effect compared with the free insulin group (Figure 15a). Moreover, a series of blood assays and in vitro/vivo assays confirmed the good blood compatibility of PBA-based nanogels. The nanogels did not cause aggregation and morphological change of the red blood cells (Figure 15b). The activated partial thromboplastin time (APPT) is used to evaluate the intrinsic and common coagulation pathways. The glucose-sensitive nanogels had no interactions with the coagulation factors and/or partial thromboplastin reagent which was confirmed by APTT. In addition, the nanogels had little effect on the extrinsic pathway of blood coagulation examined by prothrombin time (PT). Thromboelastography (TEG) assay also indicated the anticoagulant effects which were attributed to the removal of thrombin from the blood by the nanogels via electrostatic interactions. Toxicity studies further confirmed the blood compatibility of the nanogels. This work provides a strategy for the study of biocompatibility of glucose-sensitive drug carriers. Thus, compatible nanogels with good hypoglycemic effect have potential application in diabetes therapy.

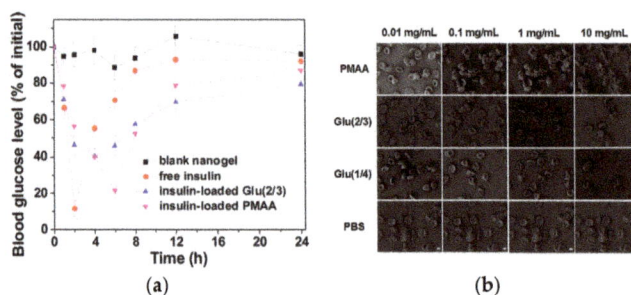

Figure 15. (a) The profiles of rats' blood glucose concentrations after injection with different samples; (b) Effect of the different nanogels on the aggregation and morphology of red blood cells. (Reprinted from ref. [84]).

5. Conclusions

PBA-functionalized hydrogels, microgels, and nanogels provide glucose-triggered drug release due to reversible swelling and shrinking induced by glucose. PBA-based glucose-sensitive hydrogels can be prepared by physical or chemical cross-linking, such as the self-aggregation of gelators, cross-linkage by cross-linkers, or reversible covalent complexation of PBA with *cis*-1,2 or *cis*-1,3-diol compounds. Even though the PBA-mediated hydrogels possess great glucose sensitivity, the poor ease of administration, i.e., they cannot be administrated by injection but probably as an implant, limit their application in diabetes therapy. In contrast, PBA-based microgels and nanogels with micro- or nano-scale dimensions exhibit excellent biocompatibility, tunable sizes, large bioconjugated surfaces, as well as convenient injection administration. For PBA-based microgels, studies are focused on the VPTT and the mechanism of glucose sensitivity. The development of PBA-based nanogels has seen significant progress in terms of glucose sensitivity, biocompatibility, and pharmacokinetic and hypoglycemic effect. PBA-functionalized nanogels with convenient administration via injection have practical application for diabetes therapy.

Even though there is considerable progress of PBA-based hydrogels, microgels, and nanogels for glucose-sensitive drug delivery, some challenges still have to be overcome to promote the clinical application of PBA-mediated platforms in diabetes therapy.

Firstly, the specificity and selectivity of glucose sensitivity are important for the design and application of glucose-sensitive platforms. Besides glucose, there are physiologically relevant saccharides, such as fructose, mannose, and galactose, which can bind to PBA and form mono(boronate) complexes. PBA and derivatives are able to form mono(boronate) complexes as well as bis(boronate) complexes with glucose. The influence of sugars on glucose sensitivity of PBA-based platforms must be considered. Additionally, the glucose sensitivity of drug carriers must be selective, that is, the platforms must timely and rapidly adjust the insulin release on-demand. The ideal drug carrier should release insulin quickly with precise dosage in response to hyperglycemic state, while not releasing glucose during normoglycemic states to keep the blood glucose levels in the normal range.

Secondly, adopting simplified preparation of PBA platforms is conducive to maintain the repeatability and controllability of the structure and property of glucose-sensitive carriers for different preparations. In addition, the glucose-sensitive platforms with high bioactivity of preloaded insulin should be administrated easily. Injection is an ideal approach because it can reach blood circulation with enhanced glucose sensitivity. The drug delivery system must be designed ingeniously with easy and repeatable preparation and convenient administration to promote its clinical application.

Thirdly, biocompatibility without long-term side effects of the platforms is a challenge for the design of PBA-based glucose-sensitive drug delivery systems. Diabetes is a lifelong chronic disease and the treatment of diabetes is a long-term process. Therefore, glucose-sensitive platforms must be nontoxic and friendly to the body without inducing inflammation. By adopting biodegradable and biocompatible materials, such as poly(acrylic acids) and polypeptides, the biocompatibility of PBA-based matrices can be enhanced.

Although some issues still need to be overcome, self-regulated drug delivery systems have promising application in diabetes treatment.

Author Contributions: The manuscript was completed through contributions of all authors. C.W., L.Z., and L.W. initiated the topic and designed the structure and main content of the article. C.W., B.L., H.Z., F.B., S.X., and G.G. wrote the article. L.Z. and L.W. revised the article.

Funding: This work was financially supported by National Natural Science Foundation of China (Grant No. 51403075, 51673190, 51603204, and 51473165) and the Scientific Development Program of Jilin Province (Grant No. 20170520153JH, 20170520152JH, 20180520217JH and 20160204015SF).

Conflicts of Interest: There is no conflict of interest regarding this manuscript.

References

1. Tripathi, B.K.; Srivastava, A.K. Diabetes mellitus: Complications and therapeutics. *Med. Sci. Monit.* **2006**, *12*, RA130–RA147. [PubMed]
2. Wild, S.; Roglic, G.; Green, A.; Sicree, R.; King, H. Global prevalence of diabetes—Estimates for the year 2000 and projections for 2030. *Diabetes Care* **2004**, *27*, 1047–1053. [CrossRef] [PubMed]
3. Renard, E. Implantable closed-loop glucose-sensing and insulin delivery: The future for insulin pump therapy. *Curr. Opin. Pharmacol.* **2002**, *2*, 708–716. [CrossRef]
4. Chou, D.H.-C.; Webber, M.J.; Tang, B.C.; Lin, A.B.; Thapa, L.S.; Deng, D.; Truong, J.V.; Cortinas, A.B.; Langer, R.; Anderson, D.G. Glucose-responsive insulin activity by covalent modification with aliphatic phenylboronic acid conjugates. *Proc. Natl. Acad. Sci. USA* **2015**, *112*, 2401–2406. [CrossRef] [PubMed]
5. Wu, Q.; Wang, L.; Yu, H.; Wang, J.; Chen, Z. Organization of glucose-responsive systems and their properties. *Chem. Rev.* **2011**, *111*, 7855–7875. [CrossRef] [PubMed]
6. Kost, J.; Horbett, T.A.; Ratner, B.D.; Singh, M. Glucose-sensitive membranes containing glucose oxidase: Activity, swelling, and permeability studies. *J. Biomed. Mater. Res.* **1985**, *19*, 1117–1133. [CrossRef]

7. Luo, R.M.; Li, H. Parameter study of glucose-sensitive hydrogel: Effect of immobilized glucose oxidase on diffusion and deformation. *Soft Mater.* **2013**, *11*, 69–74. [CrossRef]
8. Luo, J.; Cao, S.Q.; Chen, X.Y.; Liu, S.N.; Tan, H.; Wu, W.; Li, J.S. Super long-term glycemic control in diabetic rats by glucose-sensitive lbl films constructed of supramolecular insulin assembly. *Biomaterials* **2012**, *33*, 8733–8742. [CrossRef]
9. Díez, P.; Sánchez, A.; Gamella, M.; Martínez-Ruíz, P.; Aznar, E.; Torre, C.; Murguía, J.R.; Martínez-Máñez, R.; Villalonga, R.; Pingarrón, J.M. Toward the design of smart delivery systems controlled by integrated enzyme-based biocomputing ensembles. *J. Am. Chem. Soc.* **2014**, *136*, 9116–9123. [CrossRef]
10. Zhao, W.; Zhang, H.; He, Q.; Li, Y.; Gu, J.; Li, L.; Li, H.; Shi, J. A glucose-responsive controlled release of insulin system based on enzyme multilayers-coated mesoporous silica particles. *Chem. Commun.* **2011**, *47*, 9459–9461. [CrossRef]
11. Taylor, M.J.; Tanna, S.; Sahota, T.S.; Voermans, B. Rheological characterisation of dextran-concanavalin a mixtures as a basis for a self-regulated drug delivery device. *Eur. J. Pharm. Biopharm.* **2006**, *62*, 94–100. [CrossRef]
12. Tanna, S.; Sahota, T.S.; Sawicka, K.; Taylor, M.J. The effect of degree of acrylic derivatisation on dextran and concanavalin a glucose-responsive materials for closed-loop insulin delivery. *Biomaterials* **2006**, *27*, 4498–4507. [CrossRef]
13. Yoshida, K.; Hasebe, Y.; Takahashi, S.; Sato, K.; Anzai, J.-i. Layer-by-layer deposited nano- and micro-assemblies for insulin delivery: A review. *Mater. Sci. Eng. C* **2014**, *34*, 384–392. [CrossRef]
14. Makino, K.; Mack, E.J.; Okano, T.; Sung Wan, K. A microcapsule self-regulating delivery system for insulin. *J. Control. Release* **1990**, *12*, 235–239. [CrossRef]
15. Samoszuk, M.; Ehrlich, D.; Ramzi, E. Preclinical safety studies of glucose-oxidase. *J. Pharmacol. Exp. Ther.* **1993**, *266*, 1643–1648.
16. Ballerstadt, R.; Evans, C.; McNichols, R.; Gowda, A. Concanavalin A for in vivo glucose sensing: A biotoxicity review. *Biosens. Bioelectron.* **2006**, *22*, 275–284. [CrossRef]
17. Tang, Z.; Guan, Y.; Zhang, Y. Contraction-type glucose-sensitive microgel functionalized with a 2-substituted phenylboronic acid ligand. *Polym. Chem.* **2014**, *5*, 1782–1790. [CrossRef]
18. Ding, Z.; Guan, Y.; Zhang, Y.; Zhu, X.X. Synthesis of glucose-sensitive self-assembled films and their application in controlled drug delivery. *Polymer* **2009**, *50*, 4205–4211. [CrossRef]
19. Ding, Z.; Guan, Y.; Zhang, Y.; Zhu, X.X. Layer-by-layer multilayer films linked with reversible boronate ester bonds with glucose-sensitivity under physiological conditions. *Soft Matter* **2009**, *5*, 2302–2309. [CrossRef]
20. Jin, X.; Zhang, X.; Wu, Z.; Teng, D.; Zhang, X.; Wang, Y.; Wang, Z.; Li, C. Amphiphilic random glycopolymer based on phenylboronic acid: Synthesis, characterization, and potential as glucose-sensitive matrix. *Biomacromolecules* **2009**, *10*, 1337–1345. [CrossRef]
21. Shiino, D.; Murata, Y.; Kataoka, K.; Koyama, Y.; Yokoyama, M.; Okano, T.; Sakurai, Y. Preparation and characterization of a glucose-responsive insulin-releasing polymer device. *Biomaterials* **1994**, *15*, 121–128. [CrossRef]
22. Wu, W.; Zhou, S. Responsive materials for self-regulated insulin delivery. *Macromol. Biosci.* **2013**, *13*, 1464–1477. [CrossRef]
23. Nishiyabu, R.; Kubo, Y.; James, T.D.; Fossey, J.S. Boronic acid building blocks: Tools for self assembly. *Chem. Commun.* **2011**, *47*, 1124–1150. [CrossRef]
24. Yan, J.; Springsteen, G.; Deeter, S.; Wang, B. The relationship among pKa, pH, and binding constants in the interactions between boronic acids and diols—It is not as simple as it appears. *Tetrahedron* **2004**, *60*, 11205–11209. [CrossRef]
25. Matsumoto, A.; Ikeda, S.; Harada, A.; Kataoka, K. Glucose-responsive polymer bearing a novel phenylborate derivative as a glucose-sensing moiety operating at physiological pH conditions. *Biomacromolecules* **2003**, *4*, 1410–1416. [CrossRef]
26. Zhao, L.; Xiao, C.; Wang, L.; Gai, G.; Ding, J. Glucose-sensitive polymer nanoparticles for self-regulated drug delivery. *Chem. Commun.* **2016**, *52*, 7633–7652. [CrossRef]
27. Matsumoto, A.; Yamamoto, K.; Yoshida, R.; Kataoka, K.; Aoyagi, T.; Miyahara, Y. A totally synthetic glucose responsive gel operating in physiological aqueous conditions. *Chem. Commun.* **2010**, *46*, 2203–2205. [CrossRef]

28. Lee, S.; Nam, J.H.; Kim, Y.J.; Cho, Y.J.; Kwon, N.H.; Lee, J.Y.; Kang, H.-J.; Kim, H.T.; Park, H.M.; Kim, S.; et al. Synthesis of PEO-based glucose-sensitive block copolymers and their application for preparation of superparamagnetic iron oxide nanoparticles. *Macromol. Res.* **2011**, *19*, 827–834. [CrossRef]

29. Ravaine, V.; Ancla, C.; Catargi, B. Chemically controlled closed-loop insulin delivery. *J. Control. Release* **2008**, *132*, 2–11. [CrossRef]

30. Sato, K.; Yoshida, K.; Takahashi, S.; Anzai, J. pH- and sugar-sensitive layer-by-layer films and microcapsules for drug delivery. *Adv. Drug Deliv. Rev.* **2011**, *63*, 809–821. [CrossRef]

31. Ma, R.; Shi, L. Phenylboronic acid-based glucose-responsive polymeric nanoparticles: Synthesis and applications in drug delivery. *Polym. Chem.* **2014**, *5*, 1503–1518. [CrossRef]

32. Zha, L.S.; Banik, B.; Alexis, F. Stimulus responsive nanogels for drug delivery. *Soft Matter* **2011**, *7*, 5908–5916. [CrossRef]

33. Jiang, Y.; Chen, J.; Deng, C.; Suuronen, E.J.; Zhong, Z. Click hydrogels, microgels and nanogels: Emerging platforms for drug delivery and tissue engineering. *Biomaterials* **2014**, *35*, 4969–4985. [CrossRef] [PubMed]

34. Guan, Y.; Zhang, Y. Boronic acid-containing hydrogels: Synthesis and their applications. *Chem. Soc. Rev.* **2013**, *42*, 8106–8121. [CrossRef] [PubMed]

35. Mandal, D.; Mandal, S.K.; Ghosh, M.; Das, P.K. Phenylboronic acid appended pyrene-based low-molecular-weight injectable hydrogel: Glucose-stimulated insulin release. *Chem. Eur. J.* **2015**, *21*, 12042–12052. [CrossRef] [PubMed]

36. Zhou, C.; Gao, W.; Yang, K.; Xu, L.; Ding, J.; Chen, J.; Liu, M.; Huang, X.; Wang, S.; Wu, H. A novel glucose/pH responsive low-molecular-weight organogel of easy recycling. *Langmuir* **2013**, *29*, 13568–13575. [CrossRef]

37. Xu, L.; Hu, Y.; Liu, M.; Chen, J.; Huang, X.; Gao, W.; Wu, H. Gelation properties and glucose-sensitive behavior of phenylboronic acid based low-molecular-weight organogels. *Tetrahedron* **2015**, *71*, 2079–2088. [CrossRef]

38. Tao, N.; Li, G.; Liu, M.; Gao, W.; Wu, H. Preparation of dual responsive low-molecular-weight hydrogel for long-lasting drug delivery. *Tetrahedron* **2017**, *73*, 3173–3180. [CrossRef]

39. Wang, L.; Liu, M.; Gao, C.; Ma, L.; Cui, D. A pH-, thermo-, and glucose-, triple-responsive hydrogels: Synthesis and controlled drug delivery. *React. Funct. Polym.* **2010**, *70*, 159–167. [CrossRef]

40. Huang, Y.; Liu, M.; Wang, L.; Gao, C.; Xi, S. A novel triple-responsive poly(3-acrylamidephenylboronic acid-co-2-(dimethylamino) ethyl methacrylate)/(β-cyclodextrin-epichlorohydrin)hydrogels: Synthesis and controlled drug delivery. *React. Funct. Polym.* **2011**, *71*, 666–673. [CrossRef]

41. Zhang, S.-B.; Chu, L.-Y.; Xu, D.; Zhang, J.; Ju, X.-J.; Xie, R. Poly(N-isopropylacrylamide)-based comb-type grafted hydrogel with rapid response to blood glucose concentration change at physiological temperature. *Polym. Adv. Technol.* **2008**, *19*, 937–943. [CrossRef]

42. Cho, S.-H.; Tathireddy, P.; Rieth, L.; Magda, J. Effect of chemical composition on the response of zwitterionic glucose sensitive hydrogels studied by design of experiments. *J. Appl. Polym. Sci.* **2014**, *131*, 1–7. [CrossRef]

43. Yesilyurt, V.; Ayoob, A.M.; Appel, E.A.; Borenstein, J.T.; Langer, R.; Anderson, D.G. Mixed reversible covalent crosslink kinetics enable precise, hierarchical mechanical tuning of hydrogel networks. *Adv. Mater.* **2017**, *29*. [CrossRef] [PubMed]

44. Yesilyurt, V.; Webber, M.J.; Appel, E.A.; Godwin, C.; Langer, R.; Anderson, D.G. Injectable self-healing glucose-responsive hydrogels with pH-regulated mechanical properties. *Adv. Mater.* **2016**, *28*, 86–91. [CrossRef]

45. Yang, T.; Ji, R.; Deng, X.-X.; Du, F.-S.; Li, Z.-C. Glucose-responsive hydrogels based on dynamic covalent chemistry and inclusion complexation. *Soft Matter* **2014**, *10*, 2671–2678. [CrossRef] [PubMed]

46. Cai, B.; Luo, Y.; Guo, Q.; Zhang, X.; Wu, Z. A glucose-sensitive block glycopolymer hydrogel based on dynamic boronic ester bonds for insulin delivery. *Carbohydr. Res.* **2017**, *445*, 32–39. [CrossRef]

47. Dong, Y.; Wang, W.; Veiseh, O.; Appel, E.A.; Xue, K.; Webber, M.J.; Tang, B.C.; Yang, X.-W.; Weir, G.C.; Langer, R.; et al. Injectable and glucose-responsive hydrogels based on boronic acid–glucose complexation. *Langmuir* **2016**, *32*, 8743–8747. [CrossRef] [PubMed]

48. Cao, J.; Liu, S.; Chen, Y.; Shi, L.; Zhang, Z. Synthesis of end-functionalized boronic acid containing copolymers and their bioconjugates with rod-like viruses for multiple responsive hydrogels. *Polym. Chem.* **2014**, *5*, 5029–5036. [CrossRef]

49. Liu, P.; Luo, Q.; Guan, Y.; Zhang, Y. Drug release kinetics from monolayer films of glucose-sensitive microgel. *Polymer* **2010**, *51*, 2668–2675. [CrossRef]
50. Pelton, R. Temperature-sensitive aqueous microgels. *Adv. Colloid Interface Sci.* **2000**, *85*, 1–33. [CrossRef]
51. Matsumoto, A.; Kurata, T.; Shiino, D.; Kataoka, K. Swelling and shrinking kinetics of totally synthetic, glucose-responsive polymer gel bearing phenylborate derivative as a glucose-sensing moiety. *Macromolecules* **2004**, *37*, 1502–1510. [CrossRef]
52. Xing, S.; Guan, Y.; Zhang, Y. Kinetics of glucose-induced swelling of p(NIPAM-AAPBA) microgels. *Macromolecules* **2011**, *44*, 4479–4486. [CrossRef]
53. Wu, Q.; Wang, L.; Yu, H.; Chen, Z. The synthesis and responsive properties of novel glucose-responsive microgels. *Polym. Sci. Ser. A* **2012**, *54*, 209–213. [CrossRef]
54. Farooqi, Z.H.; Khan, A.; Siddiq, M. Temperature-induced volume change and glucose sensitivity of poly[(*N*-isopropylacry-lamide)-*co*-acrylamide-*co*-(phenylboronic acid)] microgels. *Polym. Int.* **2011**, *60*, 1481–1486. [CrossRef]
55. Farooqi, Z.H.; Wu, W.; Zhou, S.; Siddiq, M. Engineering of phenylboronic acid based glucose-sensitive microgels with 4-vinylpyridine for working at physiological pH and temperature. *Macromol. Chem. Phys.* **2011**, *212*, 1510–1514. [CrossRef]
56. Wang, D.; Liu, T.; Yin, J.; Liu, S. Stimuli-responsive fluorescent poly(*N*-isopropylacrylamide) microgels labeled with phenylboronic acid moieties as multifunctional ratiometric probes for glucose and temperatures. *Macromolecules* **2011**, *44*, 2282–2290. [CrossRef]
57. Hoare, T.; Pelton, R. Engineering glucose swelling responses in poly(*N*-isopropylacrylamide)-based microgels. *Macromolecules* **2007**, *40*, 670–678. [CrossRef]
58. Bitar, A.; Fessi, H.; Elaissari, A. Synthesis and characterization of thermally and glucose-sensitive poly N-vinylcaprolactam-based microgels. *J. Biomed. Nanotechnol.* **2012**, *8*, 709–719. [CrossRef] [PubMed]
59. Hoare, T.; Pelton, R. Charge-switching, amphoteric glucose-responsive microgels with physiological swelling activity. *Biomacromolecules* **2008**, *9*, 733–740. [CrossRef]
60. Liu, Y.; Zhang, Y.; Guan, Y. New polymerized crystalline colloidal array for glucose sensing. *Chem. Commun.* **2009**, 1867–1869. [CrossRef]
61. Lapeyre, V.; Gosse, I.; Chevreux, S.; Ravaine, V. Monodispersed glucose-responsive microgels operating at physiological salinity. *Biomacromolecules* **2006**, *7*, 3356–3363. [CrossRef]
62. Zhang, Y.; Guan, Y.; Zhou, S. Synthesis and volume phase transitions of glucose-sensitive microgels. *Biomacromolecules* **2006**, *7*, 3196–3201. [CrossRef] [PubMed]
63. Wu, Q.; Du, X.; Chang, A.; Jiang, X.; Yan, X.; Cao, X.; Farooqi, Z.H.; Wu, W. Bioinspired synthesis of poly(phenylboronic acid) microgels with high glucose selectivity at physiological ph. *Polym. Chem.* **2016**, *7*, 6500–6512. [CrossRef]
64. Kondo, K.; Shiomi, Y.; Saisho, M.; Harada, T.; Shinkai, S. Specific complexation of disaccharides with diphenyl-3,3′-diboronic acid that can be detected by circular dichroism. *Tetrahedron* **1992**, *48*, 8239–8252. [CrossRef]
65. Norrild, J.C.; Eggert, H. Evidence for mono- and bisdentate boronate complexes of glucose in the furanose form. Application of 1JC-C coupling constants as a structural probe. *J. Am. Chem. Soc.* **1995**, *117*, 1479–1484. [CrossRef]
66. Ancla, C.; Lapeyre, V.; Gosse, I.; Catargi, B.; Ravaine, V. Designed glucose-responsive microgels with selective shrinking behavior. *Langmuir* **2011**, *27*, 12693–12701. [CrossRef] [PubMed]
67. Alexeev, V.L.; Sharma, A.C.; Goponenko, A.V.; Das, S.; Lednev, I.K.; Wilcox, C.S.; Finegold, D.N.; Asher, S.A. High ionic strength glucose-sensing photonic crystal. *Anal. Chem.* **2003**, *75*, 2316–2323. [CrossRef] [PubMed]
68. Zhou, M.; Lu, F.; Jiang, X.; Wu, Q.; Chang, A.; Wu, W. Switchable glucose-responsive volume phase transition behavior of poly(phenylboronic acid) microgels. *Polym. Chem.* **2015**, *6*, 8306–8318. [CrossRef]
69. Zhang, X.; Lu, S.; Gao, C.; Chen, C.; Zhang, X.; Liu, M. Highly stable and degradable multifunctional microgel for self-regulated insulin delivery under physiological conditions. *Nanoscale* **2013**, *5*, 6498–6506. [CrossRef]
70. Luo, Q.; Liu, P.; Guan, Y.; Zhang, Y. Thermally induced phase transition of glucose-sensitive core-shell microgels. *ACS Appl. Mater. Interfaces* **2010**, *2*, 760–767. [CrossRef]
71. Zhang, Y.; Guan, Y.; Zhou, S. Permeability control of glucose-sensitive nanoshells. *Biomacromolecules* **2007**, *8*, 3842–3847. [CrossRef] [PubMed]

72. Wu, W.; Chen, S.; Hu, Y.; Zhou, S. A fluorescent responsive hybrid nanogel for closed-loop control of glucose. *J. Diabetes Sci. Technol.* **2012**, *6*, 892–901. [CrossRef]

73. Hamidi, M.; Azadi, A.; Rafiei, P. Hydrogel nanoparticles in drug delivery. *Adv. Drug Deliv. Rev.* **2008**, *60*, 1638–1649. [CrossRef]

74. Sharma, A.; Garg, T.; Aman, A.; Panchal, K.; Sharma, R.; Kumar, S.; Markandeywar, T. Nanogel—An advanced drug delivery tool: Current and future. *Artif. Cells Nanomed. Biotechnol.* **2016**, *44*, 165–177. [CrossRef] [PubMed]

75. Molina, M.; Asadian-Birjand, M.; Balach, J.; Bergueiro, J.; Miceli, E.; Calderon, M. Stimuli-responsive nanogel composites and their application in nanomedicine. *Chem. Soc. Rev.* **2015**, *44*, 6161–6186. [CrossRef]

76. Mauri, E.; Perale, G.; Rossi, F. Nanogel functionalization: A versatile approach to meet the challenges of drug and gene delivery. *ACS Appl. Nano Mater.* **2018**, *1*, 6525–6541. [CrossRef]

77. Li, L.; Jiang, G.; Jiang, T.; Huang, Q.; Chen, H.; Liu, Y. Preparation of dual-responsive nanogels for controlled release of insulin. *J. Polym. Mater.* **2015**, *32*, 77–84.

78. Wang, C.; Xing, Z.; Yan, J.; Li, L.; Zhao, H.; Zha, L. Glucose and temperature dual stimuli responsiveness of intelligent hollow nanogels. *Chin. J. Mater. Res.* **2012**, *26*, 44–48.

79. Zhao, L.; Xiao, C.; Ding, J.; He, P.; Tang, Z.; Pang, X.; Zhuang, X.; Chen, X. Facile one-pot synthesis of glucose-sensitive nanogel via thiol-ene click chemistry for self-regulated drug delivery. *Acta Biomater.* **2013**, *9*, 6535–6543. [CrossRef]

80. Ullah, F.; Othman, M.B.H.; Javed, F.; Ahmad, Z.; Akil, H.M.; Rasib, S.Z.M. Functional properties of chitosan built nanohydrogel with enhanced glucose-sensitivity. *Int. J. Biol. Macromol.* **2016**, *83*, 376–384. [CrossRef]

81. Zhao, L.; Xiao, C.; Ding, J.; Zhuang, X.; Gai, G.; Wang, L.; Chen, X. Competitive binding-accelerated insulin release from a polypeptide nanogel for potential therapy of diabetes. *Polym. Chem.* **2015**, *6*, 3807–3815. [CrossRef]

82. Lee, D.; Choe, K.; Jeong, Y.; Yoo, J.; Lee, S.M.; Park, J.-H.; Kim, P.; Kim, Y.-C. Establishment of a controlled insulin delivery system using a glucose-responsive double-layered nanogel. *RSC Adv.* **2015**, *5*, 14482–14491. [CrossRef]

83. Wu, Z.; Zhang, X.; Guo, H.; Li, C.; Yu, D. An injectable and glucose-sensitive nanogel for controlled insulin release. *J. Mater. Chem.* **2012**, *22*, 22788–22796. [CrossRef]

84. Zhou, X.; Lin, A.; Yuan, X.; Li, H.; Ma, D.; Xue, W. Glucose-sensitive and blood-compatible nanogels for insulin controlled release. *J. Appl. Polym. Sci.* **2016**, *133*, 1–8. [CrossRef]

molecules

Review

Hydrogels and Their Applications in Targeted Drug Delivery

Radhika Narayanaswamy and Vladimir P. Torchilin *

Center for Pharmaceutical Biotechnology and Nanomedicine, Northeastern University, Boston, MA 02115, USA;
narayanaswamy.r@husky.neu.edu
* Correspondence: v.torchilin@neu.edu; Tel.: +1-617-373-3206; Fax: +1-617-373-8886

Received: 25 December 2018; Accepted: 2 February 2019; Published: 8 February 2019

Abstract: Conventional drug delivery approaches are plagued by issues pertaining to systemic toxicity and repeated dosing. Hydrogels offer convenient drug delivery vehicles to ensure these disadvantages are minimized and the therapeutic benefits from the drug are optimized. With exquisitely tunable physical properties that confer them great controlled drug release features and the merits they offer for labile drug protection from degradation, hydrogels emerge as very efficient drug delivery systems. The versatility and diversity of the hydrogels extend their applications beyond targeted drug delivery also to wound dressings, contact lenses and tissue engineering to name but a few. They are 90% water, and highly porous to accommodate drugs for delivery and facilitate controlled release. Herein we discuss hydrogels and how they could be manipulated for targeted drug delivery applications. Suitable examples from the literature are provided that support the recent advancements of hydrogels in targeted drug delivery in diverse disease areas and how they could be suitably modified in very different ways for achieving significant impact in targeted drug delivery. With their enormous amenability to modification, hydrogels serve as promising delivery vehicles of therapeutic molecules in several disease conditions, including cancer and diabetes.

Keywords: hydrogels; applications; targeted drug delivery; drug release; hydrophobic drug delivery; clinical translation; versatile platform; administration routes; diverse therapeutic areas

1. Introduction

Hydrophilic polymeric networks that are capable of imbibing huge volumes of water and undergoing swelling and shrinkage suitably to facilitate controlled drug-release are called hydrogels. Their porosity and compatibility with aqueous environments make them highly attractive bio-compatible drug delivery vehicles. Their applications are manifold and for several biomedical needs as they are moldable into varied physical forms such as nanoparticles, microparticles, slabs, films and coatings. USA has been the largest producer of hydrogels and is expected to remain so for a few more years [1]. Hydrogels are promising, trendy, intelligent and 'smart' drug delivery vehicles that cater to the specific requirements for targeting drugs to the specific sites and controlling drug release. Enzymatic, hydrolytic or environmental stimuli often suffice to manipulate the hydrogels for the drug release at the desirable site [2]. Like the two sides of a coin, there are also the disadvantages associated with their use. The primary disadvantage in drug delivery would be the hydrophobicity of most drugs. The water-loving polymeric core is probably not very ideal to hold incompatible hydrophobic drugs, which is a challenge since many that are currently used and effective in disease therapy are hydrophobic. The tensile strength of these hydrogels is weak and this sometimes causes early release of the drug before arrival at the target site. The following review discusses on how hydrogels are being manipulated presently for improved targeted drug delivery. The modern trend in which the hydrogels are exploited for drug delivery are covered.

The attractive physical properties of hydrogels, especially their porosity, offer tremendous advantages in drug delivery applications such as sustained release of the loaded drug. A high local concentration of the active pharmaceutical ingredient is retained over a long period of time via a suitable release mechanism controlled by diffusion, swelling, chemical or based on some environmental stimuli.

Diffusion-controlled drug delivery with hydrogels uses reservoir or matrix devices that allow diffusion-based drug release through a hydrogel mesh or pores filled with water. In the reservoir delivery system, the hydrogel membrane is coated on a drug-containing core producing capsules, spheres or slabs that have a high drug concentration in the very center of the system to facilitate a constant drug-release rate. While the reservoir delivery system produces time-independent and constant drug release, the matrix system works via the macromolecular pores or mesh. This type of release is time-dependent drug release wherein the initial release rate is proportional to the square root of time, rather than being constant (Figure 1).

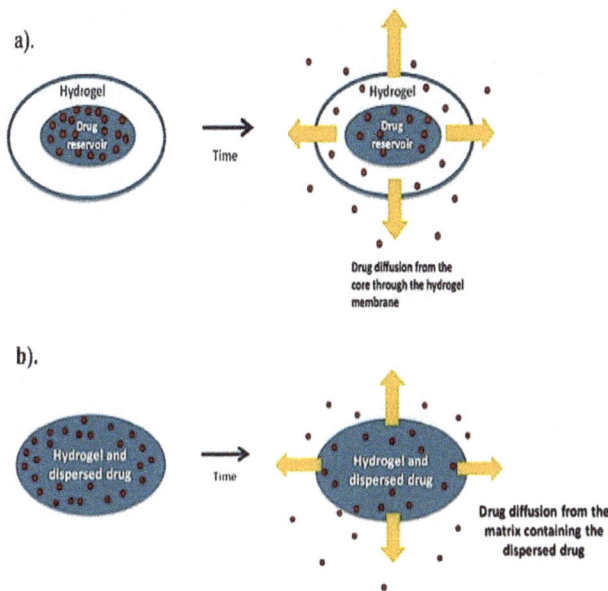

Figure 1. (a). Drug-containing core is coated with hydrogel membrane and the drug concentration is higher in the center of the system to allow constant release rate of the same in reservoir delivery system. **(b)**. Uniform dissolution or dispersion of the drug throughout the 3D structure of the hydrogel is achieved using matrix delivery (reprinted (adapted) with permission from [3]; the article is open access and the content reusable).

The swelling-controlled drug release from hydrogels uses drugs dispersed within a glassy polymer which when in contact with a bio-fluid begins swelling. The expansion during swelling occurs beyond its boundary facilitating the drug diffusion along with the polymer chain relaxation. The process, otherwise referred to as Case II transport, supports time-independent, constant drug release kinetics. Since the gradient between the dispersed drug in the hydrogel and its surrounding environment allows the active ingredient diffusion from a region of higher concentration within the hydrogel to a lower one, the process is also referred to as anomalous transport as it combines both the processes of diffusion and swelling for enabling drug release.

Ocular drug delivery carriers have been developed using hydrogels that are covalently crosslinked. These soft, biodegradable hydrogels with high swelling capacity remain in-situ in the lacrimal canal offering greater comfort for the patient. Collagen or silicone that may be used decides if the punctal-plug system could be used temporarily or permanently, respectively. Poly(ethylene glycol) hydrogels are commonly used for producing ophthalmic drug delivery systems.

Drug release in response to environmental changes would be an ideal delivery system as the release becomes very controlled and non-specific side effects at off-target sites are alleviated. Thus, sensitive drug delivery devices responsive to changes in pH, temperature, ionic strength or glucose concentration have been developed that are advantageous in the therapy of diseases such as cancer, and diabetes, characterized by local physiological changes specific to the various disease stages. The polymer composition of the hydrogel responsive to the environmental stimuli is manipulated to make it responsive to the environment [3].

Hydrogels considerably enhance the therapeutic outcome of drug delivery and have found enormous clinical use. The temporal and spatial delivery of macromolecular drugs, small molecules, and cells have greatly improved through hydrogel use for drug delivery [2]. Drug delivery using hydrogels, however, has not been free of challenges, but constant improvements are being made to identify the hydrogel design best suited for specific drug delivery purposes. Therefore, this paper discusses the recent trends in drug delivery applications using hydrogels, including their translation to the clinic and their applications to successfully deliver hydrophobic drugs.

2. Current Trend in Hydrogel Based Targeted Drug Delivery

2.1. Supramolecular Hydrogels

The supramolecular hydrogel system is composed of intermolecular interactions that are non-covalent and has two or more molecular entities held together. The non-covalent cross-linking is a very attractive aspect of these hydrogels as it helps circumvent the problems of limited drug loading potential and drug incorporation for use only as implantables which would be the only possibility with a covalently cross-linked network. Apart from offering the right physical stability for the hydrogels, these achieve drug loading and gelation simultaneously in an aqueous environment without the need for a covalent cross-linking. Recent progress has been made with supramolecular hydrogels using self-assembled inclusion complexes between cyclodextrins and bio-degradable block copolymers that provide sustained and controlled release of macromolecular drugs [4].

Natural cyclic oligosaccharides composed of six, seven or eight D-(+)-glucose units linked by D-(+)-1,4-linkages (termed α-, β- and γ-CD, respectively), are called cyclodextrins and are well-suited for use in supramolecular systems. They offer hydrophobic internal cavities with a suitable diameter and their ability to generate supramolecular inclusion complexes with various polymers make them ideal drug delivery vehicles.

A recent study involved development of a glycoconjugate prepared by amidation of homopoly-L-guluronic acid block obtained from *D. antarctica* sodium alginate with mono-6-amino-β-CD. The study was aimed at treating Chagas disease caused by *Trypanosoma cruzi*. Lipophilic non-hydroxylated coumarins were loaded into the hydrophobic core of the β-cyclodextrin to render them with trypanocidal activity. Interaction between the carboxylate groups of unconjugated α-L- glucuronate residues with calcium ions was used to produce supramolecular hydrogels of glycoconjugate of homopoly-L-guluronic block fraction (GG) with 6-NH_2-β-CD. As the *T. cruzi* parasites have only one mitochondrion, it is an ideal target for drugs to manipulate its energy process and apoptosis. Mitochondrial membrane potential studies revealed that the cyclodextrin complex with the drugs produced significant oxidative stress to destroy the parasites. The drug in the complex had increased solubility, showed improved bio-availability, controlled drug release and improved trypanocidal activity in comparison to the corresponding free amidocoumarins [5].

Cyclodextrin-functionalized polyhydrazines were used to prepare hydrogels in-situ via hydrazine bond formation with aldehyde groups on dextran aldehyde. No toxicity was observed *in vitro* with these hydrogels and they could accommodate nicardipine as hydrophobic drug into the cyclodextrin cavities. Steady release of nicardipine over 6 days was observed with the hydrogel preparation having higher hydrazine linkages. Thus, a gel capable of hydrophobic drug release in an in-situ formed device over extended periods was generated [6].

Bleeding control and wound healing by bio-adhesive hydrogels find enormous biomedical applications. In situ forming hydrogels are used to heal injured tissues based on their ability to accumulate and produce a fibrin bridge that permit fibroblast migration and collagen secretion for healing tissue injury. β-Cyclodextrins are a non-toxic adjuvant for pharmaceutical and mucoadhesive applications. Partly oxidized β-cyclodextrin was used in a recent study to exploit aldehyde groups on a hydrogel matrix for favorable reaction with amines in the tissue to result in an imine bond (Schiff's base reaction) in order to adhere to the skin and to provide improved cyclodextrin solubility in order to improve loading efficiency. Blending gelatin (the common extracellular component) with the β-cyclodextrin partly oxidized with oxidation in the presence of H_2O_2/horseradish peroxidase, resulted in very rapid formation of gelatin-β-cyclodextrin hydrogels (Figure 2). Hydrophobic drugs such as dexamethasone could be released with 2.7 fold higher efficacy when delivered in presence of the cyclodextrin relative to the gelatin-only hydrogels [7].

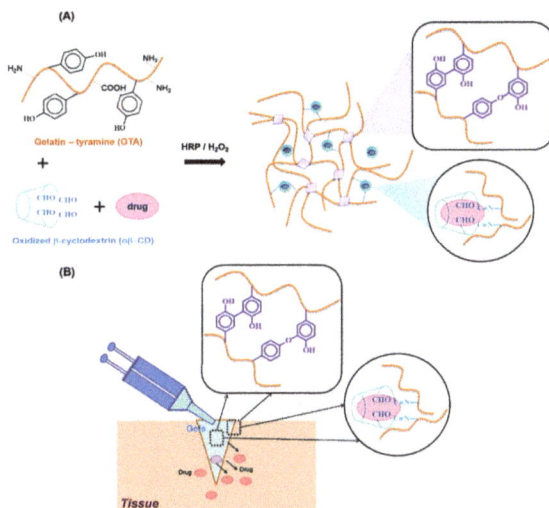

Figure 2. (**A**). Graphical representation of the methods for cross-linking to obtain gelatin-β-cyclodextrin (GTA–ob-CD) hydrogels to load hydrophobic drugs. (**B**). Schematic representation of adhesive GTA–ob-CD hydrogels in situ formed by combining HRP catalysis and the Schiff base reaction with therapeutic release (reprinted [7] with permission from the The Royal Society of Chemistry. The article is licensed by Creative Commons and the link to the license is https://creativecommons.org/licenses/by/3.0/).

Curcumin has been shown to have several therapeutic benefits and found enormous applications in conventional therapy. The challenging aspect of its delivery is the extremely low aqueous solubility. However, a glycyrrhetinic acid (GA) molecule-modified curcumin-based hydrogel has been developed to address the problem of delivery of the insoluble drug for hepatocellular carcinoma. The GA molecule-modified curcumin supplied in the pro-gelator form could produce a supramolecular hydrogel *in vitro* due to disulphide reduction by glutathione (GSH) and increase curcumin bioavailability and solubility as reported in HepG2 cells. Higher cellular uptake and potent anti-cancer activity were observed with the hydrogel *in vitro* relative to an already known curcumin-targeting compound that was tested [8].

2.2. DNA-Hydrogels

Hybrid bionanomaterials could be developed using DNA as the building block. Predictable two- or three-dimensional structures are formed from DNA molecules. Highly structured networks are formed by hybridizing complementary DNA molecules and the resultant hydrogel structures expand upon encounter with an aqueous environment that result in swelling. Not only do these materials append to any other type of nucleic acid molecules (such as siRNA, miRNA), but they can also load DNA binding drugs. High solubility, bio-compatibility, versatility and responsiveness are key features of such hydrogels. Apart from these features they can also be tagged with suitable fluorescent molecules for tracking biological studies *in vitro* [9].

An interesting application of hydrogels has been made with the development of multi-functional quantum dot (QD) DNA hydrogels. DNA hydrogels are composed of complementary strands of DNA hybridized to form a crosslinked network that swells in an aqueous environment. However, in order for biological studies to be more easily and effectively performed there has to be a tracer or a fluorescent molecule attached to the hydrogel which would then be a superlative option offering both targeted delivery and imaging. With this aspect in mind Zhang et al. recently developed a QD-based DNA hydrogel that had highly tunable size, spectral and delivery properties and bound to the DNA binding drug doxorubicin (Figure 3). The drug targeted cancer cells and the QD DNA hydrogel increased the potency of the drug *in vitro*. The single-step assembling zinc sulphide QD Doxorubicin DNA hydrogels showed increased tumor accumulation *in vitro*, high bio-compatibility, was threefold more efficacious than free DOX and served as an excellent tool for *in vivo* bio-imaging in monitoring tumor growth over time. Aptamers such as siRNA were used to target specific cell types to deliver drug specifically and modulate protein expression in various cell types [9] (Figure 3).

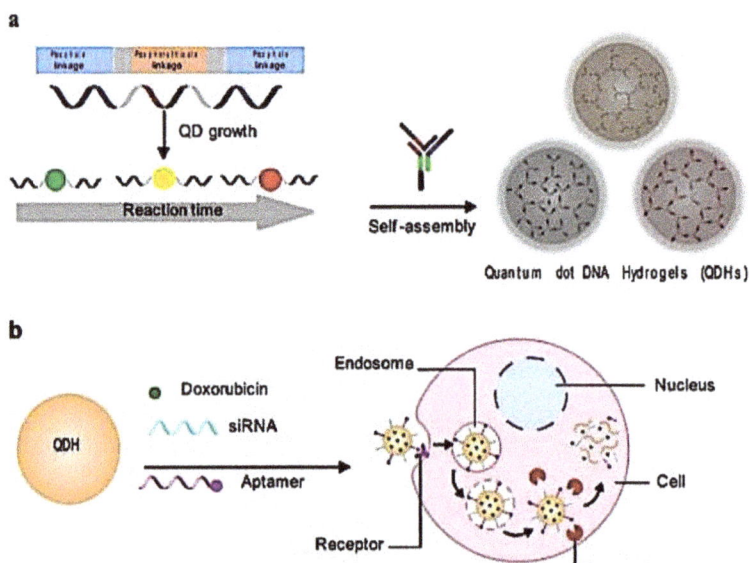

Figure 3. (a). Schematic view of the synthesis process of the DNA functionalized QDs followed by the formation of the QD hydrogel through hybridization with DNA. **(b).** Schematic view of the modification process on the QD hydrogel for specific targeting of the cell using aptamer for drug delivery. Release of Doxorubicin and siRNA happens after uptake into the cell via endocytosis (reprinted (adapted) with permission from [9]; the article is open access and the content reusable. Creative Commons International License 4.0 for the article is available from http://creativecommons.org/licenses/by/4.0/) [9].

Drug-loaded cytosine-phosphate-guanine (CpG)-DNA hydrogels have been used for cancer immunotherapy. These CpG sequences containing hydrogels elicit immune responses. As an example, these DNA hydrogels containing CpG nucleotides stimulated innate immunity through Toll-like receptor 9 and promoted immune responses of ovalbumin (OVA) incorporated into the gel by acting as an adjuvant. Adverse reactions were considerably reduced with the use of the CpG-DNA hydrogel in comparison to the OVA injected with alum or complete Freund's adjuvant [10,11].

There has been significant advancement in hydrogel modifications for enhancing the shape memory and the reversibility of the hydrogel shape after any prolonged stress on them. (example: external stimuli such as temperature). This advances the thermal responsivity behavior of the hydrogels and an example is the construction of supramolecular hydrogel with tunable mechanical properties and multi-shape memory effects. These hydrogels make use of agar that is physically cross-linked to form the hydrogel network and a supramolecular network that is cross-linked by suitable chemical bonds that fix shapes temporarily and produces a multi-shape memory effect made possible by reversible interactions (Figure 4). The supramolecular hydrogel possesses enormous bio-compatibility and bio-degradability characteristics and rely on non-covalent interactions to drive the self-assembly of small molecules in water. The formed structures have supramolecular architectures and encapsulate water [12,13].

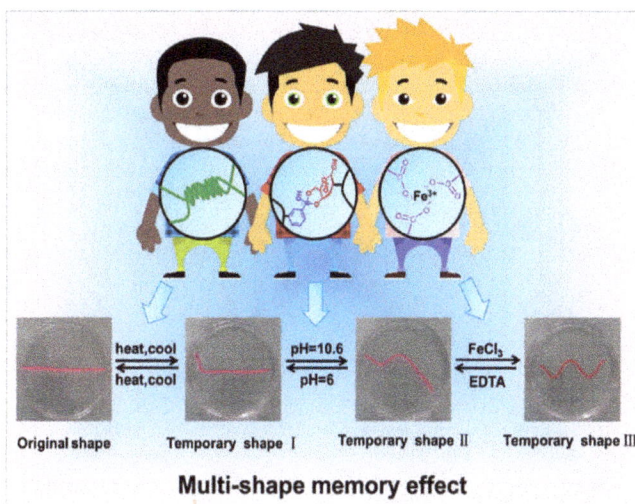

Figure 4. Schematic depiction of a novel Fe^{3+}-, pH-, thermoresponsive hydrogel with tunable mechanical properties easily changed by adjusting cross-linking densities of polymers. Moreover, several co-ordination interactions along with supporting stimuli could be exploited to stabilize temporary shapes in order to realize shape memory behavior. Programmable multi-shape memory effect can be realized by combining the 3 reversible switches (reprinted (adapted) with permission from [12]; copyright (2017) American Chemical Society).

A programmed temporary shape can turn into the memorized original shape when placed in an appropriate environment or exposed to a trigger. Such shape-memory hybrid hydrogels could also be synthesized using DNA cross-linkers. These hydrogels not only undergo phase transitions in the trigger of the stimulus, but also possess memory code to recover to the original matrix shape. Guo et al. developed a pH-controlled shape memory DNA hydrogel that was formed by co-polymerization of acrylamide residues with acrydite modified with (1) cytosine-rich sequences (forming i-motif subunits) and (2) nucleic acids exhibiting self-complementarity. Self- assembly of cytosine-rich nucleic acid strands into an i-motif structure occurred at pH 5.0 and the disassembly to a random coil form happened at pH 8.0, leading to a quasi-liquid state for the hydrogel. Re-acidification to pH 5.0 restored the original structure of the gel [14,15].

2.3. Bio-Inspired Hydrogels

A newer variety of hydrogels used for drug delivery applications are the bio-inspired hydrogels. These 3D materials recapitulate the biological micro-environment relevant to the disease condition and support studies on how the targeted drug delivery process could be optimized, how the therapy behaved *in vivo*, how the disease progressed, and so on. These are particularly useful in cancer therapy as the disease is particularly complex and normally associated with intricate cellular and physiological changes that require progressive monitoring. Engineering such microenvironments would thus be a very useful approach to promote research and study the disease condition and therapeutic process better. The stiffness of the 3D model used for studying liver cancer is a critical attribute to regulate molecular diffusivity and malignancy. The elastic moduli of the collagen gels were increased by stiffening interconnected collagen fibers with varied amounts of poly(ethylene glycol) di(succinic acid N-hydroxysuccinimidyl ester). The softer gels produced malignant cancer spheroids while the stiffer ones showed suppressed malignancy. The model provided better understanding and regulation of the emergent behaviors of cancer cells [16].

Contact lenses that are bio-inspired have been developed recently with improved drug delivery properties suited to perfectly match the eye condition in the diseased state, especially of the anterior-eye segment.

Inner layer of ethyl cellulose and Eudragit S 100 hydrogels encapsulated with diclofenac sodium salt showed sustained drug release in simulated tear film. Sustained drug levels in tear fluid for a prolonged period of time in relation to eye drops was observed. The drug timolol was released in response to lysozyme in the medium and not in PBS. Enzyme cleavable polymers in hydrogels enabled this.

Ergosterol-liposome grafted silicone materials loaded with nystatin (an anti-fungal agent) resembled a fungal infection and triggered nystatin release through a competitive mechanism. In the absence of the ergosterol (important sterol in the cell membranes of fungi) in the medium, the drug release was very negligible [17–20].

With recent progress in cancer therapy using immunology, the research has progressed in the area with the use of cells such as erythrocytes, macrophages, stem cells, dendritic cells and bacteria as building blocks to create targeted delivery systems. Stem cell membrane coated gelatin nanogels that were high tumor targeting and bio-compatible drug delivery systems were developed recently.

Those gelatin nanogels had excellent stability and tumor-targeting ability *in vitro* and *in vivo*. Targeting doxocrubicin (DOX) enhanced anti-tumor therapeutic efficacy significantly higher than the gelatin-DOX of free DOX nanogels. Evident side effects were absent in the tissues of heart, liver, spleen, lung and kidney as was revealed with histopathology staining of treated mice [21].

A similar bio-mimetic drug delivery carrier is the endosomal membrane-coated nanogel extracted from the source cancer cells for specific delivery of the small molecule drug. Hyaluronic acid nanogel composed of SiO_2/Fe_3O_4 nanoparticles formed the inner core of the hydrogel and the endosome membrane outer shell was used. The hyaluronic acid could target CD44 receptors overexpressed in a variety of tumor types. Core-shell mesoporous silica nanoparticles with Fe_3O_4 nanocrystals were used as the core to encapsulate photo-initiators and cross-linkers, followed by hyaluronic acid coating on the surface via electrostatic interaction. Photo-polymerization upon UV irradiation resulted in the in situ formed nanogel in endosomes following incubation of resultant nanoparticles with the source cells. In the presence of the loaded anti-cancer drug doxorubicin, the endosomal membrane nanogels could specifically interact with the target source cells followed by internalization to release doxorubicin. High targeting specificity, cytotoxicity and uptake of the prepared endosomal membrane coated nanogels with DOX over the bare DOX-nanogels was observed *in vitro* [22].

Detoxification of pore-forming toxins resulting from animal bites/stings using a 3D bio-inspired hydrogel matrix resembling the 3D structure of liver is a recent achievement. The 3D matrix resembling the liver is created by the hydrogel while the polydiacetylene nanoparticles installed in the hydrogel matrix serve to attract, capture and sense toxins. The 3D printed biomimetic detoxification device with installed nanoparticles would be a breakthrough innovation replacing vaccines, monoclonal antibodies, antisera to eliminate toxins from blood. Commonly used conventional antidote molecules target specific epitope structures on the pore forming toxins warranting specific treatments for the various toxins. IV-nanoparticle administration for binding and removal of toxins may cause secondary risk by poisoning from nanoparticle accumulation in liver. While challenges do not limit nanoparticle use for toxin clearance, the 3D printed *in vitro* devices have been clinically approved for toxin removal [23].

Adhesive bio-inspired gels comprising of sodium alginate, gum arabic, and calcium ions have been developed in order to mimic properties of natural sundew-derived adhesive hydrogels. Mouse adipose-derived stem cells were used in combination with the sundew-inspired hydrogel to confer interesting properties such as improved wound healing, enhanced wound closure, less noticeable toxicity and inflammation. Sundew plant's natural adhesive hydrogels were quite promising, but the researchers were quite intrigued about collecting the desirable quantity from the natural source. This inspired development of the above described hydrogel type with improved properties [24].

Magnetically actuated gel-bot (Mag-bot) is yet another recent discovery that facilitates remote control of the motion of magnetically actuated hydrogels. These hydrogels adopt a crawling motion resembling that of a maggot and move in a confined space of 3D porous media including, foams, tissues, fibres and so on, opening up broad options for targeted drug delivery. Pattern structures and lubrication effects on hydrogel mobility in the confined 3D space have been experimented in the paper [25].

It is not just the maggot that has inspired development of a new generation of hydrogels with applications in targeted drug delivery but also the drosera. The drosera-inspired model of hydrogel for targeted drug delivery with a bifunctional attribute to it has been developed recently. These hydrogels use the 'catch and kill prey' mechanism as a drosera would. They have a bottom layer functionalized with double stranded DNA that can bind the anti-cancer drug doxorubicin and there are aptamers on top that offer targeting advantages to these particles. Though there is requirement for *in vivo* studies, the sustained expression of specific aptamer on surface and the continuous killing of cancer cells by sustained drug release make these hydrogels an ideal targeted drug delivery vehicle for killing cancer cells [26].

2.4. Multi-Functional and Stimuli-Responsive Hydrogels

Hydrogels that were multi-functional and carriers of anti-cancer drugs are a typical example of the versatility of these delivery vehicles and their amenability to chemical modifications to enhance their therapeutic effects. Magnetite nanoparticles that supported increased intracellular uptake by HeLa cells also had folate ligand on them to enable targeted delivery. Moreover, the hydrogel polymers were thermally responsive and DOX loaded. Those modified hydrogels offered advantages such as increased cellular uptake and apoptotic activity *in vitro* [27].

Another stimuli-responsive hydrogel developed very recently made use of biocompatible thermally responsive polymers that facilitated rupture of cancer cells. The study was conducted *in vitro* with an external source of heat and showed successful cell rupture. Those hydrogel particles had RGD (Arginine, Glycine, Aspartate) peptides attached to their surface that could bind to cells (Figure 5). The paper discussed experiments performed to confirm firmness of the RGD peptides with the cells (ex: MDAMB21) and their rupture using external heat stimulation.

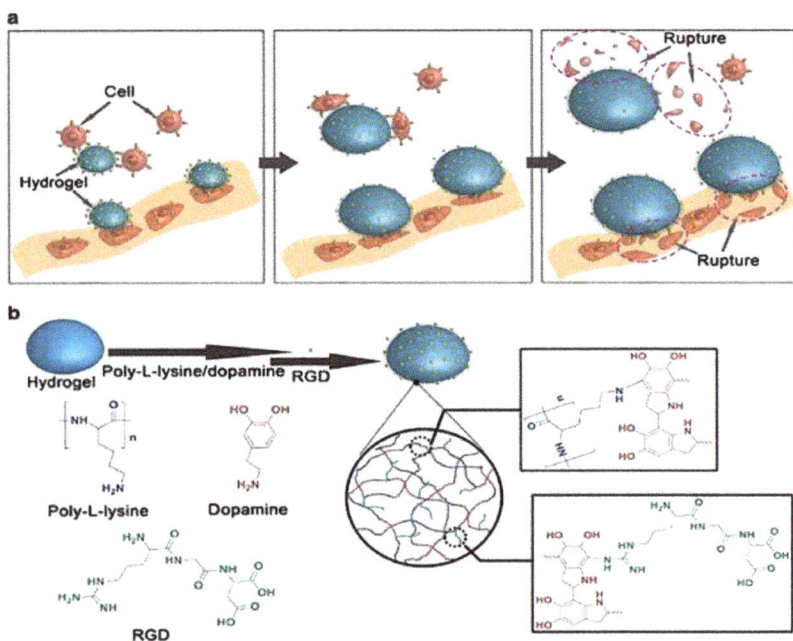

Figure 5. (a). Schematic illustration of the cancer cell attachment on stimuli-responsive hydrogel surface. External stimulus such as temperature causes expansion of the hydrogels and rupture of the cancer cells. **(b)**. Schematic depiction of the surface modification of hydrogel. The first layer of coating was polydopamine (PDA) and poly-L-lysine (PLL) followed by a layer of RGD peptides for binding cancer cells (reprinted (adapted) from [28]; the article is open access and the content reusable. Creative Commons International License 4.0 for the article is available in https://creativecommons.org/licenses/by/4.0/) [28].

Since *in vivo* conditions in tumors normally produce a higher temperature than the surrounding areas, and also that it has been proved that the physical force generated from the expansion of the cells when they were heated was sufficient to kill them, that type of thermally responsive hydrogels proved to be highly advantageous options for achieving targeted cell killing [28].

Hydrogen peroxide trigger-mediated release of drugs could be achieved using hydrogels. This has been demonstrated previously with the use of ABC type triblock copolymer poly [(propylenesulfide)-*b*-(*N*,*N*-dimethylacrylamide)-*b*-(*N*-isopropylacrylamide)] [29,30] (PPS-*b*-PDMA-*b*-PNIPAM). A model hydrophobic drug, the dye red Nile was encapsulated in the hydrogel in order to demonstrate this. The gelation to form the cross-linked hydrogel happened above 37 °C and the ambient 25 °C temperature was sufficient to encapsulate the hydrophobic drug through the formation of micelles. Thus, the temperature-responsive polymer could cross-link to be a stable hydrogel carrying the hydrophobic drug based on the temperature modification and also showed a release of the red colored dye based on the H_2O_2 release in the environment (Figure 6).

Apart from the polymer solubility switch mechanism that helped release based on the presence of the reactive oxygen species in the environment, cleavable units of polymer that underwent scission in presence of ROS have also been developed. These are also bio-degradable and help avoid any toxic build up or unwanted immune response in the system from the presence of the hydrogels [31,32].

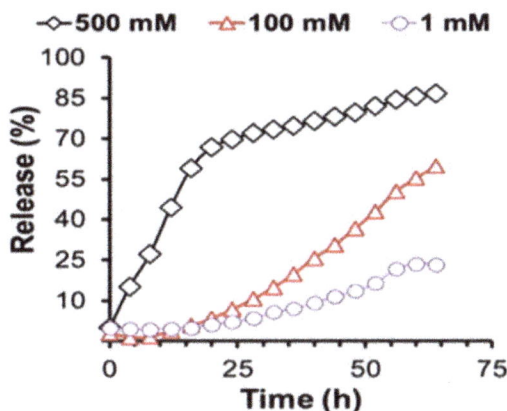

Figure 6. Drug release dependent on Reactive Oxygen Species concentration as displayed by PPS$_{60}$-b-PDMA 150-b-PNIPAAM150 triblock polymer-based thermos-responsive Nile red-loaded hydrogels. Nile red-loaded hydrogels (5 wt% triblock copolymer concentration) in PBS (pH 7.4) have been used to demonstrate Hydrogen Peroxide dependent drug release kinetics *in vitro* at 37 °C. 1, 100 and 500 mM concentrations of H$_2$O$_2$ over a 64 h time course were incubated with the hydrogel samples to study ROS dependent drug release (reprinted (adapted) with permission from [31]; copyright (2014) American Chemical Society).

Diselenide-containing block copolymer and a peptide amphiphile were used as building blocks for the construction of a UV responsive hydrogel. The gel sol transition of the hydrogel occurred in presence of gamma irradiation. The peptide amphiphile was composed of nalproxen, the drug and a hexapeptide sensitive to UV irradiation, the radiation trigger was sufficient to cause release of the drug from hydrogel. The stimuli responsive hydrogel system was thus capable of providing both chemo- and radiotherapy benefits on application [13].

Covalently cross-linked hydrogels are typically stable and elastic while the physically cross-linked hydrogels produced through ionic interactions are less stable and exhibit reduced mechanical properties. The dynamic covalent chemistry is an option to acquire stability, elasticity and also shear-thinning and self-healing characteristics for the hydrogel. Based on this, Volkan et al. reported the synthesis of hydrogel networks from reversible interactions between phenylboronic acid and *cis*-diols. The gel strength was dependent on pH and the hydrogel was evaluated for protein delivery *in vitro*. The pKa of phenylboronic acid and the pH of the environment were the determining factors for establishing the extent of hydrogel cross-linking. Soft, moldable hydrogels were produced that could be injected using standard syringe needles and were demonstrated to exhibit shear thinning and healing properties in addition to their injectable nature. Protein encapsulation was performed with the hydrogel, and the release effect monitored. Size-dependent protein release could be attributed to the mesh size of the hydrogel network. Insulin and IgG release from the hydrogel network was monitored in presence of glucose. Glucose-responsiveness of the hydrogel was confirmed, and the release kinetics could be controlled by the mesh size of the hydrogel. 3T3 fibroblast cells showed no significant toxicity as quantified using MTT assay for 24 hr. No chronic inflammation was observed *in vivo* and the materials were quite bio-compatible [33].

The temperature sensitive hydrogels respond by undergoing sol-gel transitions with changes in temperature from room ones to the physiological ones. Recently developed hydrogels were that were thermo-responsive and accommodated hydrophobic drugs efficiently were built from amphiphilic triblock co-polymers, poly(N-isopropylacrylamide)-*b*-poly(4-acryloylmorpholine)-*b*-poly (2-((((2-nitrobenzyl)oxy)carbonyl)amino)ethylmethacrylate)(PNIPAM-*b*-PNAM-*b*-PNBOC). The hydrogel carried the hydrophilic drug gemcitabine and the hydrophobic drug doxorubicin. The triblock co-polymers first assembled into micelles that had the hydrophobic and temperature responsive components formed below the lower critical solution temperature, while higher polymer concentration and temperature above the critical gelation temperature were used to form the hydrogels of physically cross-linked micellar nanoparticles. The study demonstrated the synthesis, characterization and the temperature and UV irradiation triggered synergistic release of both the hydrophobic and hydrophilic drugs *in vitro* [34,35].

Hydrogels formed by covalent cross-linking of polymers to facilitate targeted drug delivery have been achieved in recent years. The nanoparticle-hydrogel is a hybrid system that is formed in three different ways. The first method is to entrap a hydrogel within a nanoparticle, the second one is to form a 3D hydrogel network with the nanoparticles by crosslinking of the latter using hydrophobic interactions or mixing up nanoparticles of opposite charges and the final one is to covalently couple the nanoparticle with the hydrogel [36]. A liposome cross-linked hybrid hydrogel has been developed recently. Glutathione-triggered release from the stimuli responsive polymer favored drug release. Arylthioether succinimde cross-links were introduced between the Peg polymers and the liposome nanoparticles to produce the 3D hydrogel network. In presence of glutathione, the matrix was degraded, and the encapsulated drug molecules were released. Malemimide-functionalized liposomes crosslinked using peg polymers were constructed for co-delivery of doxorubicin and cytochrome-C (apoptotic cascade initiatior) and their release testing *in vitro* was monitored in presence of glutathione. The presence of glutathione is a trigger for release from the hydrogel in a reducing microenvironment such as a tumor [37].

3. Diverse Physical Attributes of Hydrogels for Drug Delivery

Structural-modification amenability of the hydrogels renders them in various shapes and sizes. This feature is particularly interesting for drug delivery applications in order to design the hydrogels as per the target sites to which the drugs have to be delivered. The hydrogel-based dosage forms can have different designs and shapes depending on the route of drug administration (Table 1).

Table 1. Different types of hydrogel products administered via different routes of administration [38–50].

Routes of Administration	Shape	Typical Dimension
Peroral	Spherical beads	1 µm to 1mm
	Discs	Diameter of 0.8 cm and thickness of 1 mm
	Nanoparticles	10–1000 nm
Rectal	Suppositories	Conventional adult suppositories dimensions (length ≈ 32 mm) with a central cavity of 7 mm and wall thickness of 1.5 mm
Vaginal	Vaginal tablets	Height of 2.3 cm, which of 1.3 cm and thickness of 0.9 cm
	Torpedo-shaped pessaries	Length of 30 mm and thickness of 10 mm
Ocular	Contact lenses	Conventional dimensions (typical diameter ≈ 12 mm)
	Drops	Hydrogel particles present in the eye drops must be smaller than 10 µm
	Suspensions Ointments	N/A
	Circular inserts	Diameter of 2 mm and total weight of 1 mg (round shaped)
Transdermal	Dressings	Variable
Implants	Discs	Diameter of 14 mm and thickness of 0.8 mm
	Cylinders	Diameter of 3 mm and length of 3.5 cm

Reprinted (adapted) with permission from [3]; the article is open access and the content reusable. The article is licensed by Creative Commons and the link to the license is available at https://creativecommons.org/licenses/by/3.0/.

4. Specific Therapeutic Areas Using Hydrogels for Drug Delivery at Present

4.1. Ophthalmic

Conventional eye drops have problems with sustained drug delivery and there is huge wastage of drug immediately following application through eye drainage. Dextenza is a very recently FDA approved (3 December 2018) ocular therapeutic hydrogel formulation for human use. This is used for ocular pain following ophthalmic surgery and is the first intracanalicular implant developed for drug delivery and developed by the company Ocular Therapeutix (Bedford, MA, USA) [51].

Thermo-responsive polymer developed by mixing poly(acrylic acid-graft-N-isopropylacrylamide) (PAAc-graft-PNIPAAm) with PAAc-co-PNIPAAm geL and incorporating epinephrine was used in the *in vitro* evaluation of ophthalmic drug release. The approach augmented the effect of intraocular pressure reduction from 8 h with the traditional drops to 36 h. The cross-linking density of the hydrogel affected the capillary network formation and offered a convenient controlled drug release method for ophthalmic drug delivery [52].

Intra-ocular pressure (IOP) elevates during glaucoma and alleviating this pressure has been quite challenging. Hydrogels could be used to resolve this problem by using them to prepare soft contact lenses composed of polymers to form networks. The highly hydrated polymer networks of hydrogels cause the drug to elute out very rapidly and this is not favorable for glaucoma therapy, which mainly uses hydrophilic drugs. However, with suitable modifications, soft contact lenses have been developed using polymers of N,N-diethylacrylamide and methacrylic acid, which delivered the hydrophilic drug timolol for about 24 h, thereby opening up ways to allow sustained hydrophilic drug delivery using hydrogels. Storing the contact lenses in a hydrated state can leach out drug and to wear them all the time are the limitations though [53].

Inner layer-embedded contact lenses have been investigated for the sustained release of highly water-soluble drug betaxolol hydrochloride on the ocular surface. Cellulose acetate and Eudragit S-100 were selected as the inner layer of the contact lenses which showed a promising sustained drug release for over 240 h in tear fluid of rabbits *in vivo* to create a controlled-release carrier of the drug in ophthalmic drug delivery [19].

Controlled drug release behavior from hydrogels was also evaluated using nepafenac as the model drug. 3D cross-linked thermos and pH sensitive hydrogel was designed that was composed of carboxymethylchitosan (CMC) and poloxamer with glutaraldehyde as the cross-linking agent. The hydrogel was found to undergo reversible sol-gel transition at temperature and/or pH alteration at a very low concentration. Sustained release of the drug nepafenac was observed in the *in vitro* model and maximum release was observed at 35 °C and pH 7.4. Cytocompatibility of the hydrogel with human corneal epithelial cells was high [54].

4.2. Oral, Intestinal

Gastroretentive drug dosage forms (GRDDFs) are particularly attractive for drugs that are absorbed in the proximal part of gastrointestinal tract. Enhancing the retention time of the drugs in the GI tract is very important in order to improve their bioavailability and enhance their therapeutic effects. These dosage forms could be exploited for their muco-adhesion to the gastric mucosa, modified to float or sink in order to prevent leaving the stomach or increase their swelling behavior and make them as large to prevent passage through pylorus for prolonged periods. Based on these ideas, polyionic complex hydrogels of chitosan with ring-opened PVP have been developed for Osteoporosis therapy. The formulation was used to release alendronate in the upper GI tract. Enhanced muco-adhesion, delayed clearance from swelling, minimal localized irritation, improved bioavailability and slower release of the active ingredients are the interesting aspects of the preparation. Also, *in vivo* experimentation showed that these hydrogels could provide optimized PK properties that maintained the drug in the therapeutic levels for a sustained period of time, minimizing fluctuations in therapeutic levels, hence also the possible side effects [55].

Inflammatory diseases such as irritable bowel syndrome have been recently treated using hydrogels. These provided safer alternatives to delivery methods that may cause systemic toxicity. Zhang et al. developed negatively charged hydrogels that preferentially accumulated in the positively charged inflamed colon and acted as carriers of the corticosteroid drug dexamethasone (Dex). The hydrogel was prepared from ascorbyl palmitate which had labile bonds responsive to inflammatory conditions and was Generally Regarded as Safe (GRAS) for administration. Enema administration to the colon of inflammation targeting (IT) hydrogel microfibers not only reached the target site but also stayed there owing to charge interaction. The formulation was therapeutically very efficacious and revealed lesser systemic drug exposure than with free Dex in the IBS mice model *in vivo* [56].

Complexation hydrogel prepared from poly (methacrylic acid-g-ethylene glycol) [P(MAA-g-EG) has been described. The targeting ligand used is the octarginine cell-penetrating peptide that causes specific delivery of insulin to the intestine. This method facilitated ideal targeting, absorption at target and allowed immediate release of insulin from absorption site. Great hypoglycemic responses were achievable and increased insulin absorption was noted from diabetic rat models used for testing. 18% glucose reduction was observed immediately on administration of the hydrogel containing insulin [57].

4.3. Cardiac Illness and Cancer

Myocardial infarction is a leading cause of death and disability in the world. Intramyocardial administration of biomaterials such as hydrogels along the perimeter region of myocardial infarction has proven to be beneficial.

Chen et al. proposed the use of a combination of curcumin (known for its anti-oxidant, anti-inflammatory and anti-oxidation properties) and nitric oxide (known as an anti-angiogenesis agent) in a hydrogel to treat myocardial infarction. The mixed component hydrogel created with the combination drugs improved therapeutic efficacy synergistically. Protective effects such as myocytic apoptotic death alleviation, reduced collagen deposition, increased vessel density (attributable to NO in the combination) and upregulated Silent Information Regulator 1 (SIRT-1), a histone deacetylase that confers resistance to the heart from ischemic injury were observed in diseased mice models *in vivo*. The hydrogel was prepared using peptide derivatives of curcumin and NO in a ratio of 4:1 and showed sustained curcumin release at a low concentration of 2.5 µg per ml per 24 h. NO was released in presence of the enzyme β-galactosidase that could break glucosidic bonds to release NO [58].

Growth factors and cytokines (paracrine factors) secreted by stem cells have been proven to be effective in repairing damaged myocardial tissue. The whole cocktail of the paracrine factors is referred to as a secretome and is isolated *in vitro*. The biomolecular composition of the secretome can be manipulated suitably by varying stem cell culture conditions. An injectable hydrogel to deliver to peri-infarct myocardium has been recently developed using secretome from human adipose derived stem cell secretome. Nano-composite hydrogel was formed from a combination of gelatin and laponite carrying the secretome and tested both *in vitro* and *in vivo* for their therapeutic effects via monitoring angiogenesis, scar formation and heart function. Significantly reduced scar area and improved cardiac function were observed *in vivo* in the secretome loaded hydrogel group in relation to the control [59].

The very recent development of a paintable hydrogel to serve as cardiac patch for treating myocardial infarction is worth mentioning in this context. The hydrogel eliminates the damage to tissue through suture or light triggered reactions as it is paintable. It has been constructed by a Fe^{3+} triggered polymerization reaction wherein the covalently linked pyrrole and dopamine undergo simultaneous polymerization with the trigger and the conductive polypyrrole produced also uniquely cross-links the network further. The functional patch is both adhesive and conductive and forms a suture-free alternative for reconstruction of cardiac function and revascularization. Bonding within 4 weeks to the beating heart boosts the transmission of electrophysiological signals with conductivity profiles equivalent to that of the normal myocardium [60].

Bio-material based immunotherapy platforms for targeted drug delivery to cancers are the latest trend observable in cancer therapy. Based on this idea, novel STINGels have been developed by Leach et al, that are peptide hydrogels to show controlled delivery of cyclic dinucleotides (CDns). Dramatic improvement in survival was observed in murine models of head and neck cancer in comparison to CDN alone or CDN delivered from a collagen hydrogel [61].

Thyroid cancer treatment using local drug delivery system formed of glycol chitosan (GC) hydrogel and doxorubicin hydrochloride (DOX·HCl) called GC10/dox has been recently developed (Figure 7). Visible light regulated the storage and swelling aspects of the hydrogel and a controlled sustained release followed the initial burst release within 18hours. Potent antitumor effects were observed *in vivo* and *in vitro* in comparison to free DOX·HCl and this is a promising research direction for thyroid cancer therapy [62].

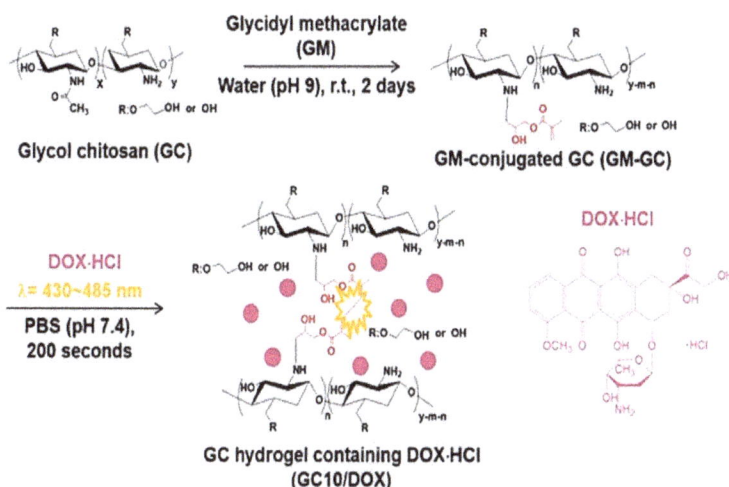

Figure 7. To glycol chitosan solution glycidyl methacrylate (GM) was added in water (adjusted to pH 9) and maintained for 2 days at room temperature. The white solid conjugate of GM was dissolved in water and riboflavin added. DOX·HCl was added and the mixture was irradiated using blue visible light (430–485 nm, 2100 mW/cm^2) for 10 minutes in order to promote hydrogelation (reprinted with permission from [62]; the article is open access and the content reusable. Creative Commons International License 4.0 for the article is available in https://creativecommons.org/licenses/by/4.0/).

Injectable hydrogels responsive to Reactive Oxygen Species that degrade in the presence of ROS and promote immunogenic tumor phenotype via local gemcitabine delivery is a recent discovery. The PVA cross-linked hydrogel with ROS-labile linkers enhance anti-tumor response with a localized release of immune checkpoint blocking antibody (anti-PD-L1 blocking antibody (aPDL-1) in *in vitro* and immunogenic *in vivo* mouse models. Tumor recurrence prevention after primary resection is the therapeutic advantage of this chemo-immunotherapy [63].

5. Translation to the Clinic

With enormous potential for therapeutic applications, several hydrogel formulations have crossed the barriers of *in vitro*/pre-clinical studies and found their way into the market. Some of them are still in the clinical study phases. Hydrogels have evolved over time to one of the best and the most versatile drug delivery platforms. Table 2 lists the widespread practical applications of the hydrogel concept that have been translated to the clinical level.

Table 2. Examples of hydrogels translated to clinical use [64–68].

Product	Type of Hydrogel	Therapeutic Application	Drug Delivered
Sericin	Dextran	Optically trackable drug delivery system for malignant melanoma	Doxorubicin
Hyalofemme/ Hyalo Gyn	Carbomer propylene glycol, Hyaluronic acid derivative	Vaginal dryness, estrogen alternative	Hyaluronic acid derivative
Dextenza	Polyethylene glycol	Intra-canalicular delivery for post-operative ophthalmic care	Dexamethasone
Regranex	Carboxymethyl cellulose	Diabetic foot ulcer	Recombinant human platelet derived growth factor
muGard	Mucoadhesive	Oral lichen planus	
	2% Poloxamer	Cervical cancer recurrence	Carboplatin

6. Conclusions

Hydrogels offer a versatile platform for the therapy of several diseases including cancer and diabetes. The water-loving nature of hydrogels and the ability to shrink and swell depending on several environmental cues or the mere presence of water is attractive for drug delivery applications. They have a high degree of porosity and the polymers building them could be cross-linked to varying degrees by adjusting their densities. With a physical structure highly amenable to modification in several ways, the hydrogel applications are not just limited to targeted drug delivery. They also find applications in hygiene products, wound dressings, contact lenses and tissue engineering.

Recent developments of hydrogels in the field of targeted drug delivery have been tremendous. They are modified with targeting ligands and diverse polymer types that confer very interesting properties on them for drug delivery. Ophthalmic drug delivery is an area seeing significant impact in therapy from hydrogels. From comfortable contact lenses to biodegradable drug delivery the applications in eye care have been enormous. They are 90% water, provide steady state drug release over days or months, deliver small molecules or large proteins, are fully absorbed in delivery and remain visible during monitoring [51].

Noteworthy is the application of pH responsive hydrogels for cancer therapy and glucose responsive hydrogels for diabetes. The use of modified stem cell membranes for targeted delivery is a very recent and attractive strategy for drug delivery. These membranes coated on hydrogels (nanogels) loaded with drugs are highly specific to the disease site in cancer and are highly bio-compatible.

Immunotherapy platforms using hydrogels are very significant in cancer therapy. Hydrogels enabling localized delivery of antibodies and other immune-regulatory molecules at cancer sites are promising drug delivery vehicles for cancer therapy. Gastro-retentive drug dosage forms (GRDDFs) are versatile drug delivery platforms for intestine and they offer the advantages of adjusting the nanoparticle size to facilitate retention of the active ingredient in the GI tract for as long as required.

Though the hydrogel-based drug delivery was originally influenced by the hydrophobicity of the drugs, several improvements have been made recently including development of cyclodextrins modified to accommodate the hydrophobic drug sufficiently. Adhesive and conductive patches developed using hydrogels are useful in cardiac repair and vascularization. Remotely controlled motility of hydrogel (mimicking motion of a magbot) and the QD DNA hydrogels are novel ideas to facilitate targeted drug delivery.

As discussed in the paper, though there are several hydrogel formulations in clinical use, there is always scope for improvement and modification of hydrogels to enhance their applications. With subtle modifications to the existing ones, the hydrogels could become superlative drug delivery vehicles surpassing the disadvantages and current limitations with the use of several conventional delivery forms and provide promising results for therapy of several illnesses.

Molecules **2019**, *24*, 603

Acknowledgments: Radhika Narayanaswamy, the first author, would like to extend my sincere gratitude to all who offered their valuable support for successful completion of the review article. I whole heartedly thank Vladimir P. Torchilin for providing me the opportunity to contribute the review to the journal *Molecules* and guiding me through the process in the midst of a super-hectic schedule. I also make use of this occasion to offer gratitude to the Guest editor for the special-issue 'Smart and functional polymers'- Jianxun Ding, for inviting us and offering a helping hand to clarify all queries patiently through the article writing and submission processes. I also extend my sincere gratitude to all the reviewers of the article who paid so much attention to detail and came up with ideas and suggestions for enhancing the content of the article. Last but not the least, I thank all my friends, family and colleagues for extending their kind support and providing constant encouragement through the different hard stages of finishing up the review article.

Conflicts of Interest: The authors declare no conflict of interest.

References

1. Coinlogitic. *Hydrogel Consumption Market Analysis by Current Industry Status and Growth Opportunities*. 2018. Available online: https://coinlogitic.com/hydrogel-consumption-market-research-report/51472/ (accessed on 11 October 2018).
2. Hoare, T.R.; Kohane, D.S. Hydrogels in drug delivery: Progress and challenges. *Polymer* **2008**, *49*, 1993–2007.
3. Caló, E.; Khutoryanskiy, V.V. Biomedical applications of hydrogels: A review of patents and commercial products. *Eur. Polym. J.* **2015**, *65*, 252–267. [CrossRef]
4. Li, J. Self-assembled supramolecular hydrogels based on polymer–cyclodextrin inclusion complexes for drug delivery. *NPG Asia Mater.* **2010**, *2*, 112. [CrossRef]
5. Moncada-Basualto, M. Supramolecular hydrogels of β-cyclodextrin linked to calcium homopoly-L-guluronate for release of coumarins with trypanocidal activity. *Carbohydr. Polym.* **2019**, *204*, 170–181. [CrossRef] [PubMed]
6. Jalalvandi, E. Cyclodextrin-polyhydrazine degradable gels for hydrophobic drug delivery. *Mater. Sci. Eng. C* **2016**, *69*, 144–153. [CrossRef] [PubMed]
7. Thi, T.T.H. Oxidized cyclodextrin-functionalized injectable gelatin hydrogels as a new platform for tissue-adhesive hydrophobic drug delivery. *RSC Adv.* **2017**, *7*, 34053–34062.
8. Chen, G. A Glycyrrhetinic Acid-Modified Curcumin Supramolecular Hydrogel for liver tumor targeting therapy. *Sci. Rep.* **2017**, *7*, 44210. [CrossRef]
9. Zhang, L. Multifunctional quantum dot DNA hydrogels. *Nat. Commun.* **2017**, *8*, 381. [CrossRef]
10. Nishikawa, M. Injectable, self-gelling, biodegradable, and immunomodulatory DNA hydrogel for antigen delivery. *J. Control. Release* **2014**, *180*, 25–32. [CrossRef]
11. Shahbazi, M.A.; Bauleth-Ramos, T.; Santos, H.A. DNA hydrogel assemblies: Bridging synthesis principles to biomedical applications. *Adv. Ther.* **2018**, *1*, 1800042. [CrossRef]
12. Le, X. Fe^{3+}-, pH-, Thermoresponsive Supramolecular Hydrogel with Multishape Memory Effect. *ACS Appl. Mater. Interfaces* **2017**, *9*, 9038–9044. [CrossRef] [PubMed]
13. Cao, W. γ-Ray-Responsive Supramolecular Hydrogel Based on a Diselenide-Containing Polymer and a Peptide. *Angew. Chem.* **2013**, *125*, 6353–6357. [CrossRef]
14. Guo, W. pH-Stimulated DNA Hydrogels Exhibiting Shape-Memory Properties. *Adv. Mater.* **2015**, *27*, 73–78. [CrossRef] [PubMed]
15. Gehring, K.; Leroy, J.L.; Gueron, M. A tetrameric dna-structure with protonated cytosine.cytosine base-pairs. *Nature* **1993**, *363*, 561–565. [CrossRef] [PubMed]
16. Liang, Y. A cell-instructive hydrogel to regulate malignancy of 3D tumor spheroids with matrix rigidity. *Biomaterials* **2011**, *32*, 9308–9315. [CrossRef] [PubMed]
17. Alvarez-Lorenzo, C. Bioinspired hydrogels for drug-eluting contact lenses. *Acta Biomater.* **2019**, *84*, 49–62. [CrossRef] [PubMed]
18. Zhu, Q. Inner layer-embedded contact lenses for pH-triggered controlled ocular drug delivery. *Eur. J. Pharm. Biopharm.* **2018**, *128*, 220–229. [CrossRef]
19. Zhu, Q. Sustained ophthalmic delivery of highly soluble drug using pH-triggered inner layer-embedded contact lens. *Int. J. Pharm.* **2018**, *544*, 100–111. [CrossRef]
20. Segura, T. Materials with fungi-bioinspired surface for efficient binding and fungi-sensitive release of antifungal agents. *Biomacromolecules* **2014**, *15*, 1860–1870. [CrossRef]

21. Gao, C. Stem Cell Membrane-Coated Nanogels for Highly Efficient *in vivo* Tumor Targeted Drug Delivery. *Small* **2016**, *12*, 4056–4062. [CrossRef]
22. Yu, J. Endosome-Mimicking Nanogels for Targeted Drug Delivery. *Nanoscale* **2016**, *8*, 9178. [CrossRef] [PubMed]
23. Gou, M. Bio-inspired detoxification using 3D-printed hydrogel nanocomposites. *Nat. Commun.* **2014**, *5*, 3774. [CrossRef] [PubMed]
24. Sun, L. Sundew-inspired adhesive hydrogels combined with adipose-derived stem cells for wound healing. *Acs Appl. Mater. Interfaces* **2016**, *8*, 2423–2434. [CrossRef] [PubMed]
25. Shen, T. Remotely Triggered Locomotion of Hydrogel Mag-bots in Confined Spaces. *Sci. Rep.* **2017**, *7*, 16178. [CrossRef] [PubMed]
26. Li, S. A Drosera-bioinspired hydrogel for catching and killing cancer cells. *Sci. Rep.* **2015**, *5*, 14297. [CrossRef] [PubMed]
27. Kim, H. Synergistically enhanced selective intracellular uptake of anticancer drug carrier comprising folic acid-conjugated hydrogels containing magnetite nanoparticles. *Sci. Rep.* **2017**, *7*, 41090. [CrossRef]
28. Fang, Y. Rupturing cancer cells by the expansion of functionalized stimuli-responsive hydrogels. *NPG Asia Mater.* **2018**, *10*, e465. [CrossRef]
29. Schlegel, P.N. Effective long-term androgen suppression in men with prostate cancer using a hydrogel implant with the GnRH agonist histrelin. *Urology* **2001**, *58*, 578–582. [CrossRef]
30. Wang, X. Vaginal delivery of carboplatin-loaded thermosensitive hydrogel to prevent local cervical cancer recurrence in mice. *Drug Deliv.* **2016**, *23*, 3544–3551. [CrossRef]
31. Gupta, M.K. Cell protective, ABC triblock polymer-based thermoresponsive hydrogels with ROS-triggered degradation and drug release. *J. Am. Chem. Soc.* **2014**, *136*, 14896–14902. [CrossRef]
32. Saravanakumar, G.; Kim, J.; Kim, W.J. Reactive-Oxygen-Species-Responsive Drug Delivery Systems: Promises and Challenges. *Adv. Sci.* **2017**, *4*, 1600124. [CrossRef]
33. Yesilyurt, V. Injectable self-healing glucose-responsive hydrogels with pH-regulated mechanical properties. *Adv. Mater.* **2016**, *28*, 86–91. [CrossRef]
34. Wang, C. Photo-and thermo-responsive multicompartment hydrogels for synergistic delivery of gemcitabine and doxorubicin. *J. Control. Release* **2017**, *259*, 149–159. [CrossRef]
35. Larrañeta, E. Hydrogels for hydrophobic drug delivery. Classification, synthesis and applications. *J. Funct. Biomater.* **2018**, *9*, 13.
36. Gao, W. Nanoparticle-hydrogel: A hybrid biomaterial system for localized drug delivery. *Ann. Biomed. Eng.* **2016**, *44*, 2049–2061. [CrossRef]
37. Liang, Y.; Kiick, K.L. Liposome-cross-linked hybrid hydrogels for glutathione-triggered delivery of multiple cargo molecules. *Biomacromolecules* **2016**, *17*, 601–614. [CrossRef]
38. Lee, P.I.; Kim, C.-J. Probing the mechanisms of drug release from hydrogels. *J. Control. Release* **1991**, *16*, 229–236. [CrossRef]
39. Ahmed, E.T.; Maayah, M.F.; Asi, Y.O.M.A. Anodyne therapy versus exercise therapy in improving the healing rates of venous leg ulcer. *Int. J. Res. Med. Sci.* **2017**, *2017*, 6. [CrossRef]
40. Hamidi, M.; Azadi, A.; Rafiei, P. Hydrogel nanoparticles in drug delivery. *Adv. Drug Deliv. Rev.* **2008**, *60*, 1638–1649. [CrossRef]
41. Bilia, A. *in vitro* evaluation of a pH-sensitive hydrogel for control of GI drug delivery from silicone-based matrices. *Int. J. Pharm.* **1996**, *130*, 83–92. [CrossRef]
42. Rahimi, M. Preemptive morphine suppository for postoperative pain relief after laparoscopic cholecystectomy. *Adv. Biomed. Res.* **2016**, *5*, 57.
43. Karasulu, H.Y. Efficacy of a new ketoconazole bioadhesive vaginal tablet on Candida albicans. *Il Farm.* **2004**, *59*, 163–167. [CrossRef]
44. Mandal, T.K. Swelling-controlled release system for the vaginal delivery of miconazole. *Eur. J. Pharm. Biopharm.* **2000**, *50*, 337–343. [CrossRef]
45. Hu, X. Hydrogel Contact Lens for Extended Delivery of Ophthalmic Drugs. *Int. J. Polym. Sci.* **2011**, *2011*, 9. [CrossRef]
46. Ludwig, A. The use of mucoadhesive polymers in ocular drug delivery. *Adv. Drug Deliv. Rev.* **2005**, *57*, 1595–1639. [CrossRef]

47. Hornof, M. Mucoadhesive ocular insert based on thiolated poly(acrylic acid): Development and *in vivo* evaluation in humans. *J. Control. Release* **2003**, *89*, 419–428. [CrossRef]
48. Brazel, C.S.; Peppas, N.A. Pulsatile local delivery of thrombolytic and antithrombotic agents using poly(N-isopropylacrylamide-co-methacrylic acid) hydrogels. *J. Control. Release* **1996**, *39*, 57–64. [CrossRef]
49. Omidian, H.; Park, K. Hydrogels. In *Fundamentals and Applications of Controlled Release Drug Delivery*; Siepmann, J., Siegel, R.A., Rathbone, M.J., Eds.; Springer: Boston, MA, USA, 2012; pp. 75–105.
50. Naveed, D.S. *Contemporary Trends in Novel Ophthalmic Drug Delivery System: An Overview*; Scientific Research: Wuhan, China, 2015; Volume 2.
51. Therapeutix, O. Engineered for Ocular Innovation. 2017. Available online: https://www.ocutx.com/about/hydrogel-technology/ (accessed on 11 October 2017).
52. Prasannan, A.; Tsai, H.-C.; Hsiue, G.-H. Formulation and evaluation of epinephrine-loaded poly (acrylic acid-co-N-isopropylacrylamide) gel for sustained ophthalmic drug delivery. *React. Funct. Polym.* **2018**, *124*, 40–47. [CrossRef]
53. Lavik, E.; Kuehn, M.; Kwon, Y. Novel drug delivery systems for glaucoma. *Eye* **2011**, *25*, 578. [CrossRef]
54. Yu, S. A novel pH-induced thermosensitive hydrogel composed of carboxymethyl chitosan and poloxamer cross-linked by glutaraldehyde for ophthalmic drug delivery. *Carbohydr. Polym.* **2017**, *155*, 208–217. [CrossRef]
55. Su, C.-Y. Complex Hydrogels Composed of Chitosan with Ring-opened Polyvinyl Pyrrolidone as a Gastroretentive Drug Dosage Form to Enhance the Bioavailability of Bisphosphonates. *Sci. Rep.* **2018**, *8*, 8092. [CrossRef]
56. Zhang, S. An inflammation-targeting hydrogel for local drug delivery in inflammatory bowel disease. *Sci. Transl. Med.* **2015**, *7*, ra128–ra300. [CrossRef]
57. Fukuoka, Y. Combination Strategy with Complexation Hydrogels and Cell-Penetrating Peptides for Oral Delivery of Insulin. *Biol. Pharm. Bull.* **2018**, *41*, 811–814. [CrossRef]
58. Chen, G. A mixed component supramolecular hydrogel to improve mice cardiac function and alleviate ventricular remodeling after acute myocardial infarction. *Adv. Funct. Mater.* **2017**, *27*, 1701798. [CrossRef]
59. Waters, R. Stem cell-inspired secretome-rich injectable hydrogel to repair injured cardiac tissue. *Acta Biomater.* **2018**, *69*, 95–106. [CrossRef]
60. Liang, S. Paintable and Rapidly Bondable Conductive Hydrogels as Therapeutic Cardiac Patches. *Adv. Mater.* **2018**, *30*, 1704235. [CrossRef]
61. Leach, D.G. STINGel: Controlled release of a cyclic dinucleotide for enhanced cancer immunotherapy. *Biomaterials* **2018**, *163*, 67–75. [CrossRef]
62. Yoo, Y. A local drug delivery system based on visible light-cured glycol chitosan and doxorubicinhydrochloride for thyroid cancer treatment *in vitro* and *in vivo*. *Drug Deliv.* **2018**, *25*, 1664–1671. [CrossRef]
63. Wang, C. In situ formed reactive oxygen species-responsive scaffold with gemcitabine and checkpoint inhibitor for combination therapy. *Sci. Transl. Med.* **2018**, *10*. [CrossRef]
64. Liu, J. Sericin/Dextran Injectable Hydrogel as an Optically Trackable Drug Delivery System for Malignant Melanoma Treatment. *ACS Appl. Mater. Interfaces* **2016**, *8*, 6411–6422. [CrossRef]
65. Chen, J. Evaluation of the Efficacy and Safety of Hyaluronic Acid Vaginal Gel to Ease Vaginal Dryness: A Multicenter, Randomized, Controlled, Open-Label, Parallel-Group, Clinical Trial. *J. Sex. Med.* **2013**, *10*, 1575–1584. [CrossRef]
66. Blizzard, C.; Desai, A.; Driscoll, A. Pharmacokinetic Studies of Sustained-Release Depot of Dexamethasone in Beagle Dogs. *J. Ocul. Pharmacol. Ther. Off. J. Assoc. Ocul. Pharmacol.* **2016**, *32*, 595–600. [CrossRef]
67. Allison, R.R. Multi-institutional, randomized, double-blind, placebo-controlled trial to assess the efficacy of a mucoadhesive hydrogel (MuGard) in mitigating oral mucositis symptoms in patients being treated with chemoradiation therapy for cancers of the head and neck. *Cancer* **2014**, *120*, 1433–1440. [CrossRef]
68. Li, J.; Mooney, D.J. Designing hydrogels for controlled drug delivery. *Nat. Rev. Mater.* **2016**, *1*, 16071. [CrossRef]

MDPI

St. Alban-Anlage 66

4052 Basel

Switzerland

Tel. +41 61 683 77 34

Fax +41 61 302 89 18

www.mdpi.com

Molecules Editorial Office

E-mail: molecules@mdpi.com

www.mdpi.com/journal/molecules

www.ingramcontent.com/pod-product-compliance
Lightning Source LLC
Chambersburg PA
CBHW051717210326
41597CB00032B/5511